The ECOLOGICAL World View

The
ECOLOGICAL
World View

Charles Krebs

University of California Press
Berkeley   Los Angeles

University of California Press, one of the most distinguished university presses in the United States, enriches lives around the world by advancing scholarship in the humanities, social sciences, and natural sciences. Its activities are supported by the UC Press Foundation and by philanthropic contributions from individuals and institutions. For more information, visit www.ucpress.edu.

Published exclusively in the United States of America, Canada, Mexico, and the Philippine Republic by
University of California Press
Berkeley and Los Angeles, California

Originally published by
CSIRO PUBLISHING
150 Oxford Street (PO Box 1139)
Collingwood VIC 3066
Australia
www.publish.csiro.au

© 2008 by CSIRO

ISBN: 978-0-520-25479-4 (pbk. : alk. paper)

Manufactured in Singapore

17 16 15 14 13 12 11 10 09 08
10 9 8 7 6 5 4 3 2 1

Set in 10/14 Adobe Palatino Roman and Optima
Cover and text design by James Kelly
Typeset by Desktop Concepts Pty Ltd, Melbourne
Printed by Imago

Dedicated to the many ecologists around the world
whose work has contributed
to the development of
ecological science

CONTENTS

PREFACE

Two views of the world dominate our thinking this century. This is a book about the Ecological World View, which contrasts sharply with the polar-opposite Economic World View to which many governments and business leaders subscribe. You are living in a century in which the Economic World View will be superseded by the Ecological World View—the signs of concern about climate change and sustainability have been reported widely in the media of 2007. As a citizen, you ought to learn something about the Ecological World View. There has been a revolution of human thinking in the last 40 years that has centered on the relationships between humans and their environment. The broader policy problems this revolution has brought about are the focus of the environment movement and the basic science behind it all is the science of ecology.

In a nutshell, ecology is concerned with the workings of the biological world within the framework of the world's environments. It is useful to learn something about ecology if you wish to understand the problems humans face in sustaining our environment and protecting the species with which we share the Earth. This textbook presents you with an outline of the science of ecology. If you understand how the natural world works, you will be better able to think with an ecological conscience.

This book presents the general principles of ecology without going into the details of ecological methods and mathematical arguments that are more essential in advanced ecology. I have strived to make this book both readable and topical—the greatest compliment any author can receive is that students think the book is readable and interesting.

Each chapter in this book raises a question about how populations and communities operate in nature, and provides examples and information for you to think about and analyse. If you would like more information on any topic, a list of suggested readings is provided as a starting point. Each chapter ends with a series of questions and problems to test your knowledge. I have not provided answers. Indeed, for many questions, the answer is not yet known.

I have tried to emphasize the historical development of ecology by adding photos of famous ecologists in most of the chapters. Science is a human activity and those scientists who have developed the science of ecology—and are still building it today—are themselves interesting characters, more worthy of recognition than the sports stars that dominate our news. Ecology can sometimes seem too removed from everyday concerns, so I have used three chapters on applied ecology to illustrate some critical problems that need action. Conservation biology focuses on practical problems that demand ecological understanding. Fisheries management is another important issue that affects the entire world. Pest management hinges on concepts of population and community dynamics that need careful thought and analysis in case we cause more problems than we solve. I have included essays in many chapters to illustrate some of the kinds of problems and questions ecologists deal with in their attempt to understand nature.

This book is my own attempt to present modern ecology as an interesting and dynamic subject. This book is not an encyclopedia of ecology, but an introduction to its problems. If there is a message in this book, it is a simple one: we make the most progress in answering ecological questions when we use experimental techniques. The habit of asking, "What experiment could answer this question?" is the most basic aspect of scientific method that students should learn to cultivate. When there is controversy, answering this question leads us to the heart of the matter.

I thank many friends and colleagues who have contributed to formulating and clarifying the material presented here. In particular I thank my university colleagues Dennis Chitty, Judy Myers, the late Jamie Smith, Carl Walters, Jim Hone, and Tony Sinclair for their assistance, and Brian Walker and the many ecologists at CSIRO Sustainable Ecosystems in Canberra who answered endless queries during this revision. CSIRO Sustainable Ecosystems and the University of Canberra have been a home since my retirement from teaching, and I am grateful for their assistance in working on this book. In particular I thank the librarians at CSIRO who provided much assistance. Briana Elwood and John Manger from CSIRO Publishing greatly facilitated the evolution of this book. Matt Lee assisted in copy editing earlier versions of many chapters. To all of these I am most grateful.

Finally I want to thank the real authors of this book, the hundreds of ecologists who have toiled in the field and laboratory to extract from the study of organisms the concepts discussed here. A person's life work may be summarised in a few sentences in this book, but we ecologists owe a great debt to our intellectual ancestors.

Charles J. Krebs
February 21, 2007

Chapter 1

AN INTRODUCTION TO ECOLOGY

IN THE NEWS

Emperor penguins in Adelie Land, Antarctica, have been declining in numbers for the last 50 years. They are now about half as abundant as they were in the 1960s. Ecologists are interested in why this drop in number is occurring. Is it part of a long-term cycle? Is global warming somehow producing this decline? Is a disease killing the adults? Where are other colonies of emperor penguins? Are other penguin species or other Antarctic species also declining? Like all scientists, ecologists begin with observations—in this case annual counts of penguins in one colony for 50 years. Next, ecologists suggest ideas to explain the observations and look for additional data to see if these ideas are correct. In this book we will discuss how to analyze ecological problems of this type.

Adelie Land

Map of Antarctica showing the location of Adelie Land

■ 1.1 INTERACTIONS BETWEEN SPECIES DETERMINE WHERE ORGANISMS LIVE AND HOW MANY LIVE THERE

Ecology is the scientific study of the interactions that determine the distribution and abundance of organisms. Some of the interactions that ecologists study, such as predator–prey interactions, are between different species. Other interactions are between organisms and factors in the physical environment, including temperature and precipitation. You probably have an intuitive feeling about the meaning of **distribution** (*where* organisms are found) and **abundance** (*how many* organisms live in a given place).

The distribution and abundance of many species vary in space. For example, the red kangaroo is common in north-eastern New South Wales, but they are rare in Western Australia and absent altogether from most of northern Australia (Figure 1.1).

What accounts for this pattern of abundance? Why is this species less abundant toward the edges of its geographic range? What limits the western and northern extension of its range? Will red kangaroos change in distribution as the climate warms? These are examples of the questions ecologists try to answer about distribution and abundance.

Ecology and Environmentalism

The word *ecology* was first used by the German zoologist Ernst Haeckel in 1866, and it came into wider use in the last half of the 19th century. But ecology was not widely recognized as an important science until the 1960s, when ecologists expressed concern about the continuing increase in the human population and the associated destruction of natural environments by pesticides and pollutants. In the public mind, the word **ecology** came to mean everything and anything about the environment, especially human impact on the environment and its social ramifications.

However, it is important to distinguish *ecology* from *environmental study*. Ecology deals with the interrelations of all organisms. While it includes humans as a very significant species by virtue of their impacts, ecology is not solely concerned with humans. **Environmental study** is the analysis of human impacts on the

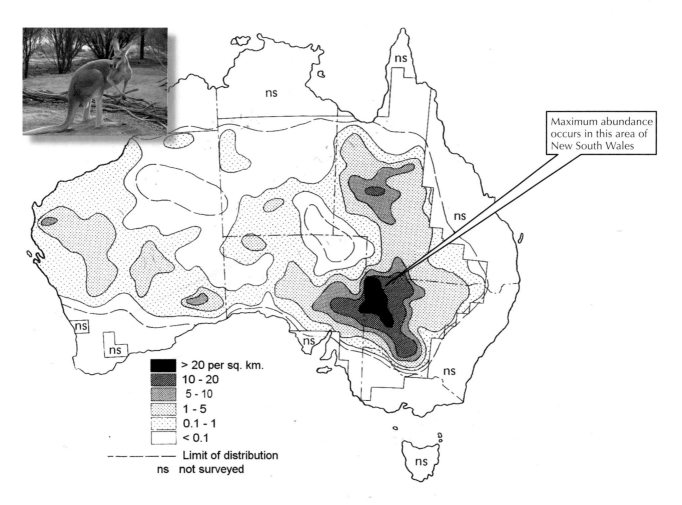

Figure 1.1. Density and distribution of red kangaroos determined from aerial surveys in 1980–82. Some coastal areas were not surveyed (ns). (After Caughley 1987b, photo courtesy of Tom Vaughan.)

physical, chemical and biological environment of Earth. As a discipline, environmental study is much broader than ecology because it integrates ecology, geology, climatology, sociology, economics, anthropology, political science and philosophy.

Environmental study has led to 'environmentalism'— a social movement whose agenda for political and social change is directed toward minimizing human impacts on Earth. Environmentalism is not ecology, although many ecologists support its goals. Ecology has much to contribute to some of the broad questions about humans and their environment, such as the unintended consequences of pest-control methods. Ecology should be the foundation for environmentalism in the same way that

physics is the foundation for engineering. Just as we are constrained by the laws of physics when we build airplanes and bridges, so we are constrained by the principles of ecology when we alter the environment. However, even if ecological research can help predict what will happen when the global temperature rises due to increased CO_2 emissions, it cannot tell us what we *ought* to do about these emissions or whether a rising global temperature is good or bad. Ecologists are not policy makers or moral authorities and, as scientists, should not make ethical judgments or political recommendations. Of course, being human, ecologists do make such judgments and most ecologists are concerned about environmental changes such as global warming. Many ecologists work hard in the political area to achieve the social goals of environmentalism.

Ecological Systems

Ecology is the science that examines the relationships between all the animals, plants, fungi and microbes on Earth. These organisms interact in ecological systems, which include all the organisms in an area as well as their non-living (abiotic) environment.

The essential message of ecology is that changing one component in an ecological system usually changes others. The science of ecology can be thought of as finding out the details. In every ecological system, such as the emperor penguins in Adelie Land, what components are connected to which others, and how strong are these connections? The devil is in the detail in every science, and nowhere is this truer than in ecology. In this book we will look at some of the details of the species interactions that occur in ecological systems so that you have a better perspective of how the ecological world works.

Humans are a dominant species in the world today, and ecology is concerned with how humans interact with all other kinds of organisms. From a practical viewpoint, we need to learn about ecological systems because our future lives will be affected by the ecological changes we cause, but another important reason for learning about ecological systems is that they are interesting too! We live in a world of nature and, like other land-dwelling organisms, we experience wind, rain, heat and cold. We can therefore understand the factors that affect ecological systems from our first-hand experience. Of all the sciences, ecology is one of the closest to our daily lives. Knowing how organisms interact is fascinating and will enrich your daily life. You will look at our world in new ways and see new interconnections.

■ 1.2 ECOLOGY'S FOUNDATION IN NATURAL HISTORY GOES BACK MORE THAN 2000 YEARS

The roots of ecology lie in natural history. Primitive tribes, which depended on hunting, fishing and food gathering, needed detailed knowledge of where and when they could find their food. The establishment of agriculture depended on an increased understanding of the practical ecology of plants and domestic animals. Early naturalists knew that a major constraint on animal populations is their food supply, and that predators and disease could also have a significant impact.

We read newspaper stories about locust outbreaks in Africa and explosions of rat populations in rice fields in Asia. These events are not new. Spectacular plagues of animals attracted the attention of the earliest writers. The Egyptians and Babylonians feared locust plagues and often attributed them to supernatural powers. Exodus (7:14–12:30) describes the plagues that God called down upon the Egyptians. In the fourth century BC, the Greek philosopher Aristotle tried to explain plagues of field mice and locusts in his *Historia Animalium*. He pointed out that the high reproductive rate of field mice enabled them to produce offspring faster than humans, or other natural predators, such as foxes and ferrets, could kill them. Aristotle stated that nothing could reduce mouse plagues except the rain, for after heavy rains the mice disappeared rapidly. Australian wheat farmers face plagues of house mice today and ask the same question: How can we get rid of these pests?

The Balance of Nature

Pests have always been a problem because they violate our feeling of harmony or balance in the universe. Ecological harmony was a guiding principle basic to the ancient Greeks' understanding of nature—the concept of the **balance of nature** is at least 2000 years old. 'Providential design,' which held that nature is designed to benefit and preserve each species, was implicit in the writings of Herodotus and Plato. This world view assumed that the numbers of every species remained essentially constant over time. When outbreaks of some populations did occur, they were usually explained as divine retribution for the punishment of evildoers. Each species had a special place in nature, and extinction did not occur because it would disrupt the balance and harmony in nature. Today, in the midst of a major extinction crisis, the balance of nature idea seems naïve, and the question of how modern extinctions will affect the functioning of natural ecological systems is an important topic of study for ecologists.

Q What is your concept of 'the balance of nature'?

Applied Ecology

Many of the early developments in ecology came from the applied fields of agriculture and fisheries, long before the word **ecology** was coined. For example, before the advent of modern chemistry, insect pests of crops were controlled with introduced predators. In 1762, the mynah bird was brought from India to the island of Mauritius to control the red locust, and by 1770 locusts were a negligible problem on the island. Around the same time, predatory ants were introduced from the mountains of south-western Arabia into date-palm orchards to control other species of ants. In subsequent years, an increasing knowledge of insect parasitism and predation led to many such introductions all over the world in efforts to fight agricultural pests—a branch of applied ecology we now call **biological control**.

Fisheries have been an important source of food for centuries, but over-fishing has led to declines in many freshwater and marine fish species. Those involved in the fishing industry around the world have long noted that the harder they fish, the fewer fish they catch, and the ones they do catch are smaller (Figure 1.2). The northern cod fishery in the North Atlantic Ocean is a good example. In 1497 John Cabot returned to England from the Grand Banks of Newfoundland and wrote about the extraordinary abundance of northern cod in these waters:

> '…the sea is covered with fish which are caught not merely with nets but with baskets, a stone being attached to make the baskets sink with the water…'

Salted cod became a delicacy in Europe, and the French, Spanish, Portuguese and English began fishing off eastern North America in the 16th century. By the 17th century, some Icelandic fisheries started to fail. But it was only during the 20th century, with the development of motor-powered fishing vessels, that industrial fishing and sustained over-fishing became rampant, leading to the collapse of the northern cod stocks and many other fish species. Fishery ecologists

Figure 1.2. Northern cod caught off Labrador in 1899. The larger fish was 1.65 m long and weighed 27 kg. After years of over-fishing, the average cod caught in 2004 was 0.5 m long and weighed 2 kg. (Photo courtesy of the National Museum of Canada.)

ask how exploitation by humans, in combination with natural processes like predation, affect the abundance of valuable marine fish species—a question we address in Chapter 18.

The development of ecology during the 20th century followed the lines drawn by naturalists during the 18th and 19th centuries and by applied ecologists working in agriculture and fisheries. The effort to understand how ecological systems work has been undertaken by a collection of colorful characters quite unlike the stereotype of scientists in white coats. From Alfred Lotka, who worked for the Metropolitan Life Insurance Company in New York while laying the groundwork of mathematical ecology in the 1920s, to Charles Elton, the British ecologist who wrote the first animal ecology textbook in 1927 and founded the Bureau of Animal Population at Oxford University—ecology has blossomed, providing an increasing understanding of our world and how humans affect ecological systems.

Charles Elton (1900–1991) carrying mouse traps on a motorbike to Bagley Woods near Oxford, England, in 1926. Vehicles were expensive, and early ecologists had to use inexpensive methods for field work. Elton was the first ecologist to conduct a field study of small rodent populations.

* **Biosphere**
* **Landscapes**
* **Ecosystems**
* **Communities**
* **Populations**
* **Individual organisms**
Organ Systems
Organs
Tissues
Cells
Subcellular organelles
Molecules

Levels of integration in biology

Figure 1.3. Levels of integration in biology. Ecology deals mainly with the six levels indicated with an asterisk.

■ 1.3 ECOLOGISTS STUDY BIOLOGICAL INTERACTIONS FROM THE LEVEL OF THE INDIVIDUAL TO THE ENTIRE BIOSPHERE

Biological systems can be analysed at different levels, ranging from molecules to ecosystems, defined largely by size. These categories are called **levels of integration** (Figure 1.3). Biological processes in each level of integration involve the level immediately below, so there is a natural hierarchy that is assumed in this pattern. In ecology, we deal primarily with the six highest levels of integration, which are marked with an asterisk (*) in Figure 1.3. **Populations** are groups of individuals of a single species, such as the emperor penguins of Adelie Land. **Communities** are groups of species that live in the same area, so they include many populations. **Ecosystems** consist of communities and their physical environment—Lake Michigan is a large aquatic ecosystem, but even a small pond can be considered an ecosystem. **Landscapes** are groups of ecosystems typically at a larger spatial scale, such as the Everglades of Florida.

Landscapes can be aggregated to include the whole-Earth ecosystem, or **biosphere**, which is sometimes called the **ecosphere**.

Scientific disciplines partly overlap in their study of different levels of integration. Ecology overlaps with physiology and behavior in studies of individual organisms, and with meteorology, geology and geo-chemistry when we consider landscapes.

Each level of integration involves a separate and distinct set of attributes and problems. For example, a population has a **density** (the number of individuals per unit area or volume)—a property that cannot be attributed to a single individual. A community has **biodiversity** (the number of species)—an attribute without meaning at the population level. In general, a scientist who deals with one level of integration seeks *explanations* from the lower levels of integration. For example, to understand why emperor penguin populations have decreased in Adelie Land during the last 50 years, an ecologist would study the numbers of births and the causes of death of individual penguins. This approach to science is said to be **reductionist**, because it reduces a problem at one level of integration to a series of problems at lower levels. An alternative approach to science is **holistic**, encompassing higher levels. An ecologist employing the holistic approach to study the decline of emperor penguin populations might analyze the consequences of this decline on the community in which the penguins live, including the fish on which the penguins feed and the predators that feed on the penguins. The example of dealing with AIDS (acquired immunodeficiency syndrome) shows why both approaches are important and should be embraced. To develop a vaccine, we must understand the action of the virus responsible for AIDS (reductionism); to stop the spread of AIDS, we must understand the social and behavioral factors that affect transmission of the virus (holism).

Much of modern biology is highly reductionist, trying to work out the chemical basis of life. Therefore, it should not surprise you that the amount of our scientific understanding varies with the level of integration. We know an enormous amount about the molecular and cellular levels of organisms, and we know very much about organs, organ systems, and whole organisms. However, we know relatively little about populations and even less about communities and ecosystems. A good example of this disparity is the Human Genome Project. This expensive and highly targeted research program has sequenced all the genes on human chromosomes, yet we do not know how many species of beetles live on Earth or how many species of trees there are in the Amazon basin.

Q **Can ecology ever be reduced to a branch of molecular biology?**

You will not find in ecology the complete theoretical framework that you find in physics, chemistry, molecular biology or genetics. It is not always easy to see where the pieces fit in ecology, and we will encounter many ideas that are well developed, but are not clearly connected to any general theory. This is typical of a young science. Many students think of science as a monumental set of facts that must be memorized, but it is more than that. Science is a search for systematic relationships, for explanations to problems in the physical world and for unifying concepts. The growth of scientific information and understanding is highly

visible in a young science like ecology. It involves many unanswered questions and much controversy.

■ 1.4 LIKE OTHER SCIENTISTS, ECOLOGISTS MAKE OBSERVATIONS, FORM HYPOTHESES, AND TEST PREDICTIONS

The essential features of the scientific method are the same in ecology as in other sciences (Figure 1.4 and Table 1.1). An ecologist begins with a problem, which

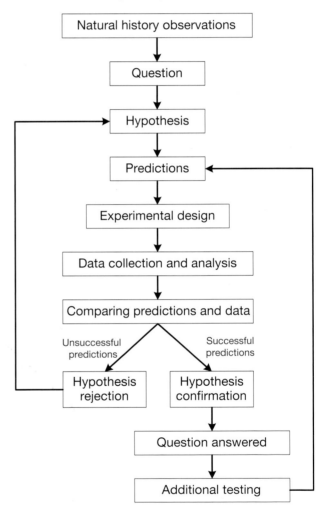

Figure 1.4. Schematic diagram of the scientific method as applied to ecological questions. Natural history observations present ecologists with patterns (like that shown in Figure 1.1) that raise questions and are the foundation of good ecology. Scientific investigation consists of an endless cycle involving hypotheses and data in an attempt to answer such questions. Critical analysis and creativity are both required for good science.

Table 1.1 Definitions of terms used in discussions of the scientific method.

Term	Definition
Scientific law	A universal statement that is so well supported by experimental observations that everyone accepts it as part of the scientific background of knowledge. There are many laws in physics, chemistry and genetics, but few in ecology.
Principle	A statement that is universally accepted because it is mostly a definition, or an ecological translation of a physical or chemical law. For example, the statement 'no population increases without limit' is an important ecological principle that must be correct in view of the finite size of Earth.
Theory	An integrated and hierarchical set of experimental hypotheses that together explain a large number of scientific observations. The theory of evolution by means of natural selection is perhaps the most frequently used theory in ecology.
Hypothesis	A statement that offers an explanation for some observation. Ecology abounds with hypotheses, as we will see in the next 20 chapters.
Model	A verbal or mathematical statement of a hypothesis. Often the words *model* and *hypothesis* are used interchangeably.
Experiment	A test of a hypothesis that involves observation of the system in question. It may also include some manipulation of the system. The experimental approach is known as the scientific method.
Fact	A specific truth pertaining to the natural world. Some observations are faulty, so an observation is not automatically a fact.

is often based on observations of natural history. For example, pine tree seedlings are not found in mature hardwood forests on the Piedmont of North Carolina. Accurate natural history is a prerequisite for all ecological studies, because if a problem is not based on correct observations, all subsequent stages will be useless. Given a problem, an ecologist suggests a possible answer. This answer, called a **hypothesis**, is basically a statement of cause and effect. In most cases, several answers might be possible, and several alternative hypotheses can be proposed to explain the observations. An example of a hypothesis is that pine seedlings do not grow in mature hardwood forests because there is too little light. An alternative hypothesis might be that the soil does not contain enough water.

Predictions and Experiments

Predictions follow logically from a hypothesis, and the more predictions a hypothesis makes, the better it is. Predictions from simple hypotheses are often straightforward: Pine seedlings will grow if you provide more light (under the light hypothesis proposed earlier), or more water (under the water hypothesis). Every hypothesis must predict something and forbid other things from happening. For example, the light hypothesis for pine seedlings predicts more seedlings if you increase the light intensity, and it forbids you finding more seedlings if you add more water. Every hypothesis should have a list of observations that are consistent with it and a list of observations that are not consistent with it (the forbidden observations).

Q **Is it better to have more hypotheses or fewer hypotheses to explain a set of observations?**

A scientist tests a hypothesis by performing an **experiment**—a set of additional observations to check the predictions. Experiments can be manipulative (for example, by providing artificial light in a mature forest), or natural (such as looking for pine seedlings in natural gaps in the forest). The protocol for the experiments and the type of data to be obtained are called the experimental design. A successful experiment produces data that allow the researcher to either

ESSAY 1.1   HOW TO READ SCIENTIFIC LITERATURE

Scientific literature fills libraries and web sites and expands every year as more and more papers are published. One way to find information on a particular topic in science is to search for it on the World Wide Web. But how can you tell whether the information you find on a web site is accurate? The most important thing is to read critically. Here are a few pointers that can help:

1. What is the question? The first and most important step in analyzing a scientific paper is to find out what question or problem the authors are addressing. This should be explicitly stated early in the paper.

2. What are the proposed explanations for this problem? List the hypotheses that are possible explanations. There must be at least two alternative hypotheses that make different predictions.

3. What observations or data would favor one hypothesis over the others? There must be some possible observations that would contradict each of the hypotheses.

4. How were data collected in the study? Scientific papers typically explain in great detail the methods used in the study. Beginning students should not worry about the validity of the methods and should assume that they correctly measure the variables under study.

5. What are the results of the study? Do they fit one hypothesis better than another? Are the results clear and conclusive, or are they inconclusive? Not all scientific papers reach a clear conclusion.

6. What should be done next? Because science is an endless process (see Figure 1.4), all research should lead to more hypotheses and more experiments, and good scientific papers end with suggestions for what should be done next.

These six pointers are no different for ecological papers than they are for other scientific papers, and they are useful ways to develop critical thinking skills.

reject or tentatively accept the tested hypothesis. A hypothesis that is tentatively accepted may be rejected later if the data from another experiment do not support it. One way to gain experience with the scientific method is to read and analyze papers in the scientific literature (see *Essay 1.1*).

Complexity in Ecology

The complexity of ecological systems often makes it challenging to apply the scientific method outlined in Figure 1.4. In some cases environmental factors operate together. For example, perhaps mature hardwood forests have too little light *and* not enough water for pine seedlings to grow. Systems in which many factors operate together are difficult to analyze, and require ecologists to formulate complex hypotheses. No matter how complex a hypothesis is, however, it must make predictions that can be tested by experimentation.

The emperor penguin population decline is a good example of this complexity. Ecologists have found that reduced survival of adult penguins is the immediate cause of the decline. But adult penguin survival is related to the extent of sea ice off Antarctica, and the amount of sea ice that forms each winter is sensitive to changes in climate. Sea ice does not directly affect the penguins—instead, it probably affects access to their

ESSAY 1.2   SCIENCE AND VALUES IN ECOLOGY

Science is thought by many people to be value-free, but this is certainly not the case. Values are woven all through the tapestry of science. All applied science is carried out because of value judgments. Medical research is a good example of basic research applied to human health that virtually everyone supports. Weapons research is carried out because countries wish to be able to defend themselves from military aggression. In ecology, the strongest discussions about values have involved conservation biology. Should conservation biologists be objective scientists studying biodiversity or should they be public advocates for preserving biodiversity? The preservation of biodiversity is a value that often conflicts with other values; for example clear-fell logging produces jobs and wood products, but reduces biodiversity. The pages of scientific journals like *Conservation Biology* are peppered with discussion about advocacy versus objectivity. Scientists have a dual role. Firstly, they carry out objective science that collects data and tests hypotheses about ecological systems. Secondly, they can also have a role as advocates for particular policies that attempt to change society, such as the use of electric cars to reduce air pollution. It is crucial to separate these two kinds of activities intellectually. Science is a way of knowing—a method for determining the principles by which systems operate. The key scientific virtues are honesty and objectivity in the search for truth. Scientists assume that once we know these scientific principles we can devise effective policies to achieve social goals. All members of society collectively decide on what social goals we will pursue, and civic responsibility is part of everyone's job, including scientists. There will always be a healthy tension between scientific knowledge and public policy in environmental matters because there are always several ways of reaching a particular policy goal. The debates over public policy in research funding and environmental matters will continue, so please join in.

food, particularly shrimp-like crustaceans called krill, and possibly the distribution of their predators. What affects the abundance and distribution of krill near Antarctica? Teams of ecologists from the United States, France, Great Britain and Australia are working to answer this and other questions.

Ecological Truth

The scientific method is often considered to be value-free, but this is not correct (see *Essay 1.2*). We expect scientists to have values and we also expect them to speak the truth. But what is the hallmark of ecological truth? The notion of truth is a profound one, which philosophers discuss in detail and scientists assume is simple.

To a scientist, truth consists of correspondence with the facts. If we say that there are 53 elephants in a certain herd in East Africa's Serengeti National Park, we are stating an ecological truth if other scientists can count the elephants and get the same number. If we had counted the same herd over 20 years and found that the numbers were continually falling, we could state that this elephant population is declining. This statement would also be an ecological truth as long as it agreed with the facts. Scientists rarely get into arguments about facts. Arguments start when the inferences are drawn from whole sets of facts. For example, suppose that we state that the elephant population is declining *because of a disease*. This statement would be an ecological *hypothesis*, not an ecological truth. We could list the predictions it makes about what we would find if we searched for a disease-causing agent in dying elephants in this herd. We might also state that elephant populations all over East Africa are collapsing because of disease. This would be a more general hypothesis and, before we could consider it an ecological truth, we would have to test its predictions by studying many more populations of elephants and their diseases. Many ecological ideas are at an incomplete stage because there has not been enough time, money or personnel to test such general hypotheses sufficiently. Like other scientists, ecologists must deal with the problem of uncertainty: How do we know if we have enough supporting evidence to regard an ecological hypothesis as an ecological truth?

The resolution to this problem in ecology and in many other areas of natural science has been the **precautionary principle,** which may be expressed as 'look before you leap' or 'an ounce of prevention is worth a pound of cure.' This principle is the ecological equivalent of part of the Hippocratic oath in medicine: 'Physician, do no harm.' The central ideas of the precautionary principle are to do no harm to the environment, to take no actions that are not reversible and to avoid risk as much as possible. Ecological truth is never obvious in complex environmental issues and emerges more slowly than we might like. We cannot wait for scientific certainty before deciding what to do about emerging problems in the environment, whether they are declining elephant populations, introduced pest species or global warming. Our knowledge obtained by the scientific method is always incomplete and science is a never-ending quest.

■ 1.5 LYME DISEASE ILLUSTRATES THE COMPLEX INTERACTIONS IN ECOSYSTEMS

To understand how ecologists use the scientific method to investigate the interactions between species, let us take a close look at the biology of Lyme disease. First described in 1977 in Connecticut, Lyme disease is now spreading from east to west across North America. The disease is caused by the transmission of bacteria, called spirochetes, from ticks to humans. When a tick infected with spirochetes bites a person, the spirochetes enter the person's bloodstream. The spirochetes do little harm to the ticks, but in humans they can cause severe damage to the nervous system, heart, eyes and joints. Because ticks transmit this disease, we clearly need to find out something about ticks and which other animals they feed on.

Figure 1.5 shows the life cycle of the deer tick, the main carrier of Lyme disease in the north-central and eastern United States. After passing through a small nymphal stage, in which they feed on the blood of small rodents and birds, deer ticks reach a size at which they feed mainly on white-tailed deer. In their second year of life, they reach sexual maturity and drop off deer to lay their eggs in the forest floor.

Thus, we must expand our model of interactions in Lyme disease to include deer, rodents and birds as well as ticks and humans. What determines the numbers of deer and rodents that live in the areas where deer ticks are found? A first guess would be their food supplies, so we next need to find out what the deer and rodents eat. One important food for both deer and rodents is the acorn—the fruit of oak trees. Oak trees do not drop the same number of acorns every year, but operate in a boom–bust cycle. During boom years, each oak tree drops thousands of acorns.

Interaction Webs

You can see that the spread of Lyme disease involves a fairly complex system with connections between several species. Ecologists typically try to summarize such connections in an **interaction web** such as the one in Figure 1.6. An interaction web is a boxes-and-arrows diagram of all the relationships between species in an ecosystem. Interaction webs have two simple rules for their construction. Firstly, arrows indicate ecological interactions, so $A \longrightarrow B$ means that species A affects species B in some way. Second, the type of interaction is indicated by a plus (+) or a minus (−) sign. Therefore, if species A affects species B positively, so that more of A leads to more of B, we can represent that interaction as

$$A \xrightarrow{+} B$$

If species A affects species B negatively (meaning that more of A leads to less of B), we would write

$$A \xrightarrow{-} B$$

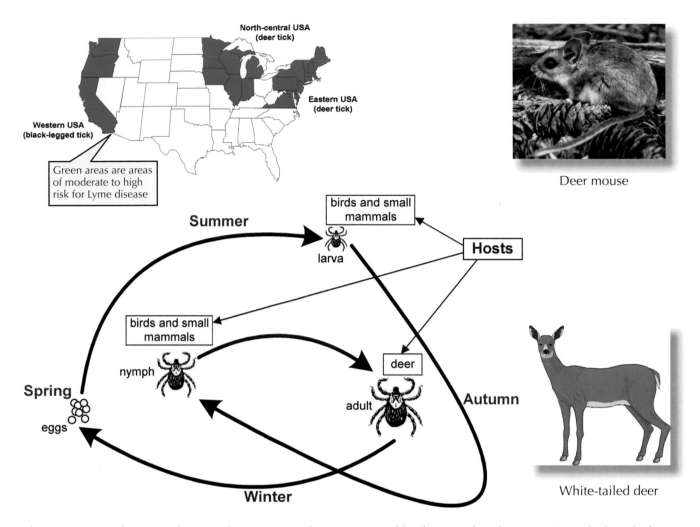

Figure 1.5 Lyme disease in the United States. Lyme disease is caused by the spirochete bacterium *Borrelia burgdorferi*. The bacterium is transmitted by several species of ticks, including the deer tick *Ixodes scapularis,* whose life cycle is shown. The deer mouse (*Peromyscus maniculatus*) and the white-tailed deer (*Odocoileus virginianus*) are two of the main hosts of deer ticks. Lyme disease is now spreading into southern Canada. (Modified after Ostfeld *et al.* 1996.)

In our example, more oak trees produce more acorns, which, in time, produce more mice and deer.

Complex Interactions in Lyme Disease

The interaction web in Figure 1.6 is a good start, but it does not tell the whole story of Lyme disease because there are many more details about the interactions that we can't summarize with + and − signs. One such detail is the *timing* of effects. Although more acorns produce more mice, this may take several months to happen. Furthermore, more acorns will not produce more deer fawns in the same year, but they will attract deer to the oak woods where acorns are abundant, so

that deer numbers will increase because of local movements. Thus, there is a time dimension to this interaction web, which is not captured in a simple diagram like Figure 1.6.

Let's investigate the natural history of the interactions that lead to Lyme disease a bit further. If mouse numbers increase, ticks have more mice on which to feed so tick numbers can increase. But mice feed on the pupal (cocoon) stage of the gypsy moth—an insect that eats the leaves of oaks and other trees in the eastern forests of North America. Consequently, more mice may increase acorn production by decreasing gypsy moth numbers. Furthermore, white-tailed deer eat tree

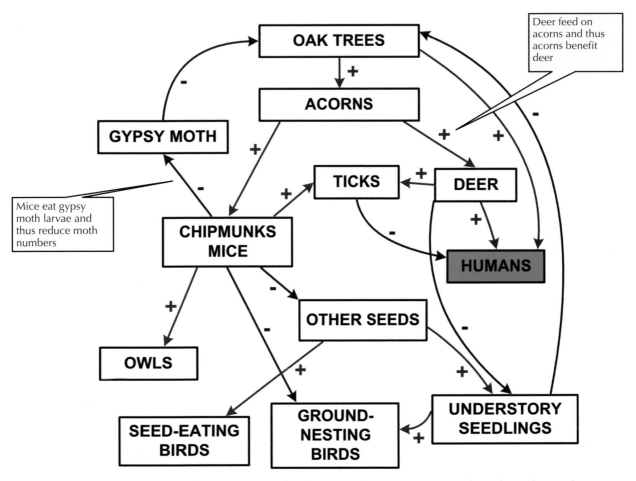

Figure 1.6 An interaction web for the key species involved in Lyme disease transmission by ticks in the north-eastern United States. Arrows represent the direction of influence between species. A plus symbol (+) and red arrow indicate that the influence increases the abundance of the species at the arrowhead. A minus symbol (–) and blue arrow indicate that the influence decreases the abundance of the species at the arrowhead. For example, more oak trees produce more acorns, and more acorns help to increase the numbers of mice and chipmunks. Interaction webs like this help ecologists unravel the complex relationships in ecological systems.

seedlings in oak forests, thereby reducing the forest understory used by many songbirds for nesting. Therefore, when acorns become abundant, more white-tailed deer are attracted to the forest in the short term, and over the long term there is a loss of forest understory, so that songbirds may produce fewer young. Now it becomes clear that we must understand the *magnitude* of these positive and negative interactions to make this interaction web more useful. For example, field studies show that years of high acorn production in New York forests result in twice the number of white-footed mice the following year. Mathematical models are particularly useful at this stage of inquiry. Ecologists need to

know by how much the numbers of organisms in interaction webs have changed, just as investors need to know by how much their bank accounts have gone up or down.

Q How might you redraw the interaction web in Figure 1.6 if you were interested in acorn production instead of Lyme disease?

Many things are connected in ecological systems, and the ecologist's job is to uncover both the direct connections and the hidden, indirect connections. Before these critical studies were done and the interaction web

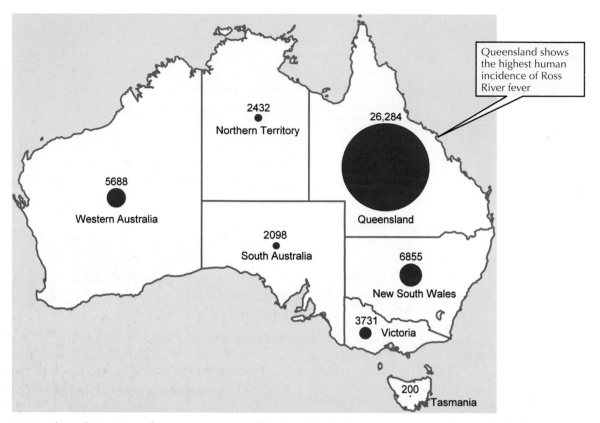

Figure 1.7 Number of Ross River fever cases reported in Australia during 1991–2000. Within individual states, cases are highly localized and depend on the numbers of mosquitoes as well as the abundance of reservoir hosts, mainly kangaroos and wallabies. Many minor infections are probably not reported, and these numbers of cases are the minimum known infections. (Data from Russell 2002.)

in Figure 1.6 was put together, no-one suspected that gypsy moths could be connected to the transmission of Lyme disease in oak forests. The best approach is to assume that there are more connections between species than are immediately obvious. Many errors in the management of agriculture, forests, fisheries and wildlife have been made by assuming too few connections between species. In ecology, it is important to recognize that we do not know everything about natural systems.

■ 1.6 ROSS RIVER FEVER ILLUSTRATES HOW HUMANS AND ECOSYSTEMS ARE INTERCONNECTED

Ross River virus is the most common mosquito-borne pathogen in Australia, affecting more than 5000 people annually. This virus is found only in Australia and

New Guinea and is common in all states (Figure 1.7). In most areas, the wet season increases the abundance of mosquitoes that carry the virus, which has been recorded from 42 different species of seven different genera of mosquitoes. The virus was first isolated in 1958 from a mosquito in Queensland, and then isolated from humans in 1972 (Russell 2002).

In humans, Ross River virus produces a variety of symptoms that can last for 6 months or more. Fever and rash are common, as well as arthritis of all the joints, resulting in persistent fatigue. Most infections occur in people between 20 and 60 years old, and are rare in children More than 47,000 cases were confirmed during the 1990s in Australia, and many more cases were probably not diagnosed. Figure 1.7 illustrates the widespread distribution of this pathogen.

Mosquitoes carry Ross River virus and the virus reservoirs appear to be mainly kangaroos and

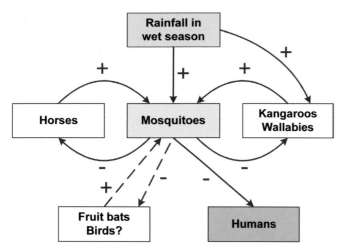

Figure 1.8 A partial interaction web for the key species involved in Ross River virus transmission in Australia. Mosquitoes are the disease vector. Arrows represent the direction of influence between species. A plus symbol (+) and red arrow indicate that the influence increases the abundance of the species at the arrowhead. A minus symbol (−) and blue arrow indicate that the influence has a negative effect on the general health of the species at the arrowhead. Dashed lines indicate that the impacts are not certain. (Data from Russell 2002.)

wallabies. Virus reservoirs are animals that carry the disease in their blood, but are often not harmed very much by the virus. Mosquitoes pick up the virus when they feed on a reservoir host, such as a kangaroo, and can then potentially carry it to people. Other species like flying foxes (fruit bats) may be carriers of the virus, but have not been adequately sampled. Horses are another reservoir for the virus, and may play an important role in maintaining the virus.

In eastern Australia up to 89% of kangaroos and wallabies tested had antibodies against the Ross River virus, indicating that they had been infected at some time in the past. The impact of this virus on macropods in Australia is completely unknown, but would appear to be relatively minor. Marsupials seem to be the main reservoir of the virus, and placental mammals like cats and dogs do not seem to be good hosts. Horses tested positive to the viral antibodies in about 30–50% of the samples taken and, because they are so closely associated with humans in many areas, they could be a major source of virus for mosquitoes to pick up. Infected humans can be significant amplifiers of

the virus, and this has caused a movement of the disease from Australia to the Pacific Islands.

The ecological significance of diseases such as Ross River fever depend on a variety of species interactions, and these interactions are rarely simple, so that the ecologist becomes a type of detective trying to unravel a tangle of interactions to find out which are highly important and which are incidental. Only by understanding the influences that species have on one another will we be able to predict the consequences of ecological changes such as those brought about by climate change.

The following chapters will explore the variety of problems and experimental efforts that characterize the science of ecology as it attempts to understand the operation of ecological systems. The results of these efforts guide the world view of ecologists. This **ecological world view** is an ethical viewpoint that is infused with an appreciation of basic ecological science, and it often differs from the view of many business people, engineers and other scientists (see *Essay 1.3*). It is part of a movement that uses scientific insights in an effort to achieve sustainable development, which is critical for all of us today as well as for future generations on Earth.

SUMMARY

Ecology is the science that examines the relationships between all organisms on Earth. The essential message of ecology is that you cannot alter just one component of an ecological system because changing one component in an ecological system usually changes others.

The roots of ecology lie in natural history. Beginning in the 17th century, ecological problems began to be discussed as modern science slowly developed. The word ecology was coined in 1866, and by the early 20th century many of the broad problems of ecology were being investigated.

The scientific method, including the testing of hypotheses by experimental observations and manipulations, is used in ecology as it is in other sciences. Ecology is a complex science in which single cause–single effect relationships are rare.

ESSAY 1.3   WHAT IS AN ECOLOGICAL WORLD VIEW?

Ecologists view the world in a particular way, and one of the reasons to study ecology is to learn about this world view. We can begin this quest by listing five principles that outline the ecological world view.

1. **You cannot alter just one component of an ecological system.** This principle is an important starting point for an ecologist, because it calls attention to the connections between species in ecological systems. If we build a dam on a river, what impact will the dam have on fish that live in the river, on the quality of water for drinking or on the plants that live on the old flood plain? On a practical level, this principle is why we require environmental impact statements for any major developments.

2. **Human actions can have long-lasting ecological impacts.** Like historians, ecologists look at events from a long-term perspective. This perspective is at dramatic odds with that of politicians who look forward a year or two, or day-traders on the stock market who take a daily view of events. Ecological processes can have effects decades or centuries in the future, so our activities today may affect our children and grandchildren. The precautionary principle encapsulates these first two principles.

3. **We can learn from history.** This general principle can be applied to most human endeavors, but it is particularly suited to ecology because we tend to think that present environmental conditions are 'normal'. What caused the extinction of many species of large mammals in North America 10,000 years ago? How has the climate changed during the last 1000 years? What impact did the introduction of the black rat 250 years ago have on Hawaiian bird life? There is a hope that if we learn from history we can avoid repeating mistakes.

4. **Conservation is essential.** Humans use a large fraction of the world's land area to grow food, reducing the amount that is available for use by the 15 million other species on Earth. Many of these species play important roles in the biosphere, and it is in our selfish interest to protect them from extinction. Conservation must therefore be part of sustainable development everywhere.

5. **Evolution continues.** Evolution is the background for all ecological interactions. When we add antibiotics to chicken feed, or use pesticides to kill insects, we select for bacteria and insects that have a natural resistance to these chemicals. Eventually, resistant populations of bacteria and insects evolve, and the antibiotics and pesticides are no longer effective.

SUGGESTED FURTHER READING

Croxall JP, Trathan PN and Murphy EJ (2002). Environmental change and Antarctic seabird populations. *Science* **297**, 1510–1514. (A detailed treatment of what is happening to penguin populations in Antarctica.)

Egerton FN, III (1973). Changing concepts of the balance of nature. *Quarterly Review of Biology* **48**, 322–350. (The definitive history of the idea that there is balance in nature.)

Ostfeld RS (1997). The ecology of Lyme-disease risk. *American Scientist* **85**, 338–346. (An overview of the problem of Lyme disease.)

Russell RC (2002). Ross River virus: ecology and distribution. *Annual Review of Entomology* **47**, 1–31. (A synthesis of the ecology of the Ross River fever virus and its geographic spread in Australia.)

QUESTIONS

1. The idea that systems like the Lyme disease interaction web have many connections has given rise to a general statement called 'the law of unintended consequences,' which holds that it is impossible to change only one thing in a complex system. Search the Web for examples of this law and discuss how widely it might be applied in other sciences.

2. An ecologist proposed this hypothesis to explain the absence of trees from a grassland area: *'Periodic fires may prevent tree seedlings from becoming established in grassland.'* Is this a suitable hypothesis? How could you improve it?

3. Describe the distribution and abundance of humans on Earth, and compare your description with the distribution illustrated in Figure 1.1 for the red kangaroo. Are there areas of Earth with no humans? What factors determine the location of areas of high human population density?

4. Debate the following position: *'The government should provide more funding for ecological and environmental research and less for space research.'*

5. How might an ecological view of sustainability differ from a medical or business view of sustainability?

6. Francis Bacon (1561–1626) is quoted as saying, 'Truth emerges more readily from error than from confusion.' How might a scientist respond to this statement? Can scientists achieve truth with the scientific method?

7. Find a distribution map of the western meadowlark (*Sturnella neglecta*) in a field guide to the birds of North America or on the Web at http://www.mbr-pwrc.usgs.gov/id/framlst/i5011id.html. Compare that map to the one for the eastern meadowlark. Do the geographic distributions of these bird species overlap? What might prevent the eastern species from moving west and the western species from moving east?

8. Does the precautionary principle mean that we should do nothing to solve environmental problems until we have achieved scientific certainty about the solutions? Discuss this question with respect to policies about global warming.

Chapter 2
GEOGRAPHIC ECOLOGY

IN THE NEWS

The red fire ant (*Solenopsis invicta*) has invaded the southern United States with devastating consequences. There are many species of fire ants in the southern states, Central America and South America, but the red fire ant is particularly dangerous because it forms large colonies that quickly swarm and sting *en masse* when an animal disturbs their mound by walking on it. Fire ant stings are painful, and more than 80 people have died as a result of multiple stings from fire ant swarms. Farm animals and wildlife species, such as ground-nesting birds, are also killed by these ants.

This map shows the range of the red fire ant in the United States in 2004, as well as its potential range. Red circles indicate sites with a high probability of reproductive success, which, for these ants, depends on high temperature and moderate rainfall. Green circles indicate sites that could possibly be colonized in years to come. It is unlikely that red fire ants will be able to live in sites marked by light blue circles under present climatic conditions.

The red fire ant was inadvertently introduced to Brisbane, Australia in 2001, presumably in a container shipped from the southern United States. There has been a concerted campaign to eradicate them, but individual colonies have kept turning up as recently as 2006, spread by moving garden mulch or potted plants around the city. Because fire ants are a serious environmental pest that could spread across the warmer regions of northern Australia, the federal and state governments have spent $175 million, and employed 650 people, over the past 6 years in an attempt to eradicate fire ants in the Brisbane area. So far 99.5% of the infestation has been eradicated and, with diligence, the remaining colonies will be located and removed.

Ecologists are interested in what limits the geographic range of this pest species, how it interacts with other ants in the environment and why it is not a particularly bad pest species in its native home in Brazil and Argentina. In this chapter we will begin to explore how ecologists analyze geographic ranges.

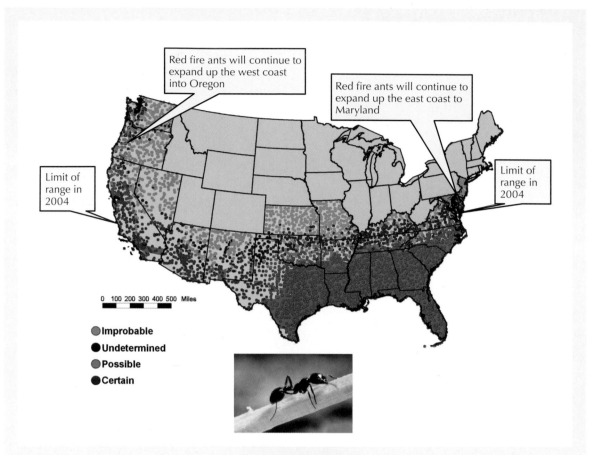

Map showing the range of the red fire ant in the United States in 2004 and its potential range

CHAPTER 2 OUTLINE

■ 2.1 ALL SPECIES HAVE A LIMITED GEOGRAPHIC RANGE

Polar bears do not live naturally in New York or Antarctica. You should not be surprised about their absence in New York because polar bears hunt the north polar ice pack for seals, and New York is too far south for the polar ice to reach. But polar bears do live in the Bronx Zoo, so clearly the climate of New York is not the restricting factor. It is more likely that their distribution is limited by the availability of seals. You may be more surprised that polar bears do not live in Antarctica, which abounds with both ice packs and seals. However, polar bears have never reached Antarctica because the tropical oceans form a barrier they cannot cross.

This example illustrates one of the two major questions that ecologists try to answer: How can we explain the geographic distributions of organisms? We will begin our analysis of geographic distributions by looking at the global distribution of organisms—a part of ecology called **biogeography.**

Biogeographical Realms

Biogeography is a discipline that spans geography, ecology and geology to provide the historical background for the present distribution of life on the continents. We can divide the land areas of Earth into six regions known as **biogeographical realms** (Figure 2.1), which tend to contain different animals and plants. Some species are typical of each biogeographical realm: we find kangaroos in Australia, lions in Africa, grizzly bears in North America and tigers in India. But

Polar bears at Churchill, Manitoba

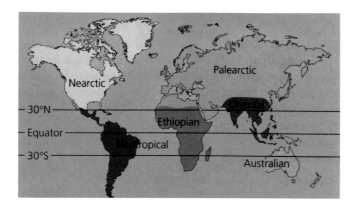

Figure 2.1. Earth's biogeographical realms. These six broad regions are a product of continental drift over the last 200 million years and barriers, such as mountain ranges, that have isolated populations. (Modified after Wallace 1876 and Brown and Lomolino 1998.)

the boundaries of these regions are not sharp, and many species live in two or more realms.

Q How would the map of biogeographical realms change if it were based on bacteria or fungi?

Biogeography originated in the 19th century as explorers began to realize that different parts of Earth had different suites of plants and animals. In 1858 Philip Sclater, a British naturalist, divided the world into the

biogeographical realms we use today. Sclater's work was based entirely on birds, yet it proved to apply to other terrestrial animals and plants. Alfred Wallace, the co-developer with Charles Darwin of the theory of evolution by natural selection, also described and analyzed biogeographical patterns. Wallace's 1876 work *The Geographical Distribution of Animals* showed that Sclater's realms could be applied to other animal groups and so set the stage for future advances in biogeography.

Biomes and Floristic Regions

Earth's surface can also be divided into **biomes**, which are major biotic communities that occupy large areas of land or water. Terrestrial biomes, such as tundra, desert and tropical rainforest, are characterized by a dominant vegetation type (Figure 2.2). The marine biome and the freshwater biome complete the global description of life on Earth.

Figure 2.3 plots the location of the terrestrial biomes along gradients of temperature and precipitation. The boundaries between the biomes are only approximate because many environmental details, such as soil type and fire frequency, are missing from this simple graph. Nevertheless, the graph reveals that the boundaries of terrestrial biomes are strongly affected by both temperature and precipitation (see *Essay 2.1*). For example, tropical rainforests require moderate to high

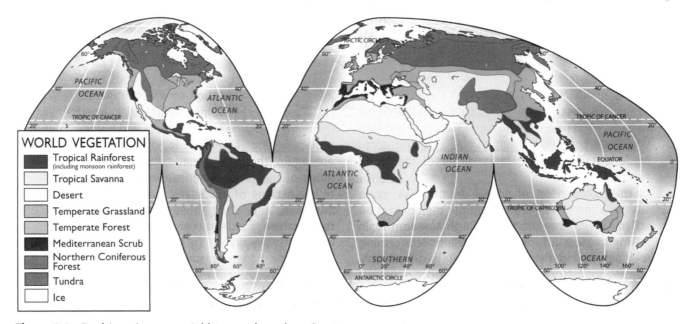

Figure 2.2. Earth's major terrestrial biomes, based on dominant vegetation. (From de Blij *et al.* 2004.)

ESSAY 2.1 GLOBAL TEMPERATURE AND PRECIPITATION

Temperature and precipitation are not uniform over the Earth. While we are familiar with our local climate, it is useful to briefly review global climatic patterns.

Temperature changes seasonally and from year to year, but average temperatures clearly vary with latitude, as shown on these maps from de Bilj (2004):

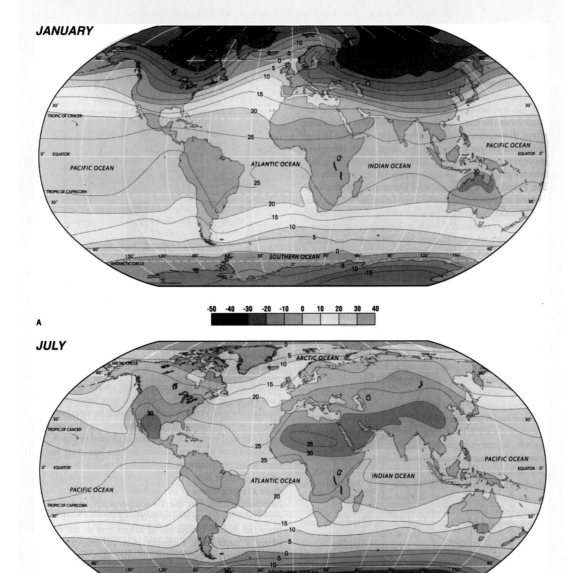

Average temperatures over the Earth in January and July

Areas near the equator are much warmer than the polar regions. The major exceptions to this general rule are the mountainous regions on all the continents.

Average annual rainfall shows a much more complex global pattern:

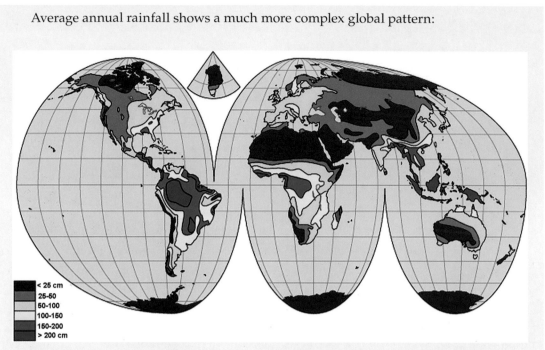

Average annual rainfall over the Earth

There is a belt of high precipitation around the equator, shown most clearly in South America, West Africa and Indonesia. Low precipitation around latitude 30° N and S is associated with the distribution of deserts around the world. Mountains and highland areas intercept more rainfall and leave a 'rain shadow' (area of reduced precipitation) on their leeward side, as can be seen in western Canada and in the United States.

These global patterns of temperature and precipitation determine the geographic distribution of biomes with their accompanying organisms.

temperatures and abundant precipitation. By contrast, tundra occurs where the temperatures are low, and deserts occupy areas with low precipitation. The important point is that we can disregard the geographic position of biomes or vegetation types on the globe and analyze the environmental factors, such as temperature and rainfall, that set their boundaries.

We can refine the biome map in Figure 2.2 to distinguish parts of terrestrial biomes, called **floristic regions**, which have characteristic vegetation. Ecologists typically recognize 14 floristic regions in North America (Figure 2.4 and Table 2.1).

Spatial Scales

Terrestrial biomes and floristic regions are typically represented on maps with a large spatial scale, but as we map geographic ranges in greater detail, we need to consider more explicitly the scale of the maps we construct for the distribution of organisms. A typical spatial analysis of geographic ranges uses a grid system with equal-area units, such as 2 km by 2 km, or 10 km by 10 km. If a species is present anywhere within a grid square, that square is considered to be occupied. To analyze the factors that control the distributions of species, we must consider distributions on a whole series of spatial scales.

Maps like those shown in Figure 2.5 provide a broad outline of the geographic ranges of particular species. For example, the giraffe is found only in the Ethiopian realm, whereas the black oak and northern red oak occur only in eastern North America in the Nearctic realm. These types of maps outline a geographic range,

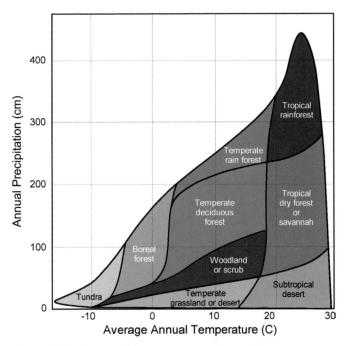

Figure 2.3. Terrestrial biomes plotted in relation to annual precipitation and average annual temperature. Boundaries between biomes are approximate. (Modified from Whittaker 1975.)

but once we look on a more local scale we need more detailed maps. If you were going to East Africa to photograph giraffes, you would want to know exactly where to find them. Giraffes live in dry, open woodlands, not in rainforests.

To determine *what limits geographic distributions*, we must carefully define the spatial resolution we seek in the answer. Figure 2.6 illustrates the ever-decreasing spatial scales we need to define a geographic range from the continental to the local scale. On a regional scale, for example, we can ask why the black oak does not grow in northern Michigan or northern Wisconsin; on a local scale, we can ask why black oaks do not grow in the active sand dunes along southern Lake Michigan, but do grow on older dunes 1 km inland.

■ 2.2 GEOGRAPHIC RANGES CAN BE VERY SMALL OR VERY LARGE

The areas occupied by a species vary by more than a million-fold. Some species, such as the Devil's Hole pupfish in Death Valley, California, are so rare that

they live in a single area occupying only 200 m^2. Other species, such as the bracken fern, are found on all the continents.

What kind of pattern would we find if we were to tally range sizes for a group of species, such as birds or mammals? Three possibilities can be imagined. The first is that few species have very large or very small ranges, while most species have ranges of intermediate size. This possibility would give a bell-shaped curve of range sizes similar to what we might get if we measured the heights of many 18-year-old males. The second possibility is that many species have large ranges (in other words, are widespread) and few have small ranges. The third possibility is that many species have small ranges and few have large ranges.

Hollow Curve Pattern

If we collect data for many different species, we find that they support the third possibility: Most species have small geographic ranges and only a few species have very large ranges. This pattern, called a hollow curve, is illustrated in Figure 2.7 for birds and mammals in North America and for vascular plants in the British Isles. It seems to apply for all groups that have been studied, from seaweeds to beetles, fossil mollusks, amphibians, reptiles, freshwater fishes, coral reef fishes and mammals, including primates.

Rapoport's Rule

In 1975, the Argentinean ecologist Eduardo Rapoport suggested that geographic ranges decrease in size from the poles to the equator. This generalization, now known as **Rapoport's rule**, is supported by studies on trees, fishes, reptiles and many mammals from several

Eduardo Rapoport (1927–) Emeritus Professor, University Nacional de Comahue, Argentina

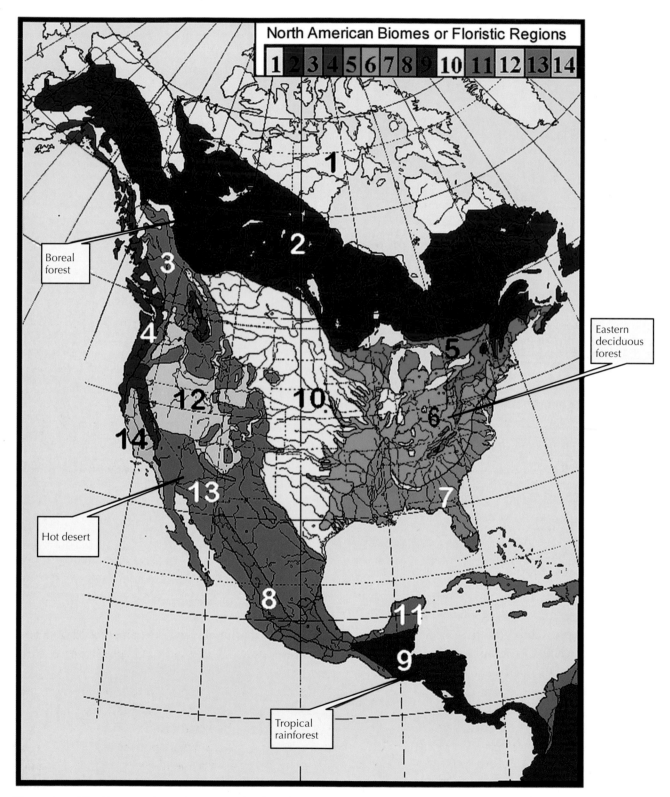

Figure 2.4. The biomes of North America broken down into 14 floristic regions, which are described in Table 2.1. The floristic regions provide a more detailed classification of vegetation types that occur within a continent. (Map courtesy of Botany Department, University of Tennessee.)

Table 2.1 Description of the 14 floristic regions of North America, which are mapped in Figure 2.4.

Map number	Name of floristic region	Alternative name	Dominant plants
1	Arctic tundra	Alpine tundra at high elevations	Dwarf willows, sedges, lichens, heath plants
2	Boreal forest	Subalpine forest (in southern mountain areas)	White spruce, black spruce, balsam fir, paper birch
3	Rocky Mountain evergreen forest	Montane forest at mid-elevation in mountains	Douglas fir, ponderosa pine, Engelmann spruce
4	Pacific coastal evergreen forest	Temperate rainforest	Western hemlock, Douglas fir, Sitka spruce, western red cedar, coast redwood, sequoia
5	Great Lakes/New England mixed forest	Northern hardwoods-hemlock	Beech, sugar maple, eastern hemlock, yellow birch, white pine
6	Eastern deciduous forest	Temperate deciduous forest	Beech, maple, tulip tree, oaks, hickories
7	Coastal Plain mixed evergreen forest	South-eastern mixed evergreen forest	Southern pines, oaks, magnolia, bald-cypress, Spanish moss, palms
8	Mexican montane forest	Tropical montane forest or Cloud forest	Many species from the Eastern deciduous forest
9	Central American rainforest	Tropical rainforest	Broadleaf evergreen trees, bromeliads, epiphytes, orchids, many diverse plant species
10	Great Plains grassland	Prairie or steppe	Big bluestem grass, various grasses and herbs
11	West Indian savannah	Tropical savannah, tropical thorn scrub, tropical dry forest	Gumbo limbo, palms, tropical grasses
12	Great Basin desert	Cool desert	Sagebrush, pinyon pine, juniper
13	Sonoran, Mojave, and Chihuahua deserts	Hot desert	Creosote bush, large cacti, yuccas, paloverde
14	Californian chaparral	Mediterranean scrub and woodland	Manzanita, ceanothus, madrone, flammable shrubs

continents, although North America shows the clearest patterns: the average geographic range of a Canadian mammal is 25 times as large as the average range of a Mexican mammal (Figure 2.8).

What is the ecological basis for Rapoport's rule? One explanation is the climatic variability hypothesis, which states that climate is more variable at high latitudes, and only organisms that tolerate this variability can live there. As a result of their tolerance, these northern species can occupy a larger range of habitats and thus have larger geographical ranges. This hypothesis predicts that the climatic tolerance of terrestrial animals and plants should increase from the equator to the poles. This prediction has not been tested.

In the ocean, temperature variation in surface waters is highest in temperate zones and much lower near the poles and in the tropics. Thus, the climatic variability hypothesis predicts that the temperature tolerance of surface marine fishes should be lowest in both polar and tropical waters. This prediction is borne out by the data in Figure 2.9. In deeper ocean waters, temperature variability is minimal, so the climatic variability hypothesis would predict no relationship

(a)

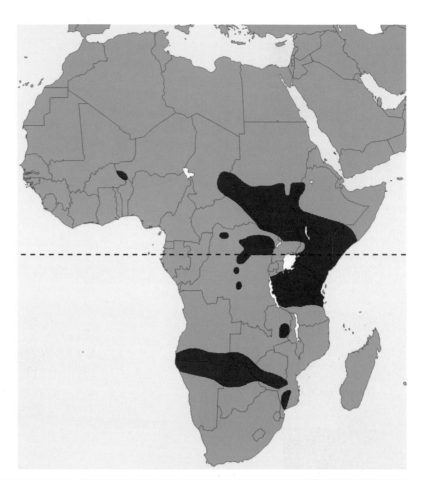

(b)

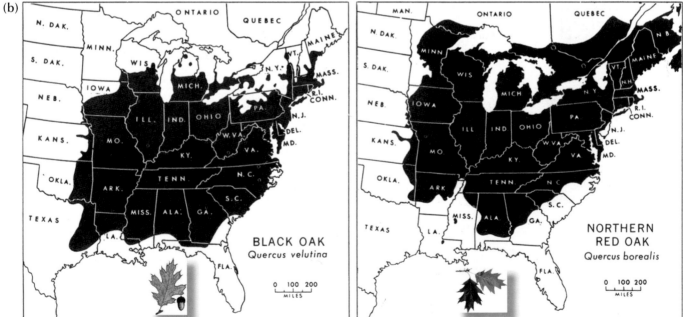

Figure 2.5. Examples of geographic range maps. (a) The giraffe (*Giraffa camelopardalis*) is common in East Africa and southern Africa, but is becoming rare in West Africa. (b) The black oak (*Quercus velutina*) and the northern red oak (*Quercus borealis*) have overlapping ranges in eastern North America. (Modified from Gleason and Cronquist 1964, photo courtesy of Calvin Jones.)

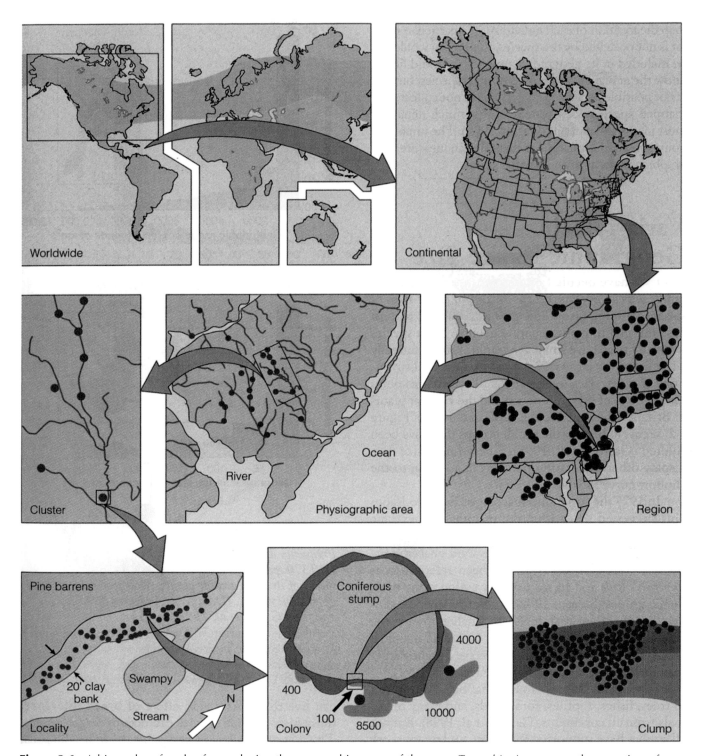

Labels in figure (top to bottom, left to right):
Worldwide
Continental
Cluster
River
Ocean
Physiographic area
Region
Pine barrens
20' clay bank
Swampy
Stream
Locality
N
Coniferous stump
400
100
8500
10000
4000
Colony
Clump

Figure 2.6. A hierarchy of scales for analyzing the geographic range of the moss *Tetraphis*. Answers to the question of what limits the range of a species may depend on the scale of the study area. (After Forman 1964.)

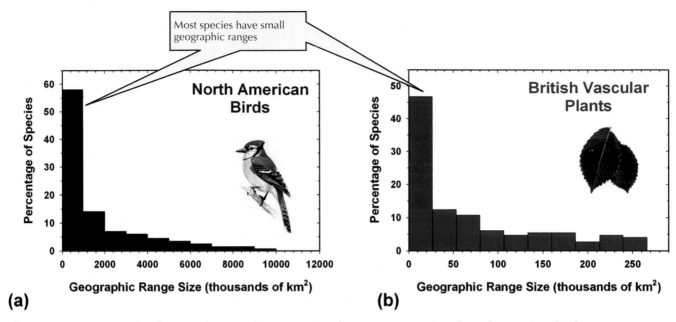

Figure 2.7. Frequency distribution of geographic range sizes for (a) 1,370 species of North American birds, (b) 1,499 species of British vascular plants. (Data from Anderson 1985 for (a), and Gaston *et al.* 1998 for (b)).

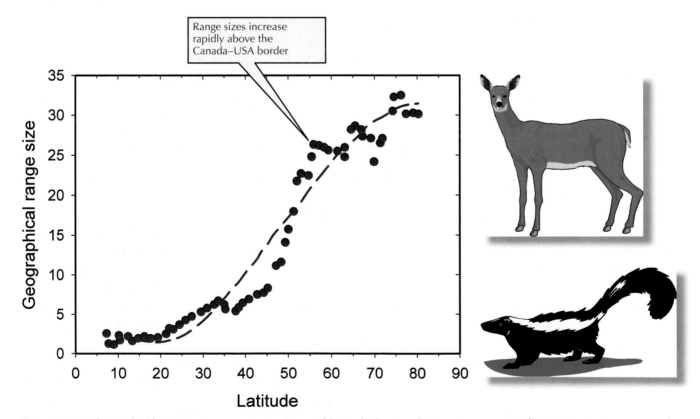

Figure 2.8. Relationship between average range size and latitude for North American mammals. Range size is expressed as the percentage of the total land area of North America, and each point represents an average for each 2° of latitude. Southern species have smaller geographic ranges than northern species, following Rapoport's rule. (From Pagel *et al.* 1991.)

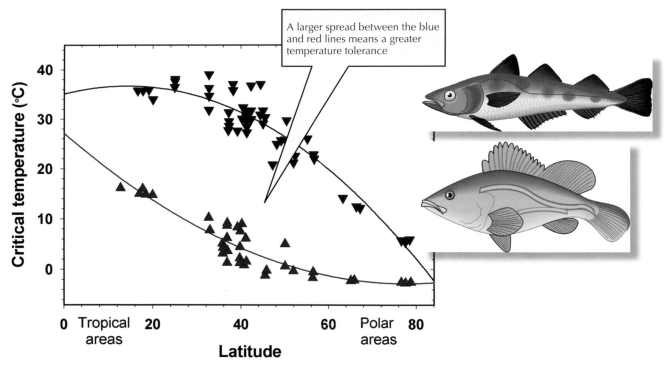

Figure 2.9. Critical temperature limits for marine fishes from shallow waters. Each dot represents data from a single species. Upper critical temperatures are in blue, and lower critical temperatures are in red. According to the climatic variability hypothesis, fish species in temperate regions should have the greatest temperature tolerance. The plotted data support this hypothesis, but more data are needed for the 0–10° latitude range. (Data from Brett 1970, fish images courtesy of Diane Rome Peebles, Florida Fish and Wildlife Conservation Commission.)

between latitude and range size for deep-sea organisms. There are no data to test this prediction.

Q Could the shape of the continents affect how well Rapoport's rule applies to the geographic ranges of terrestrial organisms?

Rapoport's rule is a good illustration of how ecologists can apply the scientific method to develop explanations for patterns in natural systems (see Chapter 1). First, ecologists describe the patterns clearly and accurately. Then, they propose hypotheses to explain these patterns, and make predictions based on each hypothesis. If the observed data agree with the predictions of a hypothesis, they may conclude tentatively that the hypothesis is supported. Then they search for exceptions to the hypothesis and try to uncover reasons why there might be exceptions.

Abundance Within Geographic Ranges

Species are not equally abundant in all parts of their geographic range. If environmental conditions within a species' geographic range become less favorable from the center to the edges of the range, we would predict the abundance of the species to be highest near the center and lowest near the edges. This prediction is supported by surveys of the western gray kangaroo in Australia (Figure 2.10a) and the eastern meadowlark, a common grassland songbird in North America (Figure 2.10b). However, many additional species will have to be tested before we can accept this prediction as an ecological generalization. If it holds true, it will have many implications for applied ecological problems. For example, it would suggest that climate changes would have the strongest effect on a species at the edge of its range, not in the center. It would also suggest that nature reserves should be placed at the center of the range to protect more individuals per unit area and that central populations should be the focus of conservation efforts.

Is there any relationship between the size of a species' geographic range and its abundance? If a species is widespread, is it always a common species? Conversely,

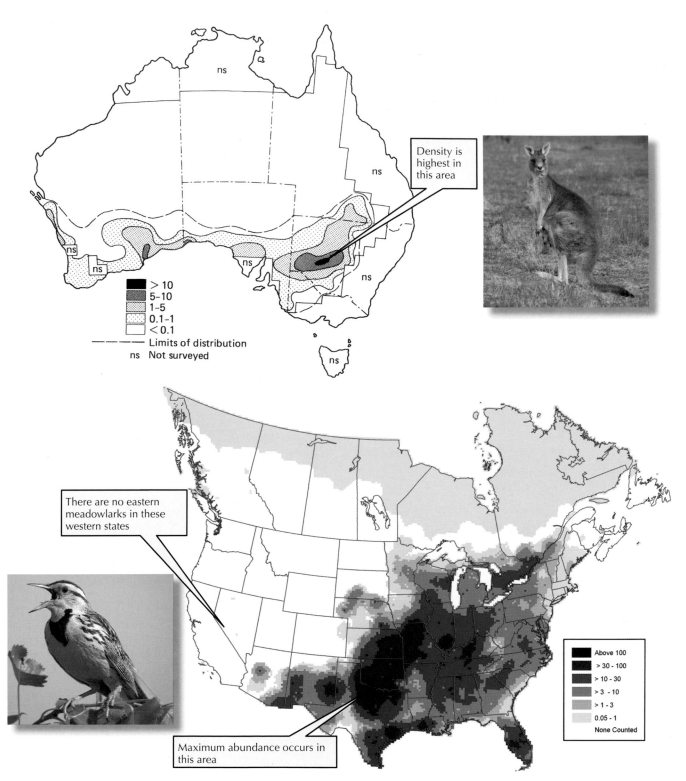

Figure 2.10. Population density maps for (a) the western gray kangaroo (*Macropus fuliginosus*) and (b) the eastern meadowlark (*Sternella magna*). For both species, population density is highest at the center of the species' range. Data in (a) were obtained from aerial surveys conducted from 1980 to 1982. (From Caughley *et al.* 1987.) Data in (b) were obtained from the North American Breeding Bird Surveys conducted from 1994 to 2003 and are expressed as the number of birds counted in standard transects of 50 stations over 39.4 km (24.5 mi). (Photo of kangaroo courtesy of Alice Kenney, photo of eastern meadowlark courtesy of Mike McDowell.)

32

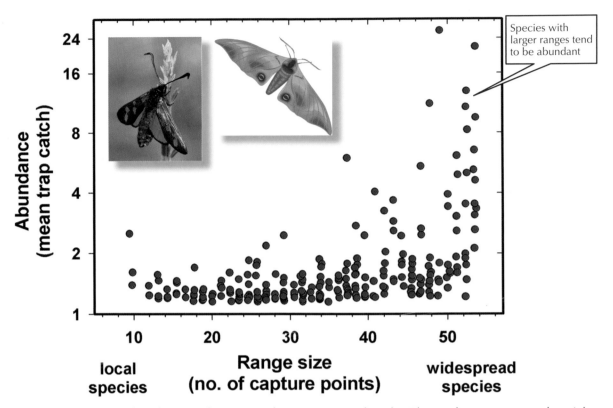

Figure 2.11. Relationship between abundance and range size for 263 species of moths. The moths were captured at night in light traps at 53 sites across the British Isles. Abundance data (mean trap catch per year) for each species were averaged over 6–14 years for all sites. (Not all light traps could be operated every year.) Range size is expressed as the number of sites occupied by a given species. Each dot represents one species. (From Gaston 1988.)

if a species is rare or threatened, does it have a small geographic range? Jim Brown, from the University of New Mexico, was the first ecologist to ask these questions and to investigate the possible relationship between range size and abundance. He and other ecologists have assembled data on a wide variety of organisms showing that species that are more widespread are typically more abundant. For example, Figure 2.11 shows data for 263 species of moths collected at 53 sites throughout the British Isles over a 14-year period. Although there is much variability in these data, they clearly show that more widespread moth species tend to be more abundant. A similar pattern has been found for frogs, birds, plants and many other groups.

One possible explanation for the correlation between abundance and range size is that species that can exploit a wide variety of resources become both widespread and common. Such species are called **generalists.** Species that exploit only a few resources, called **specialists**, have smaller ranges and are less abundant because they have strict requirements for food or shelter. This explanation is called the **ecological specialization model**. Further studies are now

Jim Brown and Astrid Kodric-Brown, Professors of Biology, University of New Mexico

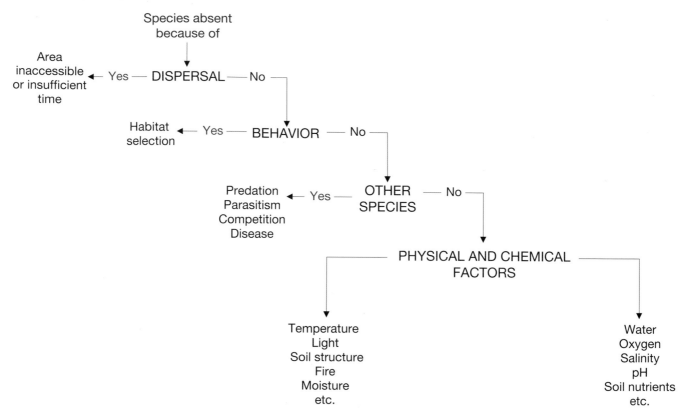

Figure 2.12. A sequence of steps for testing the factors that limit the geographic range of a species.

directed toward determining the ecological attributes of species that are widespread and abundant. Identifying these attributes may help us understand why some species are rare or endangered, and could assist in conservation efforts.

■ 2.3 A SEQUENCE OF HYPOTHESES GUIDES AN ECOLOGIST'S ANALYSIS OF WHAT LIMITS RANGES

What ecological factors limit the geographic range of a species? Ecologists are keenly interested in answering this question because it may enable us to predict the spread of pest species, such as the red fire ant discussed earlier, that have been introduced into a new area. It may also allow us to anticipate changes in the geographic ranges of species as Earth's climate warms during the next century.

To determine the factors that limit a species' geographic range, we can analyze possible factors through the sequence of steps shown in Figure 2.12. Each step in this sequence can be stated as an ecological hypothesis about one factor that may limit geographic ranges. The sequence of steps is hierarchical because you must have answered the questions at the top of the hierarchy before you can proceed to lower-level questions.

The hypothesis in the first step (DISPERSAL) is that the species is absent from an area because a barrier has prevented it from reaching the area. One way to test this hypothesis is to perform a **transplant experiment**, in which you move some individuals of that species into the area (Figure 2.13). If the transplanted individuals survive and reproduce, we may conclude that the environmental conditions in the area are adequate for the species and that its range is limited by a lack of dispersal. A proper transplant experiment should have **controls**—transplants done within the normal geographic range to reveal any effects due simply to handling and transplanting the individuals. Humans have moved many species around the globe in this way, and some of these

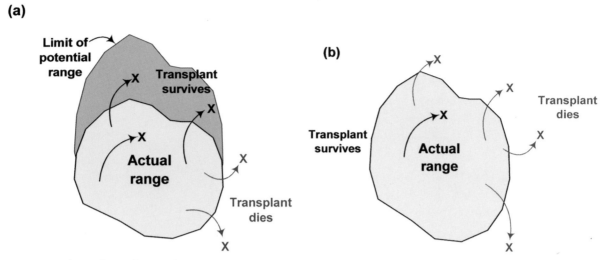

Figure 2.13. Two hypothetical sets of transplant experiments. The yellow area represents the actual current range of a species. Each arrow indicates a transplanted individual. Black arrows indicate successful transplants, and red arrows indicate unsuccessful transplants. (a) The species can potentially occupy a larger range (blue area) than it currently does. (b) The species currently occupies its potential range, so all transplants outside its current range fail. In practice, many separate transplant experiments may be needed to define the limits of a species' potential geographic range.

transplants have become serious pests, so transplant experiments must be performed very carefully.

If the species cannot survive and reproduce in the transplant areas, it is clearly not limited by dispersal, and we proceed to the next hypothesis: The species range is limited behaviorally or by habitat selection. Species may be programmed to select a smaller range of habitats than they could theoretically occupy because of evolutionary constraints. Habitats all over the world have been changed rapidly and dramatically by human agriculture and forest management, and many species have not had enough time to evolve in ways that would allow them to use these altered habitats.

Q Is habitat selection limited to animals, or can plants and fungi show this behavior as well?

More commonly, species that are not limited in their geographic range by dispersal or habitat selection may be limited by interactions with other species. These interactions may involve either the negative effects of predators, parasites, disease organisms or competitors, or the positive effects of interdependent species within the actual range. We can often determine whether interactions with other species limit the range of a species by performing a transplant experiment with a protective device, such as a cage that excludes suspected predators or competitors. For example, researchers have transplanted barnacles, which normally live on rocks in the intertidal zone, to deeper waters below the tidal zone, and used mesh cages to keep predators such as sea stars and marine molluscs away. These experiments showed that barnacles can live and reproduce in deeper waters when they are protected from their predators.

If other species do not limit the actual range of a species, we are left with the final possibility that physical or chemical factors are responsible. For example, many tropical plant species cannot withstand freezing, so they are excluded from geographic areas where frost occurs. The effects of physical and chemical factors on the survival, growth and reproduction of organisms are the subject of an entire discipline called **physiological ecology**.

Liebig's Law of the Minimum

In 1840, the German chemist Justus von Liebig laid the groundwork for physiological ecology by defining the key concept of a **limiting factor**. He postulated what is now known as **Liebig's law**[1] **of the minimum**, which states that the rate of any biological process is limited by a single factor that is least available, relative to an organism's requirements. According to Liebig's law, crop yields are limited by a single nutrient, and if one adds the limiting nutrient in fertilizer, crop production will increase. Liebig's work with artificial fertilizers was revolutionary in its time because it linked chemistry and biology and led to higher crop yields.

Q Can two factors ever limit crop yields at the same time in the same place?

The concept of limiting factors is used extensively in ecology and is particularly useful in uncovering the environmental factors that limit geographic distributions. For example, consider the question of what limits the distribution of the red fire ant in the United States. One approach to this question is to vary temperature and observe how that affects the survival and reproduction change of the species. But we must remember that the same factor may not be limiting everywhere in a species' geographic range. We might find that low temperature is limiting for the red fire ant in one area, such as northern Oklahoma, but low rainfall, or some other factor, may be limiting in another area, such as northern New Mexico. We must define the ecological tolerances of species to a whole array of potentially limiting environmental factors, a broadening of Liebig's law that was first elaborated by Victor Shelford.

Shelford's Law of Tolerance

Victor Shelford, one of the earliest North American animal ecologists, was the first to formalize the ideas of physiological ecology by applying Liebig's law to the distribution of species in natural communities. Working at the University of Illinois, Shelford developed the

Victor E. Shelford (1877–1968) Professor of Zoology, University of Illinois

major conceptual tool of physiological ecologists, Shelford's law of tolerance, which can be stated as follows: *The distribution of a species is controlled by the environmental factor for which the species has the narrowest tolerance.*

Figure 2.14 illustrates Shelford's law for two marine snails that live in the rocky intertidal zone of California. Surface temperatures never drop very low in this region, but can be quite high on exposed, rocky shores above the middle tide line. Accordingly, the upper lethal temperatures for these snails limit their ranges both locally (within the intertidal zone) and regionally (from northern to southern California).

We can conduct tolerance studies for oxygen, pH, salinity, and many other environmental factors, and build up a detailed picture of the limits of tolerance for any particular species. Figure 2.15 illustrates a hypothetical case of a plant species whose survival and reproduction is limited by two factors: moisture and temperature. To these simple Shelford tolerance graphs we can then add complications, such as variations in an organism's tolerance during its development. For example, the young stages of organisms are often the most sensitive to environmental factors. Consequently,

---

1 A more accurate name would be 'Liebig's principle of the minimum' (see Table 1.1). The same applies to Shelford's law of tolerance (see pages 36–37). However, we will continue to refer to both as laws for historical reasons.

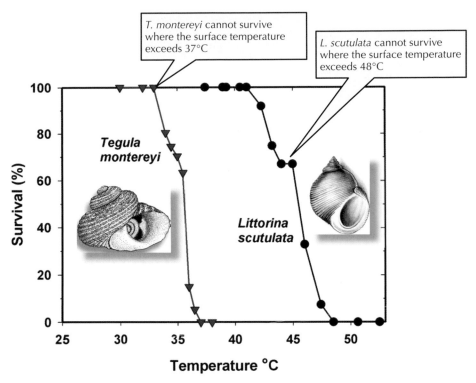

T. montereyi cannot survive where the surface temperature exceeds 37°C

L. scutulata cannot survive where the surface temperature exceeds 48°C

Figure 2.14. Upper temperature tolerance for two species of snails from the rocky intertidal zone of California. *Tegula montereyi* lives in the less exposed, lower intertidal zone, where surface temperatures are kept low by tidal water. *Littorina scutulata* lives in the upper intertidal zone, where surface temperatures are much higher and where *T. montereyi* cannot survive. (Data from Somero 2002.)

we should test the most sensitive stage when we determine a species' tolerance.

Ecologists use a great deal of natural history information when they apply tolerance studies to real-world situations. Animals are mobile and can use a variety of mechanisms to avoid lethal environmental conditions.

Arctic ground squirrels survive harsh conditions by hibernating for 6–8 months of the year.

Many birds, and some insects, escape harsh polar winters by migrating to temperate or equatorial regions. Some mammals, such as the arctic ground squirrel, hibernate during the coldest months of the year and thereby eliminate the need to feed when food is scarce. Plants become dormant and resistant to low temperatures in winter, while many insects enter a cold-tolerant stage.

In the next chapter, we will use the principles of limiting factors and physiological tolerance developed in this chapter as we consider the limits on geographic distributions in more detail.

SUMMARY

Geographic distributions can be studied on a series of spatial scales, from global to local. Different limiting factors will be critical at different scales.

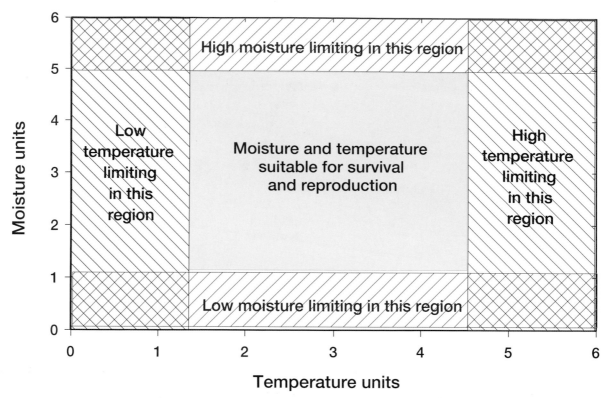

Figure 2.15. Idealized plot of Shelford's law of tolerance for two factors that limit the geographic range of a plant species. In this hypothetical example, the species cannot survive if the temperature is below 1.3 units or above 4.5 units, or if the moisture level is below 1.1 units or above 5 units. The geographic range is limited to regions where the temperature and moisture values lie within the yellow zone.

At a continental scale, most species have small geographic ranges and only a few have large ranges. Species that have large geographic ranges also seem to be the most common species locally. Such widespread species are often generalists that can use a large range of food and habitat resources. By contrast, specialists often have small geographic ranges limited by their strict requirements for food or shelter.

The geographic range of a species may be limited by a variety of factors, including dispersal, interactions with other species and physical or chemical factors such as temperature or nutrients. A transplant experiment is one way to assess the role of certain factors. For example, if a species can survive when it is transplanted to a new area outside its current geographic range, then its current range is limited by dispersal. Humans have moved many species around the globe with unfortunate consequences when pests are brought into new areas. Understanding the limitations on geographic ranges has practical uses, particularly if climate change progresses during the next century.

SUGGESTED FURTHER READINGS

Brown JH (1984). On the relationship between abundance and distribution of species. *American Naturalist* **124**, 255–279. (The classic paper on range sizes and abundance.)

Brown JH and Lomolino MV (1998). Distributions of single species. In *Biogeography*, 2nd edn. pp. 61–94. Sinauer, Sunderland, Mass. (A good discussion of geographic range limitation from the best biogeography text.)

Gaston KJ (1991). How large is a species' geographic range? *Oikos* **61**, 434–438. (An excellent discussion of how the concept of a geographic range is not simple.)

Sagarin RD and Gaines SD (2002). The 'abundant centre' distribution: to what extent is it a biogeographical rule? *Ecology Letters* **5**, 137–147. (A critical assessment of the prediction that the density of a species is highest near the center of its geographic range, and a good example of how scientists evaluate data.)

QUESTIONS

1. Find the average annual temperature and precipitation for the area where you live. Then locate the intersection of these values on the graph in Figure 2.3. What is the predicted vegetation type for your area? Does it agree with what you observe there?

2. One possible explanation for Rapoport's rule for North America is that the area available for organisms to occupy is much larger as you go north, because North America is approximately V-shaped. Hence, more northerly mammals would have larger geographic ranges. How might you test this explanation?

3. In 1846, Asa Gray, a Harvard botanist and student of evolution, pointed out that the plants of eastern North America were more similar to the plants of eastern Asia than to the plants of western North America. Suggest some possible reasons for this observation, which has been called Gray's paradox. If you want more details on Gray's paradox, see the analysis in Qian (2002).

4. Piranhas are South American predatory fish that can attack and kill large animals, including humans. Red piranhas (*Pygogcentrus nattereri*) can be bought in exotic fish stores in the United States and have been deliberately or accidentally introduced into the wild in southern states. How would you determine the factors that limit the geographic distribution of the red piranha in the United States? What you would recommend regarding the importation and sale of this fish in the United States? Bennett *et al.* (1997) evaluate this question.

5. Discuss the application of general methods for studying geographic ranges to the question of what limits the geographic range of humans, both now and early in our evolutionary history.

6. How far might the red fire ant spread in Australia if it escapes from Brisbane? Describe the general approach you might take to answering this ecological question before you read Sutherst and Maywald (2005).

Chapter 3

WHAT LIMITS GEOGRAPHIC DISTRIBUTIONS?

IN THE NEWS

San Francisco Bay supports the largest and most ecologically important expanses of tidal mudflats and salt marshes in the western United States. Many non-native species of plants and animals have been introduced to the bay, and some now threaten to cause fundamental changes in the bay's tidal areas. One such species is the Atlantic salt marsh cordgrass (*Spartina alterniflora*), which was introduced in the 1970s from the east coast. By 2003, cordgrass had colonized 1500 ha (3,700 acres) of the nearly 16,000 ha (40,000 acres) of tidal marsh and 12,000 ha (29,000 acres) of tidal flats in San Francisco Bay. This cordgrass is likely to dominate tidal marshes, cause the extinction of native tidal marsh plants, choke tidal creeks by its extensive growth and eliminate thousands of hectares of shorebird habitat. Cordgrass colonization could endanger threatened species that use existing tidal marsh habitats, including the rare California clapper rail and the salt-marsh harvest mouse.

California clapper rail

Harvest mouse

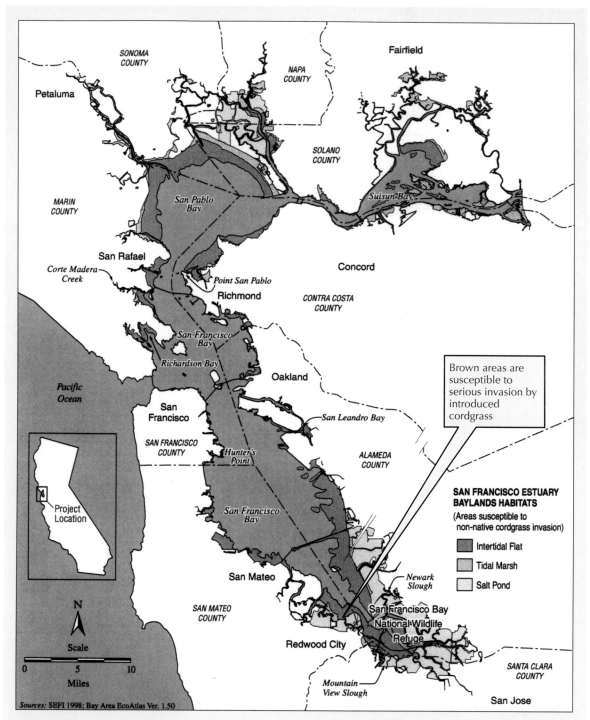

Map showing the sites in the San Francisco Bay region that are susceptible to invasion by non-native cordgrass

Once established in San Francisco Bay, invasive cordgrass could rapidly spread to other sites along the California coast through seed dispersal on the tides. California is implementing a coordinated, region-wide eradication effort called the Invasive Spartina Project (http://www.spartina.org/index.htm) to stave off this invasion at an initial cost of US$1 million per year.

Ecologists want to know how cordgrass was introduced to the San Francisco Bay area, how rapidly it might spread and what factors might limit its spread to other bays up and down the Pacific Coast.

Cordgrass can form dense stands in the tidal zone.

3.1 DISPERSAL OFTEN LIMITS RANGES ON A GLOBAL SCALE

To understand why species are found in particular places, and how their geographic ranges may shift in response to environmental changes, we need to know what factors set the limits to their ranges. In an era of global environmental change, we have a practical interest in learning how widely an introduced pest might colonize a new area and how climate change might affect the distributions of both common and endangered species. As we saw in Chapter 2, there are many factors that can limit geographic ranges. We will begin by considering continental-scale ranges and cases where a species' range is limited by its ability to move, such as on oceanic islands.

Types of Dispersal

The simplest explanation for why a species is not found in a particular area is that it has not been able to reach the area. The movement, or dispersal, of organisms can occur by two mechanisms.

1. **Diffusion** is a common form of dispersal involving the gradual spread of a population across hospitable terrain. It results from the movement of many individuals over several generations. Diffusion is well illustrated by the spread of many introduced species after their introduction to a new area.

2. **Jump dispersal** is the movement of individual organisms across large distances, usually over unsuitable terrain, to a new area where they establish another population. This form of dispersal occurs in a short time during the life of an individual. Oceanic islands can be colonized by jump dispersal. Human introductions of pest species can be viewed as an assisted form of jump dispersal.

The Spread of the Gypsy Moth, an Introduced Pest

The European gypsy moth (*Lymantria dispar*) is an example of a species that has spread by both mechanisms of dispersal. In 1868 or 1869, Ettiene Leopold Trouvelot, an amateur entomologist, brought some gypsy moth eggs from France to his home in Massachusetts. A few of the caterpillars that hatched from the eggs escaped. Trouvelot understood the seriousness of that mishap and informed the local authorities, but they were not concerned. Thus began one of the most devastating caterpillar plagues to hit North America. Gypsy moth caterpillars eat the leaves of a great variety of deciduous and coniferous trees, including apple, alder, basswood, oak, poplar, willow, birch, larch and hemlock. To curb the severe defoliation of deciduous trees caused by the caterpillars, Massachusetts initiated a control program in 1889 and by 1900 the severity of the outbreaks was reduced. Consequently, the state terminated the program. That action proved to be premature, as gypsy moths then began to spread in a wave across New England (Figure 3.1). The spread slowed around 1950, but in the 1960s gypsy moths were accidentally carried to Michigan, probably as eggs on camping equipment, initiating a new spread from that state. The gypsy moth defoliated 5.6 million ha (13.8 million acres) of forest in the United States in 1981 and another major outbreak began in 1989. In 1990, the US Forest Service started placing traps laced with a pheromone (a chemical released by female gypsy moths that attracts males) along the expanding boundary of the moth's range. These traps have slowed, but not stopped, the spread of the moth. Timber losses caused by the gypsy moth are so large that the benefits of this major research and control program outweigh its costs by more than four to one.

One factor that may affect the spread of the gypsy moth is a disease caused by the fungus *Entomophaga maimaiga*, a widespread gypsy moth pathogen in Asia. In 1904, the fungus was introduced into gypsy moth populations in North America. Little was known about the effectiveness of this introduction until 1989, when researchers discovered that *E. maimaiga* was causing extensive disease outbreaks among gypsy moths in several New England States. By 1992, the fungus was found throughout the gypsy moth's range, and recent collapses of gypsy moth populations in Pennsylvania and Virginia have been attributed to this fungus. Ecologists are now trying to predict how this fungal disease will influence the future range of the gypsy moth in North America.

The gypsy moth problem illustrates three recurring themes in geographic ecology: (1) Species moved by

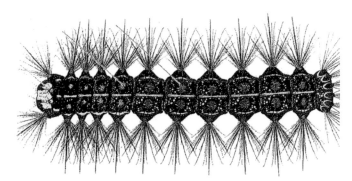

Gypsy moth caterpillar (From Michigan Department of Agriculture)

humans can become serious pests years after they are introduced; (2) governments often ignore the problem of introduced species in the early stages of introduction when the problem is often easiest to address; and (3) pests such as the gypsy moth spread by diffusion once they are established. The gradual spread of introduced species may be slowed, but cannot easily be stopped.

Aquatic Invasions

Invasions of non-native species in terrestrial habitats have long been recognized as a source of environmental problems, but much less attention has been paid to such invasions in aquatic habitats. Many aquatic invasions been assisted by humans. During the 19th century, aquatic organisms often arrived in foreign ports attached to the bottoms of ships, but this dispersal mechanism largely disappeared with the advent of antifouling paints and faster ships, which reduced the transit time during which organisms could attach to the ships' hulls.

However, the discharge of ballast water has increased dramatically as ships have become larger. A single ship can now carry 140 million kg (150,000 tons) of ballast water to maintain trim and stability. Chesapeake Bay in the eastern United States received 10 billion kg (11 million tons) of ballast water discharge in 1991, mostly from ships originating in Europe and the Mediterranean. Included in ballast water are large numbers of organisms of diverse species. For example, the ballast water in five container ships that entered Hong Kong in 1994 and 1995 was found to contain 81 species belonging to 13 phyla of animals and protists.

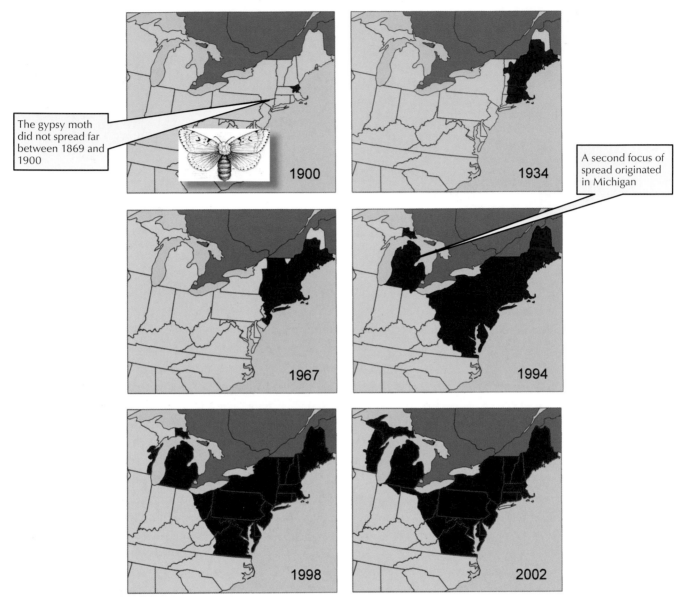

The gypsy moth did not spread far between 1869 and 1900

1900

1934

A second focus of spread originated in Michigan

1967

1994

1998

2002

Figure 3.1. Spread of the gypsy moth in the United States since its introduction in Boston in 1868–1869. Most of the forests in the eastern United States are at risk as the moth continues to spread south and west. (Map courtesy of S. Liebhold, U.S. Forest Service Northeastern Research Station.)

The biological results of ship-assisted jump dispersal are significant. Chesapeake Bay now has 116 introduced marine species and San Francisco Bay has 212. Some of these species, such as the Asian clam in San Francisco Bay, have become dominant members of the community. The zebra mussel is one of the best-known examples of an aquatic species brought to North America in ballast water. Native to Europe, the zebra mussel was found in Lake St. Clair, between Lake Huron and Lake Erie, in 1988. By 1992 it had established populations in all of the Great Lakes as well as the Mississippi, Ohio and other major rivers of the eastern United States. Zebra mussels form thick mats of up to one million animals per square meter, which can clog water intake pipes and eliminate native clam species.

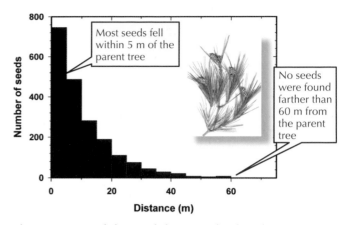

Figure 3.2. Seed dispersal distances for the Aleppo pine (*Pinus halepensis*). The average seed dispersal distance was 11.4 m. Native to the Mediterranean region, the Aleppo pine produces seeds in cones, and the seeds are wind dispersed. (Data from Nathan *et al.* 2000.)

Zebra mussels attached to a stick

Q Would you expect invasive aquatic species to be a more severe problem in fresh water or in salt water?

Introduced aquatic species can also pose health risks to humans. Single-celled marine protists called dinoflagellates are transferred worldwide in ballast water. Some dinoflagellates produce toxins that are accumulated by marine molluscs, which may be deadly to people who consume them. The cholera bacterium *Vibrio cholerae* has been found in the ballast tanks of some ships and can survive for up to 240 days in sea-water at 18°C. When released into an estuary, the bacterium can attach to a variety of organisms and thus enter the human food chain.

There are two ecological messages in these aquatic invasions. Firstly, many of these introduced species were originally limited in their global distribution by dispersal. Secondly, action is urgently needed to reduce the global transport of potentially harmful organisms in ballast water.

Tree Recolonization at the End of the Ice Age

One of the most spectacular colonizations began at the end of the last ice age, about 14,000 to 16,000 years ago. During the following 10,000–12,000 years, as glaciers retreated from Europe and North America, oaks and other temperate-zone trees expanded their range northwards by 1,000 km. In 1899 the British botanist Charles Reid calculated that this expansion would have taken a million years if the trees had spread by diffusion alone. In diffusion, the migration rate of a species depends on the species' reproductive rate and its average dispersal distance (see *Essay 3.1*). We can measure the average dispersal distance of tree seeds by placing seed traps at different distances from a tree or by mapping the locations of seedlings produced by an isolated tree. Figure 3.2 shows the results of such measurements for one species of pine. Tests on 12 other species of temperate-zone trees in the southern Appalachians revealed average seed-dispersal distances ranging from 4 to 34 m.

According to the diffusion model, if an average temperate-zone tree produces 10 million seeds in its lifetime and the seeds fall 30 m from the tree, the geographic range of trees in Europe should have expanded by only 36 km since the end of the ice age. The discrepancy between this distance and the observed expansion is called **Reid's paradox.**

Two possible explanations for Reid's paradox have been suggested. The first maintains that a species' migration rate depends more on extreme dispersal events than on average dispersal distance. Although the average dispersal distance of seeds is short, a few seeds may be carried very long distances by wind or

ESSAY 3.1   A SIMPLE MODEL OF DISPERSAL MOVEMENTS

The simplest model of species migration is called the **diffusion model**, which is based on an equation that describes chemical diffusion. Skellam (1951) showed that if trees or other organisms migrate by diffusion, the distance they migrate can be calculated using the following equation:

$$\text{Distance moved} = Dn\sqrt{\log R_0}$$

where D = average dispersal distance,

n = number of generations, and

R_0 = reproductive rate per generation.

For example, if an average spruce tree produces 20,000 seeds per year and lives as a mature tree for 50 years, the reproductive rate per generation (R_0) is 20,000 X 50, or 1 million seeds. If seeds fall an average of 35 m (D) from the tree, we can calculate how far a population of spruce trees will move in a given time, such as 1,000 years, or 20 generations (n):

$$\text{Distance moved} = 35 \times 20 \sqrt{\log(1000000)}$$
$$= 4200 \text{ meters}$$

or 4.2 km.

In practice, we must modify the diffusion model because all individual organisms do not live for the same time or reproduce at the same rate, and some seeds or spores may be transported greater distances by birds or windstorms. Nevertheless, the diffusion model forms a good starting point for estimating dispersal and colonization rates.

animals. We would classify the movement of these long-range dispersers as jump dispersal rather than diffusion. Such movements are difficult to record and measure, because they might involve fewer than one seed in 10,000.

An alternative explanation for Reid's paradox is that a few trees may have existed in isolated pockets farther north during the ice age, and recolonization of northern Europe and North America may have originated from those trees rather than from southern populations. The fossil record shows no evidence of isolated northern pockets, but if there had been only a small number of those trees, the fossils they left may be too scarce to be detected by present techniques or conditions may not have been suitable for fossilisation.

Island Colonization

The results of jump dispersal are seen most easily on islands, because all the organisms that colonize an island after it forms must do so across water. Colonization of an island by terrestrial plants and animals is determined largely by the island's distance from the nearest mainland: more remote islands generally have fewer species. (Figure 3.3). Thus, distance acts as an important filter on the species composition of islands.

Natural catastrophes can create areas of new habitat within continents, and these landlocked 'islands' provide opportunities for ecologists to study colonization by diffusion and jump dispersal. For example, during the first 6 years following the volcanic eruption of Mount St. Helens in Washington in 1980, invasion of the barren pumice plains by vascular plants was minimal, despite the proximity of seed sources (Figure 3.4). To find out why, researchers planted 16,000 seeds of various plant species native to Mount St. Helens in barren sites on the volcano. Only 1,745 seedlings emerged. Species with the largest seeds had the highest chances of growing. However, when released

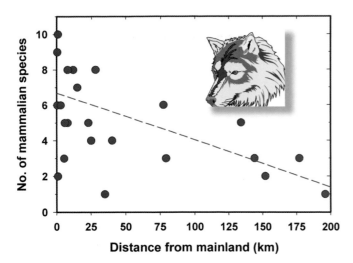

Figure 3.3. Number of terrestrial mammalian species on islands off the British Columbia coast of Canada in relation to distance from the mainland. More remote islands have fewer species. Considerable scatter in the data is caused by differences in the area of habitat on the various islands. (Data from Lawlor 1986.)

naturally, these seeds rarely got as far as 3 m from the parent plant due to their size. In contrast, plants with lighter seeds, such as *Aster,* had a greater dispersal distance, but their seeds often died from drought stress because they were small. Therefore, recolonization by plants on Mount St. Helens was patchy and slow because it was limited both by seed dispersal and seedling establishment.

Nineteen years after the eruption of Mount St. Helens, there were 95 species of vascular plants in the alpine and subalpine areas of the mountain—only one-half to one-third as many as on three nearby volcanoes. Most of the plants missing from Mount St. Helens seem to be absent because of a lack of dispersal. The mountain's subalpine zone is isolated from the subalpine zones of nearby volcanoes by 50–80 km of lowland forest, which is as great a barrier to seed dispersal as the oceans are to land mammal migration. The missing species will become re-established on Mount St. Helens only if birds, or other wide-ranging animals, carry seeds past the lowland forest barrier.

Range limitation imposed by barriers to dispersal is the basis for a country's quarantine restrictions, which are an important part of the fight to contain invading species. Quarantine measures that restrict the importation of plant pathogens are critical to

protect the agricultural crops of islands. Hawaii and New Zealand lack native snakes, and quarantine services routinely intercept snakes brought inadvertently in cargo or deliberately by people. Egg masses of gypsy moths have been found in Australia on ship cargo from Russia, and in New Zealand on imported motor vehicles. Quarantine practices remind us that pest species can move rapidly by jump dispersal.

■ 3.2 PHYSICAL OR CHEMICAL FACTORS USUALLY LIMIT RANGES ON REGIONAL AND CONTINENTAL SCALES

As you learned in Chapter 2, Shelford's law of tolerance states that the distribution of a species is controlled by the environmental factor for which the species has the narrowest tolerance. Only one environmental factor—temperature, moisture, salinity, soil pH, calcium content or something else—limits a species at any given time. An ecologist needs critical natural history information to isolate the physical and chemical factors that limit a species.

There are three steps to determining whether an environmental factor limits the range of a species:

1. Determine the stage of the life cycle that is most sensitive to the factor.
2. Determine the physiological tolerance of that stage.
3. Show that variations in the factor are within the tolerance limits of the species in its geographic range and that they exceed those limits outside its range.

Figure 3.5 shows schematically how this procedure works. If the tolerance limits of a species are exceeded in certain sites (C and D in Figure 3.5), those sites cannot be colonized. If the species is transplanted there, it will not become established and will die out. This set of procedures can be used in computer models to predict the geographic ranges of terrestrial species if temperature and rainfall are the limiting factors (see *Essay 3.2*).

The winter ranges of passerine (perching) birds in North America often correlate with minimum January temperature. One example is the eastern phoebe, which has a northern winter range limit of –4°C (Figure 3.6). Temperature limitations in temperate-zone birds are

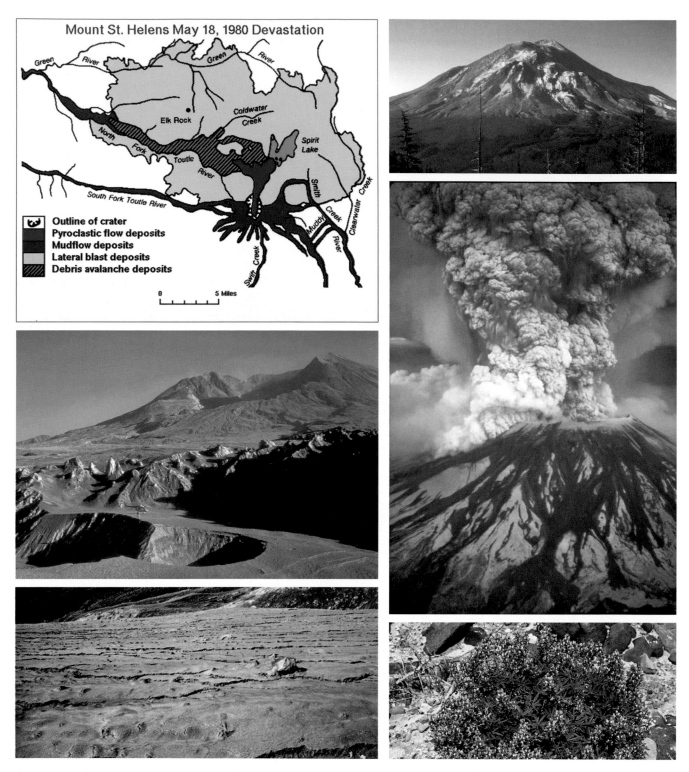

Figure 3.4. The eruption of Mount St. Helens, Washington, May 18, 1980, and a view of the mountain the day before the eruption. The center photographs show the eruption in progress and a view of the devastated area 4 months later. Note the helicopter in this photo from September 1980. The lower left photo shows an area of Pine Creek Ridge shortly after the eruption, showing the scour from mudflow and ash. The lupine (*Lupinus latifolius*) was an early colonizer. (Photos courtesy of US Geological Survey/Cascades Volcano Observatory and Roger del Moral.)

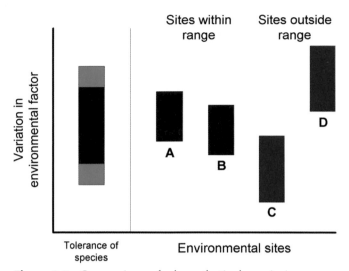

Figure 3.5. Comparison of a hypothetical species' tolerance to one environmental factor and the variation in that factor at four sites (A–D). Tolerance is measured for the stage of the life cycle that is most sensitive to that factor and is divided into an optimal zone (dark blue) and two marginal zones (light blue). The species can live in sites A and B but not in C or D.

directly linked to the higher metabolic rates needed to maintain body temperature in cold weather and the reduced amount of food available in winter to meet those metabolic demands.

Q **If bird lovers provide food for migrating birds during winter, what effect might that have on the birds' winter geographic ranges?**

The geographic limits of invertebrates and algae are often very sharply defined in the intertidal zone of rocky coastlines (Figure 3.7). In the British Isles, two barnacle species, *Chthamalus stellatus* and *Balanus balanoides*, dominate the intertidal zone. *Chthamalus* is common in the upper intertidal zone of western Britain and Ireland, but it is absent from the colder waters of eastern Britain. The lower limit of the *Chthamalus* range in the intertidal zone is often determined by competition for space with *Balanus*, which grows faster than *Chthamalus* and effectively squeezes it out of the

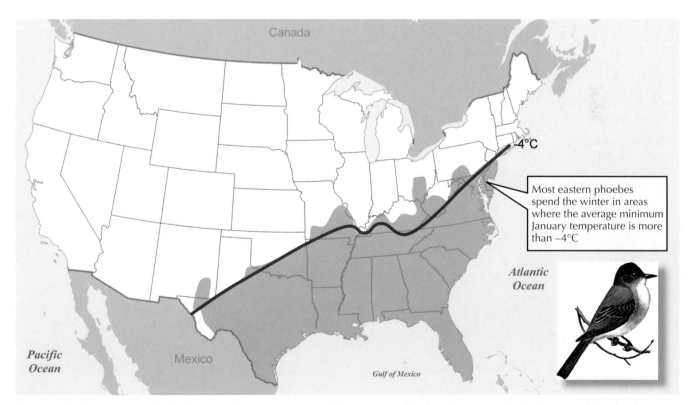

Figure 3.6. Temperature limitations on the winter range of the eastern phoebe (*Sayornis phoebe*). Most of this bird's winter range from 1962 to 1972 (green area) lies south of the –4°C isotherm of average minimum January temperature for that period (red line). The bird's winter range has moved north during the last 30 years, as average winter temperatures have risen. (Modified from Root 1988.)

ESSAY 3.2 BIOCLIM—COMPUTER SOFTWARE FOR PREDICTING POTENTIAL GEOGRAPHIC RANGES OF ORGANISMS

Well-developed computer programs, such as BioClim or CLIMEX, can be used to predict the potential ranges of species whose ranges are limited by climate. Such programs have been applied to the spread of introduced pests, including the cane toad and the fire ant in Australia (Sutherst *et al.* 1995, Sutherst and Maywald (2005).

These programs use two approaches to make their predictions. One approach, used for species whose present geographic ranges are well documented, is to construct a predictive model based mainly on rainfall and temperature readings taken inside and outside the species' ranges (Jackson and Claridge 1999). The second approach is to construct a predictive map based on the specific climatic factors that are known to limit each species' range.

The second approach was applied to the red fire ant *Solenopsis invicta*. This species' range seems to be limited in cold areas by low winter soil temperature and in dry areas by low rainfall. Korzukhin *et al.* (2001) analyzed data from 4,537 weather stations to construct the map shown at the start of Chapter 2 (page 20), which predicts the potential range of this pest species in the United States.

BioClim and other computer programs can be important resources for investigating the potential future impacts of climate change on geographic distributions that are known to be limited by temperature and rainfall.

---

middle intertidal zone (Figure 3.8). However, *Chthamalus* can survive in the middle intertidal zone if *Balanus* is removed. The lower limit of *Balanus* is set by predation, particularly by a carnivorous snail, the dog whelk (*Thais lapillus*).

The upper range limits of both barnacle species are determined by desiccation and high temperature, but because *Chthamalus* is more tolerant of these stresses than *Balanus* is, there is a zone high on the shore where *Chthamalus* can survive but *Balanus* cannot. In both species, the sensitivity of the young barnacles sets the species' upper range limits.

Q **Would you expect the factors that limit ranges on rocky shores to do so on sandy shores as well?**

The important message in the example of *Chthamalus* and *Balanus* is that different factors can be limiting in different parts of a geographic range, both on a local scale and on a regional scale. For these barnacles, physical factors (temperature and desiccation) set the upper limits, and biotic factors (competition and predation) set the lower limits. No single factor determines the geographic range.

Range Extension via Adaptation

In our analysis of range limitation by environmental factors, we have assumed that all individuals of a particular species have certain physiological tolerances built in. But, because species can undergo local adaptation, they are not genetically or physiologically uniform throughout their ranges. Darwin recognized that species could extend their ranges by adapting to local limiting environmental factors, such as temperature, but the full implications of his ideas were not appreciated until a Swedish botanist, Göte Turesson, began looking at adaptations to local environmental conditions in plants.

Turesson collected seeds from a variety of plant species growing in different areas and planted the seeds in field or laboratory plots at a single site, called a **common garden.** This technique allowed Turesson

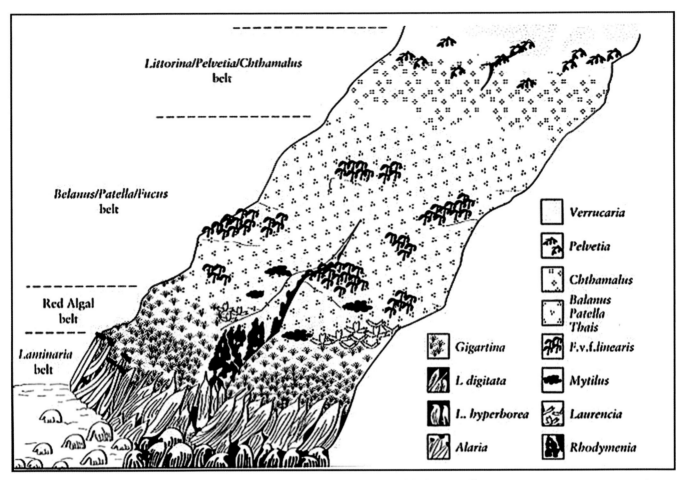

Figure 3.7. Species distributions on a barnacle-dominated, rocky intertidal slope. Different species can survive at various levels on the slope. Such slopes are very common on moderately exposed rocky shores of north-western Scotland and north-western Ireland. A similar pattern with the same or similar species can be found on the east and west coasts of North America. Similar patterns can be found on the rocky intertidal zones of Australia and New Zealand. (Modified from Lewis 1972.)

Göte Turesson (1892–1970) Professor of Systematic Botany and Genetics, Uppsala, Sweden

to separate the **phenotypic** (environmental) and **genotypic** components of variation among plants of the same species growing in different environments. If all of the variation within a species is phenotypic, then plants of that species should have the same characteristics when grown in a common garden. On the other hand, if all of the variation is genotypic, then plants should have the characteristics typical of the habitat where the seeds were obtained. A combination of phenotypic and genotypic determination should produce intermediate characteristics.

Turesson discovered that some of the normal variation within a species was maintained in the common garden, indicating a combination of phenotypic and

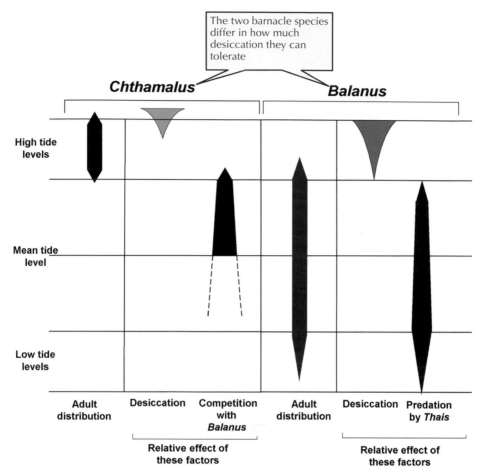

Figure 3.8. Intertidal-zone ranges of the barnacles *Chthamalus stellatus* and *Balanus balanoides* on rocky coastlines in Scotland, and the main factors that limit those ranges. For each factor, the width of the shaded area indicates the importance of that factor in the corresponding part of the intertidal zone. (Modified from Connell 1961a, 1961b.)

genotypic determination. For example, the sea plantain (*Plantago maritima*) grows as a tall, robust plant in marshes along the coast of Sweden and also as a dwarf plant on exposed sea cliffs in the Faeroe Islands. When plants from marshes and from sea cliffs were grown side by side in the common garden, they showed a significant, although reduced, difference in height (Figure 3.9). In 1922 Turesson coined the word **ecotype** to describe genetic varieties within a single species. Turesson's early studies on ecotypes helped create the research field of ecological genetics.

The perennial herb yarrow (*Achillea lanulosa*) provides a classic example of ecotypes. In the Sierra Nevada in California, yarrow grows at elevations below about 3,000 m (10,000 ft). The average winter temperature decreases with increasing elevation, and

yarrow plants at the higher elevations undergo winter dormancy and are smaller. To determine how much of this height variation is genetically determined, ecologists collected seeds from yarrow plants at several elevations and planted them in a common garden at Stanford University in California (Figure 3.10). The plants that grew showed height variations that correlated with the elevations where the seeds were collected, indicating that height variation in yarrow plants is partly genetically determined.

Q **Would you expect all plant species to have ecotypes?**

Many species have expanded their geographic ranges during the last century, but, in nearly all cases,

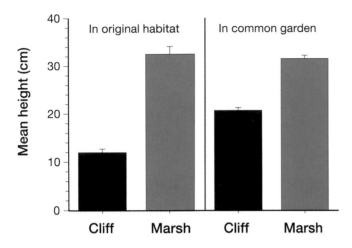

Sea plantain growing in a cliff environment

Sea plantain growing in a marsh environment

Figure 3.9. Mean heights of two ecotypes (cliff and marsh) of sea plantain (*Plantago maritima*) grown in their original habitat and in a common garden. (Data from Turesson 1930, photos courtesy of Digital Flora of Newfoundland and Labrador and Flora Czechoslovakia.)

we do not know if genotypic changes accompanied these expansions. If we could study a species as it expanded its range, we might gain some insight into how organisms can extend their limits of tolerance. We should have more opportunities to do that as Earth's climate warms.

Impacts of Rapid Climate Change

As you've learned in this chapter, climatic factors, such as temperature and moisture, limit the geographic ranges of many species of terrestrial plants and animals. Earth's climate is warming and is expected to become more variable over the next century. One way to predict how these climate changes may affect species' geographic ranges is to examine what has happened in temperate regions since the end of the last ice age.

Recall that temperate-zone trees in the northern hemisphere started migrating northwards when the continental glaciers began to retreat about 14,000 to 16,000 years ago. A detailed record of these migrations is captured in fossilized pollen deposited in lakes and ponds. Margaret Davis and her students at the University of Minnesota have been leaders in deciphering this record. Their research indicates that different tree species migrated independently in North America. Oaks, maples and beeches moved rapidly toward the north-east from the lower Mississippi Valley, while hickories advanced more slowly. Hemlocks and white pines moved toward the north-west from refuges along the Atlantic Coast. Thus, New Hampshire saw the arrival of sugar maples 9,000 years ago, hemlocks 7,500 years ago and beeches 6,500 years ago.

Figure 3.11 shows the current distribution of forest types in the eastern United States and the distribution predicted by climate change models if atmospheric CO_2 levels double. According to the models, southern forest types will move north as the potential ranges of all temperate-zone trees expands northwards. These changes will happen only if trees can colonize new areas with sufficient speed. For example, the American beech must move north by 7–9 km per year to stay within its climatic tolerance limits. However, this species has migrated at a rate of only 0.2 km per year since the end of the last ice age. If these kinds of predictions are even approximately correct, then slowly colonizing species of

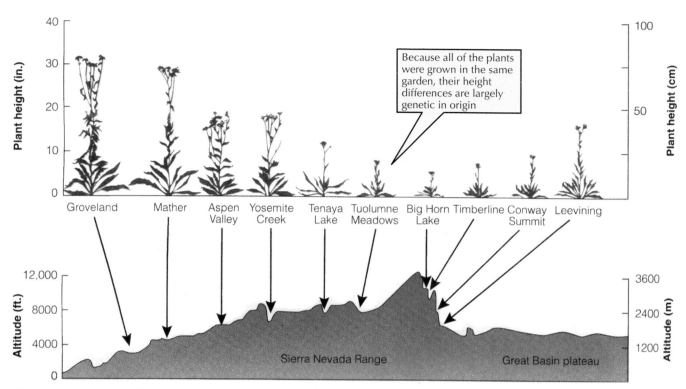

Because all of the plants were grown in the same garden, their height differences are largely genetic in origin

Figure 3.10. Height differences among yarrow plants (*Achillea lanulosa*) in a common garden. The plants were grown from seeds collected at sites in the Sierra Nevada indicated in the profile at the bottom. (The horizontal distance across the profile (800 km) is compressed relative to elevation.) (Modified after Clausen *et al.* 1948.)

trees will require human assistance to move into their new ranges. Climate change is happening too quickly for trees to disperse by natural mechanisms alone.

The effects of climate change will not appear immediately. Long-lived species, such as trees, will survive for many years their present ranges. As the climate changes, they will reduce their seed production and finally be unable to produce viable seedlings.

Q **What kinds of ecological processes introduce uncertainty into long-term predictions of range shifts due to climate change?**

Many lines of evidence link global climate shifts during the last 50 years with a variety of biological changes. The largest body of data relates to the timing of life-cycle events that are strongly affected by temperature, such as the start of egg laying in birds, the time of first flowering in plants and the time of emergence of mosquitoes. For most plants and animals that have been studied, these events have been occurring progressively earlier over the last half-century (Figure 3.12). Over the same period, the geographic ranges of 99 species whose ranges are well known have shifted toward the poles an average of 6 km per decade. The impact of climatic warming is established without a doubt for many species, with the largest effects occurring in temperate and polar regions.

Margaret B. Davis (1931–) Regents' Professor of Ecology, University of Minnesota

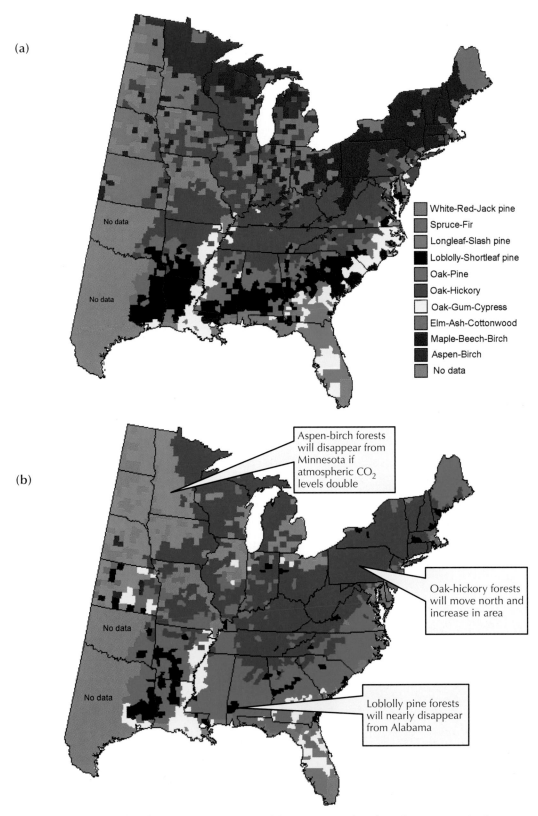

Figure 3.11. Present geographic ranges (a) and predicted future ranges (b) of ten forest types in the eastern United States. The predicted ranges are the result of five climate-change models that assume a doubling of atmospheric CO_2 levels. (From Iverson *et al.* 1999a, b; see http://www.fs.fed.us/ne/delaware/atlas/index.html)

Climate change will also have large impacts in mountainous regions. Most mountains have a *tree line*—an upper limit to the distribution of forests. Above tree line is alpine tundra—a community of low-growing plants that can survive the harsh climatic conditions. As global warming proceeds, trees will colonize the higher levels and alpine tundra species may be lost if the zone disappears entirely. Reduced precipitation in mountains will affect aquatic organisms by reducing stream flows and raising water temperature. Changes in cloud cover in mountain regions will dramatically affect tropical cloud forests, whose flora and fauna depend on high humidity.

■ 3.3 PREDATORS, DISEASES, PARASITES, AND COMPETITION CAN LIMIT RANGES ON A LOCAL SCALE

If a species can disperse to and colonize an area, its distribution will be limited on a local level by either biological agents or the physical–chemical environment. A species may not be able to live in a local area because predators, diseases, parasites or competitors prevent successful colonization. We will focus on the first two of these factors in this section.

Limitation by Predators

The best illustrations of range limitation by predators are cases in which introduced predators have severely affected native prey species. For example, the European red fox (*Vulpes vulpes*) was introduced to Australia in the 19th century and has contributed to the extinction, or near-extinction, of several marsupials, including a small rat-kangaroo called the burrowing bettong (*Bettongia lesueur*; Figure 3.13). Once distributed across much of the Australian mainland, by 1990 the burrowing bettong was restricted to three islands off the Western Australia coast, where red foxes had not reached. In 1992, conservationists reintroduced the burrowing bettong to a peninsula in Shark Bay, Western Australia, after erecting an electrified fence

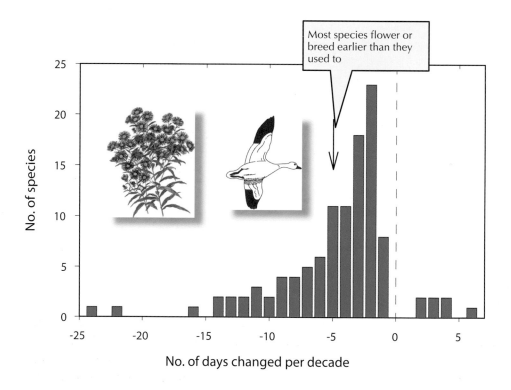

Figure 3.12. Change in the timing of temperature-related life-cycle events (such as the time first flowering time) for 694 species studied over an average of 30 years. The overwhelming majority of species experienced a shift toward earlier events, consistent with climatic warming. The blue arrow marks the mean change of 5 days earlier per decade. (Data from Root *et al.* 2003.)

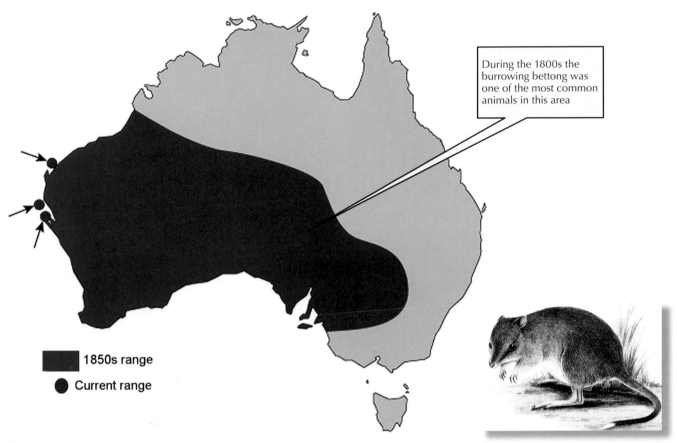

During the 1800s the burrowing bettong was one of the most common animals in this area

■ 1850s range

● Current range

Figure 3.13. Current and former geographic ranges of the burrowing bettong *Bettongia lesueur,* a small (2 kg) marsupial. It is now confined to three islands off Western Australia, and attempts are being made to re-introduce it to the mainland. (Modified from Ovington 1978.)

across the neck of the peninsula to exclude foxes. Foxes already on the peninsula were killed off with meat bait containing a compound extracted from Western Australian plants (poison peas, *Gastrolobium* spp.) that is lethal to non-native predators, but is harmless to native species. The reintroduction of bettongs was successful, unlike many reintroduction programs in Australia that have failed because of heavy predation.

Other examples of predator-related range changes have resulted from the elimination of predators. White-tailed deer expanded their range in eastern North America after wolves were eliminated in the late 19th century. Coyotes similarly expanded their range into northern Canada as wolf populations there were reduced or eliminated. Wolves kill coyotes and are a good example of how predators can potentially affect the distribution of other predators as well as that of their prey.

Limitation by Diseases

A large proportion of the bird species that were once endemic (found nowhere else) to the Hawaiian Islands have become extinct in historical times. One major factor in their extinction is introduced diseases. The extinction of many low-elevation bird species before 1900 may have resulted from avian pox, in conjunction with extensive habitat clearing for agriculture and the introduction of rats, cats and pigs. A second wave of extinction in Hawaiian birds began in the early 1900s and was probably the result of avian malaria. At that time, native birds were relatively common only above 1500 m elevation, as they are today. Most introduced bird species occupy lower elevations, and so does the main malarial vector, the mosquito *Culex quinquefasciatus.* However, native Hawaiian birds are much more susceptible to malaria than the introduced birds are. Consequently, avian malaria occurs most often at intermediate elevations,

The Hawai'i Oo (*Moho nobilis*), once a common bird of Hawaii, last seen in 1934

where the ranges of the vector and the most susceptible hosts overlap. The geographic range of many native bird species has been reduced to the highest-elevation forests, where *C. quinquefasciatus* mosquitoes are rare and avian pox does not occur.

The roles of predators and diseases in the geographic ecology of plants and animals have been studied far less than their potential importance warrants (see *Essay 3.3*). We should be aware that current geographic ranges of many species may be the result of limitations imposed by predators and diseases in the past.

Q What data would you need to demonstrate that the existing range of a species is limited by a predator or a disease?

SUMMARY

To understand why species are found in particular places, we need to know the factors that set the limits to their geographic ranges. Many species are limited on a global scale by their inability to move across barriers such as oceans. Humans move many pest species to new locations, both deliberately and accidentally.

On regional and continental scales, the constraints on geographic ranges are often produced by physical and chemical factors, such as temperature, moisture, salinity and pH. The continental ranges of many species correlate closely with climatic variables.

The geographic ranges of species are dynamic and, as Earth's climate changes, species will move into new areas that become climatically suitable, if time permits. Ecologists are concerned that the speed of climate change may be too great for slowly colonizing species to move and that genetic adaptation to local temperature and rainfall patterns may be lost in some species. Another concern is that pest species, such as the mosquitoes that carry West Nile virus, will migrate far beyond their current ranges.

On a local level, it is usually biotic interactions and physical–chemical factors, rather than dispersal ability, that limit geographic ranges. A combination of predation, disease, parasitism and competition between species can prevent many species from colonizing new areas.

ESSAY 3.3   WHAT WAS THE SOURCE OF THE WEST NILE VIRUS?

The emergence of new infectious diseases is one of the major problems of our time, and a good example is the West Nile virus. This virus was isolated and described in 1937 in the West Nile district of Uganda. West Nile virus is carried by mosquitoes and infects birds as its primary host. It was first detected in the Western Hemisphere in New York City in June 1999, when an unusually large number of dead and dying crows were found. By autumn 1999, 62 people had become infected with West Nile virus in and around New York City. In humans, the virus causes symptoms ranging from mild illness to encephalitis and meningitis. It is lethal in up to 14% of cases. Prior to 1999, there had been only a few sporadic outbreaks of infection by the virus in humans, including in Israel in the 1950s and in South Africa in 1974.

Since the mid-1990s, West Nile virus has become more virulent—killing birds and horses in eastern Europe, Israel and now North America. A new variant of the virus seems to have evolved and is responsible for these recent severe outbreaks (Kramer and Bernard 2001). The variant that is spreading through North America is structurally most similar to a variant isolated in Israel in 1998, which suggests that the North American variant migrated from the eastern Mediterranean area. The way in which the virus moved to North America is uncertain—it could have been carried by an infected mosquito, a bird, a human or some other vertebrate host.

West Nile virus has spread very rapidly in North America. By 2000 it had been found in 12 eastern states, from the Canadian border to North Carolina—a distance of 900 km. By 2001 it had spread west to Missouri and Iowa and by 2002 it was in all states except Arizona, Utah, Nevada, Oregon, Hawaii and Alaska. By early 2003 in the United States there were more than 4,000 cases of West Nile virus infection in humans, and 263 deaths were attributed to the virus, many in elderly people (see http://www.cdc.gov/ncidod/dvbid/westnile/index.htm). There have been only a few cases of West Nile virus detected in birds in Canada, and most of the human cases were Canadians who had visited parts of the United States known to harbor the virus.

The West Nile virus story is typical of many newly emerging diseases: a sudden dispersal of a known pathogen from an area where it has existed for many years causing relatively little damage to people or other animals, followed by a burst of mortality in new hosts that are not adapted to the pathogen. Humans typically acquire these diseases as incidental hosts, rather than as the primary target of the pathogen. The speed of spread of the pathogens shows how easily the combination of human transport and natural movements of animals can overcome migration barriers with devastating consequences.

Part of the solution to the problem of newly emerging diseases lies in preventing jump dispersal. Quarantine procedures prevent many of these undesirable introductions, but, because of increased travel and trade, countries are being bombarded with possible introductions of potential pathogens. Both diagnosis and quarantine procedures need to be strengthened to assist containment.

SUGGESTED FURTHER READINGS

Ayres DR, Smith DL, Zaremba K, Klohr S and Strong DR (2004). Spread of exotic cordgrasses and hybrids (*Spartina* sp.) in the tidal marshes of San Francisco Bay, California, USA. *Biological Invasions* **6**, 221–231. (A good discussion of a current marine plant invasion in California.)

Baskin Y (2002). *A Plague of Rats and Rubbervines: the Growing Threat of Species Invasions.* Island Press, Washington, D.C. (A riveting account of the problems of invasive species around the world.)

Clark JS, Fastie C, Hurtt G, Jackson ST, Johnson C, King GA, Lewis M, Lynch J, Pacala S, Prentice C, Schupp EW, Webb TI and Wyckoff P (1998). Reid's paradox of rapid plant migration. *BioScience* **48 (1)**, 13–24. (How did plants move back north after the glaciers disappeared?)

Dauphin G, Zientara S, Zeller H and Murgue B (2004). West Nile: worldwide current situation in animals and humans. *Comparative Immunology, Microbiology and Infectious Diseases* **27**, 343–355. (A summary of how the latest human viral disease has spread around the world.)

Parmesan C, Ryrholm N, Stefanescu C, Hill JK, Thomas CD, Descimon H *et al.* (1999). Poleward shifts in geographical ranges of butterfly species associated with regional warming. *Nature* **399**, 579–583. (How butterfly ranges in Europe have changed in the last 100 years.)

Pielou EC (1991). *After the Ice Age: The Return of Life to Glaciated North America.* University of Chicago Press, Chicago. (A beautifully written, general summary of the movements of plants and animals during the last 10,000 years in North America.)

QUESTIONS

1. Find information about the quarantine services of your country on the Web and make a list of the species or plants, animals and microbes that are of the most concern as potential invaders. Some possible web sites:

    – for Australia, http://www.aqis.gov.au/

    – for Canada, http://www.inspection.gc.ca/english/sci/surv/pesrave.shtml

    – for New Zealand, http://www.biosecurity.govt.nz

    – for the United States, http://www.cdc.gov/ncidod/dq/index.htm

2. Deep-sea hydrothermal vents are hot-water springs in volcanically active areas of the ocean's floor. The prokaryotes surrounding these vents can live at high temperatures and harness the chemical energy in sulfides released in the hot water. Given that these vents appear and disappear on a short timeframe separated by various distances, discuss how organisms adapted to living near them might be able to disperse through the surrounding cold ocean and colonize new vents. Would you expect Atlantic, Pacific and Indian Ocean vents to have different species associated with them? Van Dover *et al.* (2002) discuss this question.

3. Since about 1950, the ticks that transmit Lyme disease have been spreading south from the northeastern and north-central United States, carrying the disease into new regions (see Rich *et al.* 1995). How could you determine what limits the geographic distribution of these ticks?

4. Elfinwood (or krummholz) trees are prostrate, low-growing shrub-like forms of trees found near the tree line on mountains. How would you test the idea that elfinwood trees are ecotypes of trees of the same species that grow at lower elevations?

5. Male and female birds of the same species do not always migrate together. For example, adult male dark-eyed juncos (*Junco hyemalis*) remain farther north in winter than females and juveniles (Ketterson and Nolan 1982). Suggest two possible explanations for this observation, and describe how someone could test your explanations.

Chapter 4

BEHAVIORAL ECOLOGY: EVOLUTION IN ACTION

IN THE NEWS

If a newspaper headline read 'Babysitters critical to survival,' everyone would sit up and take notice. In the case of African wild dogs (*Lycaon pictus*), such a headline would be true. These dogs were once common across much of southern Africa, but, because of habitat loss and extermination by ranchers, their numbers have decreased to fewer than 1,000. Biologists have been trying to determine why wild dog populations continue to decline despite active measures to reduce mortality.

The key seems to be a change in the dogs' behavior. Wild dogs are highly social, living in large groups that hunt and care for young collectively. Hunting success increases with the number of hunters, so a large group is critical to maintaining the food supply. However, a wild dog cannot hunt and care for young at the same time. In groups that have fewer than five adults, all adults hunt and none guard the pups, which are then vulnerable to predation by lions and hyenas. This trade-off between hunting and pup guarding leads to very low reproductive success in small groups. As the population of wild dogs has declined, the average group size has decreased from 40 to 100 individuals to fewer than 10. Thus, wild dogs appear to be trapped in a positive-feedback cycle in which population decline results in low reproductive success, which, in turn, causes further population decline. Baby-sitters may be critical for this social species' survival.

An African wild dog (Photo courtesy of Calvin Jones)

CHAPTER 4 OUTLINE

■ 4.1 BEHAVIORAL ECOLOGISTS ANALYZE THE ECOLOGICAL AND EVOLUTIONARY CONTEXTS OF BEHAVIORS

The ecology of a species is ultimately determined by interactions between *individuals* and their environment. The environment includes other individuals of the same species as well as members of other species, such as predators. The environment also includes physical factors, such as temperature, rainfall and wind. The ways that organisms respond to each other, and to particular cues in the environment, are called *behaviors*. In this chapter we will focus on the decisions made by animals as they interact with their food resources, mates and other members of their social group. How does a hummingbird decide where to feed? How does a male lion achieve reproductive success? These are some of the questions we will address.

All animal behaviors are generated through a complex set of physiological and neurological reactions triggered by environmental stimuli. However, the mechanisms that produce a behavior and how they develop are the focus of such disciplines as psychology, physiology, neurobiology and developmental biology. Behavioral ecologists are more interested in understanding the ecological and evolutionary contexts of behavior. They want to learn how an individual's behavior is shaped by its social and physical environment, both past and present, and how specific behaviors affect its chances of surviving and reproducing.

An example of the kinds of questions behavioral ecologists ask is, 'Why do black-headed gulls remove eggshells from their nests after their chicks hatch?' Nikolaas Tinbergen, one of the founders of behavioral ecology, asked this question in the early 1960s. Tinbergen observed that the bright white lining of hatched eggs is much more visible than the outside of unhatched eggs, which are khaki colored. He hypothesized that, by removing eggshells from the nest, gulls reduce the risk of egg predation by crows. In one test of this hypothesis, he measured the predation rates on white and khaki-colored chicken eggs (Figure 4.1). As he had predicted, the white eggs were more likely to be eaten by crows.

Figure 4.1. Nikolaas Tinbergen (1907–1988), one of the first biologists to analyze the adaptive value of animal behavior. In this photo, he is painting chicken eggs to mimic gull eggs for a field experiment. Tinbergen, Konrad Lorenz and Karl von Frisch shared the 1973 Nobel Prize in physiology or medicine for their research on individual and social behavior patterns.

Tinbergen asked 'why' questions. 'Why' questions concern the current function and evolution of behaviors. To appreciate the ecological approach to studying behavior, we need to understand how evolution shapes the behaviors and other traits of species.

Adaptation Through Natural Selection

Evolution is the change in the characteristics of species over time. Many characteristics of organisms are clearly not random, but are well suited to the organisms' particular lifestyles. The process that generates the fit between the characteristics of organisms and their environment is **natural selection.** Two English naturalists, Charles Darwin (1809–1882) and Alfred Russel Wallace (1823–1913), independently and almost simultaneously proposed the theory that evolution occurs through natural selection. Because Darwin supported this theory with much more evidence, he is generally given most of the credit for its development. Darwin formulated many of his ideas on natural selection while traveling around the world as the ship's botanist aboard HMS *Beagle* (Figure 4.2). During his travels, he was impressed by the enormous diversity of animals and plants and by their close relationships on a local scale. In his book *The Origin of Species* (1859), Darwin observed:

'Nevertheless the naturalist, in travelling, for instance, from north to south, never fails to be struck by the manner in which successive groups of beings, specifically distinct, though nearly related, replace each other.'

These patterns in the diversity and distribution of species led Darwin to argue that new species evolved over time from a common ancestor by a process of descent with modification.

Natural selection is the evolutionary link between organisms and their environment. It is critical for all ecologists to understand how it operates. We cannot predict how organisms will respond to changes in their environment without knowing something of the factors that helped shape the traits of those organisms. Many of these traits are the outcome of natural selection that has acted over thousands of generations, and continues to act to this day.

Natural selection can occur in a population if the following conditions are met:

Variation: Individuals in a population show heritable (genetic) variation in traits such as their morphology, physiology and behavior.

Competition: The number of individuals increases faster than the available resources, leading to competition.

Differential reproduction: Individuals that are best able to compete for resources in an environment will produce the most offspring.

The logical outcome of these conditions is that individuals whose traits make them better at finding food or mates or avoiding predation will produce more young. If those traits are heritable, the individuals that have them will become a larger proportion of the population in the next generation. Thus, any heritable trait that leads to high reproductive success in an environment will be *selected* for and will become increasingly common in a population. Over time, this selection can result in a fit between the traits of a species and its environment. This fit is called **adaptation.**

Figure 4.2. The voyage of the ship HMS *Beagle* around the world from 1832 to 1836. The drawing shows the HMS *Beagle* in the Straits of Magellan, Tierra del Fuego, with Mount Sarmiento in the distance. Charles Darwin's observations on this voyage laid the foundations of his theory of evolution by natural selection. (Drawing courtesy of the British Museum.)

Behavioral Changes via Natural Selection

The results of descent with modification can be clearly observed in a group of closely related Hawaiian birds called honeycreepers (family Drepanidae). The Hawaiian Islands are volcanic islands that formed between 5.5 and 0.5 million years ago. Because the islands were never connected to a continent, the birds that live on the islands must have descended from birds that flew across the ocean—an example of jump dispersal. Genetic data corroborate this conclusion: all 28 of the living or recently extinct honeycreeper species appear to have descended from a single species of cardueline finch that arrived on the Hawaiian Islands approximately 3.5 million years ago.

Variation in bill shape within the Hawaiian honeycreeper family is as great as that observed across whole orders of other birds (Figure 4.3). Why do honeycreepers have such varied bill shapes? The answer appears to lie in adaptation of the bill for specialized food sources. Because relatively few animal species colonized the Hawaiian Islands, honeycreepers were able to diversify and feed on many different plants with little competition from other animals. Specialized feeding requires a specialized feeding apparatus, and a single individual cannot make use of all types of food. For example, a thick bill suitable for shelling seeds cannot be used to harvest nectar efficiently. Over time, different bill shapes evolved—each adapted for certain foods.

Thus, thick-billed species such as the Kona grosbeak (Figure 4.3a) feed on large seeds, hooked-billed species such as the Maui parrotbill (Figure 4.3b) feed on insect larvae, and slender-billed species such as the i'iwi (Figure 4.3c) have curved bills and brush-like tongues specialized for feeding on nectar. Despite having evolved from a single ancestral species, Hawaiian honeycreepers show a range of feeding adaptations that mimic those of birds on continents.

We can see how changes in feeding behavior can lead to adaptive changes in bill shape by examining a single species of Hawaiian honeycreeper, the i'iwi (Vestiaria coccinea). Historically, i'iwi fed on nectar from the long, tubular corollas of lobelioid flowers (Figure 4.4), but the disappearance of these flowers, and the extinction of several closely related honeycreeper species,

Figure 4.3. Variation in bill shape in a single group of closely related birds, the Hawaiian honeycreepers. Natural selection over many generations has favored bill shapes that are adapted for different food sources. Three species, the Kona grosbeak (*Chloridops kona*, (A) the Maui parrotbill (*Pseudonestor xanthophyrs*, (B), and the i'iwi (*Vestiaria coccinea*, (C) are discussed in the text. (From Freed *et al.* 1987, drawings used with permission of H. Douglas Pratt.)

has led to a recent shift in the feeding behavior. Present day i'iwi feed almost exclusively on the nectar in flowers of the ohia tree (*Metrosideros polymorpha*), flowers that lack a tubular corolla and have a very different shape than lobelioid flowers. By comparing the bills of live i'iwi with those of museum specimens collected more than 100 years ago, ecologists were able to observe

Figure 4.4. An example of how bill shape in one species of Hawaiian honeycreeper matches the original feeding behavior of the species. (a) An i'iwi near lobelioid flowers, historically the main nectar source for this species . Note that the curvature of the flowers' tubular corollas matches that of the i'iwi's bill. (Photo courtesy of Jim Denny.) (b) An i'iwi next to ohia flowers, now the main food source, but the bill shape does not match the flower shape. (Photo courtesy of Jack Jeffrey.)

a modification of bill shape that coincided with this change in diet. They found that the average length of the upper mandible (top half of the bill) is smaller in present-day i'iwi, primarily because there are fewer individuals with extremely long upper mandibles (Figure 4.5). This is exactly the pattern you would predict if the diet change selected a more efficient adaptation for feeding on ohia flowers: a shorter bill. Evolutionary changes in structure are linked to changes in behavior, which in turn affect reproductive success in these birds.

Adaptive Behavior

As the Hawaiian honeycreeper example illustrates, behaviors at the population or species level can correlate with morphological traits. Because both morphological and behavioral changes occur slowly in evolutionary time, however, it is not possible to study them directly. Even though we can sequence the DNA in individuals, this will not help us understand the adaptive value of behavior because no complex behavior is under the control of a single gene. Instead we must adopt an indirect approach to analyze why a particular kind of behavior is adaptive.

Q **Is it possible for a behavior not to be adaptive?**

This approach asks what benefits individuals might gain from behaving in certain ways. To answer that

question, behavioral ecologists examine the decisions that animals make when faced with environmental options such as where to feed, what to eat, where to live and which individuals to mate with. An animal's

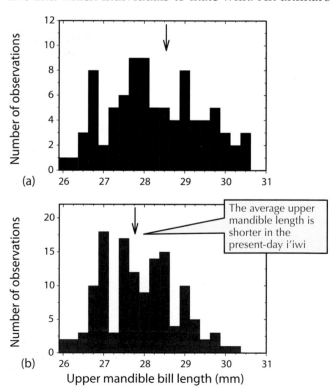

Figure 4.5. Lengths of the upper mandibles of adult i'iwi males. (a) Museum specimens collected about 100 years ago. (b) Present-day birds. Arrows indicate the average upper mandible length in each population. (From Smith *et al.* 1995.)

decisions translate into differences in survival, fecundity or mating success and therefore are shaped by natural selection.

Can we apply this kind of analysis to human behavior? Many people have argued that behavioral ecology can add little to our understanding of human behavior because so much of our behavior is learned or culturally derived. Cultural evolution has triumphed over biological evolution in humans, so this argument goes. But more data suggest that this view is not always correct. Although human behavior is influenced by many factors, behavioral ecology can identify the kinds of factors that may influence human behavioral decisions.

One good example is parental care, which is a major investment lasting many years in humans. Like all animals, humans must divide limited resources between reproduction and other activities as well as between different offspring. The choices involved require **trade-offs**, which are compromises between two desirable, but incompatible, activities. From an evolutionary point of view, how much should parents invest in their offspring? Evolutionary theory suggests that animals are more likely to invest in parental care when the recipients of this care are close relatives. Variation in relatedness explains, in part, why households with stepchildren have lower male parental care, whether measured as food provided in the Hadza, a nomadic hunter-gatherer society, or as the amount of money spent on postsecondary education in the United States. Behavioral ecology can help explain resource allocation, mate choice and factors that influence conflict in human societies.

■ 4.2 ALL BEHAVIORS HAVE COSTS AND BENEFITS

All organisms are constrained by time and energy. Time spent engaged in one activity cannot be spent on another, and energy expended in doing one thing will not be available to do something else. For this reason, organisms must choose between different activities, and the nature of their choices will be shaped by evolution. We can analyze some of the choices that a species makes by comparing the costs and the benefits of alternative activities. This kind of assessment, called a **cost–benefit analysis**, is commonly used in economics to determine whether the financial cost of a project is less than the economic benefit that can be expected from the project. In behavioral ecology, costs are typically measured in terms of energy consumed, the probability of injury or the probability of being killed by a predator. Benefits are usually measured in terms of a gain in energy or an increase in reproductive success.

Q Do ecologists need a common unit of measurement to compare the costs and benefits of behaviors?

Behavioral ecologists assume that natural selection favors aspects of an individual's behavior that maximize the net benefit. For example, individuals that make better decisions about where to feed should have a higher net energy intake and be in better condition. Therefore, they should be better able to avoid predators and diseases, attract mates and produce many young. Thus, natural selection should favor any behavioral attribute that consistently leads to good feeding decisions.

An ecologist can use a mathematical formulation called an *optimality model* to predict which set of behaviors will maximize an individual's reproductive success in a given environment. Optimality models make explicit the relationships between costs and benefits of behaviors under various conditions. They are most useful in circumstances where it is clear that making the right decision maximizes some payoff, such as survival rate, reproductive success (number of young produced), feeding efficiency (energy gained per unit time) or mating success (number of matings per unit time).

Territorial Defense
We can examine how an optimality model works by considering territorial defense in animals. An animal's **territory** is any defended area. Many mammals, birds, lizards and fishes defend a feeding area against other individuals of the same species. How big a territory should an individual defend? To answer this question, we need to think about the costs and benefits of defending a territory. The costs are time, energy and risk of injury. The total cost will increase with the size of the

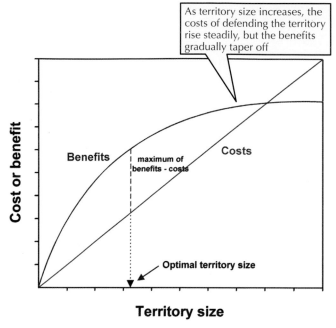

As territory size increases, the costs of defending the territory rise steadily, but the benefits gradually taper off

Figure 4.6. A hypothetical relationship between the size of a feeding territory and the costs (red curve) and benefits (blue curve) of defending that territory. The optimal territory size is the size where the difference between the benefits and costs is greatest (indicated by the dashed black line). Different individuals may have different cost and benefit curves and thus different optimal territory sizes.

territory, and, for simplicity, we will assume that the relationship between cost and territory size is linear (Figure 4.6). The benefit of defending a territory is exclusive access to food, which increases with the size of the territory. Since an individual can consume only a certain amount of food, however, the benefit curve gradually levels off as the territory becomes larger. Above a certain territory size, there is no further increase in benefit (Figure 4.6). The optimal territory size is the one that maximizes the *relative benefit*, which is the difference between the costs and benefits. In the hypothetical example shown in Figure 4.6, the relative benefit would be greatest at the territory size indicated by the arrow. Clearly, the optimal territory size is determined by the shapes of the cost and benefit curves, which vary with the species, habitat and an individual's age or mating status.

If an animal does not behave as predicted by an optimality model, we should ask whether we have assessed the costs and benefits of the behavior correctly or whether additional factors should be considered. For example, the optimality model in Figure 4.6 assumes that cost and benefit curves are constant over time. Suppose instead that the shape of the cost curve varies from year to year. Should an animal change its territory size each year in response to these variations, or should it maintain a territory size that is optimal in average years?

One difficulty with optimality models is that they consider only one or two behaviors at a time, whereas individuals must simultaneously optimize all aspects of their behavior. We assume, however, that if a behavior such as territorial defense is directly linked to survival or reproductive success, then natural selection favors the ability of an individual to optimize that behavior.

Optimal Foraging

For most animals, their food is patchily distributed in time and space. Consequently, acquiring food involves many behavioral decisions, such as what type of food to consume, where and how to search for food and, once food is located, how much to eat. Since animals must acquire food at a certain rate to maintain their physiological functions, the efficiency with which they can find and eat food is also important. Thus, we can assume that natural selection favors **optimal foraging**, which is any method of searching for and obtaining food that maximizes the relative benefit (the difference between costs and benefits). Foraging provides an excellent opportunity to examine the factors that influence behavioral decisions because its benefits and costs are relatively easy to define, measure and manipulate. Much of the research on foraging has been carried out on birds.

Hummingbirds obtain most of their food energy from the nectar in flowers. Nectar is a resource that occurs in tiny amounts in individual flowers. It consists mostly of water and some dissolved sugars and varies highly in availability. Hummingbirds have very high energy requirements for their body weight due to their small size, high body temperature and use of hovering flight. To maintain a sufficient rate of food intake, a hummingbird must choose flowers that are full of nectar and avoid those that provide little nectar.

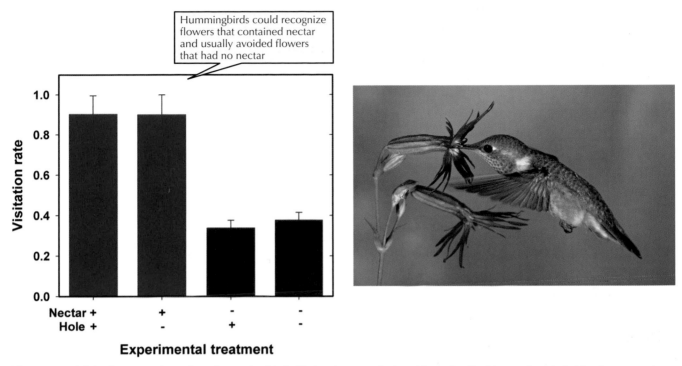

Figure 4.7. Visitation rates by rufous hummingbirds (*Selasphorus rufus*) and broad-tailed hummingbirds (*S. platycercus*) on scarlet gilia flowers (*Ipomopsis aggregata*). Researchers manipulated some flowers by removing nectar, adding holes, or both. The holes mimicked those created by bumblebees (*Bombus occidentalis*), which puncture flowers near the base and remove the nectar without pollinating the flower. In this experiment, hummingbirds consistently identified flowers that contained nectar, regardless of whether the flowers had holes. How they were able to do this is unknown. (From Irwin 2000, photo courtesy of Alan Murphy Photography.)

Experiments in which researchers manipulated the nectar content and appearance of flowers showed that hummingbirds are extremely good at identifying flowers with a high nectar content, even after a potential cue to nectar content has been altered (Figure 4.7). Foraging decisions by hummingbirds increase their rate of food intake.

For hummingbirds, and many other animals, food is distributed in a series of discrete patches across the landscape—some patches contain more food than others. If an animal forages optimally, it should prefer patches where the difference between benefits and costs is high. The benefits of foraging can be measured in terms of the amount of food obtained in each patch, and the costs can be measured in terms of the time taken and the probability of injury or predation. How will a forager respond when the costs of feeding in different patches are varied? We can answer this question by providing the same amount of food (a fixed benefit) in experimental patches that differ in their risk of predation or level of competition (varied costs). We can then determine how animals respond to changes in the costs of foraging by measuring how much food they eat in each patch. This approach was first used by Joel Brown in 1988 to investigate the foraging behavior of small mammals in desert habitats. He predicted that if the food levels are equal in two patches, a forager should stay longer and eat more food in the patch where the costs of foraging are lower.

One animal on which Brown's hypothesis has been tested is the gerbil (*Gerbillus spp.*). Gerbils are nocturnal, seed-eating rodents that live in sandy burrows. Their major predator is the barn owl—a rodent specialist. Ecologists studied the foraging behavior of gerbils by placing seed trays in open areas and under bushes in experimental enclosures. Some enclosures were illuminated; others were dark. Some enclosures contained owls; others did not. If predation is the major

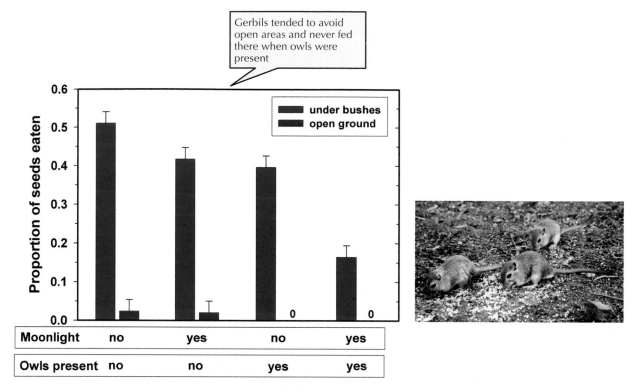

Figure 4.8. Proportion of seeds eaten by gerbils under bushes (green bars) and in the open (red bars) in the Negev Desert. Trays of seeds were set out in experimental enclosures in which the presence of moonlight and the presence of a predator, the barn owl, varied. (Data from Kotler *et al.* 1991, photo courtesy of Günther Eichhorn.)

cost of foraging by gerbils, they should eat more seeds under bushes and spend more time foraging there, especially in enclosures that are illuminated or contain predators. That is exactly what the researchers found. As Figure 4.8 shows, gerbils fed primarily at trays under bushes and reduced their overall feeding on bright nights, particularly when owls were present. They fed in open areas only when owls were absent. The results indicate that these desert rodents make choices based on the benefits of easily available food and the costs due to predation, and that the risk of predation influences their foraging behavior. If we were managing populations of gerbils, this study could tell us what kinds of habitat alterations might improve or decrease their survival and breeding success.

Optimal foraging studies support the conclusion that animals are finely adapted to searching for food in ways from which they achieve maximum relative benefit. Because there is always a chance that a predator will catch a foraging animal in a random encounter,

natural selection continues to favor the most efficient foraging traits.

■ 4.3 NATURAL SELECTION FAVORS GROUP LIVING IN SOME SPECIES

Many animals live in groups. Grazing herbivores form large herds, fish school together, carnivores form hunting groups, birds breed in large colonies and some animals live in extended family groups. If natural selection favors individual interests over group interests, why should animals ever associate, much less cooperate, with others to hunt or raise young? We can start to understand the factors that drive the evolution of group living by evaluating its benefits and costs (Table 4.1).

Benefits of Group Living

If food is sparsely distributed and difficult to locate, living in a group can increase an individual's foraging

Table 4.1. Potential benefits and costs of group living in animals.

Potential benefits	Potential costs
Increased foraging efficiency	Competition for food Increased risk of disease or parasites
Reduced predation	Attraction of predators
Increased access to mates	Loss of paternity Brood parasitism
Help from kin	Loss of individual reproduction

success by allowing it to obtain information about the location of food from successful foragers. This idea, which was first proposed by Paul Ward and Amotz Zahavi in 1973, seems to explain why some birds nest in colonies.

For example, ospreys (*Pandion haliaetus*) are fish-eating birds of prey that can nest alone or in groups. Hunting for fish is energetically expensive and time consuming—ospreys can spend up to 8 hours per day hunting. Individuals that are successful at foraging are highly conspicuous because they return to the nest with a whole fish in their talons. A study of a nesting colony in Nova Scotia, Canada, found that ospreys were more likely to depart in the same direction as individuals that returned to the nest with fish. But they based their decision about departure direction on the type of prey that had been caught. They were more likely to follow the direction of an individual that returned with a schooling fish, such as alewife, pollock or smelt, than one that returned with a non-schooling flounder (Figure 4.9a). This foraging behavior was adaptive because individuals that followed the flight path of previously successful foragers increased their foraging efficiency: They spent less time searching before returning with a schooling fish than did individuals that nested alone (Figure 4.9b). Thus, for prey species that are highly clumped in schools and hard to locate, information transfer can be a major benefit of group living.

A second potential benefit of group living is a reduced risk of predation. For example, if a predator takes a single individual as prey, each individual's risk of predation would drop from 10% in a group of 10 to 1% in a group of 100, if all other factors are equal. This 'dilution effect' is a passive benefit of larger groups. But this benefit must be balanced against the higher probability that a predator will find a large group than a small group or an individual. Animals in a group can also actively lower their risk of predation by being vigilant for predators. Increasing group size can make vigilance more effective and less costly, since many eyes increase the probability of detecting predators and reduce the time each individual must spend being vigilant. Less time spent being vigilant should translate into more time for other activities, such as foraging.

A balance between foraging and vigilance for predators can be observed in the behavior of western gray kangaroos (*Macropus fuliginosus*), which feed in social groups on grasslands across southern Australia (Figure 4.10). Kangaroos are vulnerable to predation by dingoes (wild dogs), which use the cover of trees and bushes when stalking individuals. By observing kangaroos at a site with numerous dingoes and another site with few dingoes, ecologists found that the balance between foraging and vigilance varied with group size and the rate of predation. At the site with numerous dingoes (and a high predation rate), individuals benefited from associating in a group, because the percentage of time they spent foraging increased, and the percentage of time they spent being vigilant decreased, as the group size increased (Figure 4.11a). At the site with few dingoes (and a low predation rate), there was no relationship between foraging or vigilance and group size (Figure 4.11b). Nevertheless, kangaroos at the low-predation site continued to forage in groups, which suggests that there may be additional advantages to group living in western gray kangaroos besides minimizing predation.

From an evolutionary point of view, success is measured in terms of the number of copies of one's genes in future generations (Dawkins 2006). An individual can increase its evolutionary success, or fitness, *directly* by producing its own young, and *indirectly* by increasing the survival or reproductive success of close relatives, which have some of the same genes. Helping, and being helped by, relatives is one benefit of group living in some animals, so cooperation in these animals has an evolutionary explanation (see *Essay 4.1*).

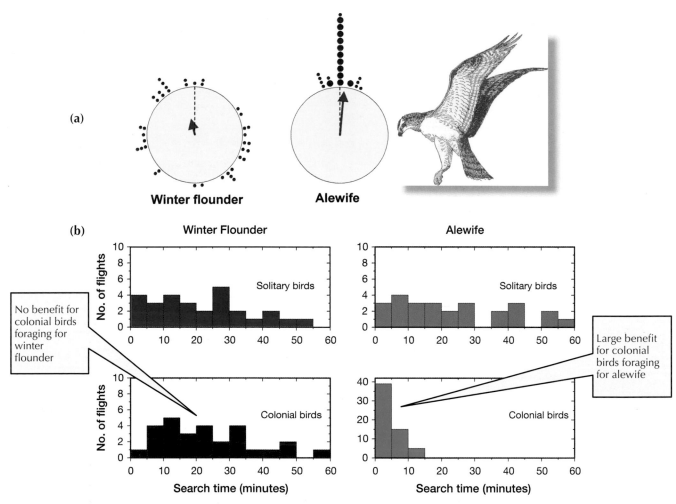

Figure 4.9. Benefits of group living in ospreys. (a) The departure directions of ospreys that left a colony after another osprey arrived with a fish. The direction of arrival is indicated by the dotted line pointing to 12 o'clock. Red arrows inside the circles indicate the average flight path taken by the arriving osprey for each fish species. Blue dots outside the circles indicate the direction of the departing birds; the largest dots represent 10 flights. Ospreys followed the path of foragers that returned with alewife, which form schools, but did not follow the path of foragers that returned with flounders, which do not school. (b) Search times for winter flounder and alewife by solitary ospreys and by ospreys that lived in a colony and watched returning birds. Birds in the colony caught alewife more quickly than did solitary birds. There was no significant difference between colonial and solitary birds in their search times for winter flounder. (Data from Greene 1987.)

Belding's ground squirrels (*Spermophilus beldingi*) provide an example of apparent cooperation in group-living animals. These rodents live in burrows in alpine and subalpine meadows in western North America. Although both sexes disperse from the burrow where they are born, males move much farther than females. This difference in dispersal distance leads to neighborhoods where the females are closely related, but the males are not. Belding's ground squirrels produce loud alarm calls when predators—chiefly coyotes, pine

martens and long-tailed weasels—are in the area. Alarm calls serve as an early warning for other ground squirrels living nearby, but they provide no immediate benefit to the caller. In fact, they may increase costs for the caller by attracting predators to it. Why then should any individual produce alarm calls? Paul Sherman addressed this question by studying a population of individually tagged Belding's ground squirrels over several years. He found that females were far more likely to give alarm calls than males (Figure 4.12a).

Figure 4.10. Western gray kangaroos. These grazers typically forage in groups—a behavior that increases protection from predators, such as dingoes. However, group living may also increase competition for food. (Photo courtesy of A.J. Kenney.)

However, females differed in the frequency with which they called: Females with relatives nearby, even young females that had no offspring of their own, called more often than females that had no relatives in the area (Figure 4.12b). Thus, Belding's ground squirrels are more likely to call when doing so may benefit the survival of their close relatives. The evolution of traits that increase the survival, and ultimately the reproductive success, of an animal's relatives rather than the animal itself is termed **kin selection.**

Costs of Group Living

Living in a group has costs as well as benefits (see Table 4.1). The magnitude of these costs limits the extent to which a species forms groups and explains why some groups are larger than others. Not

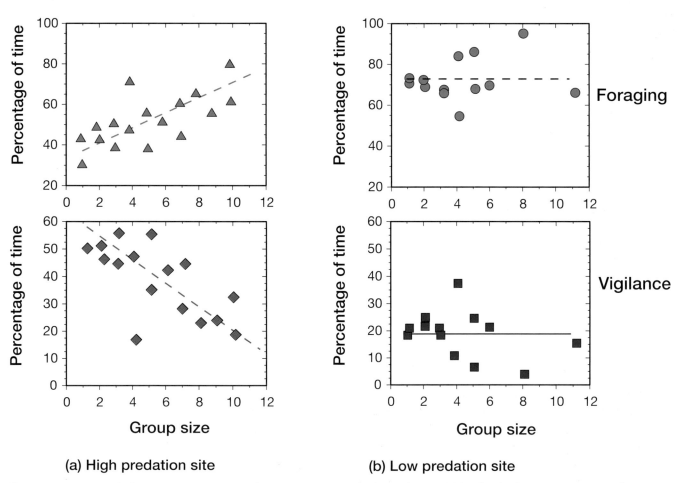

(a) High predation site

(b) Low predation site

Figure 4.11. Group behavior in western gray kangaroos. (a) In a high-predation site, individuals spent more time foraging and less time being vigilant as the group size increased. (b) In a low-predation site, the percentage of time kangaroos spent foraging and being vigilant did not vary with group size. (After Blumstein and Daniel 2002.)

ESSAY 4.1   DO INDIVIDUALS ACT FOR THE GOOD OF THE SPECIES?

Natural selection occurs due to the reproductive advantages of some individuals. This view of the world implies that all individuals are in competition with each other and will behave to further their own interests. From a philosophical viewpoint, the idea that the world is full of selfish individuals clashes with many of the values we hold for human societies, such as cooperation, community spirit and selflessness. Does the variety of behaviors that we observe in animals, even the apparently cooperative ones, really arise from the interactions of selfish individuals? Can traits evolve that favor the larger interests of a group or society? Does evolution lead only to selfishness? These are key questions that interest social scientists, philosophers and biologists.

It is easy to imagine that populations of selfish individuals might overexploit the available resources and become extinct, whereas populations that have evolved social behaviors preventing over-exploitation of resources might have better long-term survival prospects. Natural selection for traits that favor groups rather than individuals is termed *group selection*. The idea that groups of animals could evolve self-regulating mechanisms that prevent over-exploitation of their food resources was first argued in detail in 1962 by V. C. Wynne Edwards, an ecologist at the University of Aberdeen in Scotland. Despite its intuitive appeal, group selection is not considered very important in producing changes in species. Group selection operates much more slowly than individual selection, making it a much weaker selective force in most circumstances.

Imagine, for example, a species of bird such as the puffin that lives in large colonies and lays only a single egg. Could laying a single egg have evolved in puffins by group selection to limit population growth and maintain an adequate food supply for the long-term good of the puffin colony? The answer is no because any mutation that increased the number of eggs laid, would lead to those individuals leaving more copies of their genes to the next generation, compared with birds laying a single egg. Consequently genes for laying two or more eggs would spread rapidly through the population, long before food limitation set in. Short-term advantages to selfish individuals will accrue much more quickly than long-term advantages to the group, so it is difficult to see how traits favored by group selection can be maintained in a population.

But this does not mean that all behavior must be selfish and altruism is dead. To understand apparently cooperative behaviors that benefit the group or society, we need to look for benefits accruing to individuals. Individual selection can produce behaviors that are a benefit for the group.

A horned puffin

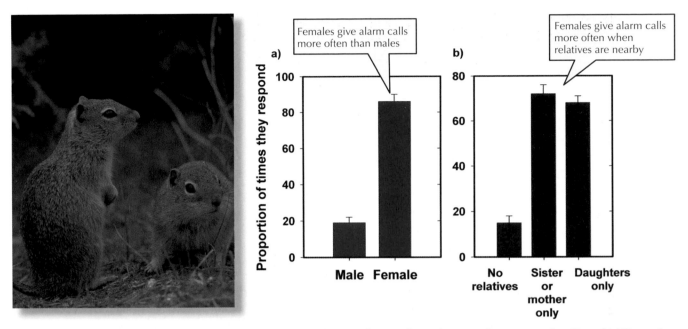

Figure 4.12. Patterns of alarm calling by Belding's ground squirrels. (a) Effect of sex on frequency of calling. (b) Effect of type of nearby relatives on frequency of calling by females. In both (a) and (b), the vertical axis is the percentage of time that squirrels produced alarm calls when a predator approached. (Data from Sherman 1977, photo courtesy of Don Baccus.)

surprisingly, living in large groups leads to competition for resources, such as food or mates. For example, Magellanic penguins (*Spheniscus magellanicus*) form breeding colonies of up to 200,000 birds on subantarctic islands. Colony size in this species appears to be limited by competition for food, which consists of squid and pelagic schooling fishes, including anchovies. Adults and chicks in small colonies ingest more prey of high energy content than do individuals in large colonies (Figure 4.13), and fledglings in small colonies are healthier and therefore more likely to reach adulthood.

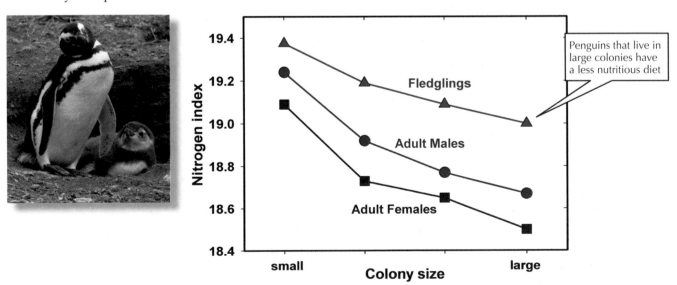

Figure 4.13. Relationship between nitrogen index and colony size in Magellanic penguins. The nitrogen index is based on the ratio of stable nitrogen isotopes in blood samples and is an indicator of food quality. A higher nitrogen index reflects a diet of more nutritious prey, such as anchovies. (Data from Forero *et al.* 2002, photo courtesy of Matt Mueller.)

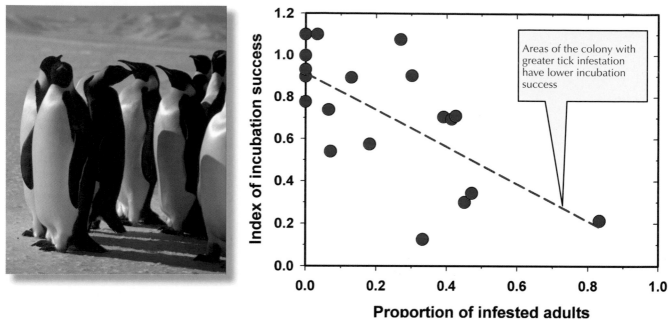

Figure 4.14. Relationship between incubation success and level of tick infestation in a king penguin colony in the Crozet Archipelago. The index of incubation success is the ratio of the number of nests at hatching in February divided by the number of nests at laying in November. Each point on the graph represents data from one small area in the colony, which contained 25,000 birds in total. (Data from Mangin *et al.* 2003; photo courtesy of Don Siniff.)

Breeding in large colonies can also increase the transmission of diseases and parasites. Another species of penguin, the king penguin (*Aptenodytes patagonicus*), breeds in Antarctica in colonies of up to 500,000 individuals. In large colonies, adults and chicks become infested with ticks (*Ixodes uriae*). High rates of tick infestation reduce the incubation success of adults (Figure 4.14).

Another important cost of group living is loss of parentage. Breeding in a group increases the chance that an animal will raise another individual's offspring. Cliff swallows (*Petrochelidon pyrrhonata*) are small songbirds that nest in colonies ranging from a few birds to thousands. Although male–female pairs of cliff swallows cooperate in building nests and raising young, both males and females engage in behaviors that can increase the number of offspring raised outside the pair bond. Males attempt additional copulations, called *extra-pair copulations,* with other females in the colony. The frequency of extra-pair copulations increases with colony size (Figure 4.15). Females engage in *brood parasitism*—placing extra eggs in other females' nests—a behavior whose frequency also increases with colony

size. Extra-pair copulation and brood parasitism increase the cost for some individuals while providing a benefit for the individuals that gain extra offspring. Thus, the potential costs and benefits of group living can vary among the members of a group.

Group Living in African Lions

Ecologists have been studying the social behavior of lions (*Panthera leo*) for more than 40 years in eastern and southern Africa. Lions are the most social member of the cat family—forming groups called prides composed of 1–7 males, 2–18 females and their young. Prides are relatively small in arid areas, such as the Kalahari, and relatively large in areas that have more abundant large prey, such as the Serengeti Plain. In this section we will examine the costs and benefits of different pride sizes and try to understand the benefits of group living for lions. Why do lions live in prides and why do pride sizes vary from place to place?

Male and female lions behave in very different ways, and these differences influence the costs and benefits of group living for each sex. Females almost never leave the pride in which they were born. They

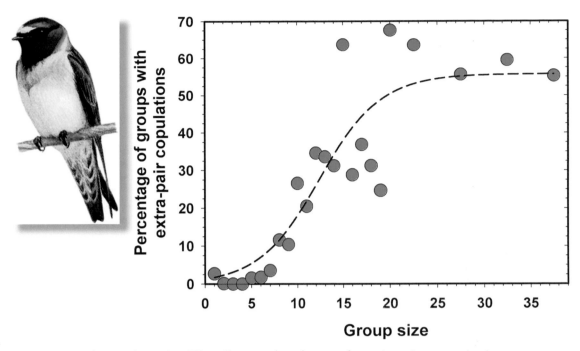

Figure 4.15. One cost of group living in cliff swallows: reduced parental certainty. As group size increases, extra-pair copulations become more common. (Data from Brown and Brown 1996.)

cooperate with their mothers, sisters and other female relatives in hunting, raising young and defending territory. In contrast, male lions are highly transient. They leave the pride in which they were born when they reach 2 to 3 years of age and roam widely in search of a new pride. Males that do not belong to a pride often form groups with related or unrelated males, forming coalitions that challenge males in existing prides for breeding positions. These challenges may result in the death of one or more of the participants. Once in a pride, the males do little hunting and instead spend most of their time defending their territory by patrolling, scent marking, and roaring. Due to frequent challenges, males rarely retain control of a pride for more than 2 years.

Because of the behavioral differences between male and female lions, we will consider the benefits of male–male groups and female–female groups separately. For males, the major benefit of grouping is straightforward. Single males rarely succeed in obtaining a breeding position within a pride. Large coalitions are more likely to take over a pride and are more effective at repelling challenges from other males. Consequently, an individual male's reproductive success

increases with the number of males in a coalition (Figure 4.16). The longer a coalition can remain in control of a pride, the more cubs those males can produce and the greater the cubs' chances of survival are. Although male reproductive success increases with coalition size, individual breeding success becomes more variable in the largest coalitions: Some males mate often, whereas others rarely mate. As a result, very large coalitions tend to self-destruct owing to male–male competition for matings.

For female lions, the benefit of group living—as measured by reproductive success—is greatest in groups of 3–10 females (Figure 4.17). This appears to be the **optimal group size**—the size that results in the largest relative benefit. How can we explain this observation? Careful calculations have shown that very small prides (and even solitary lionesses) have the highest rates of food intake. Thus, hunting success seems to decrease as group size increases. In contrast, larger groups facilitate territorial defense, which is important in preventing male takeovers. When new males take over a pride, they typically kill all the young cubs. That causes the females in the pride to rapidly enter estrus, allowing the new males to father offspring

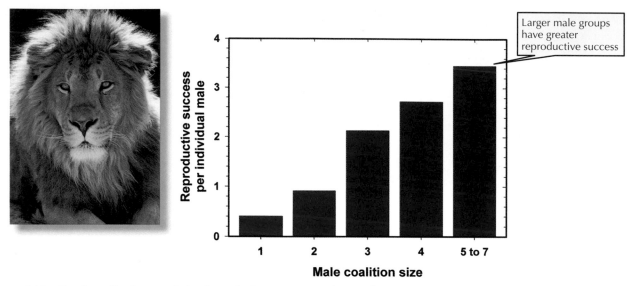

Figure 4.16. One benefit of group living in male lions: increased reproductive success. (Data from Packer *et al.*, 1988, photo courtesy of Mikale Pilgrim.)

quickly. Cub survival is higher in larger female groups because larger groups are better able to save young cubs from infanticide. Thus, the optimal group size in female lions may represent a balance between hunting success and territorial defense.

Q Would you expect all lion groups to have exactly the same optimal size? Why or why not?

As our example of African lions illustrates, understanding which factors favor group living in a species can be complex (Table 4.2). Although we can easily identify potential costs or benefits of group living, to single out the important factors, we must determine how this behavior affects the survival and reproductive success of an *individual*. Doing this successfully requires detailed data on individuals from groups of different sizes or carefully designed field experi-

ments, or both. The relative benefit of group living may vary with habitat type and other environmental conditions, making long-term studies especially important. In many species, the costs and benefits of group living differ between the sexes, which can lead to conflict between males and females over the optimal group size.

SUMMARY

To survive as species, all animals must obtain food, avoid predators and disease, and produce offspring. They achieve these goals through a variety of behaviors, which must be appropriate for their particular environment. Natural selection is the force that achieves the fit between how individuals behave and their subsequent survival. Many animals must make

Table 4.2. Specific benefits and costs of forming male or female groups in African lions.

Sex	Benefits of grouping	Costs of grouping
Male	Increased ability to gain control of a pride (access to mates) Increased ability to maintain control of a pride (higher survival of offspring)	Sharing of paternity with coalition members
Female	Preferential feeding of close kin (help from kin) Territorial defense (increased female and offspring survival)	Lower rate of food intake

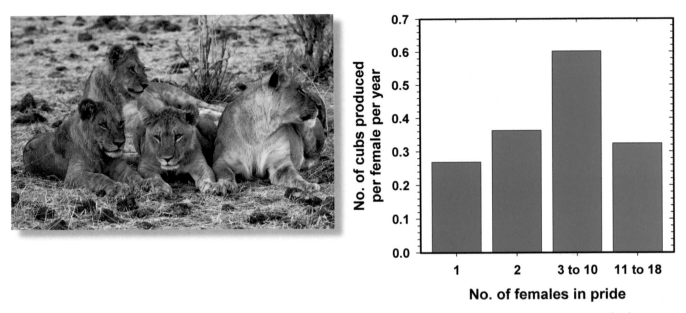

Figure 4.17. Reproductive success of female lions in prides of differing size. The production of cubs is maximal when prides contain 3 to 10 females. (Data from Packer *et al.* 1988, photo courtesy of Calvin Jones Photography.)

decisions about where to forage, which individuals to mate with, how large a territory to defend, and which habitat to select for nesting.

The key to understanding the behavior of individuals is to determine the costs and benefits of these decisions in terms of the number of offspring an individual produces. This approach has been particularly successful for foraging behavior, and we can identify foraging rules by which animals optimize their food intake rates. A cost–benefit analysis can also help us identify the factors that affect the social structure of a species, such as its optimal group size and how large a territory it defends. Understanding the factors that influence the behavior of individuals may allow us to predict how different species will respond to habitat loss and reductions in population size.

SELECTED REFERENCES

Anderson KG, Kaplan H and Lancaster J (1999). Paternal care by genetic fathers and stepfathers I: Reports from Albuquerque men. *Evolution and Human Behaviour* **20**, 405–432. (A quantitative assessment of human investment in related and unrelated children.)

Forero MG, Tella JL, Hobson KAM, Bertellotti M and Blanco G (2002). Conspecific food competition explains variability in colony size: a test in Magellanic penguins. *Ecology* **83**, 3466–3475. (A good example of how costs rise in large nesting groups of birds.)

Hamilton WD (1971). Geometry of the selfish herd. *Journal of Theoretical Biology* **31**, 295–311. (The classic paper that uses simple models to show the costs and benefits of group living.)

Krebs JR and Davies NB (1993). *An Introduction to Behavioural Ecology.* 4th ed. Blackwell Scientific Publications, Oxford, UK. (The classic text on behavioral ecology synthesizing theory and examples for senior students.)

Packer C and Pusey AE (1997). Divided we fall: cooperation among lions. *Scientific American* **276**, 52–59. (A clear presentation of how ecologists studied lions to find out the advantages and disadvantages of group living.)

Sherman PW (1977). Nepotism and the evolution of alarm calls. *Science* **197**, 1246–1253. (A classic paper explaining how alarm calls help relatives.)

QUESTIONS

1. The species recovery committee for African wild dogs is considering transplanting some wild dogs to other sites. Would you recommend this action? Why or why not? What other actions would you recommend?

2. Lemmings are arctic rodents that show large variations in population density every 3–4 years. Some people have suggested that when lemming numbers are high, lemmings reduce pressure for resources by running into the sea in great numbers. Would individual selection favor this behavior? Could it evolve by means of group selection? Why would this behavior be likely or unlikely to evolve?

3. Altruism—personal sacrifice on behalf of others—is difficult for evolutionary biologists to explain because natural selection favors the interests of individuals. Nevertheless, altruistic behaviors towards relatives are observed in many animal societies. Is there any way that altruism among non-relatives can evolve in animal societies? How might altruism arise in human societies if it is based on self-interest? Gintis *et al.* (2003) discuss this question.

4. Many birds form groups in which only one female breeds and other birds act as helpers at the nest. Discuss the general conditions that might favor such breeding groups. Why might an individual choose to stay as a helper in a group rather than move away and breed elsewhere? Heinsohn and Legge (1999) discuss this problem of cooperative breeding.

5. In Scotland, female offspring of red grouse disperse to surrounding areas, while male offspring take up a territory next to their father. A male's territory is always occupied exclusively by one bird. Describe how the aggression associated with territorial defense might differ if a male is surrounded

by his sons or by unrelated males. Mougeot *et al.* (2003) describe this system and some experiments on this issue.

6. Infanticide is observed in many mammals, birds and insects. Female infanticide is surprisingly common in human cultures. Read the following report from *The Lancet* (2003, 362: 1553), a prestigious journal for medical research, and answer the questions below:

> 'The ratio of boys to girls in India is becoming increasingly skewed in favour of boys, as more and more girls are being selectively aborted or killed. Consolidated data released by the United Nations Population Fund (UNFPA) on Oct 29 revealed a shocking decline in the number of girls compared with boys during the past 10 years, mainly as a result of abortion of female fetuses and killing of newborn girls.'

> 'While India's population rises, the ratio of girls to boys is in steep decline. Between 1991 and 2001, the number of girls per 1000 boys fell from 945 to 927, for the 0–6 years age group…'

Using the approaches discussed in this chapter, (a) formulate two hypotheses to explain infanticide in humans, (b) design an experiment or describe the data you would collect to test your hypotheses, and (c) discuss the proposition that infanticide is adaptive in humans.

Chapter 5

POPULATION DYNAMICS—ABUNDANCE IN SPACE

IN THE NEWS

Grizzly bears in North America represent two views of how humans fit into natural ecosystems. On the one hand, grizzly bears are hunted by sportspeople, whose activities are an important economic benefit to outfitters and other businesses in rural areas of Alaska and western Canada. On the other hand, grizzlies are a conservation icon—the embodiment of a wilderness ideal that many North Americans want to preserve for their children. In 2001–2003, these two views were on opposite sides of a heated public controversy over the abundance of grizzly bears in British Columbia. One side argued that there were at least 14,000 grizzlies in the province, so some hunting could be sustained. The other side maintained that the grizzly bear population was only 6,000, so all hunting should be stopped. The controversy boiled over in newspapers, and both sides put pressure on the government to accurately estimate the abundance of grizzly bears in the province.

One common method for estimating the abundance of a population of animals is the mark–recapture technique, which we will discuss in section 5.1. This method has not traditionally been applied to grizzly bears because they are relatively secretive animals that live at low density in thick forests, so capturing them is difficult. Recently, however, wildlife ecologists have used DNA fingerprinting to identify individual bears based on hair samples. The samples are collected from a barbed-wire fence surrounding a pile of logs soaked in rancid fish oil, which attracts the bears. This variation of the mark–recapture technique allows the ecologists to recapture evidence from specific individuals without ever seeing or handling them.

Population estimates based on DNA fingerprinting indicate that grizzly bears are more numerous in British Columbia than either side in the controversy previously thought—the whole province contains about 15,000 grizzlies. Therefore, it is likely that hunting can continue in areas where the bears are common without threatening the grizzly bear population with extinction.

Female grizzly bear

The Ecological World View

■ 5.1 TO ANALYZE POPULATIONS, WE MUST MEASURE ABUNDANCE

As you learned in Chapter 1, the abundance of a species in a given habitat is simply the number of individuals of that species that live in that habitat. Abundance varies both in space and in time. In this chapter we will consider only the spatial dimension. In Chapter 6 we will look at how abundance in a particular area rarely stays constant for long.

How do ecologists determine abundance? The most direct way is to count all of the individuals in the area in question. This is similar to a census as carried out by different countries every 5 or 10 years to determine the current size of their human population. It is fairly easy to count all the trees in a particular area or locate all members of a species of birds in which the males sing to defend territory or to attract females. Large animals that live in small areas can sometimes be counted using aerial surveys, while other animals, such as the northern fur seal, may be counted when they are all gathered in breeding colonies.

For the vast majority of species, however, obtaining a direct count of all the individuals in an area is impractical or impossible. For these species, the investigator must be content to count only a small proportion of the population and to use this sample to estimate the total. There are two general methods of making such estimates: quadrat counts and mark–recapture.

Quadrat Counts

Organisms that do not move quickly can be counted in small sections, called **quadrats**, within a larger area and the numbers from these sections can then be extrapolated to the whole area of study (Figure 5.1). Quadrats can be square, rectangular or even circular. As an example of how quadrat counts are carried out, suppose you counted 120 individuals of a forest herb species in ten quadrats, each measuring 2 m by 2 m. The average density of the species in those quadrats would then be 120 plants/40 m^2, or 3 plants/m^2. (Recall from Chapter 1 that the density of a species is the number of individuals per unit area or volume.) If the whole area occupied 10,000 m^2, then we could estimate that it contained 30,000 plants of that species. Quadrat

ESSAY 5.1   COUNTING TREES

Plants are relatively easy to count because they do not move around like animals. The simplest way to conduct a census for plants is to set up a quadrat and count all the plants within the boundaries. To illustrate how this is done, let us look at maps of two tree species in Lansing Woods, an old-growth forest in southern Michigan that has been designated as a park to preserve the beech–maple forest growing there. On one 7.9-ha (20-acre) site in Lansing Woods in Michigan, there are 135 black oak trees, so the density is 17 trees/ha (7 trees/acre):

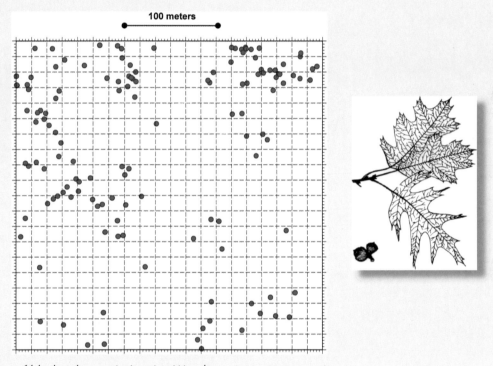

Map of black oak trees in Lansing Woods

On the same site, there are 514 sugar maples, so their density is 65 trees/ha (26 trees/acre):

Many similar quadrats have been constructed for other forests. One of the most ambitious is a 50-ha site on Barro Colorado Island in Panama, where Stephen Hubbell, Robin Foster and many assistants mapped the locations of 215,000 small trees (1–10 cm in diameter) and about 21,000 larger trees, belonging to more than 200 species. They have followed these individually marked trees for more than 30 years to measure the dynamics of tropical rainforest trees. Marking individual trees enables the researchers to obtain a great deal of information about the trees' life histories.

Spatial maps of trees are useful for determining density, but they also contain considerable information about the spacing of individuals, which is of wide interest to plant

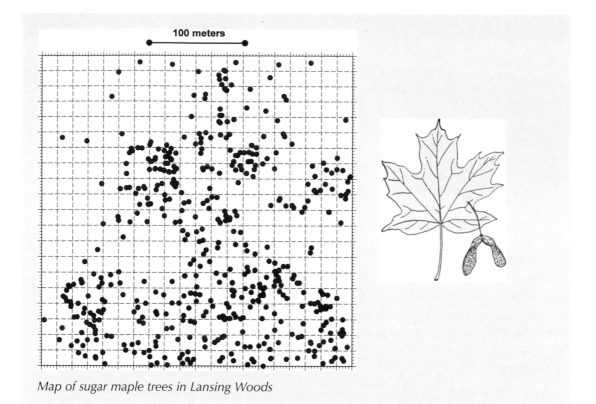

Map of sugar maple trees in Lansing Woods

ecologists who wish to understand why patterns of clumping occur. In this way, simple quadrat counts and quadrat maps can lead to a variety of interesting questions about how forests are sustained.

counts have been used extensively in plant ecology and are the most common method for sampling plants (see *Essay 5.1*).

Figure 5.1. Quadrat sampling in beach areas of Rocky Bay, Washington State in 2002. A 0.25 square meter quadrat is used to count the density of invertebrate burrows and to estimate the plant cover. (Photo courtesy of US Environmental Protection Agency.)

The reliability of abundance estimates obtained by quadrat counts depends on three conditions: (1) The population in each quadrat must be determined exactly, (2) the area of each quadrat must be known and (3) the quadrats must be representative of the whole area of study. The last condition is usually satisfied by sampling many randomly selected quadrats. The more quadrats you count, the more accurate your estimate will be.

Mark–recapture

Mark–recapture techniques involve the capture, marking, release and recapture of mobile animals. Over the last 50 years, a variety of clever methods have been developed for marking individual animals, including paints and dyes on feathers and fur, numbered ear tags and leg bands, collars with radio transmitters and, as you read at the beginning of this chapter, even an animal's DNA fingerprint (Figure 5.2). The key feature of

Figure 5.2. Two methods of marking birds. (a) Tasmanian native hen with plastic and metal numbered leg bands. (b) Greylag geese in Sweden. One adult is marked with a lettered plastic neck band. (Photos courtesy of Elsie Krebs and Nordic Greylag Goose Project, Sweden.)

these methods is that they allow researchers to identify animals when they recapture them. Mark–recapture techniques have been an important method of study for many animal species.

The simplest mark–recapture technique to estimate population abundance is the **Petersen method**[1], which was developed by the Danish fisheries scientist C. G. J. Petersen in 1898. The Petersen method follows this line of reasoning: If you capture, mark and release animals into the general population, then a later sample taken from that population will have the same proportion of marked animals as the whole population has. We can represent this reasoning with the following equation:

$$\frac{\text{No. of marked animals in second sample}}{\text{Total no. of animals in second sample}} =$$

$$\frac{\text{No. of marked animals released}}{\text{Total population size}}$$

The time interval between the release of marked animals and taking the second sample is critical. On the one hand, it must be long enough to allow marked individuals to become evenly distributed throughout the whole population. On the other hand, it must be short enough that no new animals are added to the population and no marked animals die.

The Norwegian fisheries scientist Knut Dahl was one of the earliest scientists to use of the Petersen method. In 1912 he marked 109 trout (*Salmo fario*) in

small Norwegian lakes to estimate the size of the population that was subject to fishing. A few days after releasing the marked fish, he caught 177 trout, of which 57 were marked. Inserting these data into the above equation gives

$$\frac{57}{177} = \frac{109}{\text{Total population size}}$$

We can rearrange this equation to solve for the total population size of 338 trout.

The Petersen method can be used to analyze trends in populations that have been threatened and are now recovering. One example is the humpback whale (*Megaptera noveangliae*), whose numbers were greatly reduced by the whaling industry until 1966, when the species became protected. Humpback whales range in length from 10 to 15 meters, weigh from 27,000 to 45,000 kg (30 to 50 tons) and are found in all of the world's oceans. They become reproductively mature at 4–8 years of age and mate during their winter migration to warmer waters. The female gives birth to a single calf a year later.

For the past 40 years, biologists have been using the Petersen method to estimate the abundance of these whales. Each whale has a unique set of markings on its large tail flukes, so researchers do not have to capture the whales or otherwise interfere with their movements to identify them (Figure 5.3). In one study off the central coast of Ecuador, 79 humpback whales were photographed and cataloged in 1996. The following year, 72 whales were photographed in the same area, of which 3 were identified from the 1996 photographs;

---

1   Wildlife ecologists refer to the Petersen method as the *Lincoln method* because it was first used on ducks by F. C. Lincoln in 1930.

Figure 5.3. Tail flukes of two humpback whales photographed off the Oregon coast. The markings on a whale's flukes are as distinctive as a person's fingerprints. (Photos courtesy of Oregon State University, Marine Mammal Program.)

the remaining 69 had not been cataloged in 1996. With these data, we can estimate the humpback whale population migrating off central Ecuador:

$$\frac{3}{72} = \frac{79}{\text{Total population size}}$$

Solving for total population size gives an estimate of 1,896 whales. By carrying out similar surveys in other parts of the world, whale researchers estimated that the world humpback population in 2003 was about 15,000 individuals, which is 10–15% of the population size before the advent of whaling.

Other Methods for Comparing Population Sizes

Ecologists use a variety of sampling methods that enable them to compare the *relative* sizes of populations without providing an estimate of population size. Such methods might allow you to say that there are more individuals of a certain species in area X than in area Y, but they would not tell you how many individuals there are in either area. There are a great many of these methods, but we will list only a few:

1. *Traps.* Traps include mousetraps spread across a field, light traps for night-flying insects, pitfall traps in the ground for beetles, suction traps for flying insects and plankton nets (Figure 5.4).
2. *Number of fecal pellets or mass of feces.* This measure has been used for snowshoe hares, deer, field mice, rabbits and caterpillars.
3. *Vocalization frequency.* The number of calls heard per 10 minutes in the early morning can be used to gauge the size of bird populations. The same method is used for frogs, crickets and cicadas.
4. *Number of artifacts.* This count is useful for animals that leave evidence of their activities, such as bird tracks in sand plots, mud chimneys made by burrowing crayfish, tree squirrel nests and pupal cases from emerged insects.
5. *Questionnaires.* Questionnaires filled out by hunters, fishers and trappers provide estimates of population changes in economically important animals.
6. *Cover.* The percentage of the ground surface covered by individuals is an especially important method for comparing the relative sizes of many plant populations. This method is also used by ecologists studying invertebrates in the rocky intertidal zone.
7. *Roadside counts.* The number of birds or other highly visible organisms observed while one drives a standard distance has been used as an index of abundance.

Q **How would you test any particular index of abundance to determine if it is reliable?**

As we conclude our discussion of techniques for measuring and comparing abundance, two points are important to remember. Firstly, detailed, accurate information is easily obtained for many plant species,

■ 5.2 GOOD AND POOR HABITATS ARE DEFINED BY RESOURCE LEVELS

Every birdwatcher, mushroom picker, hunter or fisher knows places where they can be sure of finding a particular species of animal or plant, and other places where they would be very unlikely to find that species. Nature is a mosaic of good, poor and impossible habitats for any organism. Conservation programs to protect species, such as the mountain pigmy possum and the orange-bellied parrot in Australia or bald eagles, grizzly bears or caribou in North America, can succeed only if we know what comprises a good habitat for these species. Similarly, pest control programs require us to know what makes habitats poor for undesirable species.

Why are some habitats good and others poor for a particular species? We can take two approaches to answer this question. One is to look for general attributes of species that are related to abundance. For many animals, the most important of these attributes is body size. A second approach is to describe the variation in abundance from site to site for particular species and correlate this variation with environmental parameters, such as temperature, rainfall and soil type. Both approaches begin with a thorough description of the pattern of abundance and are followed by experiments designed to test hypotheses that explain the pattern. Let's look at a few examples of these approaches.

Abundance in Relation to Body Size

One of the general rules of animal abundance is that small animals are usually more abundant than large animals. Figure 5.5 shows the relationship between population density and body weight for mammals and birds, which have been studied more extensively than other animal groups. This relationship allows us to predict the population density of any mammalian or bird species of a given body weight—a useful starting point if we want to know whether that species is abnormally common or rare (see *Essay 5.2*). Although there is a great deal of variation around the trend lines in Figure 5.5, the figure reveals two clear patterns. Firstly, for a given body weight, birds are generally less abundant than mammals. For example, the graph

Figure 5.4. Plankton sampling in the Southern Bering Sea with a bongo plankton net. The lead weight on the bottom allows the two nets to sink to the desired sampling depth, and they are then towed obliquely to the surface. The two nets provide duplicate samples. The two current meters above the nets measure the volume of water filtered through the nets, which typically have a mesh size of 333 to 500 microns. Many variations of these methods are used for sampling zooplankton and small larval stages of fish species. (Photo courtesy of U.S. National Oceanic and Atmospheric Administration.)

but we have only rough estimates of abundance for most animal species. Secondly, because of the difficulty of obtaining accurate data on animals, a disproportionate amount of research has been done on the more easily studied animal groups, particularly birds and mammals, and on species of economic importance. This research imbalance introduces an obvious bias into the discussions that follow.

ESSAY 5.2   CALCULATION OF EXPECTED POPULATION DENSITY FOR A HERBIVOROUS MAMMAL

We can use the linear relationship shown in Figure 5.5 to calculate the expected abundance of a mammal or bird that we are studying. For example, consider a Yukon chipmunk that weights 85 g. According to Silva *et al.* (1997) and Peters (1983), the slope (*b*) of the red line in Figure 5.5 is −0.66, and the y-intercept (*a*) is 1.30. Consequently:

$$\log(\text{population density}) = a + b(\log[\text{body weight}])$$
$$= 1.30 - 0.66(\log[0.085])$$
$$= 1.30 + 0.707 = 2.007$$

The estimated population density is the antilog of 2.007, or $10^{2.007}$, which is 102 individuals per km^2. A similar calculation for a larger mammals like the 1250 g rufous hare-wallaby from Australia would estimate a population density of 2.4 individuals per km^2.

For an 85-g bird, the expected population density would differ because the blue line in Figure 5.5 has a different slope (−0.54) and a different y-intercept (0.22):

$$\log(\text{population density}) = a + b(\log[\text{body weight}])$$
$$= 0.22 - 0.54(\log[0.085])$$
$$= 0.22 + 0.578 = 0.798$$

The estimated population density for this bird is the antilog of 0.798, or 6.3 individuals per km^2. Similarly, for a hypothetical 1250 g bird, the estimated population density would be 1.5 individuals per km^2.

These calculations predict that an 85-g bird species will have a lower population density than a mammalian species of the same body weight, in agreement with the data shown in Figure 5.5.

predicts a density of 1 individual/km^2 for a 1-kg bird and 20 individuals/km^2 for a 1-kg mammal. Secondly, larger animal species have lower population densities. Thus, we would expect a density of about 60 individuals/km^2 for 200 g squirrels but only about 0.5 individual/km^2 for 250 kg elk. This inverse relationship between population density and body size also applies to plants and invertebrate animals. The greater abundance of small plants and animals affects the structure of ecological communities and ecosystems, a topic we will discuss further in Chapter 15.

Q **What factors might affect the scatter of points around the lines shown in Figure 5.5?**

Abundance in Relation to Limiting Environmental Factors

We need to know a great deal about the biology and natural history of a species before we can assess the factors that limit its abundance. While it is true in general that resource levels limit abundance, the problem is to determine which particular resource or combination of resources is critical for each species. A resource may be nutrients, light or water for plants, food for animals, warmth for plants or animals, space free from competitors or predators, or space that has been changed by disturbances like fires or flooding. There is no general theory that allows us to predict exactly what resource is limiting for a particular species, so we

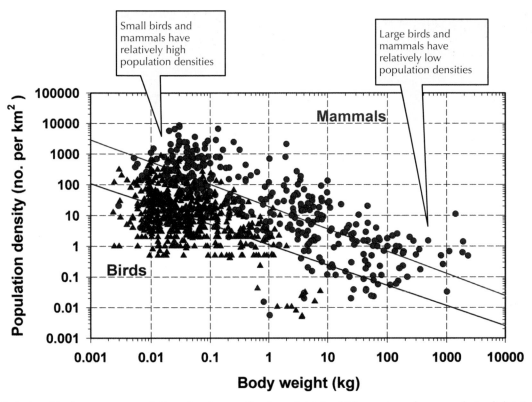

Figure 5.5. Relationship between population density and body weight for 350 species of mammals (red dots) and 552 species of birds (blue triangles) from around the world. Average trend lines are shown for each group. Note that the scales are logarithmic. (Data from Silva *et al.* 1997.)

proceed by a series of case studies to illustrate possible limitations on abundance.

Many species show large variations in abundance in different parts of their geographic range. One example is the wood pigeon, whose range includes most of Europe as well as parts of Asia and North Africa. Because wood pigeons eat seeds, they can be a serious agricultural pest in areas that are used to grow grain crops, such as oats and rapeseed. In the British Isles, wood pigeons are most abundant in eastern England, where the most extensive agricultural areas are located, but they are relatively uncommon in Ireland, where grain crops are grown in only a few areas (Figure 5.6). Based on this distribution pattern, we may predict that food supplies limit the abundance of wood pigeons and that food limitation is most dramatic in winter. Data are needed to test these predictions.

Limitation by Physical Factors

The abundance of many plants can be directly linked to physical factors such as temperature, rainfall and soil nutrients. For example, the scribbly gum, a species of Australian eucalypt, grows only in areas where the mean annual temperature is between 9°C and 14°C (Figure 5.7). Within this range, its abundance increases with temperature until the upper critical temperature is approached. Temperature and moisture limitations on plant abundance typically operate on the seedling stages of trees. Once trees are established, they can tolerate a much wider range of these factors.

The availability of cavities in trees appears to limit the local abundance of cavity-nesting birds. Carl Bock and David Fleck counted birds in 54 small plots for two years in two Colorado forests. Then they added nest boxes to half of the plots (chosen randomly) and counted birds in all the plots for an additional two years (Figure

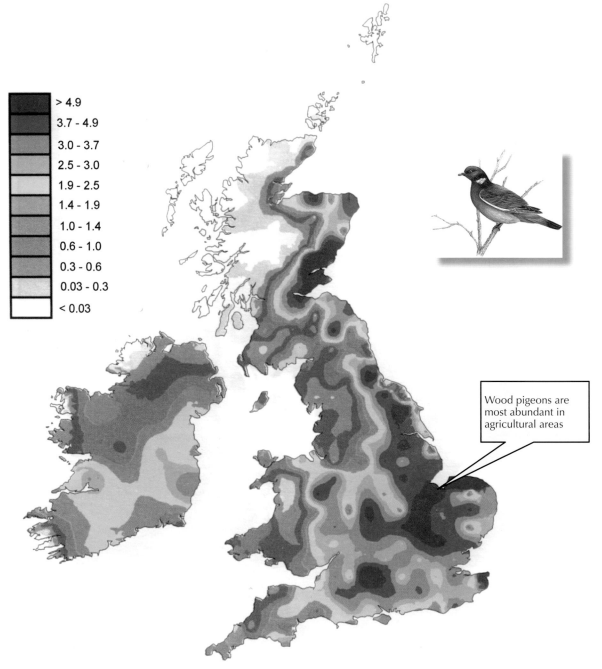

Figure 5.6. Abundance of the wood pigeon (*Columba palumbus*) in the British Isles. The abundance scale is relative and is based on point counts (a type of quadrat sampling) conducted from 1988 to 1991. (Data from Gibbons *et al.* 1993, image courtesy of the Royal Society for the Protection of Birds.)

5.8). They found that the abundance of cavity-nesting birds, such as the western bluebird (*Sialia mexicana*) and the mountain chickadee (*Parus gambeli*), increased more than threefold in the plots with added nest boxes, while the abundance of birds that do not nest in cavities did not change. Experiments like this are the best way of identifying limiting factors for abundance. If we continued to add nest boxes to an area, bird numbers would reach the point where nest sites are no longer limiting, and some other limiting factor would come into play.

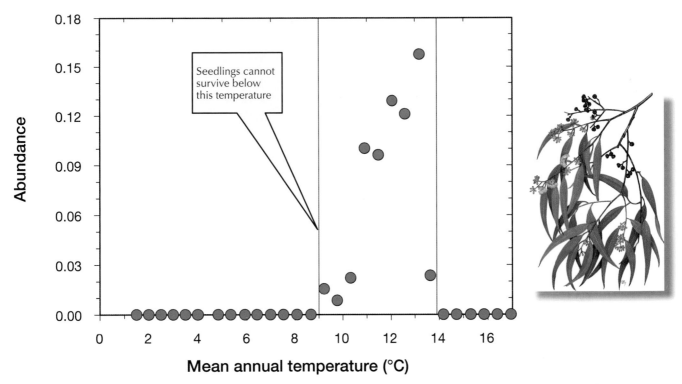

Figure 5.7. Abundance of the scribbly gum (*Eucalyptus rossii*) in relation to mean annual temperature in south-eastern Australia. The vertical lines mark the tolerance limits for this tree. The abundance scale is relative. (Data from Austin *et al.* 1994.)

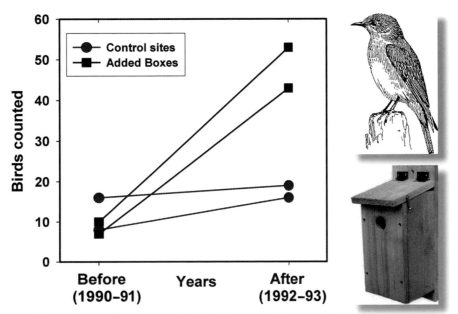

Figure 5.8. Abundance of five species of cavity-nesting birds in 54 study areas in Colorado. Between the first (1990–1991) and second (1992–1993) survey periods, researchers added nest boxes to 27 of the study areas. The number of birds increased significantly only in the areas that had next boxes (blue bars). (Data from Bock and Fleck 1995.)

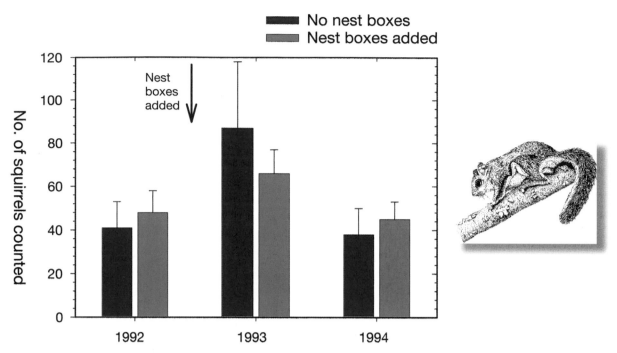

Figure 5.9. Abundance of southern flying squirrels in six study areas in South Carolina. After the 1992 survey period, researchers added nest boxes to three of the study areas. The addition of the nest boxes did not affect abundance in this species, as the number of squirrels in all six areas increased in 1993 and decreased in 1994. These results contrast with those in Figure 5.8, demonstrating that all populations are not limited by the same factor. (Data from Brady *et al.* 2000, drawing courtesy of Robert Savannah, U.S. Fish and Wildlife Service.)

Can we extend the conclusions of Bock and Fleck's nest-box experiment to all cavity-nesting species? The southern flying squirrel (*Glaucomys volans*) is a small mammal that uses cavities for storing nuts and denning, and it competes with cavity-nesting birds for access to tree cavities. Matthew Brady and his colleagues tested the hypothesis that the number of cavities limits the abundance of southern flying squirrels in South Carolina. In the fall of 1992, they added 100 nest boxes to each of three forested areas, increasing the number of available cavities by 66%. Over the next two years, they counted flying squirrels in those areas and in three similar areas without nest boxes. Surprisingly, they discovered that adding nest boxes had no significant effect on the number of squirrels (Figure 5.9). Clearly, some factor other than the availability of cavities—perhaps food supply—must limit flying squirrel numbers in this part of South Carolina.

Limitation by Predators

The abundance of the large kangaroos in Australia has increased since the British colonized the continent. The

paradoxical increase in kangaroo populations has occurred in spite of intensive shooting programs— kangaroos are considered pests by ranchers and are harvested for meat and hides. The reason seems to be that ranchers have improved the habitat for large kangaroos in three ways. Firstly, water has been made available for sheep and cattle, removing the impact of water shortage for kangaroos in arid environments. Secondly, kangaroos feed on grass, and ranchers have cleared timber and produced grasslands for livestock. Thus, both the water supply and the food supply have increased. Thirdly, predation by the dingo (*Canis familiaris dingo*) on kangaroos has been removed. To protect their sheep, ranchers shot and poisoned dingoes in south-eastern Australia, and the Australian government constructed a 5,400-km-long fence to keep dingoes from recolonizing that part of the country (Figure 5.10a). These measures led to a spectacular increase in red kangaroo (*Macropus rufus*) populations in places where dingoes were eliminated: The density of red kangaroos is 166 times as high in New South Wales as in parts of South Australia that are north or west of the

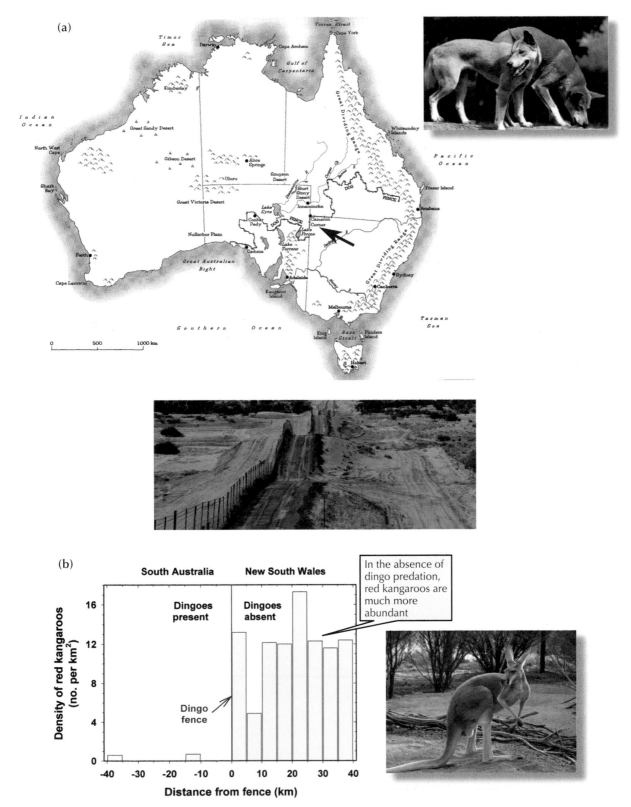

Figure 5.10. (a) Map of the 'Dingo Fence,' designed to keep dingoes out of sheep-grazing country in south-eastern Australia. (b) Density of red kangaroos along a transect across the Dingo Fence on the New South Wales–South Australia border in 1976. The arrow in part (a) marks the location of the transect. Predation by dingoes limits the abundance of red kangaroos. (Map modified from Woodford (2003); data from Caughley *et al.* 1980, photo courtesy of Tom Vaughan.)

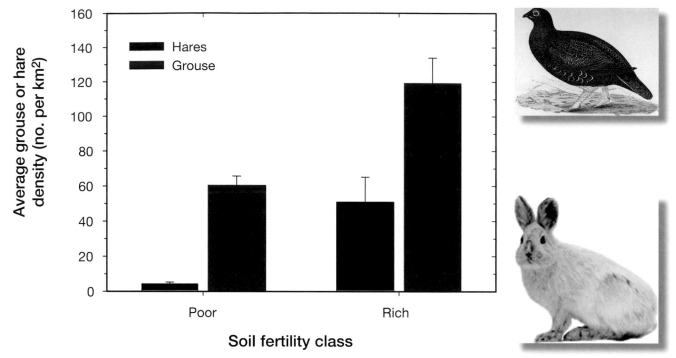

Figure 5.11. Density of mountain hares and red grouse on eight moors in Scotland in relation to soil fertility, which depends on the underlying rocks. Rich soils support more nutritious heather, which translates into higher average densities of hares and grouse. (Data from Watson *et al.* 1973.)

fence (Figure 5.10b). Thus, the elimination of one problem for sheep ranchers (predation of sheep by dingoes) has worsened another problem (competition for grazing land by red kangaroos).

Limitation by Food Supply

Much of wildlife and fisheries management operates on the assumption that increasing a species' food supply can turn a poor habitat into a good one. Red grouse (*Lagopus lagopus*) and mountain hares (*Lepus timidus*) are good examples. Robert Moss and Adam Watson were part of a team of ecologists who investigated why populations of these two species had declined on Scottish moorland after 1930. They found that the density of grouse and hares was high on some moors and much lower on others. These differences in species density were correlated with differences in the nutritional quality of heather—the main food of both mountain hares and red grouse. Heather that grows on rich soil contains more nitrogen and phosphorus and supports higher densities of these species than does heather that grows on poor soil (Figure 5.11).

Thus, this example supports the generalization that better food increases abundance. The research by Moss and Watson also showed that land management that includes periodic burning can turn poor moors into good moors for grouse and hares.

Deer populations in western North America respond in a dramatic way to differences in the quality of nutrition. On good feeding grounds, some does produce twins, and fawn survival is high. On poor feeding grounds, fewer does have twins and fawns grow slowly, entering their first winter at a smaller size. Slow growth also increases the age at sexual maturity: Most does produce their first fawns at two years of age on good range, but does on poor range do not give birth until they are three years old. Parasite loads and the incidence of disease may also be higher in undernourished deer. Male deer have a higher mortality rate than female deer, and this difference is accentuated on poor range, causing females to greatly outnumber males. Deer management thus aims to provide habitats with an abundance of food plants of high nutrient content. However, this aim is not easy to accomplish because,

Robert Moss

Female moose in Alaska

by feeding selectively, deer reduce the abundance of high-quality food plants.

In the eastern United States, the eradication of wolves and bears has released white-tailed deer populations from growth limitations set by predators. At the same time, agricultural practices have increased the availability of high-quality food for deer. The resulting 'overabundance' of white-tailed deer is now one of the most serious wildlife management problems in this part of the country. If deer are allowed to reach high numbers, they can rapidly turn a good range into a poor one.

Limitation by Disturbances

Forest fires may cause habitat alterations that favor some species and harm others. In North America and Scandinavia, moose (*Alces alces*) feed on many of the broadleaved shrubs and small trees that colonize burned areas—as the dense forest matures, these plants are eliminated and moose decline in numbers. Species such as moose are sometimes called *fugitive species* because they are continually on the move and depend on the opening of new habitats by fire, logging or other disturbances. Fugitive species can be very successful. Moose are extremely abundant in Sweden—nearly ten times as abundant as in the best North American habitats—because logging continually provides open habitats containing the shrubs moose prefer to eat and because wolves and other predators have been largely eliminated.

Adam Watson

The attainment of high population densities in some plants also depends on fire. The most famous example is the giant sequoia (*Sequoiadendron giganteum*) of California. Because of their very thick, fire-resistant bark, large sequoia trees are not damaged by ground fires that are fatal to many other conifers. Sequoia seedlings also germinate best on the bare soil that is left after ground fires have removed forest-floor litter. In the absence of fire, sequoias are replaced by other conifers. Therefore, early conservation attempts to protect the remaining sequoia forests by preventing or extinguishing all forest fires could have doomed sequoias to disappear. However, since the 1960s the U.S. National Park Service has used fire as a part of sequoia forest management (Figure 5.12).

Limitation by Habitat Structure

Changing the structure of a habitat can affect the number of animals that use the habitat, even if the food supply is unchanged. For example, young salmon (fry) spend the first year or more of their lives in streams, where they defend feeding territories against intruders. The size of the territories depends on several factors, including the topography of the stream bottom and the current. When large stones are placed on the bottom, the fry become more visually isolated from one another, territories become smaller, and population density increases as a result (Figure 5.13). Increasing the current has the same effect because higher currents cause the fry to keep closer to the stream bottom, again increasing visual isolation. Thus, both territorial behavior and physical features of the habitat limit population density among salmon fry.

Q **Which of the environmental factors discussed in this section limit the human population?**

■ 5.3 POPULATIONS ARE NOT CONTINUOUSLY DISTRIBUTED IN SPACE

We have been discussing populations (see *Essay 5.3*) as though they were continuously distributed in space across a relatively uniform environment. But if a species lives in a patchy habitat its spatial distribution will also be patchy. Patches of habitat may occur naturally

Figure 5.12. Prescribed burning in the Redwood Mountain Grove of giant sequoias in Kings Canyon National Park, California, in 2003. Fire consumes the accumulated forest fuels, leading to the recycling of nutrients and preparing a seedbed for sequoias. (Photo courtesy of US National Park Service.)

or as the result of human actions. For instance, in areas where forests have been largely cleared for agriculture, open fields separate small patches of forest. The wheat belt of Western Australia is a classic example of such an area (Figure 5.14).

Populations and Metapopulations

When a species is divided into small populations because of a patchy habitat, and occasional migration between patches is possible, the small populations collectively are called a **metapopulation**. Metapopulations are often distributed over 'islands of habitat' that are surrounded by a sea of unsuitable habitat.

The bay checkerspot butterfly in central California is one example of a metapopulation. The caterpillars of this butterfly feed on plants that grow only in soil

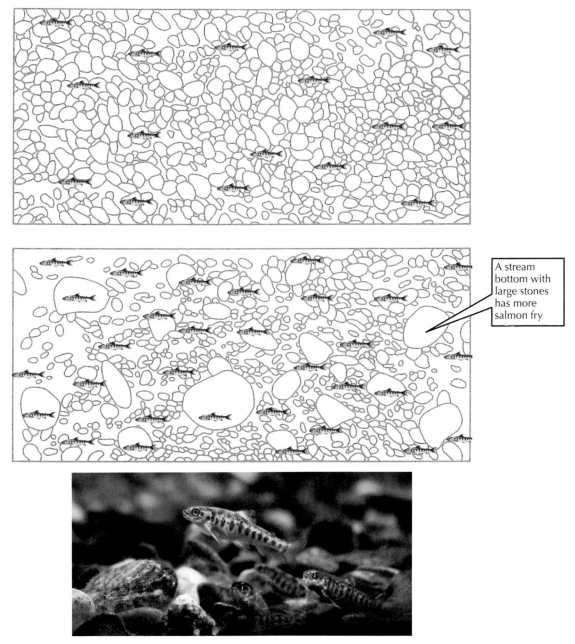

Figure 5.13. The position of Atlantic salmon fry within territories in two sections of an artificial stream. In the lower section, which has large stones on the stream bottom, the territories are smaller and the population density is therefore higher. (Modified after Kalleberg 1958, photo courtesy of Peter Steenstra, US Fish and Wildlife Service.)

A stream bottom with large stones has more salmon fry

containing serpentine, a green rock that is toxic to most plants because of its low nitrogen and phosphorus content and its high levels of nickel, chromium and magnesium. The plants that sustain the checkerspot are adapted to the peculiar chemistry of serpentine and, therefore, the distribution of both the plants and the checkerspot are confined to serpentine areas. Near San Jose, California, Morgan Hill is a large serpentine outcrop that supports a source population of checkerspot butterflies that colonise many of the small patches of serpentine that are nearby (Figure 5.15). These small patches contain checkerspot populations in some years, but not in others, so there is continual extinction of local populations followed by

ESSAY 5.3   WHAT EXACTLY IS A POPULATION?

A population is defined by ecologists as a *group of organisms of the same species that occupy a particular space at a particular time*. Thus we may speak of the mule deer population of Glacier National Park, the mule deer population of Montana, the human population of Tasmania or the human population of Australia. The key point is that a population is always a single species. The ultimate constituents of the population are individual organisms. For sexually reproducing organisms, a population may be subdivided into **local populations**. Individuals in a local population have a common gene pool and have the potential to breed with one another. Consequently, the mule deer of Montana would not constitute a local population because deer from eastern Montana would never be able to breed with deer from the western part of the state—the two groups are isolated by distance. The boundaries of a population both in space and in time are not precise and in practice are usually fixed somewhat arbitrarily by the investigator.

Although a population is composed of individuals, it has group characteristics, which are statistical measures that cannot be applied to individuals. The basic characteristic of a population that we are often interested in is its **density**. We can also measure the **reproductive rate** (the birth rate or the rate of egg, seed or spore production) and the **death rate** of a population. In addition to these measures, we can derive secondary characteristics of a population, such as its age distribution, sex ratio and genetic composition. All these parameters result from a summation of individual characteristics, but are statistical attributes of the population.

Ecologists began a serious quantitative study of populations only in the 1920s, and since that time considerable information on the population dynamics of organisms has accumulated as one part of our ever-growing ecological knowledge.

recolonization from the source population. At any given time, there are some suitable serpentine patches that are not inhabited by the butterfly because they have not been recolonized.

Q Do all populations that are distributed in habitat patches constitute a metapopulation?

Metapopulations form a link between the factors that limit distributions and those that limit abundance. If a species has good dispersal powers, it can recolonize scattered habitats quickly, and the average density of the smaller populations in an area will be high. On the other hand, if a species has poor dispersal powers, it may leave many suitable sites unoccupied, and its average density will be low.

Habitat Suitability Models

If the suitability of a habitat for a particular species can be defined, it should be possible to predict the average density the species could reach in that habitat. With this knowledge, we should be able to manage endangered and economically important species better. For example, before we reintroduce a species such as the wolf into an area, we need to know if the area still contains suitable habitat in spite of changes humans have made to the area. Defining habitat suitability has been a goal of wildlife and fisheries management agencies for the past century and has led to the creation of many **habitat suitability models.** A habitat suitability model is a mathematical model that explains quantitatively how specific ecological factors affect the abundance of a species.

Figure 5.14. Agricultural landscape with scattered woodlots (dark areas) in the wheat belt area of Western Australia, a patchy habitat created by agricultural clearing. Narrow corridors along roads connect some of the woodland patches. (Photo courtesy of Denis Saunders.)

the water. We can enter these characteristics into a habitat suitability model and then use the model to predict whether the habitat in another area is suitable. Let us consider two examples to illustrate this process.

Juvenile Atlantic salmon live in streams in eastern Canada and New England. The decline in the Atlantic salmon population during the last 25 years might be partly explained by a reduced suitability of freshwater streams due to agriculture and urban development Salmon prefer sections of streams that are 30–60 cm deep with a current of 60–70 cm/sec and a substrate that consists of gravel averaging 4–5 cm in diameter. They avoid fast-flowing sections of streams and substrates that are muddy or composed of large rocks. With this information, fish ecologists constructed a habitat suitability model that predicts the local density of juvenile Atlantic salmon with high precision (Figure 5.16). The model can suggest actions that are likely to increase the amount of suitable habitat in a river, such as removing or adding woody debris to adjust the flow rate.

The Eurasian lynx (*Lynx lynx*) was almost extinct around 1900, but has been slowly recovering in Europe since then. Lynx are now well established in Switzerland and the Czech Republic, but they do not live in large parts of their former range in Germany (Figure 5.17). Which regions of Germany are now suitable for lynx reintroduction? The simplest way to answer this question is to describe the present lynx habitat in Switzerland and the Czech Republic and look for similar habitats in Germany. German ecologists found that the best lynx habitat contains connected forest areas within a radius of 5 km, the size of a female lynx home range in Switzerland. By analyzing satellite images of Germany, the researchers concluded that about 24,000 km^2 (8% of the country's land area) is suitable for lynx reintroduction. That amount of habitat could support about 370 lynx.

We said earlier that habitat suitability models are based on the assumption that species live only in suitable habitats. However, this does not necessarily mean that a habitat is unsuitable for a species just because that species is not presently found there. If the distribution of a species is limited by dispersal, there may be large areas of suitable habitat that are not occupied.

A great deal of natural history information must go into the design of habitat suitability models, and this information is vital to understanding how a species' abundance varies with habitat quality. If we assume that species live only in suitable habitats, we can use their observed distribution and abundance to define the habitat characteristics that affect their abundance. Habitat characteristics for terrestrial animals might include the amount of forested landscape, the distance to water, the average temperature and rainfall and the abundance of predators. For stream-dwelling invertebrates, the important characteristics might include the depth of the stream, the composition of the stream substrate, and the temperature, turbidity, and flow rate of

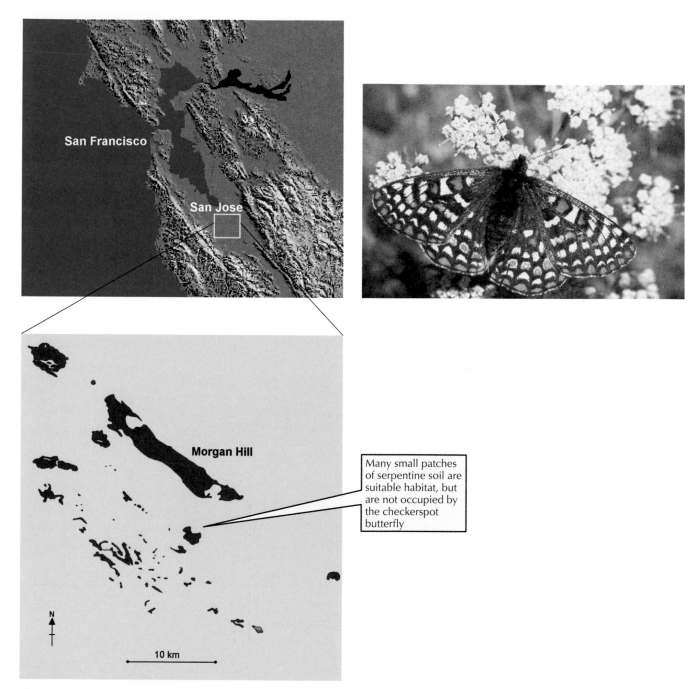

Figure 5.15. Distribution of serpentine soils (brown areas on enlarged map) in central California. The whole region contains a metapopulation of the bay checkerspot butterfly (*Euphydryas editha*), which is found only in the serpentine patches. The large (1,500 ha) outcrop called Morgan Hill just south of San Jose is a source of butterflies for the smaller patches. (From Harrison 1991; photo courtesy of Richard Arnold.)

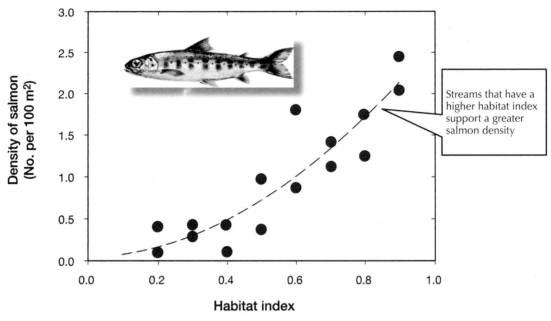

Figure 5.16. Habitat suitability model for juvenile Atlantic salmon (5–10 cm in length, 1–2 years old) in streams in eastern Canada. The habitat index is a measure of habitat suitability based on water depth, flow rate and gravel size in the streambed. An optimal habitat would have an index of 1.0. (From Guay *et al.* 2000.)

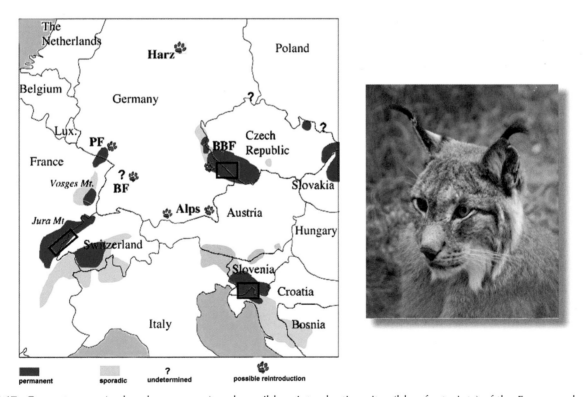

Figure 5.17. Current range (red and gray areas) and possible reintroduction sites (blue footprints) of the European lynx in central Europe. Key potential sites for reintroduction in Germany include the Harz Forest, the Black Forest (BF) and the Alps. Black rectangles enclose areas where lynx have been studied intensively by European ecologists. (After Schadt *et al.* 2002, photo courtesy of Ralf Schmode.)

In these cases, we may underestimate the number of places where the species could thrive.

SUMMARY

Ecologists have developed an array of methods for estimating abundance. Quadrat counts are often used for plants and less mobile animals, while mark–recapture techniques can be used for mobile animals. Once we have estimated abundance, we can begin to ask what causes its spatial variation—what comprises good and poor habitats.

For animals, food is often the environmental factor that limits abundance, but predators or nesting sites can be limiting as well. For plants, the limiting factors are often temperature, moisture and soil nutrients. There is, as yet, no general theory of habitat quality and much experimental work remains to be done.

If we can determine the habitat characteristics that are favorable and those that are unfavorable for a species, we can use this knowledge to identify possible areas for the reintroduction of endangered species and to manage habitats to reduce the abundance of pest species.

SUGGESTED FURTHER READINGS

Begon M, Mortimer M and Thompson DJ (1996). *Population Ecology: A Unified Study of Animals and Plants*. 3rd edn. Blackwell Science, Oxford, UK. (An excellent textbook on population ecology for senior students by leading British ecologists.)

Newton I (1994). Experiments on the limitation of bird breeding densities: a review. *Ibis* **136**, 397–411. (An excellent overview of the factors that limit bird densities.)

Silva M, Brown JH and Downing JA (1997). Differences in population density and energy use between birds and mammals: a macroecological perspective. *Journal of Animal Ecology* **66**, 327–340. (Why birds and mammals differ in their use of food energy.)

Sutherland WJ (Ed.) (2006). *Ecological Census Techniques: A Handbook*. Cambridge University Press, Cambridge, UK. (A good practical guide to many of the methods ecologists have developed to census plants and animals.)

QUESTIONS

1. Suggest some reasons why birds generally have lower population densities than mammals of the same size, as shown in Figure 5.5. Would you expect this rule to apply to flightless birds as well? Explain your reasoning.

2. During three days of trapping, an ecologist captured, marked and released 46 European rabbits. The following week, she trapped the same area and caught 37 rabbits, of which 27 were marked from the previous week. Use the Petersen method to estimate the size of this rabbit population.

3. An animal's DNA fingerprint can serve as a mark in the mark–recapture method of estimating population size. What possible errors might occur with this approach that would not occur with other ways of marking individuals?

4. Rats and mice can be serious pests in urban areas around the world. Suggest two possible approaches to changing an urban block from a good habitat for rats and mice to a poor habitat.

5. The density of juvenile Atlantic salmon in a river can often be predicted by a habitat suitability model that considers the water depth, flow rate and gravel size of the river (Figure 5.16). If you applied this model to a river and obtained results that did not agree with the prediction, what other factors might you add to the model?

6. It is illegal to hunt female grizzly bears with cubs in the United States and Canada. What might be the justification for this wildlife-management decision?

7. What problems might you encounter if you tried to build a habitat suitability model for a species like the bay checkerspot butterfly (Figure 5.15) that lives in scattered habitats as a metapopulation?

Chapter 6

POPULATION DYNAMICS—ABUNDANCE IN TIME

IN THE NEWS

*House mouse (*Mus musculus*)*

Outbreaks of rodents occur periodically in many parts of the world, destroying agricultural crops and stored food supplies and at times transmitting diseases to humans. During the Black Death of the 14th to 16th centuries, almost half the population of Europe died from the disease now known as bubonic plague, which is caused by bacteria transmitted from rat fleas to humans. In south-eastern Australia, the introduced house mouse (*Mus domesticus*) undergoes periodic variations in abundance. In most years it is difficult to find even a few house mice in the cereal-growing areas, but over the past century a major outbreak has occurred about every 10–15 years, with smaller outbreaks in between. During outbreaks, mice can reach densities exceeding 1,000 per ha, destroying grain crops and vegetables and infesting houses, hospitals and schools in rural areas. In the record outbreak of 1917, half a million mice were captured in four nights trapping on one farm alone.

For the past 25 years, ecologists have studied house mouse populations in south-eastern Australia to obtain more detailed information on how and why these destructive outbreaks occur. Their investigations reveal that house mouse abundance changes spectacularly from year to year:

Why were there extended periods of little or no population growth—for example, from 1985 to 1987 and from 1998 to 2001? Why did

Half a million house mice captured on a single farm over four nights. (Photo courtesy of Grant Singleton.)

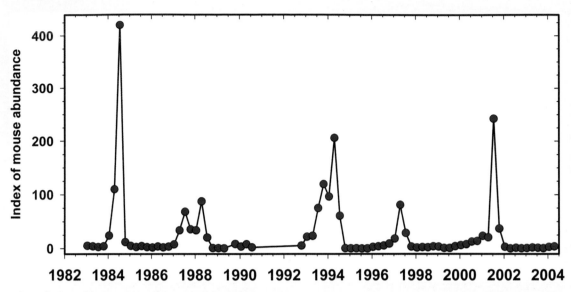

Abundance of house mice from 1982 to 2004

the population explode in 1984, 1993 and 2001? A key factor in the answers to these questions is the weather. Good rains during autumn and winter encourage grass growth and seed production, which starts the mouse breeding season early with large litter sizes. In these years of good food supplies, mortality caused by moisture stress is low, and the population explodes under the joint influence of good survival and high reproduction. Predators are too scarce to affect mouse numbers once an increase begins, but the population stops growing when food becomes scarce and epidemic diseases spread under the crowded conditions.

There are two main options for reducing the impact of these outbreaks: sterilize females to reduce the reproductive rate early in the season, or use poisons to increase the death rate. Because poisons can harm desirable species living in agricultural areas, the former option is preferred.

■ 6.1 NO POPULATION CAN INCREASE WITHOUT LIMIT

Populations are dynamic. They grow and decline, and one important goal of ecology is to understand why. Growing populations are highly desirable if they are crop plants or fish that we want to harvest, but they can also be undesirable if they are pest species or disease organisms such as the anthrax bacterium. Population dynamics is a quantitative subject—it requires us to calculate the rate at which populations change. But we can use graphs to illustrate the principles of population change without delving into the underlying mathematics. Let us start our analysis with a simple model of how a population might increase over time. We will then look at patterns of population change over time in nature.

Geometric Population Growth

Imagine a population that has been released into a favorable environment and begins to increase in numbers. What forms will this increase take, and how can we describe it quantitatively? To begin, let us consider the simple case in which generations are separate, as in annual plants, which have a single breeding season and a lifespan of one year. What happens to this population will depend on the number of female offspring that survive and reproduce the following year (see *Essay 6.1*). The pattern of population change is called **geometric population growth** (or exponential growth) because, in the simplest case, the population grows like the geometric series 1, 2, 4, 8, 16, 32, 64, and so on.

Figure 6.1 shows some examples of geometric population growth in which different numbers of female offspring are produced in each generation. As you would expect, the more female offspring that are produced, the more rapidly the population increases. This is exactly the way money grows in a bank account with a constant annual rate of interest.

However, populations do not continue to grow geometrically as in Figure 6.1. If they did, the world would be stacked high with elephants and oak trees. Aristotle pointed this out 2,300 years ago, and Darwin repeated it 150 years ago. Therefore, we must modify the

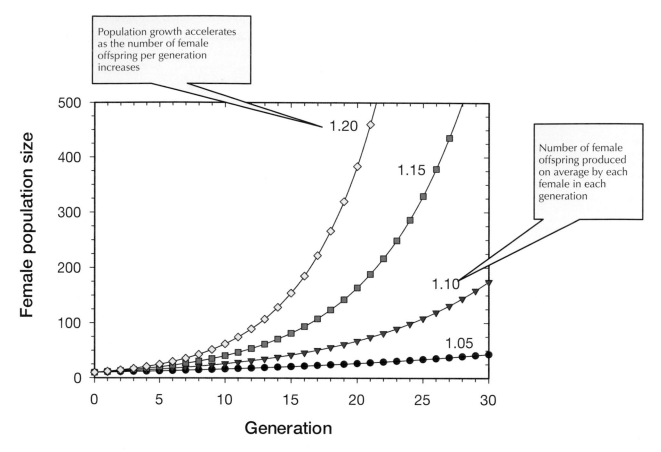

Population growth accelerates as the number of female offspring per generation increases

Number of female offspring produced on average by each female in each generation

Figure 6.1. Geometric population growth in a hypothetical species of annual plant that has a constant reproductive rate. The four curves show how the growth rate depends on the average number of female offspring produced per generation (1.05, 1.10, 1.15 and 1.20). Each population begins with 10 females at generation 0.

geometric model to take into account the fact that all populations eventually stop growing.

A Model for a Regulated Population

To make a model in which a population stops growing, we must allow the reproductive rate of the population to vary with the size of the population. In small populations, the birth rate will be high and the death rate from diseases and natural enemies will be low. In large populations, either the birth rate will decrease or the death rate will increase from a variety of causes, such as food shortage or epidemic disease (in animals) or competition for nutrients or light (in plants). The simplest of these models assumes there is a negative linear relationship between population size and reproductive rate: As the population grows, fewer female offspring are produced in each generation

(Figure 6.2). At the population size where the number of female offspring per generation drops to 1.0, the birth rate equals the death rate and the population is in equilibrium.

The patterns of population change predicted by the simple linear model depend on the slope of the blue line in Figure 6.2. If the slope is gradual, the population moves toward the equilibrium size in a smooth, sigmoid (S-shaped) curve and remains there forever as a stable population (Figure 6.3a). This type of population growth is called **sigmoid growth.** However, a slight increase in the slope of the blue line causes the population to fluctuate up and down forever, a pattern called a **limit cycle** (Figure 6.3b). A further increase in the slope introduces **chaotic fluctuations**, in which the changes in population size are irregular and unpredictable (Figure 6.3c).

ESSAY 6.1 CALCULATION OF GEOMETRIC POPULATION GROWTH FOR AN ANNUAL SPECIES

Our model species has a single annual breeding season and a lifespan of one year. Let us assume that all of the female offspring produced one year survive and breed the following year. We can then write the simple equation

$$N_{t+1} = R_0 N_t$$

where N_t = the number of females in generation t,

N_{t+1} = the number of females in generation $t + 1$, and

R_0 = the net reproductive rate (the number of female offspring produced per female per generation).

The net reproductive rate (R_0) is the fractional change in the population each year. Its value determines how the population changes over time. Consider the simplest case, in which R_0 is constant. There are three possibilities:

1. If $R_0 = 1$, the population is constant.

2. If $R_0 > 1$, the population is increasing geometrically without limit. For example, let $R_0 = 1.5$ and $N_t = 10$ when $t = 0$:

Generation	Population Size (N_t)
0	10
1	(1.5)(10) = 15
2	(1.5)(15) = 22.5
3	(1.5)(22.5) = 33.75

In this example, the population increases by 50% each generation.

3. If $R_0 < 1$, the population is declining because it is not replacing itself. For example, let $R_0 = 0.9$ and $N_t = 100$ when $t = 0$:

Generation	Population Size (N_t)
0	100
1	(0.9)(100) = 90
2	(0.9)(90) = 81
3	(0.9)(81) = 72.9

In this example, the population decreases slowly to extinction, losing 10% each generation.

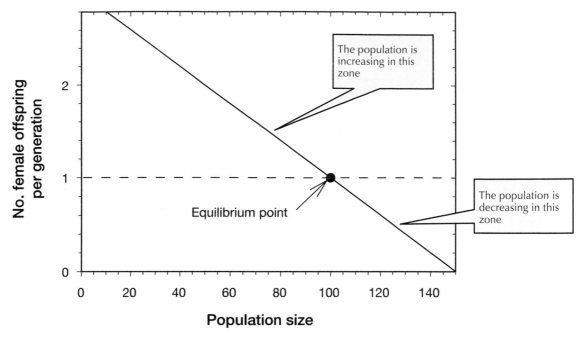

Figure 6.2. Relationship between population size and number of female offspring produced per generation. In this hypothetical example, the reproductive rate varies linearly with population size. When the population is 100 (the equilibrium point), the number of female offspring per generation is 1.0, so each female just replaces herself and the population size does not change. The slope of the blue line may be more or less steep in different populations.

Much of this model of population growth was clarified and elaborated by the mathematical ecologist Robert May, who worked at Princeton University and later at Oxford University. The fact that such a simple model can produce a diversity of population growth patterns is one of the most surprising results of mathematical ecology in the past 50 years. In the next section, we will consider how well these patterns match those of real-world populations.

Q Would this model of population growth make different predictions if the reproductive rate depended on the population size one or two generations earlier, rather than on the present size?

Robert May (1936–) Professor of Zoology, Oxford University

■ **6.2 NATURAL POPULATIONS ARE RARELY STABLE**

What factors limit the growth of populations? To answer this question, we must first realize that there are four processes that can cause the size of a population to change: reproduction (addition of individuals through birth or seed production) and immigration (movement of individuals into a population) increase population size, whereas death and emigration (movement of individuals out of a population) decrease

116

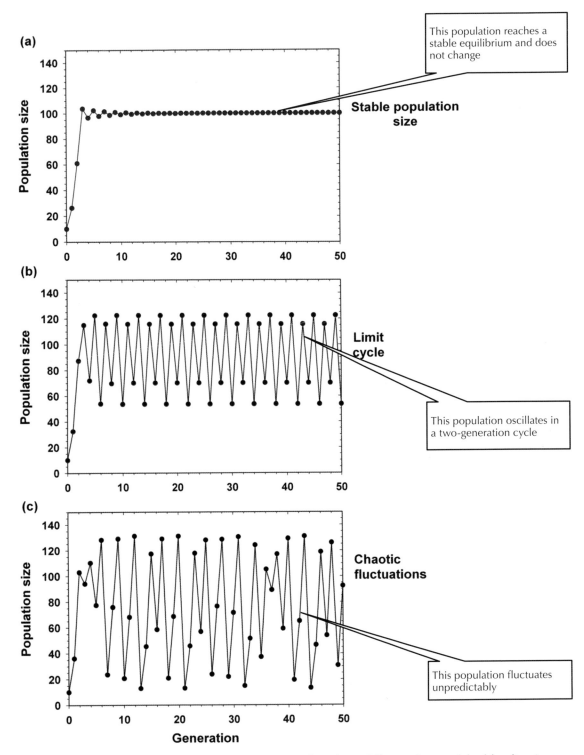

Figure 6.3. Hypothetical examples of regulated population growth, based on three different slopes of the blue line in Figure 6.2. In each example, the beginning population size is 10, and the equilibrium size is 100. (a) If the slope is shallow, population growth slows as the equilibrium is approached, resulting in a curve that plateaus to an equilibrium. (b) If the slope is moderate, the population oscillates around an average size of 90 in a two-generation cycle. (c) If the slope is steep, the population fluctuates in an irregular pattern that never repeats itself. Thus, knowing the past trajectory allows you to predict the future exactly in (a) and (b) but tells you nothing about the future in (c).

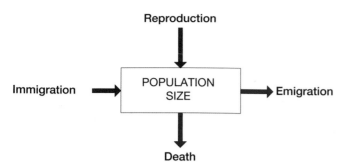

Figure 6.4. Four processes that contribute to population change. Blue arrows indicate processes that increase population size; red arrows indicate processes that decrease it.

population size (Figure 6.4). In many populations, natural controls balance these four processes, causing the population size to stay fairly constant over long periods. In the short term, however, we see little evidence of this balance. Birds that are common one winter may be rare the next. Garden pests that are bothersome one summer are nowhere to be seen the following year. Fluctuations in natural populations are quite frequent.

We can use a systematic, hierarchical analysis to determine why a population is increasing or decreasing (Figure 6.5). The first step is to measure the rates of reproduction, death, immigration and emigration. Once we have identified a change in the rate of one of these processes, we examine the agents that control them, which fall into four broad categories: weather, nutrients, other organisms (including predators, pathogens and parasites) and habitat. For any particular population, one of these categories may be of overriding importance. Let us look at a few examples of population changes to illustrate this framework of analysis.

Desert Locust Plagues

Locust plagues have been a recurrent disaster for agricultural societies for thousands of years. Several species of locusts migrate in swarms, which concentrates their destructiveness (Figure 6.6). Each locust eats approximately its own weight in food per day, so a swarm leaves little green vegetation remaining after passing through an area.

One of the most destructive locust species is the desert locust (*Schistocerca gregaria*), whose range includes Africa north of the equator, the Middle East and India (Figure 6.7). Swarms of this locust probably constituted one of the plagues of Egypt described in the Old Testament. Since 1860 there have been ten major plagues of the desert locust at irregular intervals (Figure 6.8). High populations last for 7–13 years and then collapse to low numbers for up to 6 years.

The most important requirement for the development of a desert locust plague is moisture. The locusts' main breeding areas in northern Africa have scant, erratic rainfall, so moisture is often limiting. To survive, locust eggs must be laid in moist soil. Moisture is also needed for the growth of plants on which both nymphal and adult locusts feed. Plagues collapse when dry weather returns, hindering locust reproduction and reducing the availability of vegetation.

Q **Would you expect there to be much genetic variation among local populations of the desert locust?**

Desert locusts undergo a remarkable transformation before they swarm: They change from a pale, short-winged, solitary form to a dark, long-winged, gregarious form (Figure 6.9). Boris Uvarov, the Russian entomologist who discovered the two forms in 1913, called the pale form the **solitaria** phase and the dark form the **gregaria** phase. Uvarov observed that an individual locust could change from the solitaria phase to the gregaria phase during its development, thus demonstrating that locusts in both phases belong to the same species.

Phase transformation in desert locusts is predominantly a behavioral change. Solitaria locusts are repelled by other locusts, while gregaria locusts actively aggregate. In the laboratory, forcing solitaria locusts together in small cages produces some of the changes involved in phase transformation, but we do not fully understand the ecological events that bring about this transformation in field populations. Aggregation may be assisted by a pheromone produced in the intestine of immature locusts. This pheromone causes immature solitaria locusts to be more gregarious and to change color towards that of the gregaria phase. The effects of crowding may be cumulative

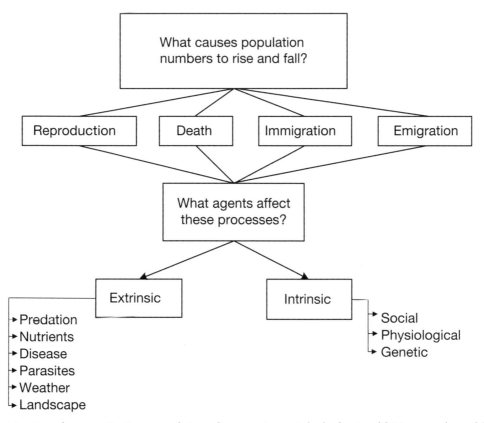

Figure 6.5. A decision tree for investigating population changes. A great deal of natural history study and field experimentation is needed to answer these questions.

Figure 6.6. A goat herd runs away from a swarm of desert locusts near Kaedi, Mauritania. Livestock are in competition with the insects for available grazing land. (Photo courtesy of FAO/G. Diana.)

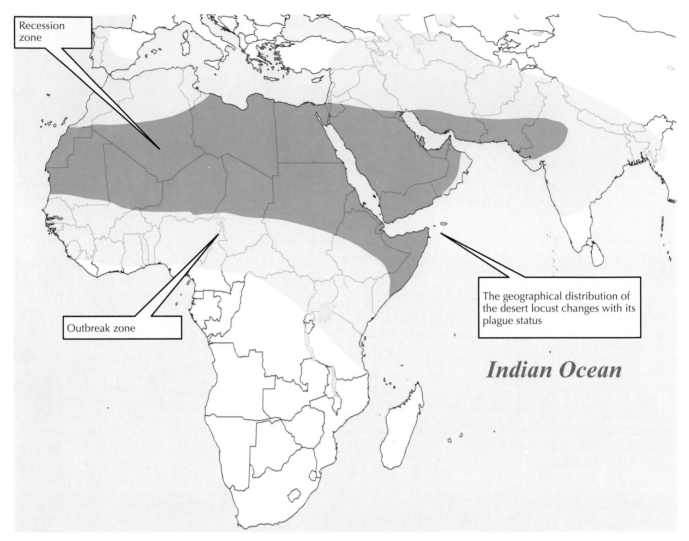

Figure 6.7. Geographic range of the desert locust. Locusts are normally present at low density in semi-arid areas within the recession zone, indicated in green on this map. Swarms of locusts that arise in hot spots within the recession zone may move into the outbreak or plague zone (yellow). Populations in the outbreak zone die out when the plague collapses. (Modified after Symmons and Cressman 2001.)

from generation to generation, but the mechanism of inheritance is not known.

You might suppose that the more striking coloration of the gregaria phase is a visual signal that directs other desert locusts to aggregate. This behavioral response would be adaptive if locusts in groups were more likely to survive than isolated locusts. In laboratory experiments, however, immature locusts that could see a highly colored gregaria individual, but could not detect its pheromone, did not aggregate, which indicates that the gregaria coloration by itself does not cause aggregation. One alternative hypothesis is that the gregaria col-

oration warns potential predators that these locusts are distasteful, but this hypothesis has not been tested.

Locust plagues are particularly difficult to control because swarms can move thousands of kilometers quickly and cross political boundaries. Until recently, the aerial spraying of insecticides was the only way to control swarms, but the desire to stop using these toxic chemicals has stimulated interest in biological pest control—a topic we will discuss in Chapter 19. The Food and Agricultural Organization of the United Nations provides up-to-the-minute information on desert locust outbreaks at the FAO website.

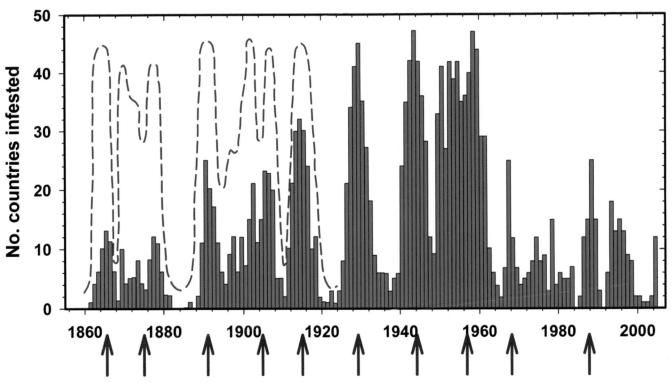

Figure 6.8. Outbreaks of the desert locust since 1860. The dashed line indicates the probable extent of outbreaks during the period 1860–1920, for which the data are incomplete. Red arrows mark major plagues. (Data from Symmons and Cressman 2001.)

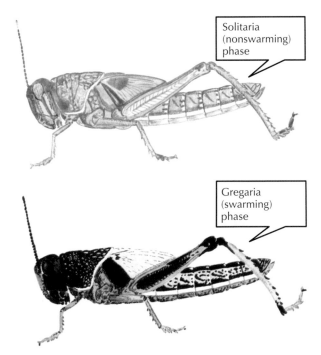

Solitaria (nonswarming) phase

Gregaria (swarming) phase

Figure 6.9. Two phases of the desert locust (*Schistocerca gregaria*). (Image courtesy of FAO Locust and Other Migratory Pests Group.)

The Yellowstone Elk Population

In 1968 Yellowstone National Park instituted a hands-off management policy called 'natural-regulation management,' which permitted populations of elk and other large mammals to reach unmanipulated levels within the park. Previously, elk had been culled to prevent perceived overpopulation (Figure 6.10). The natural-regulation management policy predicted that the elk population would increase and reach a stable equilibrium due to a decline in the birth rate and a rise in the death rate. Data from a particularly well-studied elk population in the northern part of the park agree with this prediction. This population began to increase in 1968, accompanied by a slight reduction in the percentage of pregnant females, an increase in the age at sexual maturity and a major decline in calf survival during the first year of life.

The major factors that limit populations of large mammals are disease, predators, food shortage and weather. The only serious disease that affects elk is brucellosis, but it is of minor importance to the elk in

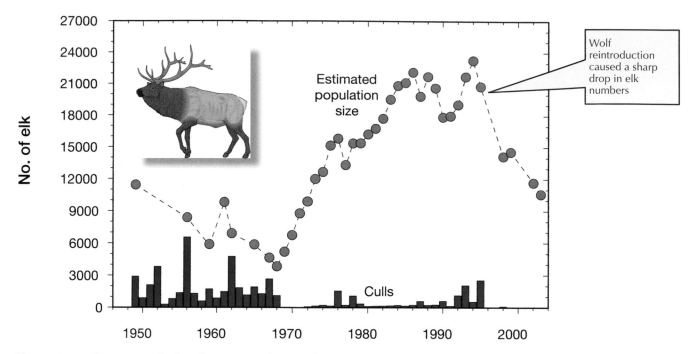

Figure 6.10. Changes in elk abundance in northern Yellowstone National Park from 1949 to 2003. When the culling of elk inside the park was stopped in 1968, the elk population immediately increased. (Culling of elk that roamed out of the park continued after 1968.) The reintroduction of wolves to the park in 1995 caused the elk population to decline by 6% per year. (Data from Houston 1982 and Yellowstone National Park files, courtesy of Glenn Plumb and Francis Singer.)

Yellowstone. Predators of elk include grizzly bears, black bears, coyotes and golden eagles, which, combined, kill about one-third of elk calves before they reach one month of age. Grizzly bears also kill adult elk, as do mountain lions and wolves. Wolves were a major predator throughout most of North America before Europeans arrived, but were eliminated from the Yellowstone ecosystem during the 1920s. They were reintroduced in 1995.

Predators combined with food shortage and weather drive population changes in Yellowstone elk. Losses of elk calves during winter are greater when the population is larger (Figure 6.11a) and when the winter is more severe (Figure 6.11b). Some of these losses are due to predation. In addition, summer rainfall increases summer plant production, which provides more food and increases elk calf survival. Thus, during the years when wolves were absent from the park, the northern Yellowstone elk population was controlled largely by other predators and by the effect of variable weather. Weather affects both summer grazing conditions and winter snow levels, which

affect the nutrition and subsequent mortality of calves in their first year of life.

Q What would happen to the Yellowstone elk herd if all predator species were eliminated?

In the decade since wolves were reintroduced to Yellowstone, the additional predation by wolves has caused the elk population in the northern part of the park to decline by about half (see Figure 6.10). This population has not yet reached equilibrium and may never do so because of variation in the weather as well as human influences in the areas surrounding the park.

Water Fleas

Many organisms show strong annual fluctuations in abundance and, for them, the pattern of population growth can be observed every year. One example is the water flea *Daphnia*. This crustacean, which is only 1–2 mm in length, is common in the plankton of many temperate lakes and ponds, especially those that lack

(a) Population size

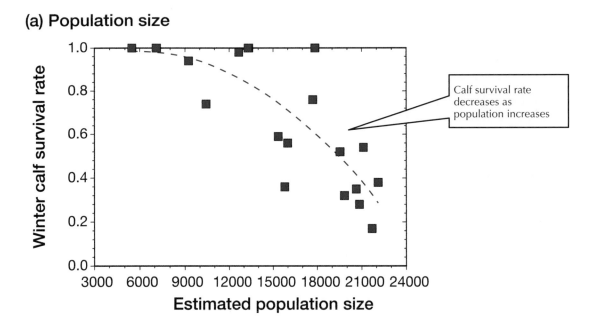

(b) Winter weather

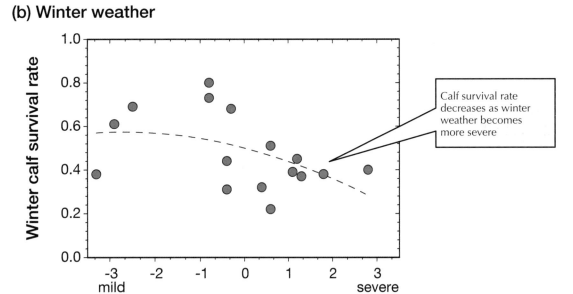

Figure 6.11. Winter survival of elk calves in Yellowstone National Park in relation to (a) the size of the elk population, and (b) the severity of winter weather. The winter weather index is a composite of snow depth and minimum temperature and ranges from −4 (mildest) to +4 (most severe). (Data from Singer *et al.* 1997.)

fish predators. *Daphnia* populations undergo a geometric increase in density each spring followed by a decline in autumn; both the duration and the magnitude of this population explosion vary from year to year (Figure 6.12). The geometric growth and later collapse of *Daphnia* populations result from changes in food availability. *Daphnia* feed on small algae whose populations rise in spring with increasing water temperature and abundant nutrients, such as phosphorus. Algal numbers decrease in autumn as the water temperature falls and nutrients become less abundant. In lakes without fish predators, the growth of *Daphnia* populations thus mirrors the rise and fall of their food resources.

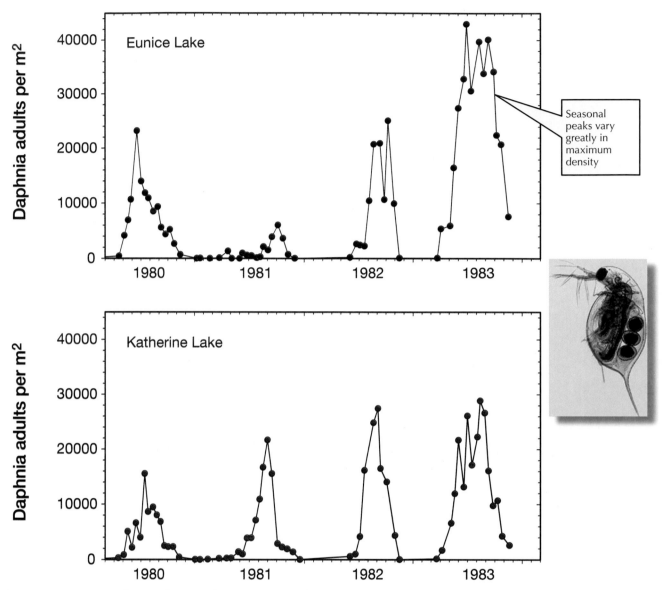

Figure 6.12. Density of water flea (*Daphnia rosea*) populations in Eunice Lake and Katherine Lake, British Columbia, from 1980 to 1983. These temperate lakes show strong seasonal dynamics that vary from year to year. (Data from Walters *et al.* 1990, photo courtesy of Michael Lynch, Indiana University.)

■ 6.3 HUMAN POPULATION GROWTH MUST ALSO BE LIMITED

The human population worldwide has been growing throughout most of recorded history, and probably for more than two million years before that. Figure 6.13 shows the growth of the human population over the last 2,000 years. In 2005 there were an estimated 6.45 billion people on Earth, and the population was growing at a rate of 1.15% per year, or 203,000 people per day. Because no population can continue to increase without limit, Figure 6.13 immediately raises the question, what will limit human population growth?

Demographic Transition

The human population can achieve stability when the birth rate equals the death rate. There are two extreme forms of this stability, one with high birth and death rates, and one with low birth and death rates. Many regional human populations have undergone a change

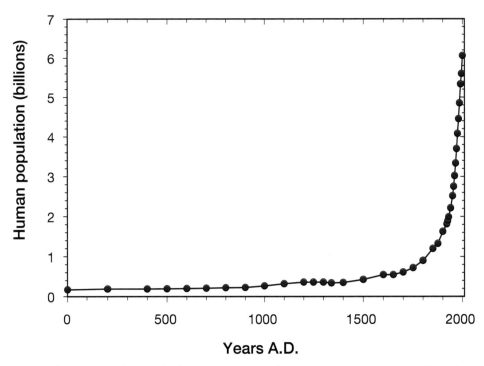

Figure 6.13. Human population growth over the last 2,000 years. The increase appears approximately geometric (compare with Figure 6.1). (Data from Cohen (1995) and the Population Reference Bureau.)

from the former extreme to the latter. This change, called the **demographic transition**, is illustrated in Figure 6.14 for Sweden and Mexico. The demographic transition is a descriptive observation that is not well understood, rather than a law of human population growth (see *Essay 6.2*). During the 20th century, death rates declined rapidly in most countries, but birth rates have declined in a more variable manner. The decrease in birth rates has been dramatic in China but slow and irregular in India. In much of Africa, the transition to lower birth rates is just beginning.

Most human population growth is occurring in developing countries, where about 80% of the world's people now live. The majority of developed countries, in contrast, have populations that are close to equilibrium, with reproductive rates near replacement (an average of 2.1 children per family). Some developed countries, such as Canada and the United Kingdom, have reproductive rates below 2.1, and their populations will eventually decline if there is no immigration and the reproductive rate does not change. Most developed countries, however, are still increasing in population without immigration because the birth rate

exceeds the death rate. This increase will stop in about 30 years.

The United Nations projects a world population in 2050 that might range from 7.4 to 8.9 billion people. The range reflects different assumptions about future changes in fertility and mortality. If this projection is accurate, then at least 1 billion people will be added to the population in the next 45 years. Will Earth be overpopulated if there are 7.4 billion people? Is it already overpopulated? How many humans can the biosphere support?

Carrying Capacity of Earth

Ecologists use the concept of **carrying capacity** to answer these kinds of questions. Carrying capacity is the maximum population of a given species that can be supported indefinitely in a particular habitat without permanently impairing the productivity of that habitat. This concept has come from range management, in which both the rancher and the ecologist ask how many cattle or sheep can graze on a given area of pasture without degrading the land. Degraded pastures are readily noticed because of immediate

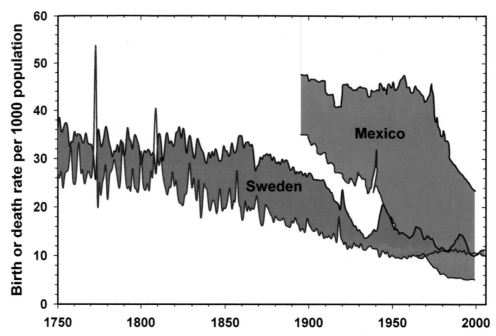

Figure 6.14. Demographic transition in the human populations of Sweden and Mexico. The transition took 150 years in Sweden but about half that time in Mexico. When the birth rate exceeds the death rate, the population grows (shaded areas), if emigration is low. In 2005 Mexico had a crude birth rate of 21 and a crude death rate of 4. In Sweden at the same time the comparable rates were 11 and 10. (Data from the Population Reference Bureau 2006.)

economic feedback to ranchers, and much research has been devoted to maintaining pastures in good condition for grazing animals for human consumption.

Q **Might the concept of carrying capacity be viewed differently by a cattle rancher and a wildlife biologist?**

What is the carrying capacity of Earth for humans? Scientists have tried to answer this question for more than 300 years. Figure 6.15 shows the range of estimates for Earth's carrying capacity, from the first attempt by the Dutch scientist Antoni van Leeuwenhoek in 1679 to the present day. Two features are striking about this graph. Firstly, the estimates vary widely, from less than 1 billion to over 1 trillion. Secondly, they do not converge over time but seem to increase in variability, although the median has remained around 10–12 billion. Why are these estimates of carrying capacity so variable?

Carrying capacity for humans is difficult to estimate, and the scientists who produced the estimates in

Figure 6.15 used quite different methods to get their answers. Some used theoretical population growth curves to predict the future maximum of the human population. Others multiplied the population density in areas that were assumed to be maximally populated by the area of habitable land on Earth. A different approach to estimating carrying capacity is to focus on a single population constraint, such as food or water. For example, if food is the limiting factor, we could in principle calculate the carrying capacity if we knew how much land is available for growing crops, the average yield of those crops, how much energy each unit of crop would provide, and the amount of energy each person requires:

$$\text{Carrying capacity} = \frac{(\text{ha land})(\text{yield per ha})(\text{kJ per crop unit})}{\text{no. kJ needed per person per year}}$$

But because there are many crops, different soils, variable yields, losses to pests and differences in assumed standards of living, estimates based on this equation are subject to a great deal of variability.

ESSAY 6.2 THE DEMOGRAPHIC TRANSITION: AN EVOLUTIONARY DILEMMA

The demographic transition involves two puzzling aspects of human fertility. Firstly, the increased availability of resources is accompanied by a large decrease in the number of children that parents produce. This change was shown very clearly in Europe during the 19th century. Secondly, rich families reduce their fertility earlier than the rest of the population and often have fewer than the average number of children. Thus, there is a negative correlation between wealth and fertility. If we assume that humans should follow the principles of Darwinian natural selection, these observations are surprising. When other organisms have more resources, they produce more offspring.

Three hypotheses have been proposed to explain why humans with access to plentiful resources have lower fertility:

1. Low fertility is optimal because of the competitive environment in which offspring are raised. This idea is the classic view of evolutionary anthropologists, who suggest that high levels of parental investment are critical to a child's success and are costly to the parents. According to this argument, there is a trade-off between offspring quality and offspring number, and maximal fitness is achieved by having fewer offspring with higher parental investment.

2. **Low fertility is a consequence of Darwinian selection on non-genetic mechanisms of inheritance.** Cultural selection through imitation drives the demographic transition in fertility because people see that successful people have fewer children. This hypothesis is attractive because it postulates that ideas, rather than economic resources, can drive fertility rates, but it suffers from some serious flaws. For example, it doesn't explain why rich and successful people should reduce their fertility in the first place.

3. **Low fertility is maladaptive—a by-product of rapid environmental change.** This hypothesis suggests that having lower fertility is indeed an inappropriate evolutionary response to greater resource availability and is not favored by natural selection. The use of contraceptives is often cited as one explanation for the prevalence of this maladaptive situation in human societies, but it may be difficult to apply this idea to events in the 19th century. This hypothesis remains intriguing but also vague and untestable until the precise environmental changes can be identified.

There is currently much interest in applying the principles of evolutionary ecology to human behavior. However, a shortage of data on human societies at the individual level makes it difficult to find out, for example, whether individuals who have fewer offspring in fact have more grandchildren. At the present time, the demographic transition can be best explained by the first hypothesis, in which people can have more descendents in the long term by pursuing wealth at the cost of immediate reproductive success, as long as wealth is inherited.

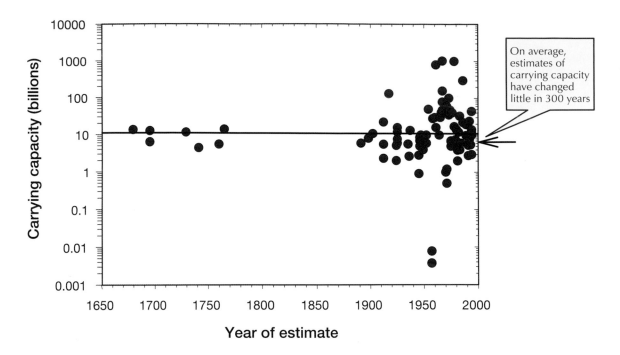

Figure 6.15. Estimates of Earth's carrying capacity for humans as suggested by various scientists since 1679. Both minimum and maximum estimates are plotted. The blue line is the median of these estimates, and the arrow indicates the human population in 2005. (Data from Cohen 1995, Appendix 3.)

Any discussion of carrying capacity must involve the concept of *sustainability*, which refers to the use of resources in a manner that can be continued indefinitely in the future. Sustainability is a critical issue for this century because humans are running into limits that are set by the biological resources of Earth. We will consider sustainability further in Chapter 21.

Like all populations, human populations are subject to the rule that critical resources limit population growth sooner or later. The next four chapters explore how these limitations operate in plant and animal populations.

SUMMARY

Populations change over time, and a variety of population growth models have been proposed to quantify these changes. The most basic model describes geometric growth, which can occur for only a short time. Therefore, most realistic models account for regulated population growth. The most surprising result of pop-

ulation models is that a great variety of population changes—from stability to unpredictable fluctuations—can arise from very simple assumptions. Data from natural populations illustrate that few populations are stable and wide fluctuations are common.

No population can increase without limits. Many studies have analyzed the environmental factors that stop population growth and cause population decline. Reproduction, death, immigration and emigration are all involved in changing population size. Pest control and the conservation of declining species are two practical examples of the need to understand the causes of changes in populations in size over time.

The human population is increasing rapidly, and this increase must come to an end during this century, because humans are not exempt from ecological laws. Sustainability is essential for all populations, including humans. The carrying capacity of Earth for humans may already have been exceeded, and the transition from a growing human population to a stable one is one of the most important problems of this century.

SUGGESTED FURTHER READINGS

Berryman AA (1981). *Population Systems: A General Introduction.* Chapter 2. Plenum Press, New York. (A clear discussion of a simple population regulation model.)

Cohen JE (1995). *How Many People Can the Earth Support?* W.W. Norton, New York. (A brilliant, detailed discussion of the carrying capacity of Earth with background history.)

Levin SA (Ed.) (2001). *Encyclopedia of Biodiversity. Volume 4, Population dynamics.* Academic Press, San Diego. (A crisp synopsis of the ideas and concepts of population changes.)

Moss R and Watson A (2001). Population cycles in birds of the grouse family (Tetraonidae). *Advances in Ecological Research* **32**, 53–111. (An important overview of the detailed studies that have been carried out to determine why grouse numbers fluctuate.)

Silvertown JW and Charlesworth D (2001). *Introduction to Plant Population Biology.* 4th edn, Chapters 4 and 5. Blackwell Science, Oxford, UK. (A view from two leading plant ecologists of how plant populations rise and fall.)

QUESTIONS

1. An annual plant reproduces after one year of growth and a biennial plant reproduces after two years of growth. Would a population of a biennial plant that produces 200 seeds per plant have the same potential geometric rate of increase as a population of an annual plant that produces 100 seeds per plant? Why or why not?

2. The widespread application of the insecticide DDT in the mid 1900s reduced the population of bald eagles in the United States. After the use of DDT on crops was banned in 1972, the bald eagle population began to rise. From 1986 to 2000, the population increased at an average rate of 2% per year. If this rate of increase continues, how long will it take for the bald eagle population to double in size? Explain why the answer is *not* 50 years (2% per year × 50 years).

3. Many developed countries now have an average family size that is smaller than the replacement level of 2.1 children per family. Why, then, are all of these countries still increasing in population, even if you discount immigration?

4. Review some of the controversy over the reintroduction of wolves into Yellowstone National Park in 1995. (If you wish, you can start at http://www.nps.gov/yell/naturescience/wolfrest.htm). What general principles should guide the reintroduction of animals into national parks?

5. How would you set up an experiment to determine whether the color of the gregaria phase of the desert locust (see Figure 6.9) is an adaptation that warns predators that this phase is distasteful?

6. Determine the annual birth rate, death rate and percentage increase in the human population of your country. How do immigration and emigration affect the percentage increase? These data can be obtained from the World Population Data Sheet published each year by the Population Reference Bureau in Washington, DC (http://www.prb.org/).

∴ The simple model of population growth illustrated in Figures 6.2 and 6.3 assumes that for any population size there is a fixed number of female offspring. In animal and plant populations, however, there is variation in the number of offspring produced. How might the predictions of a population growth model differ if it included variability in reproductive rates rather than a fixed rate (for example, if the number of female offspring in Figure 6.2 at a population size of 50 might be 1, 2 or 3 instead of exactly 2)?

Chapter 7

NEGATIVE SPECIES INTERACTIONS— PREDATION, HERBIVORY AND COMPETITION

IN THE NEWS

No predator has been persecuted more in recent times than the coyote. Coyotes are major predators of both sheep and cattle in the western part of North America. They also prey on cats and dogs in the suburbs of many North American cities, to the dismay of pet owners. Western newspapers have frequent articles about the need for more coyote control to protect livestock and pets as well as wildlife species, such as pronghorn. The response to the coyote problem has been to eradicate them. Coyotes are trapped, poisoned and shot from helicopters and from the ground. In 2002 the U.S. Department of Agriculture spent more than $13 million on coyote control and killed 86,000 coyotes—$150 per coyote killed.

Ecologists can bring a different insight into the issue of coyote control. They recognize that coyotes have become a problem because wolves and bears have been reduced in number, or eliminated, in many western areas. Coyotes have a high reproductive rate, and their numbers rebound quickly if only part of a population is killed. Coyote females begin reproducing at younger ages and have larger litters when population density is reduced. In spite of much killing, coyotes now are as abundant as ever in the western states. Eradicating these predators is clearly extremely difficult.

If we cannot eradicate a predator that is causing losses, we must learn to live with it—sheep and cattle ranchers have developed an array of management actions to cut down on losses. For example, sheep can be kept in buildings at lambing time instead of being left out on the range where coyotes can attack the lambs. Guard dogs or llamas can be kept with the flock for protection.

What are the consequences of coyote control for their prey species and for other predators in the ecosystem? To answer this question, researchers removed coyotes from a shortgrass prairie ecosystem in western Texas for two years. They found that both prey (rodents), and other predators (including badgers, bobcats and gray foxes) increased in abundance. The results of this study reaffirm a basic principle of ecology you encountered in Chapter 1: *You cannot change just one thing in an ecosystem.* Removing coyotes in some ecosystems may cause worse problems than learning to coexist with them.

Coyote

CHAPTER 7 OUTLINE

■ 7.1 INTERACTIONS BETWEEN SPECIES CAN BE NEGATIVE OR POSITIVE

Populations of a single species are not found alone in nature. Rather, they exist in a matrix of many populations of other species. Although some species in an area will be unaffected by the presence or absence of one another, in some cases two or more species will interact. The evidence for such interactions is quite direct: A population of one species changes when a population of a second species changes. We are concerned here only with direct interactions between two species. Ecologists classify interactions on the basis of their effects, which may be negative or positive.

In negative interactions, one or both of the species involved suffer some loss, either in population size, reproduction or mortality. There are five types of negative interactions between individuals of different species:

- **Predation:** one animal species eats all or part of a second animal species.
- **Herbivory:** one animal species eats all or part of a plant species.
- **Competition:** two species use the same limited resource or harm each other while seeking a resource.
- **Infection:** a microorganism lives in or on a host and impairs the physiological function of the host.
- **Parasitism:** two species live in a physically close and obligatory association in which one (the parasite) depends metabolically on the other (the host).

Positive interactions are grouped into two large categories:

- **Mutualism:** Two species live in close association with each other to the benefit of both.
- **Commensalism:** Two species are closely associated—one draws a benefit while the other is unharmed.

We will discuss infection and parasitism in Chapter 8 and mutualism and commensalism in Chapter 9. In this chapter we will focus on how the first three negative interactions—predation, herbivory and competition—affect the distribution and abundance of the interacting species.

■ 7.2 PREDATION IS THE PRIMARY FACTOR LIMITING THE ABUNDANCE OF MANY POPULATIONS

Predation is an important process from two points of view. Firstly, predation may restrict the distribution of a species (as we saw in Chapter 3) or reduce its abundance. If the prey is a pest, we may consider predation useful because it makes the pest less common. However, if the prey is a valuable resource, such as caribou or domestic sheep, we may regard predation as undesirable because it causes an economic loss or a conservation problem. Secondly, predation is a major selective force in evolution, and many adaptations we see in organisms, such as warning coloration, have their explanation in the evolution of predator–prey systems.

An obvious experiment to determine how predators affect the abundance of their prey is to compare prey populations in areas where specific predators are present with similar populations in areas where those predators are absent. One such experiment was done on the wood pigeon (*Columba palumbus*)—a common species in city parks and other urban areas in Europe. Wood pigeon nests are subject to predation by a large number of avian and mammalian predators, including the hooded crow, the goshawk and the pine marten. Figure 7.1 contrasts wood pigeon populations in two urban parks in south-west Poland: Slowacki Park, which was colonized in 1972 by hooded crows (*Corvus corone*), and Legnica Park, which has remained mostly free of nest predators since 1972.

The density of wood pigeon nests in Slowacki Park declined 90–95% when hooded crows became established, while the density in Legnica Park remained high. Urban parks in Poland without nest predators have an average of 71 wood pigeon pairs per 10 ha, whereas open farmland with the normal contingent of nest predators has only 0.5–1.5 pairs per 10 ha. Nest predation clearly reduces the number of breeding wood pigeons.

Doomed Surplus Concept

Can we conclude from the experiment on wood pigeons that predation generally reduces the abundance of prey species? Not necessarily. Paul Errington

Paul Errington (1902–1962) Professor of Animal Ecology, Iowa State University

of Iowa State University argued that you cannot measure the effects of predation simply by counting the numbers of prey killed; you must determine the factors that make certain individuals vulnerable to predation while other individuals are protected. Errington studied predation on muskrat (*Ondatra zibethicus*) populations in the marshes of Iowa for 25 years. He found that mink were a primary cause of muskrat deaths, but he hypothesized that mink were removing only muskrats that had been driven out of their territories by competition for space with other muskrats. According to Errington, mink were merely acting as the executioners for animals excluded by the social system—animals that would have died from disease or exposure if not from predation. Errington introduced the important concept that in some systems predation may remove only the 'doomed surplus' from populations. This concept also appears to apply to the Serengeti Plains of eastern Africa, where a variety of predators—lions, leopards, cheetahs, wild dogs and spotted hyenas—seem to have little impact on the populations of their large mammalian prey. Most of the prey taken by these predators are old, injured or diseased animals. Therefore, in each predator–prey system, predator impacts should be inferred only from experiments

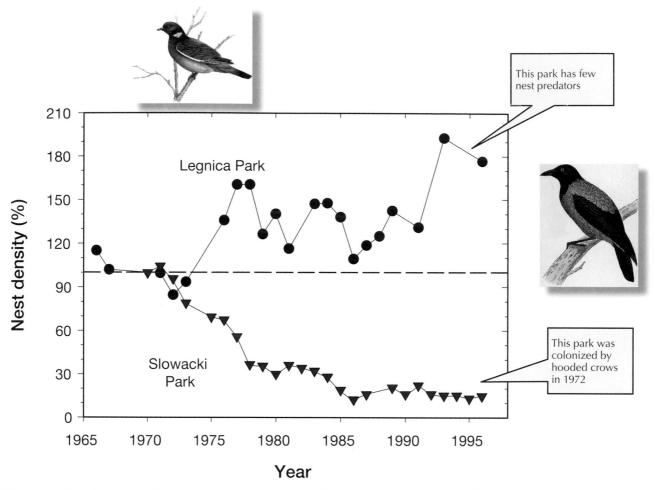

Figure 7.1. Density of wood pigeon nests in two urban parks in south-western Poland from 1966 to 1996. Because the parks are of different sizes, nest density is presented as the percentage of the density in each park in 1970 (indicated by the blue dashed line). (Data from Tomialojc 1999.)

that compare unmanipulated areas with areas from which predators have been removed.

Q What factors might make it difficult to carry out large-scale predator manipulation experiments?

Predator Control

Wildlife managers have relatively few methods for increasing desirable wildlife populations. Among the most popular of these methods is predator control, which can be highly effective in some cases. In one experiment in North Dakota, duck nesting success increased substantially when striped skunks were removed from waterfowl nesting areas (Figure 7.2).

However, a variety of other predators raid duck nests, including foxes, coyotes, ground squirrels, badgers and raccoons. In the prairies of North America, the red fox is a major predator of duck eggs, but coyotes reduce the abundance of foxes by killing them or excluding them from coyote territories. Nest success in North Dakota averaged 32% in areas where coyotes were common and 17% in areas where the red fox predominated. Thus, interactions between predators can affect the survival and reproductive success of prey in complex predator–prey systems.

One of the classic controversies regarding predator control in North America has centered on wolves and moose. If wolf predation affects the abundance of moose, then removing wolves should increase moose numbers.

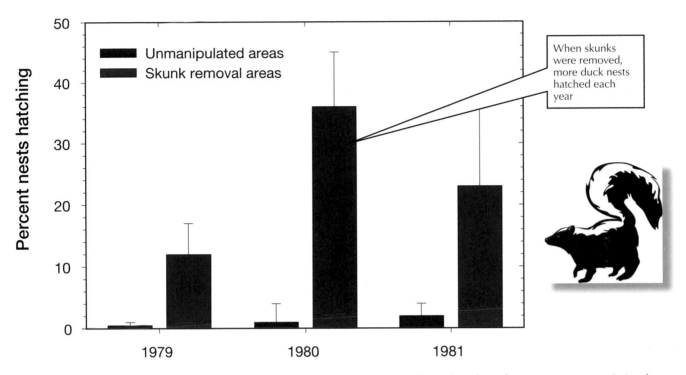

Figure 7.2. Mean hatching rates of upland duck nests in waterfowl areas of North Dakota from 1979 to 1981. Striped skunks (*Mephitis mephitis*) were removed from some of the areas (red bars) during the April-to-July nesting season. Skunk removal dramatically improved duck nesting success in these areas compared with that in areas where skunks were present (blue bars). (Data from Greenwood 1986, Table 3.)

This hypothesis has been tested experimentally five times in Canada and Alaska. Wolf removal increased moose calf survival in three of the five experiments, but only one experiment produced a significant increase in the moose population. Predation may affect moose numbers, but it is the combined predation by grizzly bears, wolves and other predators that is most important, not predation by wolves alone. When wildlife ecologists put radio collars on moose calves in the Yukon, they found that 58% of all calf deaths were caused by grizzly bears and 25% by wolves. In these kinds of multi-predator systems, removing one predator may not change prey abundance very much. For predator control to be most effective, it is important to identify the most significant predator species in each system.

Wolves also prey on caribou, and there has been considerable argument about how much predation reduces caribou abundance and whether predator control would allow more hunting of caribou. Although many wolf-control programs have been implemented, few of them have used a proper experimental design

that includes unmanipulated populations for comparison. Circumstantial evidence suggests that a combination of predation on young caribou and human hunting of adult caribou holds caribou populations in check (Table 7.1).

Consequently, you cannot manage caribou populations by considering *only* hunting or *only* natural predators. In a world without human hunters, predators alone would hold caribou populations to a density of about 0.4 per km^2. Herds with few predators can reach densities that exceed 20 per km^2, at which point the food supply becomes limiting and some individuals, particularly calves, starve.

Caribou, moose and other large species may represent an interesting example of populations whose reproduction curves have two stable equilibrium points (Figure 7.3), instead of the single equilibrium point we discussed in Chapter 6 (see Figure 6.2, page 116). Such populations may temporarily evade the factor that sets the lower equilibrium and then increase in size to an upper equilibrium set by a second factor,

Table 7.1. Effect of predation and hunting on caribou. Each value in the table is the average rate of increase or decrease in the population of 40 herds. (Data from Bergerud 1980.)

Hunting mortality	No predators	Few predators	Many predators
None	0.28	0.11	0.02
< 5% per year	0.08	0.05	−0.01
> 5% per year	−0.11	−0.19	−0.13

such as food shortage. When the lower equilibrium is set by predation, the region to the left of the threshold is called a **predator pit.** The key experimental test for a predator pit is to remove predators and observe whether the prey population increases to a new equilibrium size that is set by a different limiting factor.

Introduced Predators

Instances in which humans have accidentally introduced a new predator provide clear evidence of the influence of predators. A striking example is the near elimination of lake trout (*Salvelinus namaycush*) from the Great Lakes by the sea lamprey (*Petromyzon marinus*). Adult sea lampreys, which range from 30 to 50 cm in length, live in the Atlantic Ocean and migrate into North American streams to spawn. They use their sucker-like mouth and rasping tongue to attach themselves to the side of a fish, tear a hole in the fish's skin, and suck out body fluids. Few fish survive after being attacked by a lamprey. Niagara Falls presumably

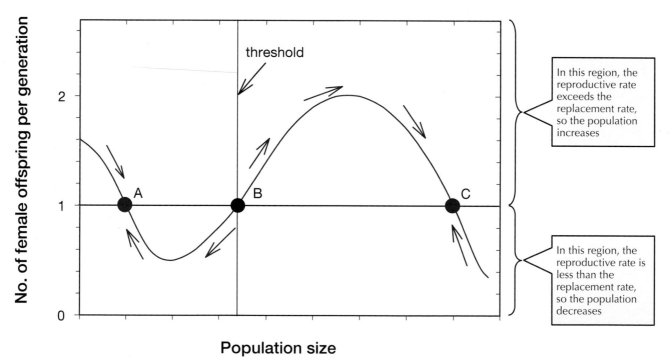

Figure 7.3. Population model with two stable equilibrium points (A and C). In this hypothetical population, the blue line marks the replacement reproductive rate (1.0 female offspring per generation). When the reproductive rate is greater than the replacement rate, the population increases (red arrows point to the right). When the reproductive rate is less than the replacement rate, the population decreases (red arrows point to the left). Point B is an unstable equilibrium that marks a threshold from which populations always increase or decrease. This type of model might be appropriate for caribou populations in which the lower equilibrium is set by predators and the upper equilibrium by food shortage. Compare with Figure 6.2 on page 116.

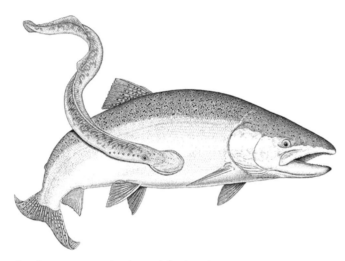

Sea lamprey attached to a lake trout

blocked the passage of lampreys to the upper Great Lakes before the Welland Canal was built in 1829. The first sea lampreys were found in Lake Erie in 1921, in Lake Michigan in 1936, in Lake Huron in 1937 and in Lake Superior in 1938. Lake trout catches decreased to virtually zero within about 20 years of the lamprey invasion in each lake (Figure 7.4). Control efforts have been applied to reduce the lamprey population since 1951, and lampreys are now rare in the Great Lakes. Attempts have been made to rebuild the Great Lakes fishery by releasing trout bred in hatcheries, and lake trout populations have increased somewhat in Lake Superior. However, the fish are still rare in all the other Great Lakes, where restoration has been hampered by a loss of genetic diversity, a shortage of spawning areas, the presence chemical contaminants and the introduction of exotic species, such as Pacific salmon.

Q Would you expect evolution to favor traits of predators that make them more efficient at hunting prey?

Escape from Predation

Many prey species have adaptations that enable their populations to escape from the limiting impact of predation. Some species, such as caribou, wildebeest and snow geese, migrate away from their predators. On the Serengeti Plains, the vast majority of large mammalian prey species are migratory, whereas most of the predators are resident. Lions, for example, seem to be limited in numbers by the resident prey species available in the dry season when the migratory ungulates are elsewhere. Other species escape from predation limitation by being very large (elephants), by breeding much faster than their predators (house mice) or by hibernating (ground squirrels).

What can we conclude about the impact of predators on prey species? Efficient predators may have strong effects on prey populations, but not all predators can be assumed to be efficient. In general, the impact of predators is greatest when the potential rates of increase of predators and their prey are nearly equal, and when predators are highly specialized on one, or a few, prey species. Most severe predator impacts have arisen when humans have introduced predators, such as the sea lamprey, into new ecosystems.

■ 7.3 PLANTS HAVE DEFENSES THAT REDUCE HERBIVORY

Plant–herbivore interactions constitute a major part of the interactions between species in terrestrial ecosystems. Even though plants cannot escape from herbivores by moving, most herbivores do not completely eliminate their food plants. Herbivores can be important selective agents on plants. Some herbivore species have evolved mechanisms, such as territorial behavior, that prevent them from destroying their food supply. Other mechanisms, such as predation, may limit herbivore abundance, allowing some plants to escape being eaten. But the major reason why plants thrive despite the threat of herbivory is that most plants have evolved an array of defenses against herbivores and are not easily eaten.

Q Is it possible for a herbivore species to evolve traits that would conserve its food plants?

Secondary Plant Substances

Plants contain a variety of chemicals whose functions have long puzzled plant physiologists and

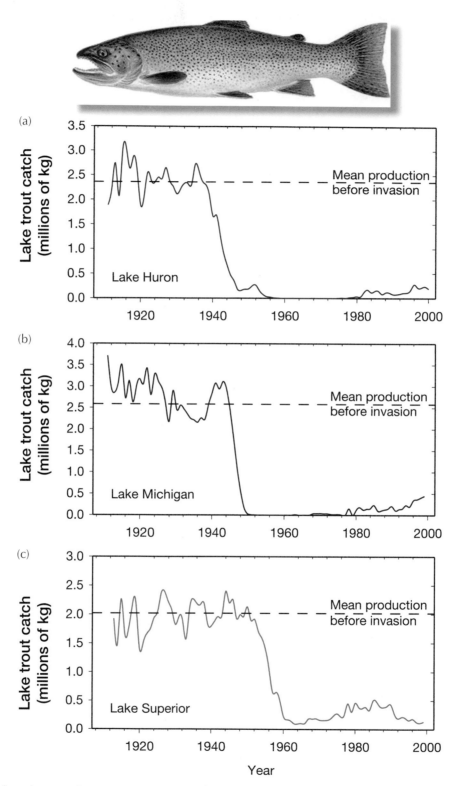

Figure 7.4. Effect of sea lamprey invasion on commercial production of lake trout in (a) Lake Huron, (b) Lake Michigan, and (c) Lake Superior. Dashed lines indicate the mean production over the 50 years before lamprey invasion. Lake trout populations in the Great Lakes have not recovered to pre-invasion levels despite more than 50 years of sea lamprey control. (Data from the Great Lakes Fishery Commission, 2005.)

biochemists. These chemicals, called **secondary plant substances**, are by-products of the primary metabolic pathways in plants and are not directly involved in growth or photosynthesis. Secondary plant substances include:

- the chemicals that give the spices nutmeg, cinnamon and clove their characteristic tastes and aromas
- terpenoids, which are found in peppermint oil and catnip
- a diverse group of compounds called alkaloids.

Among the better-known alkaloids are:

- nicotine, which is a product of tobacco plants
- morphine, which is derived from the juice of the opium poppy
- cocaine, which is a substance contained in the leaves of the coca plant of South America
- caffeine, which is found in more than 60 species of plants, including tea, coffee and cacao (the source of chocolate).

How do secondary plant substances help plants survive? One view is that they defend plants by making the plants distasteful to herbivores. According to this view, well-defended plants grow faster and produce more offspring. However, synthesizing secondary substances has a metabolic cost for the plant because it requires that energy and nutrients to be diverted from other needs, such as growth and seed production. If plant defense characteristics are inherited, then all the elements needed for natural selection are present, and we can make four general predictions:

1. Plant species will evolve more defenses if they suffer severe damage from herbivory, and they will evolve fewer defenses if the cost of defense is high.
2. Within a plant, more defenses will be allocated to those valuable tissues that are at risk.
3. Defense mechanisms will be reduced when herbivores are absent and will be increased when plants are attacked.
4. Because of their cost, defense mechanisms cannot be maintained if plants are severely stressed by other environmental factors.

These four predictions are supported by much evidence.

Both the type and the amount of defense used by plants depend on the vulnerability of the plant tissues. Plants that are attacked by many herbivores typically invest more heavily in the defense of growing tips and young leaves, which are more valuable to the plants than mature leaves. Tannins, resins, alkaloids and other secondary substances are concentrated at or near the surface of the plant, thereby increasing their effectiveness.

Some plants put much of their energy into anti-herbivore defense, while others spend very little on defense. How can we explain this difference? The answer lies in a plant's ability to replace tissues taken by herbivores. Fast-growing plants typically have leaves that are shed relatively quickly, whereas slow-growing plants retain their leaves much longer. Because each leaf represents a relatively greater investment for slow-growing plants, they are more vulnerable to herbivores and thus invest more in defensive chemicals. This explanation predicts that there would be a trade-off between using the products of photosynthesis for growth and using them for defense against herbivores. Figure 7.5 illustrates this trade-off.

Inducible Plant Defenses

Plants can defend themselves at all times or only when they are attacked by herbivores. Defensive mechanisms that are activated only when they are needed are called *inducible defenses*. Induction times for defensive reactions by plants have been studied for only a few species and vary from 12 hours to more than a year. Rapid defensive responses in plants are the subject of intensive interest in plant–herbivore research.

If defenses are costly, we would expect a reduction in inducible defenses when herbivores are absent (prediction #3 on page 139). Experiments on thorny plants support this assumption. Thorns, spines and prickles are produced by many terrestrial plants and act as physical defenses against some herbivores. The Mediterranean shrub *Hormathophylla spinosa* is heavily browsed by goats and other ungulates, which typically eat 80% of the flowers and fruits. When plant ecologists fenced out goats for two years, they found that thorn density on plants inside the fence dropped by

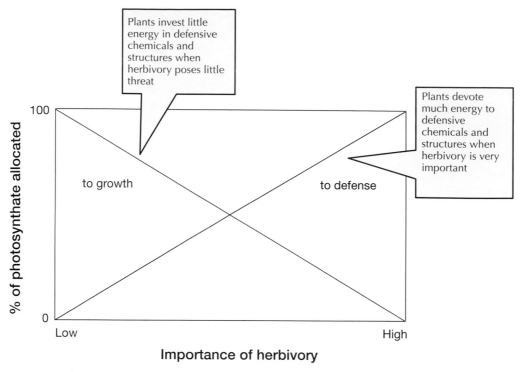

Figure 7.5. Trade-off between using energy captured in photosynthesis for growth or using it for defense. When herbivory is significant, we would expect plants to use more of their energy for defense. This energy allocation would reduce the potential growth of leaves, shoots, seeds and roots. (Modified from Stamp 2003.)

nearly 50% (Figure 7.6). In Tanzania, goats that feed on *Acacia tortilis* trees induce the trees to produce thorns, while trees that are protected from grazing do not grow thorns. The more thorns a tree has, the fewer shoots it loses to goats. In the Negev Desert of Israel, acacias in areas exposed to goats have more thorns than acacias that have been fenced from large herbivores for more than 10 years.

Herbivores have also evolved mechanisms that enable them to circumvent plant defenses to some extent. In some insect species, for example, the larvae or nymphs, which are the main feeding stages of the life cycle, are produced at the same time their food plants are growing many new leaves, which are initially low in secondary substances. Some herbivores also have enzymes that detoxify secondary plant substances. While thorns on acacia trees may reduce goat browsing, they do not deter insect herbivores. Plants must have a variety of adaptations for coping with different types of vertebrate and invertebrate herbivores.

Q **Is the interaction between plants and herbivores similar to a military arms race? Explain.**

Ideas about herbivory have had important consequences for human land use (see *Essay 7.1*). Cattle grazing is a significant land use throughout the world—60% of the world's agricultural land is grazing land. About 96% of Earth's natural grasslands and savannahs are now used for cattle and sheep grazing. If the economic object is to maximize revenue from grazing, there are important issues of sustainable land use that require study by agricultural scientists and ecologists. This is a critical issue to which we will return in Chapter 21.

If animal populations are not limited by predation, food supplies are typically the next most likely limiting factor. Because of the complex interactions between herbivores and their food plants, ecologists must perform careful experiments to determine whether food supplies are limiting in specific situations.

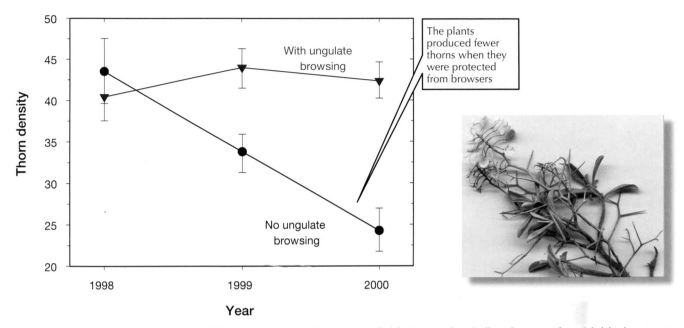

Figure 7.6. Changes in the number of thorns on the Mediterranean shrub *Hormathophylla spinosa* in fenced (black curve) and unfenced (red curve) plots. There is a trade-off between thorn production and fruiting in this plant: The fewer thorns it produces, the more fruit it can bear. (After Gomez and Zamora 2002.)

■ 7.4 COMPETITION OCCURS WHEN SPECIES HAVE SIMILAR RESOURCE REQUIREMENTS AND RESOURCES ARE IN SHORT SUPPLY

We read about sports competition and business competition every day in the newspapers. Species can also compete for resources in nature, but we must distinguish between two different types of competition, defined as follows:

- **Scramble competition** occurs when a number of organisms use common resources that are in short supply. All competing individuals are equally affected—there are no 'winners' or 'losers.'
- **Interference competition** occurs when the organisms seeking a resource harm one another in the process (for example, by fighting), even if the resource is not in short supply. Some individuals acquire resources at the expense of others, so there are 'winners' and 'losers.'

Note that both types of competition may be between two or more different species or between members of the same species. We discuss here only competition between different species.

A variety of resources may become the center of competitive interactions. For plants, light, nutrients and water may be important resources, but plants may also compete with other plants for pollinators or for space. Water, food and mates are possible sources of competition for animals. Competition for space also occurs in some animals and may involve many types of specific requirements, such as nesting sites, wintering sites or sites that are safe from predators. Thus, resources are diverse and complex. The first rule of competition is that species must require at least some of the same resources before they can be potential competitors.

Competition may be a subtle process. There is no need for animals to see or hear their competitors. A species that feeds by day on a plant may compete with a species that feeds at night on the same plant, if the plant is in short supply. Conversely, many or most of the organisms that an animal does see or hear will not be its competitors. This is true even if the organisms use the same resources, if those resources are superabundant. Oxygen, for example, is a resource used by most

ESSAY 7.1   HERBIVORY, ECONOMICS, AND LAND USE

Ecological ideas about herbivory collide with economic ideas about land use in the western United States, where about 70% of the land area—including that inside wilderness areas, wildlife refuges, national forests and some national parks—is grazed by livestock. Ecologists ask two questions about grazing impacts: (1) What are the ecological costs of grazing these areas? (2) Is grazing in its current form sustainable? Economists ask about the costs and benefits of grazing, but rarely look at the ecological costs that are not expressed in dollars. The result has been an ongoing and acrimonious controversy over land use in this part of the country—a controversy with many facets. The key issue is whether western rangelands are degraded by excessive cattle grazing.

How might an ecologist analyze this controversy? An experimental ecologist would like to measure the ecological impacts of grazing on populations of plants and animals by comparing similar grazed and ungrazed areas. But there is almost no ungrazed land available in this part of the United States for such comparisons. Much of the land left that is left ungrazed is located on steep slopes or in rocky areas that differ dramatically from the surrounding habitats. One solution is to use livestock exclosures to create ungrazed areas. But this approach also has problems because most exclosures are small and are on land that was previously grazed. Small exclosures may not include all the species in a community, especially the rare ones, and, if initial grazing impacts are the most severe, historical carry-over will affect even long-term exclosure studies on sites that were previously grazed because seed banks will be exhausted and some plants will be locally extinct. Therefore, exclosures will underestimate the true impact of grazing on plants and animals. These problems may be addressed by constructing larger exclosures and by monitoring them for a longer time.

Some ecologists and land managers argue that livestock are essential for the ecological health of western grazing lands. Support for this argument has come from ecologists who have suggested that plants may increase their productivity if they are grazed. Before Europeans settled North America, bison, pronghorns, deer and elk grazed in the West, and cattle have largely replaced these species. There is little ecological evidence for the hypothesis that grazing is good for western rangelands, but the idea keeps coming up as one justification for the current grazing system.

The use of public lands for grazing must be balanced with the needs of conservation and recreation. In particular, all those concerned must work out sustainable land-use practices that will satisfy these diverse economic and ecological needs. Western rangelands should not be exclusively national parks, nor should they be exclusively overgrazed plant communities. Science can show us what policy goals can be achieved by good land management.

terrestrial organisms, yet there is no competition for oxygen among these organisms because oxygen is so abundant. Competition in plants usually occurs among individuals that are rooted in position and therefore differs from competition among mobile animals, which have the option to move away from competitors.

Ecological Niches

To investigate competition between species, we must first determine the range of environmental values in which each species can survive and reproduce. This range is the **fundamental niche** of the species—the set of resources it can utilize in the absence of competition and other biotic interactions. For example, the fundamental niche of a plant species might be a specific range of temperature and soil moisture. Because of competition from other species, predation, disease or a failure to disperse to suitable areas, no species occupies its entire fundamental niche. In nature, the actual environmental range in which a species lives, called its **realized niche**, is a subset of its fundamental niche (Figure 7.7).

Q **Would you expect the realized niche of a species to vary over the species' geographic range?**

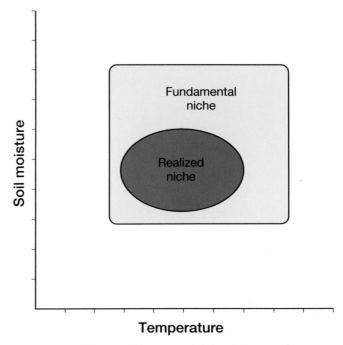

Figure 7.7. Schematic illustration of the difference between the fundamental niche and the realized niche of a plant species. The fundamental niche is defined by the species' tolerance for specific environmental factors; the realized niche is the range of factors in which the species is actually found. Only two factors—soil moisture and temperature—are illustrated here, but more factors may be involved.

In measuring niches, we concentrate on those resources that are the subject of competition. For example, research on finches on the Galápagos Islands has revealed how competition for food can shift the realized niches of bird species. As Figure 7.8 illustrates, species of ground finches with different bill sizes consume different sizes of seeds, thereby minimizing competition for food and increasing their chances of survival during the dry season, when seeds are scarce.

Species may be restricted by competition or predation to a small part of their fundamental niche. Figure 7.9 shows how this might happen in evolutionary time. For example, a bird species A in Figure 7.9a may be restricted in its feeding niche to a narrow range of food sizes because two similar species, B and C, harvest smaller and larger foods more efficiently. Alternatively, a plant species (D in Figure 7.9b) may grow best in soils that have an intermediate level of nitrogen but may be

restricted to poorer soils by a superior competitor (E in Figure 7.9b).

If two species have overlapping realized niches, they are potential competitors. How much does competition shape the realized niches of such species? The best way to answer this question is to remove one of the species and measure the survival, reproduction and growth of the other species (see *Essay 7.2*). Another approach is to observe introduced species that displace native species through competition. Studies on snakeweed and ants illustrate these two approaches to studying competition.

Snakeweed (*Gutierrezia microcephala*) is a widely distributed shrub in the south-western United States that grows to a diameter of 1 m and lives for up to 20 years. Two plant ecologists, Matthew Parker and Amy Salzman, studied the impact of competition on the

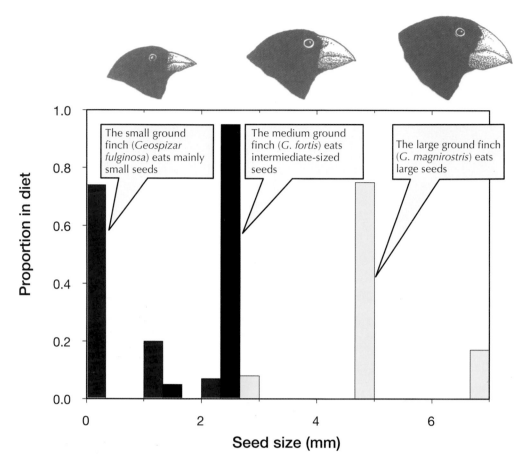

Figure 7.8. The realized feeding niches of three of Darwin's finches on the Galápagos Islands. The separation of feeding niches in these species reduces possible competition for food. (Data of Schluter 1982.)

survival of 2-year-old snakeweeds by removing adult snakeweeds, grasses or all plants within 0.5 m of each test plant. They found that competition reduced the survival of young snakeweeds, but only when herbivores (grasshoppers) were present in the study plots (Figure 7.10). When grasshoppers were excluded, all of the test plants had similar survival rates. Thus, both competition and herbivory affect snakeweed survival.

The Argentine ant (*Linepithema humile*) is native to South America, but was introduced to California in 1907. Since then, it has been spreading throughout temperate and subtropical regions of North America, displacing all native ants in the areas it invades. What explains the strong competitive ability of Argentine ants? They are not more aggressive than other ant species: In tests of aggressiveness, sometimes native ants won and sometimes Argentine ants won. Moreover, the

chemical defenses of Argentine ants are no more effective than those of native ants at repelling other ant species. The key to Argentine ant success seems to be their ability to form supercolonies consisting of multiple nests. Whereas native ants form individual colonies and defend the territory around their colonies from other ants, Argentine ants move freely between different nests without any territorial defense. The number of worker ants is much larger in supercolonies, and this numerical advantage gives Argentine ants a competitive edge for finding food. In one experiment, researchers in northern California placed food at fixed distances from ant colonies. Argentine ants found the food within 4 minutes, but native ants required 10–35 minutes to do so. Thus, Argentine ants can secure most of the food resources in an area and drive native ant species to extinction by scramble competition for food.

(a)

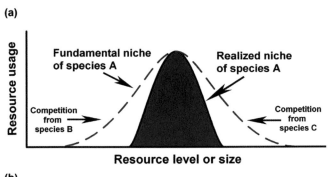

(b)

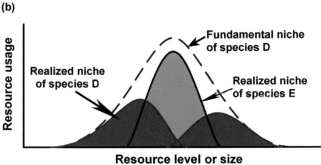

Argentine ant

Figure 7.9. Effect of competition on species' realized niches. (a) Three species competing for a common resource. Species *A* has a wide fundamental niche (dashed red line), but its realized niche (red) is narrower because of competition from species *B* and *C* (whose niches are not shown here). (b) Two other species, *D* and *E*, are competing for a common resource. Species *E* (green) is the superior competitor. Species *E* (green) forces species *D* (blue) out of its optimum into the peripheral parts of its fundamental niche, causing species *D* to have a bimodal, asymmetrical realized niche. Plants competing for nitrogen in the soil could show this pattern. (Modified from Austin 1999.)

Evolution and Competition

If two species are competing for a resource that is in short supply, both would benefit by evolving traits that reduce competition, as suggested in Figure 7.11. The benefit is a higher average population size for each species and, presumably, a lower chance of extinction. But in many cases, evolution may simply alter the terms of competition without reducing it. For example, if evolution reduces competition between two species (*A* and *B*) by causing A to use smaller food items than *B*, it may increase competition between A and a third species, *C*, which also feeds on small food. The realized niche of each species may be set by a web of other possible competitors, so evolution may not completely eliminate competition.

r-selection and *K*-selection

Species can differ dramatically in their ability to win in competition. How might evolution produce such differences? The concept of competitive ability in animals is intuitively clear, but difficult to define. To understand how competitive ability might evolve, we need to look at life-history strategies more broadly. In some environments, species exist near the maximal density or carrying capacity of the habitat (symbolized by *K*) for much of the year, and these species are subject to **K-selection.** In other environments, the same species may rarely approach the carrying capacity, but remain on the rising portion of the population growth curve for most of the year; these species are subject to **r-selection,** where *r* is the rate of population increase. Robert MacArthur and E. O. Wilson suggested in 1967 that as a population initially colonized an empty habitat, *r*-selection would predominate for a time, but ultimately the population would come under *K*-selection.

Species that are *r*-selected should be under little pressure from competition because they generally do not live in crowded conditions. Hence, they should be unlikely to evolve strong competitive abilities. These species should have an array of traits characteristic of rapid population growth—high reproductive potential, high dispersal powers and short lifespans. These are all traits of colonizing species and weedy plants.

By contrast, species that are *K*-selected live in crowded conditions and therefore should be subject to

ESSAY 7.2 STUDYING PLANT COMPETITION WITH REPLACEMENT SERIES

One method for studying the effect of one plant species on a competing species is to use *replacement series*, either in the field or in the greenhouse. A replacement series can be viewed schematically as an array of plots with different combinations of the two species. In the diagram below, green dots represent species A and red dots represent species B:

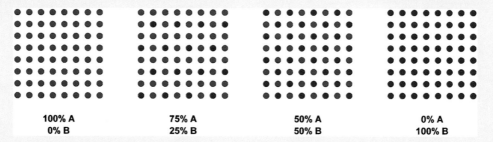

| 100% A | 75% A | 50% A | 0% A |
| 0% B | 25% B | 50% B | 100% B |

In this type of replacement series, the density of plants is kept constant and only the percentage composition is changed. The variables of interest are the yield of each species and the combined yield of both species. Competition between the two species, as well as competition among individuals of the same species, determines the yield. Replacement series were pioneered by the Dutch ecologist C. T. de Wit nearly 50 years ago and have been an important technique for investigating competition between plant species.

The results from one replacement series involving perennial ryegrass (*Lolium perenne*) and white clover (*Trifolium repens*) were as follows (Jolliffe 2000):

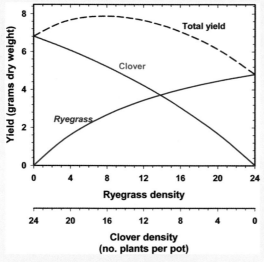

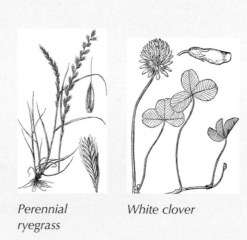

Perennial ryegrass

White clover

Graph showing a replacement series involving perennial ryegrass and white clover

The total yield was maximal when clover comprised about 65–75% of the mixture, indicating that the mixture was more productive than either single-species monoculture in this experiment. This study has been repeated with combined densities greater than 24 plants per plot to explore how these relationships vary with overall plant density.

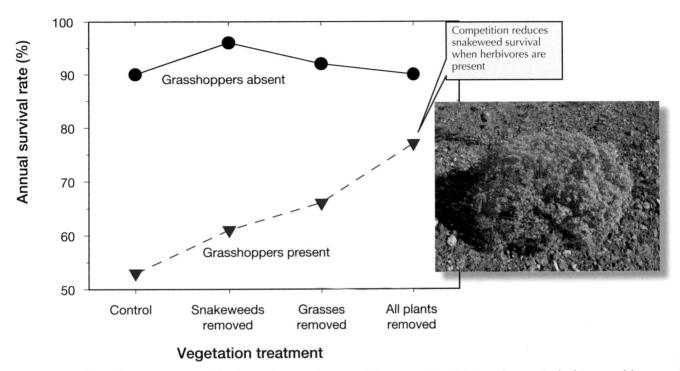

Figure 7.10. Effect of competition and herbivory by grasshoppers (*Hesperotettix viridis*) on the survival of 2-year-old snakeweed plants. Control plants were not manipulated. Herbivory has the largest impact on snakeweed survival, but competition from other plants also reduces survival rates. (Data from Parker and Salzman 1985, photo courtesy of Dr Clinton Shock, Oregon State University.)

strong competitive pressures, which should push these species to use their resources efficiently. A species that can convert limiting resources into reproductive adults the fastest is usually the superior competitor. *K*-selected species should show traits characteristic of stable, high-density populations—low reproductive rates, limited adaptations for dispersal and long life spans. These general traits can be seen in many long-lived species.

C-S-R Model of Plant Strategies

Plant ecologists have taken a slightly different view of competition because plants cannot move. Vascular plants face two broad categories of ecological factors that affect their growth and reproduction. One category includes shortages of resources, such as light, water or nitrogen, temperature stresses and other physical-chemical limitations. The second category includes losses due to grazing, diseases, windstorms, frost, erosion and fire. Philip Grime called the first category *stresses* and the second category *disturbances*. Grime examined the four possible combinations of

these two categories in 1979 (Table 7.2). If stress and disturbance are too severe, no plant can survive, so the combination of high stress and high disturbance is biologically impossible. Grime suggested that the other three combinations, or strategies—labeled C, S, and R—form the primary focus of plant evolution and that individual plant species have tended to adopt one of these life-history strategies.

The C, S, and R strategies can be illustrated as a triangle (Figure 7.12), which emphasizes that they represent trade-offs in life-history traits. The critical assumption is that a plant species cannot be good at all three strategies, but must trade off one set of traits

Table 7.2. Grime's C-S-R plant strategies.

Intensity of disturbance	Intensity of stress	
	Low	**High**
Low	Competitive strategy (C)	Stress-tolerant strategy (S)
High	Ruderal or weed strategy (R)	None possible

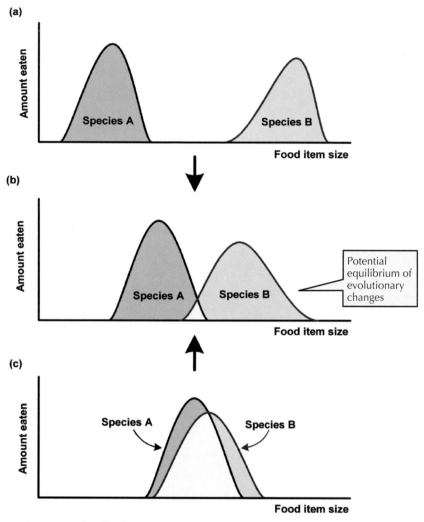

Figure 7.11. Evolutionary changes in food-utilization curves for two species (*A* and *B*) that are potential competitors. Food of particular sizes is the resource for which competition may occur in this hypothetical situation. In (a) some foods are not being used, and so both species could increase their abundance by moving their feeding niches into the middle region and using the resources that are presently not used. Natural selection will favor individuals that can use the food items in the middle region. In (c) the feeding niches of the two species overlap extensively; *A* and *B* could benefit from eating smaller and larger food items, respectively, which would reduce competition. Arrows indicate the direction of evolutionary pressures toward case (b).

against another. Grime's concept has been called the C-S-R model. Competitive (C) plants, including many perennial herbs, shrubs and trees, show characteristics of *K*-selection: dense leaf canopies, high growth rates, low seed production and relatively short life spans. Stress-tolerant (S) plants often have small, evergreen leaves, low growth rates, low seed production and long life spans. Ruderals (R) are weeds; they are small, grow rapidly, typically live only one year and devote much

of their resources to seed production. Ruderals are the *r*-strategists of the plant world.

Q **Why are there three plant life-history strategies, but only two animal strategies?**

The evolution of competitive ability, although approached differently by botanists and zoologists, has achieved a convergence of ideas. Both MacArthur

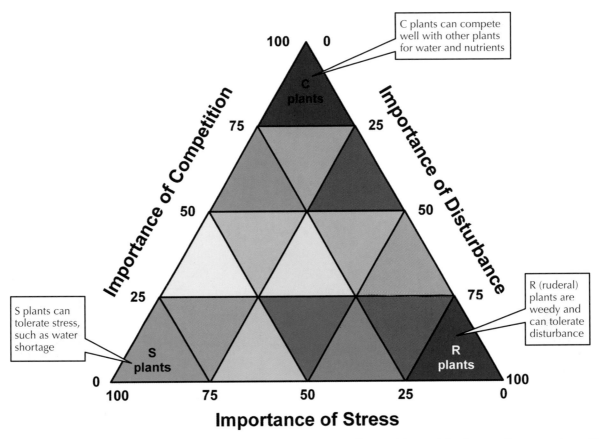

Figure 7.12. Grime's C-S-R model of plant life-history strategies. According to this model, plant attributes evolve within this triangle depending on the relative importance of competition, stress and disturbance. (Modified from Grime 1979.)

and Wilson's *r*-selection and *K*-selection model and Grime's C-S-R model describe well the trade-offs species must face in evolutionary time. Species cannot be good at everything and adaptations are always a compromise between conflicting goals.

SUMMARY

Species interact negatively when the presence of one changes the distribution, or reduces the abundance, of the other by lowering its birth rate or survival rate. Predation, herbivory and competition are major negative interactions between species and are responsible for many of the changes in distribution and abundance that occur in natural communities.

Some prey species escape from their predators through adaptations such as migration, so not all predators reduce the abundance of their prey. The most spectacular examples of predator impacts have come from introduced predators, such as the sea lamprey in the Great Lakes.

The interaction between plants and herbivores has strong impacts on the characteristics of plants and the feeding habits of animals. Plants have evolved an array of secondary substances and structural defenses that deter herbivore feeding. Many spices, and some drugs, are secondary plant substances.

Competition is one of the best-studied negative interactions between species. Competition occurs over resources such as food, water, or nitrogen. Species must require the same resources to be considered as competitors. Plant competition differs from most animal competition because individual plants cannot move from place to place to avoid competition. Competitive ability evolves by natural selection—evolutionary pressures from competition shape the life histories of plants and animals.

SUGGESTED FURTHER READINGS

Bergelson J (1996). Competition between two weeds. *American Scientist* **84**, 579–584. (Detailed studies of how two weeds interact by one of the leaders in the field of plant competition.)

Côté IM and Sutherland WJ (1997). The effectiveness of removing predators to protect bird populations. *Conservation Biology* **11**, 395–405. (A comprehensive review of whether predator-removal programs work to save species.)

Henke SE and Bryant FC (1999). Effects of coyote removal on the faunal community in western Texas. *Journal of Wildlife Management* **63**, 1066–1081. (A well-designed experiment to determine the effect of removing coyotes on prey species and other, smaller predators.)

Holway D A (1999). Competitive mechanisms underlying the displacement of native ants by the invasive Argentine ant. *Ecology* **80**, 238–251.(A detailed analysis of why the invasive Argentine ant wins in competition.)

Lima SL (1998). Nonlethal effects in the ecology of predator–prey interactions. *Bioscience* **48**, 25–34. (A review of how predators can affect their prey without killing them.)

Strauss SY, Rudgers JA, Lau JA and Irwin RE (2002). Direct and ecological costs of resistance to herbivory. *Trends in Ecology and Evolution* **17**, 278–285. (A review of the cost to plants of becoming resistant to herbivores.)

QUESTIONS

1. In *The Origin of Species* (1859, Chapter 3) Charles Darwin made the following assertion about competition between species:

    'As the species of the same genus usually have, though by no means invariably, much similarity in habits and constitution, and always in structure, the struggle will generally be more severe between them, if they come into competition with each other, than between the species of distinct genera.'

    Discuss the idea that competition will be more severe between closely related species.

2. The birds, lizards and mammals of Guam—an island in the western Pacific Ocean—have been driven to extinction or to low numbers by an introduced predator: the brown tree snake (*Boiga irregularis*). Is it adaptive for a predator to drive its prey to extinction? Read the discussion in Rodda *et al.* (1997) and evaluate the rarity of this situation.

3. Insects consume 10–50% of the leaves of Australian eucalyptus trees. Yet these trees contain very high concentrations of essential oils and tannins (Morrow and Fox, 1980). Discuss how such a high level of herbivory could occur if these oils and tannins are defensive chemicals against herbivores.

4. Plants that are native to islands without large mammalian herbivores have no evolutionary history of being grazed. Therefore, they may be vulnerable to damage, and possible extinction, if such herbivores are introduced to the islands. How would you test the hypothesis that island plants lack defenses against large mammalian herbivores?

5. If competition occurs over resources, how would you define what a resource is for a particular species of plant or animal? Suppose you discovered a new species of insect in South America: how would you determine what its possible competitors might be?

Chapter 8

NEGATIVE SPECIES INTERACTIONS— INFECTION AND PARASITISM

IN THE NEWS

Influenza in its many forms is one of the most common diseases affecting humans each year, and many people underestimate its potential for harm. Avian influenza, or 'bird flu,' is caused by type A strains of the influenza virus. First identified in Italy more than 100 years ago, avian influenza occurs worldwide, and all species of birds are thought to be susceptible to the 15 forms of this virus. However, some birds are more resistant than others. Ducks and geese are the natural reservoir of the avian influenza viruses, but they are also very resistant to the viruses, which produce only a mild illness in them. By contrast, chickens and turkeys are extremely susceptible to bird flu and typically die from the highly pathogenic forms of this virus. Transmission of bird flu virus seems to occur when domestic flocks of chickens or turkeys have contact with waterfowl. Quarantine of infected chicken farms and destruction of exposed flocks are the standard means of preventing spread of the disease.

Because the influenza virus is unstable and genetically labile, a relatively harmless flu virus can mutate into a highly pathogenic virus. That happened during a 1983–1984 outbreak of bird flu in the United States. The subtype H5N2 flu virus, which initially caused a mild disease in chickens, mutated within 6 months to a form that killed 90% of infected chickens. Poultry farmers had to destroy 17 million chickens to end this epidemic.

The global concern is that these lethal bird viruses will cross over and infect humans. Influenza viruses can shift hosts by an exchange of genetic material between two subtypes. For example, the human flu virus in a person could acquire genetic material from the avian strain of the virus in a chicken. Highly pathogenic avian flu viruses can remain infective on farm equipment, cages or clothing, especially when temperatures are low. Pigs can be infected with both avian flu and human flu strains, so they may serve as a medium for the shift of highly pathogenic avian strains into forms that will readily infect humans.

The ability of the flu virus to jump from birds to humans was responsible for the 'Spanish flu' pandemic[1] of 1918–1919, which killed at least 40 million people. Influenza pandemics have occurred every 10–50 years during the last 300 years:

---

1 A pandemic is an epidemic that affects a large number of individuals on a global scale.

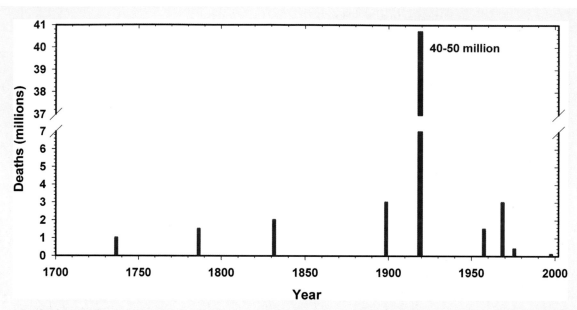

Deaths from influenza pandemics for the past 300 years

Outbreaks in 1957 and 1968 claimed 4.5 million lives. In 1997, chicken flu caused about 6 million birth deaths in Asia and within three days all of Hong Kong's poultry had to be killed. Given the close contact between large numbers of people and domesticated animals in developing countries and the speed of international travel, conditions are now favourable for another major pandemic. However, we do not have the ability to predict where or when new flu virus strains will appear. Vaccination is the major weapon we have against the flu virus, but, because the virus can mutate so rapidly, immunity gained to one strain will not confer immunity against a new strain. Flu vaccines take 4–8 months to produce and are expensive. Consequently, medical response teams are constantly racing to stay ahead of the ever-changing targets for vaccination. Newly emerging diseases such as bird flu are now one of the most pressing medical threats to humans.

■ 8.1 PATHOGENS AND PARASITES HAVE NEGATIVE IMPACTS ON SPECIES

Dealing with disease resulting from infection has been one of the great preoccupations of humans throughout our recorded history, from the Black Death of the 14th century to the smallpox plague of the 19th century and the Spanish flu pandemic of 1918–1919. Today's scourges include AIDS, drug-resistant tuberculosis, West Nile virus and mad-cow disease. Generations of children succumb to common diseases such as measles, and each winter brings another flu epidemic.

Infection is a negative interaction in which a pathogen (disease-causing agent) lives on or in a host organism, to the benefit of the pathogen and the detriment of the host. Pathogens include microorganisms, such as some bacteria and fungi, as well as viruses and prions (protein bodies), which are non-living. Parasitism has much in common with infection as a negative biotic interaction, but it differs from infection mainly because parasites are often large, multicellular organisms, such as tapeworms. Large parasites are called **macroparasites.** Some parasites, such as the spirochete bacteria that cause syphilis, are also pathogens. Pathogenic microorganisms are called **microparasites.**

The **virulence** of a pathogen depends on the intensity of the disease it causes and is measured by host mortality. Although people are often very concerned about the lethal effects of pathogens and parasites, the **sublethal effects**—any effects that reduce well-being without causing death—are probably more important for plants and animals in ecological settings. Infected or parasitized animals may produce fewer offspring, be captured more easily by predators, or be less tolerant of temperature extremes. Infection and parasitism can thus interact with competition and predation in affecting population dynamics. Almost every individual plant and animal harbors both pathogens and parasites.

■ 8.2 COMPARTMENT MODELS ARE USEFUL FOR ANALYZING HOW DISEASES AFFECT POPULATIONS

One way to begin to understand a disease is to build a simple model that describes how the disease spreads

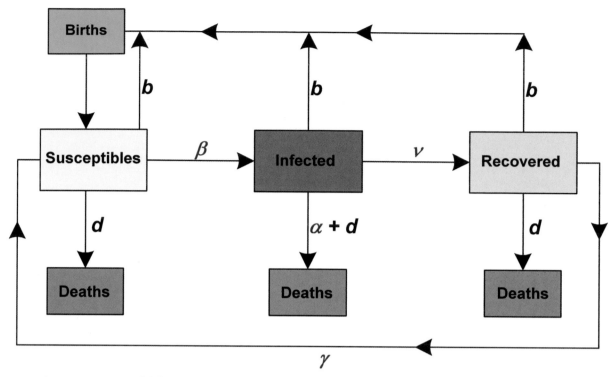

Figure 8.1. Compartment model for a directly transmitted microparasitic disease such as syphilis. Each box represents a number of individuals and, in each time step in which the model is run, individuals move between the boxes at the rates given by the symbols alongside the arrows. (Modified from Anderson and May 1979.)

among individuals and what happens to individuals that become infected. Many types of models have been used in the study of disease. **Compartment models** are box-and-arrow models that show simplified population dynamics, and they are a good starting point for learning to think about epidemics. The boxes in these models represent groups of individuals in the population, and the arrows represent the movements of individuals from one group to another (Figure 8.1). For example, a susceptible individual may move into the infected category if it contracts the disease, or an infected individual may die or recover from the disease. In their simplest form, compartment models assume a constant host population, and because this assumption is valid for the human population in the short term, these models have been used extensively for exploring human diseases.

Parameters of Compartment Models

We will use microparasites that are directly transmitted between hosts and that reproduce within the host to illustrate a compartment model. Microparasites have short generation times and thus have very high repro-

duction rates. Hosts that recover from infections typically acquire some immunity against reinfection, sometimes for life. In many cases, the duration of the infection is short relative to the lifespan of the host.

For microparasitic infections, we can divide the host population into three groups of individuals: susceptible, infected and recovered. As shown in Figure 8.1, the host population has a natural birth rate (b) and death rate (d). The rates of four processes determine the progression of a disease caused by a microparasitic infection:

1. the rates at which infected individuals die from the disease (α)
2. the rates at which infected individuals recover from the disease (v)
3. the rate of transmission between infected and susceptible individuals (β)
4. the rate at which recovered individuals lose their immunity to the disease (γ).

This model is relatively simple because it does not consider the abundance of the microparasite in the host (individuals are either infected or not infected) or

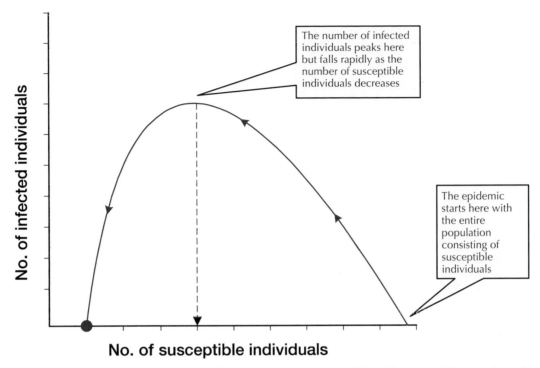

The number of infected individuals peaks here but falls rapidly as the number of susceptible individuals decreases

The epidemic starts here with the entire population consisting of susceptible individuals

No. of infected individuals (y-axis)

No. of susceptible individuals (x-axis)

Figure 8.2. Trajectory for an epidemic described by the compartment model in Figure 8.1. The number of infected individuals rises to a peak and then falls to zero at the threshold of transmission (red dot). (Modified after Heesterbeek and Roberts 1995.)

variations in host susceptibility due to genetic or nutritional differences.

Q How would a compartment model for macroparasites differ from that shown in Figure 8.1 for microparasites?

Compartment models are useful for answering questions about the stability of the host–microparasite interaction. Will the disease persist in a population or will it disappear? How do the proportions of susceptible and infected individuals change as the disease moves through a population? The answers to these important questions are the keys to understanding the effects of disease in populations.

Epidemics

The general course of an epidemic described by a simple compartment model is shown in Figure 8.2. An epidemic spreads through a population until the number of susceptible individuals falls below a threshold of transmission. At the threshold, the susceptible population becomes so small that it is unlikely an infected individual will contact a susceptible individual. Individuals can become immune to the disease, or die from it, as the epidemic progresses. In populations that are growing due to births or immigration, the disease may persist as new susceptible individuals come into the population.

Culling and vaccination are two methods for reducing the size of the susceptible population below the threshold of transmission. The more transmissible a pathogen or parasite is, the higher the fraction of individuals in a population that must be culled or vaccinated to prevent an epidemic. For example, the virus that causes German measles is highly transmissible among children, and a vaccination rate of 95% is needed to prevent an epidemic of this disease.

Q Which method—culling or vaccination—would be preferable for preventing an epidemic in a wild population of animals?

Simple compartment models can be elaborated upon to account for the specific details of different diseases. We turn now from these models to consider the

155

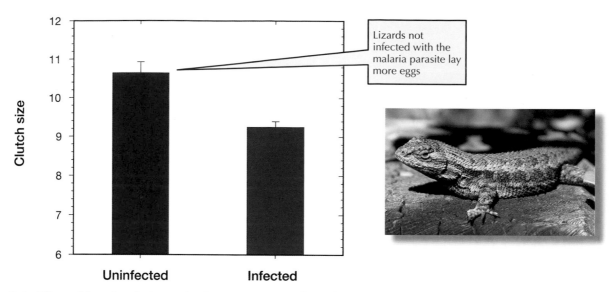

Figure 8.3. Effect of lizard malaria on clutch size in western fence lizards from 1978 to 1982. (Data from Schall 1983, photo courtesy of Bob Dyer.)

impacts that pathogens and parasites have on individual plants and animals and on populations.

■ 8.3 PATHOGENS AND PARASITES AFFECT INDIVIDUAL ORGANISMS BY REDUCING REPRODUCTIVE OUTPUT AND INCREASING MORTALITY

Individual hosts are like islands or patches of habitat that must be colonized by pathogens or parasites if a disease is to spread. Much more is known about the impact of infection and parasitism on domestic animals and humans than about their impact on wild animals and plants. One reason few studies on this topic have been carried out in the wild is that many biologists have assumed that evolution causes pathogens and parasites to become relatively harmless to their hosts, and so we would not expect infection and parasitism to have severe impacts on wild organisms. But this assumption is probably not correct—an increasing number of studies are finding that parasites and infections have significant effects on the reproduction, survival and growth of organisms.

Effects on Reproduction

Because food is not unlimited and most organisms have a limited amount of energy available, it is not surprising to find that pathogens and parasites can reduce reproductive output. For example, about 25% of western fence lizards (*Sceloporus occidentalis*) in California are infected with *Plasmodium mexicanum*—the parasite that causes lizard malaria. Malaria-infected lizards lay clutches of eggs that are about 20% smaller than those produced by uninfected lizards (Figure 8.3). Infected lizards store less fat in the summer and therefore have less energy available the following spring to produce eggs.

Effects on Mortality

Bird chicks are often attacked by parasites and, if parasite infestation is severe, the chicks may die in the nest. For example, ticks may reduce chick survival by causing blood loss and by transmitting pathogens. One study on a colony of cattle egrets in Queensland, Australia, found that chicks infested with ticks (*Argas robertsi*) grew more slowly and were more likely to die than those without ticks (Figure 8.4). The degree of tick infestation was sensitive to fluctuations in the tick population due to environmental factors. In the 1991–1992 breeding season, tick infestations were severe—an average of 24 ticks per chick—and chick mortality was significant. During the 1992–1993 breeding season, however, rainfall was low and ticks did not survive well. There were fewer ticks (an average of 5 per chick) that season, and chick mortality was much lower. This type of result is common in studies of parasitic diseases,

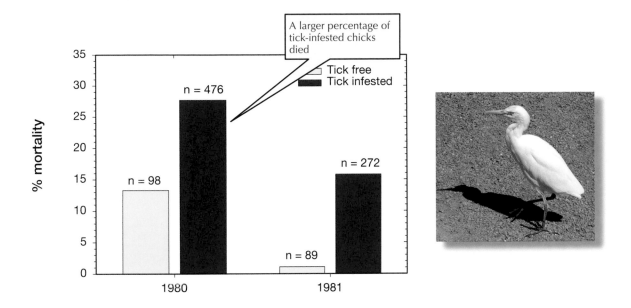

Figure 8.4. Effect of tick infestation on mortality of cattle egret (*Bubulcus ibis*) nestlings in Queensland during two breeding seasons. The numbers of chicks studied are given above the bars. (Data from McKilligan 1987, photo courtesy of David Cook.)

which may have strong effects on host populations in some years and insignificant effects in others.

No one doubts that diseases kill animals, and numerous veterinary examinations of dead animals have suggested that a parasite or pathogen was the immediate cause of death. However, a population ecologist needs to know more. What fraction of mortality in a population is caused by disease? To answer this question, we must examine disease outbreaks in natural populations.

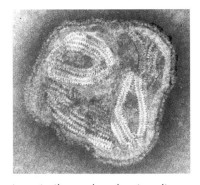

A paramyxovirus similar to the phocine distemper virus. The ribonucleoprotein capsid has a 'herring-bone' appearance. These viruses are single-stranded RNA viruses. This family includes the viruses that cause measles and mumps in humans.

In the spring of 1988, harbor seals (*Phoca vitulina*) in the North Sea began to die in large numbers. Dead seals were first noticed in the waters between Denmark and Sweden, but within a few months they were found near the Netherlands, Great Britain, and Ireland (Figure 8.5). Harbor seals are distributed throughout the North Atlantic and North Pacific, and before 1988 there were approximately 50,000 harbor seals in European waters. The massive epidemic in 1988 claimed an estimated 60% of the harbor seal population in the Baltic Sea, and the deaths accumulated very rapidly (Figure 8.6). The cause of the deaths was not clear initially, but a virus was suspected because the dying seals showed symptoms—including aborted pregnancies—similar to those of canine distemper—a viral disease. Eventually, the causative agent was indeed identified as a virus similar to canine distemper virus; it was named phocine distemper virus (PDV). Within 2 weeks of infection by the virus, seals developed symptoms and typically died of pneumonia and associated secondary bacterial and viral infections.

The incidence of infection for this epidemic could not be measured directly, but indirect estimates suggest that that 95% of the harbor seals in the North Sea were infected with the virus. Deaths from PDV seemed

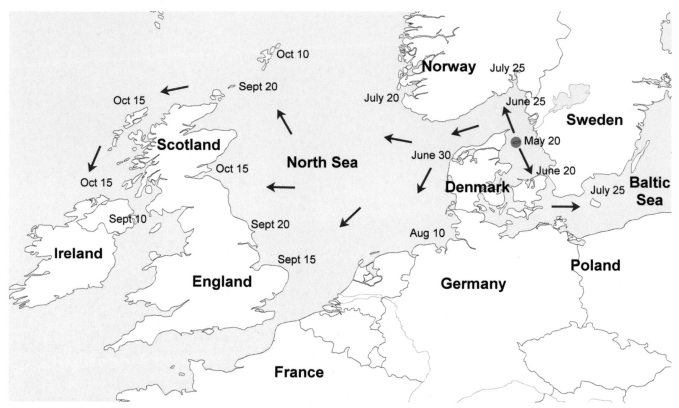

Figure 8.5. Map of the spread of harbor seal deaths in the North Atlantic in 1988. These deaths were part of the first well-documented epidemic among free-ranging marine mammals. (Data from Swinton *et al.* 1998.)

to be more common in males than females, although both sexes were infected. The disease spread so rapidly that dispersal of infected seals between colonies must have been frequent. Small colonies were as likely to become infected as large colonies, and the main predictor of a colony's chance of becoming infected was its proximity to other colonies. Harbor seal colonies in northern Norway and Iceland escaped the epidemic, presumably because no infected seals dispersed to these distant colonies.

Could this viral disease persist in the harbor seal population? Infected individuals that recover are immune for life, but, because births occur each year, there is a continual source of susceptible seals in the population. In the Baltic Sea, pups constitute about 20–22% of the harbor seal population. Using this information, ecologists constructed a compartment model of the 1988 epidemic, which showed that PDV could not be maintained in harbor seal populations as a persist-

ent infection. Then how did the disease originate in the North Sea harbor seals in 1988? It may have been introduced by another species of seal—the harp seal (*Phoca groenlandica*)— in which the virus causes a relatively innocuous disease. Harp seals are normally rare in southern waters, but in 1987 and 1988 harp seals moved south in large numbers from northern Norway into the North Sea. PDV may have crossed species boundaries at this time, triggering the 1988 epidemic among the more susceptible harbor seal population.

Despite the large number of harbor seal deaths in 1988, the harbor seal populations of Western Europe were only temporarily affected and quickly returned to their former numbers. Another less severe epidemic reduced the harbor seal population in the North Sea in 2002. The seal epidemics of 1988 and 2002 raise the general question of how often a disease can exert a long-term effect on a population—a question we turn to next.

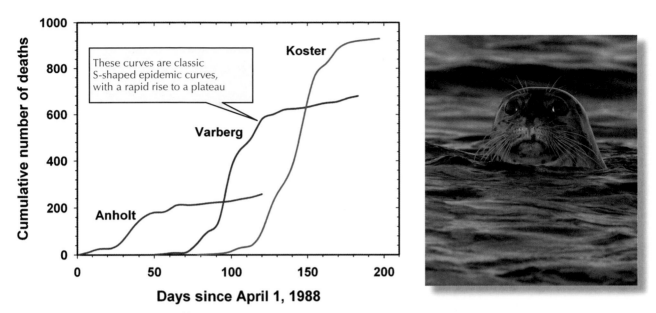

Figure 8.6. Cumulative number of harbor seal deaths in three colonies between Denmark and Sweden in 1988. The epidemic started at the small Anholt colony in April, at the larger Varberg colony in mid-May and at the Swedish Koster colony in late June. On average, an estimated 60% of the seals died in each colony. (Data from Heide-Jørgensen and Härkönen 1992, photo courtesy of Tom Grey.)

■ 8.4 DISEASES CAN REDUCE POPULATIONS

There are few studies of plant or animal diseases in which the history of each individual in a closely monitored population is known. Most often the available data are estimates of seroprevalence—the percentage of individuals in a population that have antibodies to a particular pathogen. Seroprevalence measures how widespread a disease has been in a population in the recent past. Consequently, the impact of a disease on a population is usually not well understood. Three examples illustrate the range of problems ecologists face when they try to measure the population impacts of disease.

Brucellosis in Ungulates

Brucellosis is a highly contagious disease of ungulates (hoofed mammals) caused by the bacterium *Brucella abortus*. The disease is prevalent in cattle throughout the world and, because it manifests itself in pregnant females by causing abortion, it is commonly called 'contagious abortion'. The livestock industry has made a huge effort to eradicate brucel-

losis, but the disease persists in the western United States in wild populations of bison and elk, which can transmit it back to domestic cattle. Transmission occurs most readily through exposure to an aborted fetus or other birth materials. As Figure 8.7a shows, about 50–60% of adult bison in Yellowstone National Park have antibodies to *Brucella*, indicating that they must have had the disease and have become immune. There is considerable controversy over whether brucellosis is a native disease of bison or was introduced into North America by cattle (it is most likely that brucellosis was not present in bison before 1917 and was contracted from domestic cattle).

Q Could social organization in herding animals affect disease transmission rates?

Andy Dobson and Mary Meagher constructed a simple model of the spread of brucellosis through bison populations to determine whether the disease could be eliminated by a culling program. Brucellosis has a sharply defined threshold of transmission equal to about 200 bison, above which the proportion of

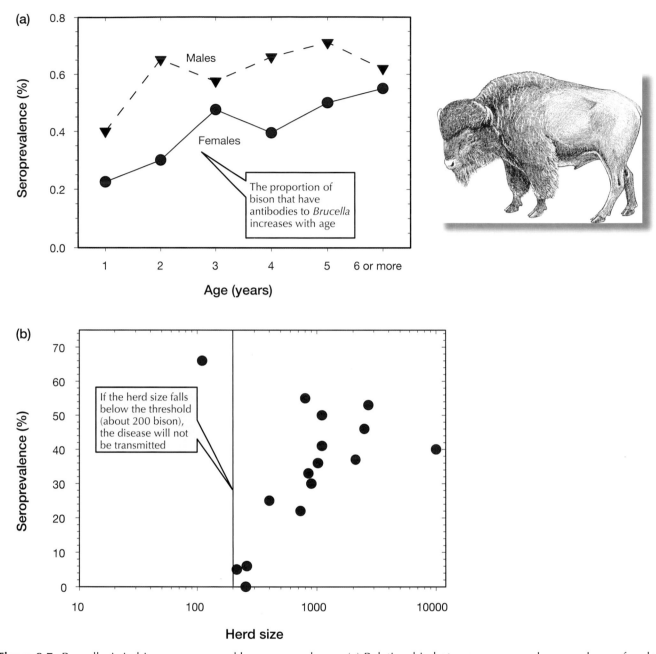

Figure 8.7. Brucellosis in bison as measured by seroprevalence. (a) Relationship between seroprevalence and age of male and female bison in Yellowstone National Park in the winter of 1990–1991. (Data from Pac and Frey 1991.) (b) Relationship between seroprevalence and size of bison herds in six national parks in Canada and the western United States. (After Dobson and Meagher 1996.)

infected individuals rises smoothly with population size (Figure 8.7b). There are now about 4,000 bison in Yellowstone National Park. While it would be possible to reduce the population to 200 through culling, doing so would put Yellowstone bison in danger of extinction and would undoubtedly be very unpopular.

Therefore, it is unlikely that culling will be a viable strategy for eliminating brucellosis in Yellowstone bison. An alternative strategy—vaccination—works well in cattle, but is not very effective in bison.

Brucellosis reduces the growth rate of bison populations, and herds with brucellosis are smaller than herds

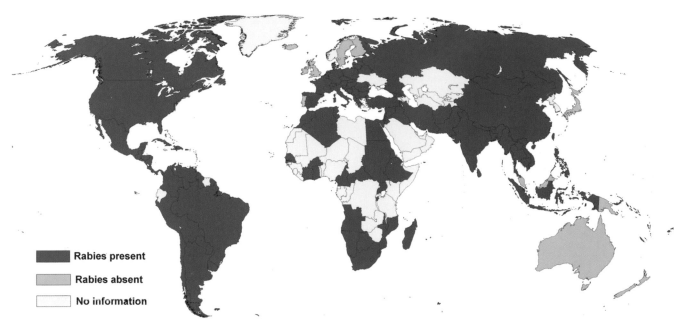

Figure 8.8. World distribution of rabies. There are only a few countries in which rabies is absent. The annual number of deaths worldwide caused by rabies is estimated to be 55,000, mostly in rural areas of Africa and Asia. An estimated 10 million people receive post-exposure treatments each year after being exposed to rabies-suspect animals. (Data from World Health Organization for 2001–2006.)

that are infected with the disease. The detrimental impact of brucellosis on elk and bison populations has been studied much less than the transmission of the disease between domestic cattle and wildlife in areas surrounding national parks, such as Yellowstone.

Rabies in Wild Mammals

Rabies is one of the oldest known diseases, and one of the most terrifying for humans. Democritus described rabies around 500 BC and, 200 years later, Aristotle wrote about rabies in his *Natural History of Animals.* Rabies is a directly transmitted disease of the central nervous system caused by a number of viruses of the genus *Lyssavirus* in the family Rhabdoviridae[1]. All mammals are susceptible to the disease, but carnivores are the viruses' essential hosts. The viruses are usually transmitted in saliva by the bite of an infected animal, although a few cases of aerosol transmission from bats in caves have been reported. Rabies is widespread in the world (Figure 8.8) and only a few coun-

tries are free of this disease. About 55,000 people contract rabies each year, mostly in India and the Far East. In many parts of the world, rabies infects humans via domestic dog bites, but vaccination of dogs has cut this link to humans in North America and Europe. The incubation period in humans is highly variable, ranging from less than 10 days to more than 6 years. There is no cure for rabies, which is almost always fatal once symptoms appear. Between 1994 and 2002, 28 people died in the United States from rabies, and 79% of the confirmed cases in humans were caused by viruses carried by bats.

Among wild animals, rabies is particularly common in foxes, wolves, coyotes, skunks, raccoons, bats and jackals. In Europe, the red fox is the main reservoir for rabies; in North America, the main reservoirs are raccoons, skunks, foxes and bats. The major vectors of rabies vary in different regions of the United States (Figure 8.9), and these vectors carry a diverse set of rabies viral genotypes. Foxes accounted for most reported cases of rabies in the United States before 1960, skunks from 1960 to 1990, and raccoons since 1990 (Figure 8.10). Little is known about the

---

1 Virologists have established a classification system for viruses that includes orders, families, genera, and species. This system is separate from the one used to classify organisms.

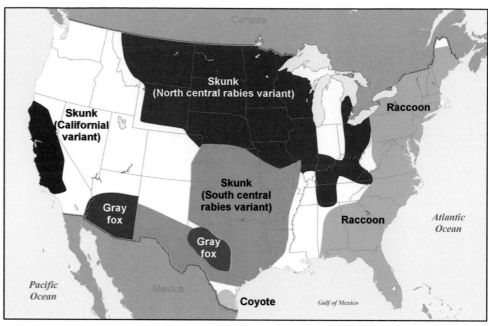

Figure 8.9. Major reservoirs of rabies in different regions of the United States. Other mammals serve as minor reservoirs of the disease in each region. The geographic ranges of these five species are much wider than the areas shown here. There are several variants of the rabies virus that are spread by specific mammals. (Modified from Krebs *et al.* 2005.)

quantitative incidence of rabies in bats or the impact of rabies on bat populations.

An epidemic of rabies in eastern North America began in the 1970s in Virginia and has been spreading for more than 30 years (Figure 8.11). This epidemic probably started when humans brought diseased raccoons into the area from the south-eastern states. Raccoon rabies crossed the border into Ontario, Canada,

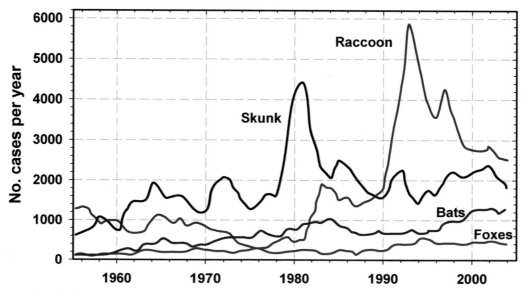

Figure 8.10. Number of rabies cases reported to the Centers for Disease Control in the United States from 1955 to 2004. The rise in the number of raccoon rabies cases since 1980 has resulted from an epidemic that spread through the eastern United States. (Data from Krebs *et al.* 2005.)

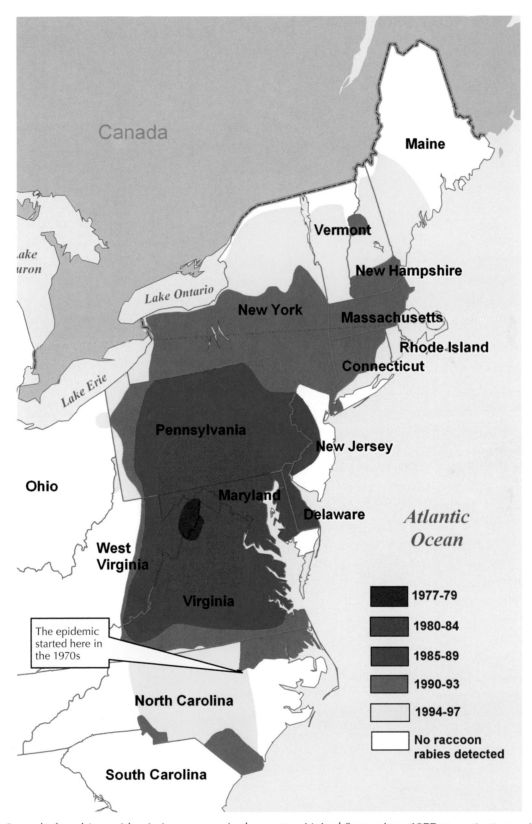

Figure 8.11. Spread of a rabies epidemic in raccoons in the eastern United States since 1977. (From the Centers for Disease Control, courtesy of John W. Krebs.)

in 1998 and has also spread south, meeting another epidemic coming north from Florida.

Because raccoons are so common, particularly around human habitations, rabies in raccoons has been particularly targeted by control agencies in the United States and Canada in recent years. Health officials are attempting to reduce the incidence of raccoon rabies by vaccinating wild raccoons, using a recombinant virus vaccine approved in April 1997. Some raccoons are trapped, injected with the vaccine, and released. However, most of the vaccine is added to bait consisting of small cubes of fish oil and wax polymer, which the raccoons eat. Millions of baits have been distributed annually in Massachusetts, New York, New Jersey, Vermont and Ontario, but it is not yet clear how effective this program has been.

- Could a disease like rabies ever die out naturally in wild populations?

A major epidemic of rabies came out of Russia in the 1930s and was first reported in Poland in 1939. It has moved westward at a rate of 20–60 km per year, reaching the northern coast of France in the late 1980s (Figure 8.12). More than 70% of the infected animals have been red foxes. One feature of this epidemic is that it appears to show a 4–5 year cycle, rather than being a constant disease problem.

Roy Anderson and his colleagues have developed a simple model of a rabies epidemic that encapsulates much of the ecology of this disease. We can use this model to address a critical management question: can we eliminate rabies from a fox population by culling or vaccination? Foxes have high rates of reproduction and dispersal, which makes culling an unsuccessful strategy for controlling the disease unless the foxes are in poor habitat or the rate of culling is extremely high.

In contrast, vaccination directly reduces the size of the susceptible pool and is much more effective in the control of rabies. The proportion of foxes that would have to be vaccinated to break the transmission cycle and eradicate the disease varies with the density of the fox population. At a density of 2 per km^2, about 50% of the population would need to be vaccinated. Extensive programs of vaccination of wild foxes with baits have

been carried out in Switzerland, Austria, Hungary, France, Belgium and Germany. These programs have been highly successful: by 1999 rabies was much reduced in Western Europe and Switzerland was rabies free.

At present we have no quantitative data on the impact of rabies on mammalian host populations. The amount of mortality rabies causes in natural populations of foxes, raccoons and skunks is believed to be high, and the resulting population declines are often commented on, but have not been measured rigorously. Arctic foxes and wolves in northern Canada have periodic outbreaks of rabies that reduce their populations to low numbers. All of these animals have high reproductive rates so, even if rabies kills a large fraction of their populations, they recover from epidemics fairly quickly.

Myxomatosis in the European Rabbit

The European rabbit (*Oryctolagus cuniculus*) was introduced into Australia in 1859, and within 20 years it had reached very high densities throughout the continent. Beginning in 1950, the Australian government attempted to reduce rabbit numbers by releasing the myxoma virus—a pox virus of the genus *Leporipoxvirus*. The European rabbit had no prior evolutionary exposure to this virus, whose original host is the South American jungle rabbit (*Sylvilagus brasiliensis*). The disease is transmitted passively by the bites of arthropod vectors, principally mosquitoes and fleas. The virus does not replicate in the vectors.

Frank Fenner (1914–) Professor of Microbiology, Australian National University, Canberra. Dr Fenner was instrumental in studying the evolution of the myxoma virus and the European rabbit in Australia.

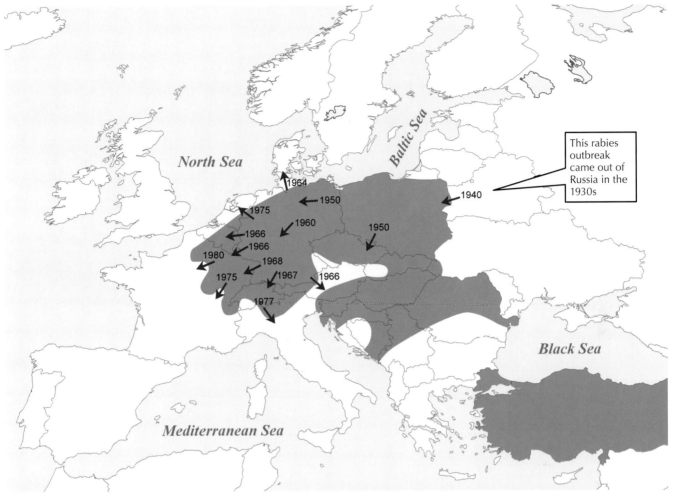

Figure 8.12. Spread of a rabies epidemic in Europe. There has been no westward movement of the epidemic since 1983, probably due to oral vaccination of foxes with baits. (After Macdonald and Voigt 1985.)

The myxoma virus causes myxomatosis, a disease that rarely kills South American jungle rabbits, but is highly lethal to European rabbits. After its introduction into Australia, it killed more than 99% of the rabbits it infected. Figure 8.13a shows the precipitous crash in rabbit numbers that followed the release of myxomatosis in one area in south-eastern Australia in 1951. Myxomatosis was also introduced to France in 1952, from where it spread throughout Western Europe. In Great Britain, 99% of the entire rabbit population was killed in the first epidemics from 1953 to 1955. This type of extreme mortality is common when pathogens infect new host species, and introducing the pathogen can be a useful strategy if the host species are pests.

Both the myxoma virus and the European rabbit have been evolving since the virus was introduced into Australia and Europe. More virulent strains of the virus have been replaced by less virulent strains, which kill fewer rabbits and take longer to cause death. Because the host remains alive longer while infected with less virulent strains, the virus has a higher probability of being spread by vectors than with more virulent strains. At the same time, rabbits have become more resistant to the more virulent strains of the virus (Figure 8.13b).

What impact do these changes in the virus and the rabbits have on the population dynamics of the rabbits? Because myxomatosis causes less mortality now than it did immediately after the virus was

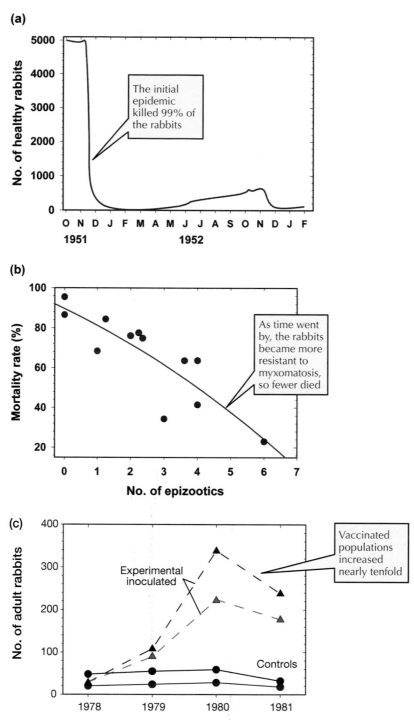

Figure 8.13. Effects of myxomatosis and vaccination on European rabbits. (a) Crash of the rabbit population at Lake Urana, New South Wales, after the myxoma virus was introduced in 1951. Numbers of healthy rabbits were counted on standardized transects. (After Myers *et al.* 1954.) (b) Decline in mortality rates of wild rabbits near Lake Urana as a function of time since the myxoma virus was introduced. Mortality was measured after infection with a virulent strain of the virus. (After Fenner and Myers 1978.) (c) Effect of vaccination on the numbers of adult rabbits in four fenced areas in south-eastern Australia. Rabbits in two areas (triangles) were vaccinated with an attenuated strain of the myxoma virus that produced immunity to virulent strains. Rabbits in the other two areas (dots) were inoculated with a virulent strain. (Data from Parer *et al.* 1985.)

introduced, you might hypothesize that the disease has little impact on rabbit numbers at present. One way to test this hypothesis would be to compare rabbit populations with and without myxomatosis, but it is impossible to find a field population of rabbits that does not already have the disease. Two groups of researchers have tried an alternative approach: reducing the impact of myxomatosis by making rabbits immune or by eliminating transmission by vectors. In Australia, Ian Parer and his colleagues compared four fenced populations of rabbits, two inoculated with an attenuated strain of the virus (to produce immunity with little mortality) and two inoculated with a virulent strain. They found that the immune populations increased approximately ten-fold over the populations that had been inoculated with the virulent strain (Figure 8.13c). A similar experiment in England reduced the numbers of rabbit fleas (the main vector) using insecticides: this produced a two- to three-fold increase in rabbit numbers. These results show clearly that myxomatosis still limits rabbit populations despite the virus's reduced virulence in field populations. Myxomatosis is a good example of the strong impact that an introduced disease can have on a wild population.

■ 8.5 PATHOGENS CAN BECOME MORE OR LESS VIRULENT THROUGH EVOLUTION, AND THEIR HOSTS CAN EVOLVE RESISTANCE

How do pathogens and their hosts evolve? This simple question has become critical now that we know that some pathogens, such as the viruses that cause AIDS and bird flu, appear to have moved into the human population from other animals. One idea is that pathogen–host systems become stable as they evolve because natural selection changes the characteristics of both pathogens and hosts in a direction that produces population stability. In particular, the conventional wisdom about these systems is that virulence is selected against, so that pathogens and parasites become more benign and thus persist. But does natural selection really work that way? What can we say about the evolution of virulence?

Evolution of Virulence

One way to study the evolution of pathogen–host systems is to perform a serial passage experiment, in which a pathogen is transferred from one host to another in a chain. Host properties are held constant so that evolutionary changes in the pathogen can be monitored. Because the pathogen is propagated under defined laboratory conditions, the biological attributes of pathogen individuals at the end of the experiment can be compared with those of individuals that started the experiment. Serial passage experiments were developed for vaccine studies, but can be used very effectively for investigating the evolution of virulence. A general result of such experiments with many viral, bacterial, fungal and protozoan pathogens is that the pathogens become more virulent with each generation in their native host species. Figure 8.14 illustrates this result for the bacterium *Salmonella typhimurium*.

The increase in pathogen virulence observed in serial passage experiments does not occur in most natural disease systems, presumably because natural host populations are genetically more variable. According to the **Red Queen hypothesis**, genetic variation is beneficial because it hinders pathogen adaptation (see *Essay 8.1*). The Red Queen hypothesis states that any selective advantage that one species might obtain through evolutionary change may be offset by evolutionary changes in other species in the community. If hosts are genetically variable, pathogens will be on average less virulent than if hosts are genetically uniform (as is the case for some crops).

Coevolution in Disease Systems

The evolution of pathogens and their hosts is a classic example of **coevolution**, or reciprocal evolutionary change. For example, the evolution of resistance to the myxoma virus in European rabbits is easily explained by selection operating at the individual level: rabbits that are more resistant to the virus leave more offspring. It is more difficult to explain the evolution of reduced virulence in the virus, however. Virulence in a virus is related to fitness because more virulent viruses make more copies of themselves. But if more virulent viruses kill rabbits more quickly, there will be less time available for transmission of the virus by mosquitoes

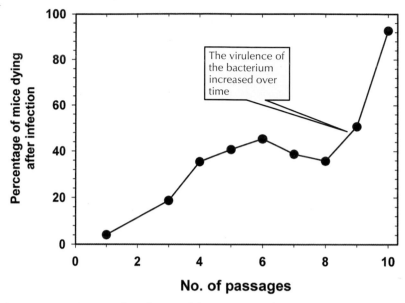

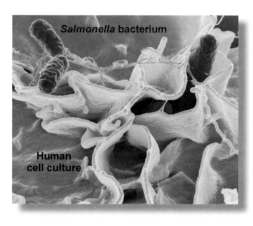

Figure 8.14. Change of virulence of the mouse typhoid bacterium *Salmonella typhimurium* after ten serial passages in laboratory mice. *Salmonella typhimurium* (red, rod shaped bacterium) is a common cause of food poisoning in humans. It uses its whip-like flagella to sense a cell and then attach to the cell wall. (Data from Ebert 1998, photomicrograph courtesy of National Institutes of Health.)

or fleas. Thus, there is a trade-off between virulence and transmissibility in the myxoma virus, and natural selection favors a moderate level of virulence, which maximizes the reproductive rate of the virus.

We do not know if the myxoma virus–rabbit system has reached a stable equilibrium or if continuing evolution will allow the rabbit population to slowly recover to its former levels. There is some evidence that the interaction between the virus and the rabbit is changing in Britain, and the size of the rabbit population seems to be slowly increasing there. Evolutionary changes in the myxoma virus–rabbit system in Australia have been complicated by the introduction in 1997 of another viral disease, rabbit hemorrhagic disease, which has further reduced the rabbit's average population density. There is still much to be learned about how pathogens and their hosts interact and evolve.

SUMMARY

Pathogens and parasites are involved in negative interactions with host organisms in which the host loses and the pathogen or parasite gains. Much of our understanding of the dynamics of pathogen-caused diseases has come from studies of human diseases, such as influenza and measles.

Compartment models can be used to visualize the interactions between pathogens or parasites and hosts. The host population is usually broken down into susceptible, infected and recovered individuals. These simple models are characterized by a few rates that define the outcome of the interactions. Pathogen–host systems have a threshold of transmission below which the disease will be eliminated from the population. The objective of much of the study of applied disease ecology is to determine how best to move the host population below this threshold. In general, vaccination has been more effective than culling in meeting this objective for wild animals.

Pathogens and parasites can affect the reproductive rate or the mortality rate of their hosts. Many studies show impacts on mortality, but few measure how large these impacts are in nature or indicate whether the average density of the host species has been reduced. Pathogens and parasites often have debilitating effects on their hosts and make them more susceptible to predators, bad weather or food shortage.

ESSAY 8.1   WHAT IS THE RED QUEEN HYPOTHESIS?

In Lewis Carroll's *Alice in Wonderland*, there is a scene in which Alice and the Red Queen have to run as fast as they can to get nowhere because the world is running by at the same speed. Lee Van Valen used this metaphor in 1973 to illuminate biological evolution. Any selective advantage that one species might obtain through evolutionary change may be offset by changes in other species in the community. For example, if a prey species evolves the ability to run faster, it will gain no advantage in surviving predation if the predators also evolve this ability. Therefore, pathogen–host systems, plant–herbivore systems and predator–prey systems may continually undergo evolutionary change without affecting the overall balance in the system. Increasing fitness in one species is always balanced by increasing fitness in all other species. The characters run and run and run, but get nowhere. This idea is called the Red Queen hypothesis.

"A slow sort of country !" said the Queen.
"Now, here, you see, it takes all the running
you can do, to keep in the same place.
If you want to get somewhere else, you
must run at twice as fast as that."

Evolution can be much faster in pathogens and parasites whose generation times are short relative to those of their hosts. If the success of a pathogen or parasite is determined by its ability to infect a host, the main selection pressures will come from the most common host genotypes. A host species whose genotype changes over time can present a moving target that the pathogen or parasite cannot catch. This is one possible reason for the evolution of sexual reproduction, in which recombination in each generation presents a new array of host genotypes to the coevolving array of pathogens and parasites. The Red Queen hypothesis thus predicts continually changing dynamics in the evolution of pathogens, parasites and hosts—not a stable equilibrium in which there is one winner and one loser.

The evolution of virulence has progressed from the idea that well-adapted pathogens and parasites are benign to a more dynamic view. Virulence will increase through evolution if that enables a pathogen to increase its fitness by producing more copies of itself. One of the main factors limiting disease virulence is host genetic variability; genetically uniform populations are particularly susceptible to virulent disease outbreaks. Much remains to be done to link epidemiology and ecology in disease studies.

SUGGESTED FURTHER READINGS

Diamond, JM (1999). *Guns, Germs, and Steel: The Fates of Human Societies.* W. W. Norton & Company, New York. 480 pp. (A best-selling book that discusses the role of diseases in European colonization of the world.)

Ebert D and Bull JJ (2003). Challenging the trade-off model for the evolution of virulence: Is virulence management feasible? *Trends in Microbiology* **11**, 15–20. (A good discussion of whether humans can manipulate the evolution of diseases toward harmless ends.)

Ewald PW (1995). The evolution of virulence: a unifying link between parasitology and ecology. *Journal of Parasitology* **81**, 659–669. (An evaluation of the idea that diseases become benign in their hosts through evolution.)

Grenfell BT and Dobson AP (Eds) (1995). *Ecology of Infectious Diseases in Natural Populations.* Cambridge University Press, Cambridge, UK. (A good overview of the effects of disease on populations.)

Harvell CD, Mitchell CE, Ward JR, Altizer S, Dobson AP, Ostfeld RS and Samuel MD (2002). Climate warming and disease risks for terrestrial and marine biota. *Science* **296**, 2158–2162. (A discussion of how global warming will affect disease transmission.)

Weiss RA (2002). Virulence and pathogenesis. *Trends in Microbiology* **10**, 314–317. (An evaluation of the critical question of why viruses cause disease.)

Wills C (1996). *Yellow Fever, Black Goddess: The Coevolution of People and Plagues.* Addison-Wesley Publishers, Reading, Massachusetts. (A timely discussion of how and why some human diseases rise and fall.)

QUESTIONS

1. Anthrax is a disease caused by the bacterium *Bacillus anthracis*, which is lethal to most mammalian herbivores. Within a few months in 1983–1984, an anthrax epidemic wiped out 90% of the impala population of Lake Manyara National Park in Tanzania. Discuss the biological mechanisms that would permit an epidemic of this type to appear suddenly in a population and then disappear for decades. Prins and Weyerhaeuser (1987) discuss this particular epidemic.

2. Rabies is a disease with interesting spatial spread patterns (Figures 8.11 and 8.12). Foxes defend discrete, non-overlapping territories. How might territorial behavior affect the spatial dynamics of rabies spread?

3. The myxoma virus has been introduced into European rabbit populations on islands to eradicate the rabbits for conservation purposes. What factors might affect the success of this eradication program? Flux (1993) considers this issue.

4. Male bison show a higher prevalence of antibodies to brucellosis than do female bison (Figure 8.7a). Suggest two conditions under which this situation might be produced.

5. McNeill (1976), Wills (1996) and Diamond (1999) discuss the idea that the introduction of diseases by Europeans made possible their conquest of aboriginal peoples in the Americas, Africa and Australia. Review the arguments supporting this idea for an area of the world that interests you and discuss their validity.

Chapter 9

POSITIVE INTERACTIONS BETWEEN SPECIES—MUTUALISM AND COMMENSALISM

IN THE NEWS

Fungi are never in the news and yet they are among the most important organisms on earth, so they should be. Plants must take up nutrients from the soil to grow and almost all plants have fungi, called mycorrhizae, growing on or in their roots. These fungi help the plant by taking up inorganic nutrients, such as phosphorus, from the soil and donating them to the plant in exchange for carbohydrates, such as sugars, that the fungi obtain from the plant roots. These kinds of win–win interactions are called mutualisms because they benefit both of the species involved.

Ecologists and agricultural scientists discovered the importance of mycorrhizae by observing what happens to plants that do not have mycorrhizae. In Oregon, a Douglas fir tree nursery was started in the Willamette Valley in 1961 in old agricultural fields. Because the foresters were concerned about root diseases and reducing weeds, they fumigated the sandy soil before sowing the first crop and

Douglas fir tree seedlings in an Oregon nursery. Only some patches of trees picked up mycorrhizal fungi that colonized the seedling roots and improved nutrient uptake. (Photo courtesy of Jim Trappe.)

killed all the soil organisms—good as well as bad. The photo on page 171 shows the Douglas fir seedlings in their third growing season. Most seedlings are stunted, off color and deficient in nutrients, especially phosphorus. However, some tree seedlings became inoculated with mycorrhizal fungi—presumably via air-borne spores—and started growing normally. Once the mycorrhizae were established, the fungi grew outwards to colonize adjacent seedlings, which then also began to grow normally. The fungi spread during the growing season, resulting in patches of better growth, with the largest seedlings in the middle. When the stunting syndrome appeared, the soil in the nursery was heavily fertilized, but the seedlings did not respond as they could not pick up the soil phosphorus without the help of the mycorrhizal fungi. Without phosphorus, the plants could not extend their root systems to acquire other nutrients.

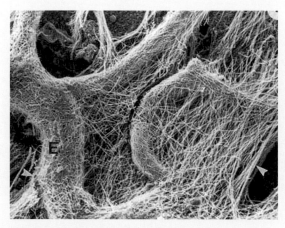

*A scanning-electron-microscope photo of the roots of Australian jarrah trees (*Eucalyptus marginata) *with a fungal mycelium attached*

Many types of fungi act as mutualists to help higher plants acquire soil nutrients. They are able to do this by forming a mantle around roots. The scanning-electron-microscope photo at left shows the roots of Australian jarrah trees (*Eucalyptus marginata*) with an abundant web-like fungal mycelium attached.

The mycorrhizal association is a true mutualism because both the plant and the fungus benefit. In infertile soils, nutrients taken up by the mycorrhizal fungi lead to improved plant growth. Plants with mycorrhizae are more competitive in infertile soils and better able to tolerate environmental stresses than plants without mycorrhizae.

There are many different species of mycorrhizal fungi, and a species of plant may be colonized by a variety of different fungal species. Ecologists are just now beginning to appreciate how this complex community of soil organisms interacts with trees, shrubs, grasses and herbs. Mycorrhizal fungi are essential for modern agriculture and forestry, but go largely unnoticed in the news.

■ 9.1 POSITIVE INTERACTIONS BETWEEN SPECIES CAN BENEFIT ONE OR BOTH OF THE SPECIES INVOLVED

Positive interactions (also called **facilitation)** are defined as interactions between two species that benefit at least one of the species and do not harm the other. Positive interactions between species have never been as prominent in the ecological literature as negative interactions, such as predation, competition, and disease. This might be because most interactions between species are negative ones or because less research has been done on positive interactions. The balance is slowly changing and much interest now focuses on species interactions that are not harmful to the species involved. When both species derive benefits from the association, it is called a **mutualism**[1]—this is a win–win interaction, rather than the win–lose interactions we discussed in the last chapter. In some cases positive interactions between two species are win–neutral, so that one species gains while there is no harm done to the other. These associations are called **commensalisms**.

Many **mutualisms** have been known for hundreds of years. Bees pollinate flowers and gain by obtaining pollen as food, while the plants gain by gene flow (through movement of pollen) and seed fertilization. Nitrogen-fixing bacteria and mycorrhizal fungi inhabit the roots of plants and gain protection and carbohydrates from the plant while supplying nitrogen or other soil nutrients to the plant in exchange. But we should always remember that there are costs to mutualisms as well as benefits, and we need to determine how the benefits exceed the costs for positive interactions. Ecologists first wish to describe these mutualisms and then ask how they might affect the distribution and abundance of species in nature (*Essay 9.1*).

■ 9.2 MUTUALISTIC INTERACTIONS OCCUR WHEN ANIMALS POLLINATE AND DEFEND PLANTS

While it is tempting to think of mutualisms in their simplest form as a two-species interaction—for example,

---

1 The term **symbiosis** is sometimes used as equivalent to mutualism.

ESSAY 9.1   WHY ARE CORALS BLEACHING?

Coral reef bleaching has increased dramatically in many tropical areas around the globe in the last 20 years. Corals are animals that contain symbiotic algae within their cells, and this symbiotic relationship is one of the most important mutualisms in the biosphere. The symbiotic algae provide color to the corals and undertake photosynthesis, thus contributing to coral growth—a positive relationship. When corals bleach they lose their symbiotic algae, and therefore their color, and often die. Widespread bleaching can cause the death of whole coral reefs. The primary cause of coral bleaching is thought to be elevated sea surface temperatures. Many reef-building corals live very close to their upper lethal temperatures, and small increases of 0.5 to 1.5°C over a few weeks, or larger increases of 3–4°C over several days, can kill corals (Huppert and Stone 1998).

What might cause increased sea surface temperatures? In the Pacific Ocean, the 3–7 year cycle of El Niño's warm and cool phases is a major suspect. In the Pacific, El Niño is characterized by broad-scale warming of ocean water along the equator. The nearly cyclic nature of El Niño events suggests that if we can predict these temperature changes, we should be able to predict episodes of coral bleaching. Part of the difficulty in accepting the simple temperature model of coral reef bleaching has been the observation that not all corals bleach in a given area, and areas outside the Pacific, such as the Caribbean, are also affected. One possible explanation is that these large-scale oceanographic events have worldwide climatic repercussions, and are thus not confined to the traditional El Niño regions of the Pacific and Indian Oceans.

Bleaching in corals on Great Keppel Island during the 2002 mass bleaching event. This was the worst coral bleaching event in the history of the Great Barrier Reef Marine Park in Australia. (*Photo courtesy of Dr Ove Hoegh-Guldberg, University of Queensland.*)

If the temperature explanation for bleaching is correct, why should bleaching have increased dramatically in the last 20 years? Three factors may be involved. Firstly, global warming is a likely candidate. Secondly, degradation of coral reefs from pollution may have reduced the general resilience of reefs to bleaching damage. Finally, increased ultraviolet radiation could be combining with increased temperatures to induce bleaching. El Niño years are often associated with clear skies and calm seas, which leads to high penetration of UV radiation into the sea water.

The ecological consequences of the collapse of coral reefs with their rich biodiversity are large, and this problem deserves global attention (Knowlton 2001, Sheppard 2003).

between a particular pollinator and a particular plant species—in natural ecosystems specialized, two-species partnerships are rare. We must remember to think instead of multi-species systems in which, for example, many insects pollinate a particular plant, and a single pollinator may use pollen from several different plant species. For simplicity, most natural history studies concentrate on two-species interactions, from which we can gradually build a more complex picture. Let us consider two examples of mutualisms.

Bees and Coffee

Pollination of plants by animals is one of the oldest described mutualisms. But if a plant species is self-pollinating, botanists have tended to assume that the plant does not need pollinators. The African shrub *Coffea arabica* has become one of the most important tropical plants for commercial agriculture—the source of coffee beans. Because the coffee shrub is self-pollinating, agriculturalists have assumed that it could gain nothing from the presence or absence of insect pollinators. In Central America coffee flowers are visited by a variety of native insects, but also by introduced African bees. These pollinators live in patches of forest that surround the coffee plantations. Is there any evidence that coffee yields are affected by insect pollinators?

African honeybees colonized Panama in 1985 and became major pollinators of coffee that was grown near forests. Ripe coffee berries on open branches were both heavier and more abundant than berries on branches that were covered with fine cloth mesh to keep pollinators out. In areas where African honey bees were the exclusive pollinators, berries were 25%

heavier than controls with no insect pollination. If these small-scale results are correct, pollinators may control coffee harvests on a large scale. Is there any further evidence for this?

Figure 9.1 shows the trends in coffee yields from African sites where intensive cultivation has eliminated almost all native forest in the coffee growing regions. The 20–50% decline in coffee bean yields from these countries suggests loss of pollinators as a possible cause.

Coffee bush (*Coffea arabica*)

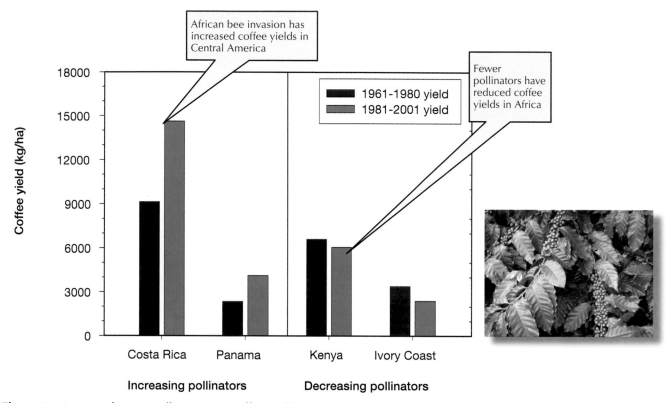

Figure 9.1 Impact of insect pollination on coffee yields in Central America (Costa Rica and Panama), where African honey bees have become important pollinators since 1985, and in Kenya and the Ivory Coast, where land clearing has reduced forest habitats where pollinators live. Coffee is a self-pollinating plant, and no one had expected that insect pollinators could affect yields. (Data from Roubik 2002, photo courtesy of Alice Kenney.)

By contrast, at the same time in Latin America coffee yields have increased 25–30% in the last 20 years because African bees have become major pollinators in these countries. This increase in yield seems to arise from more outcrossing (or hybrid vigor)—African bees bring pollen from other trees to coffee flowers. The implication for coffee growers is that it is important to preserve habitat for pollinators, and that crops such as coffee that self-pollinate can benefit from insect pollination. We do not know how many other self-pollinated plants would set more seed or fruit if more pollinators were present.

Q Must all plant–pollinator interactions be mutualistic?

Ants and Acacias

A mutualistic system of defense has been achieved by the swollen-thorn acacias and their ant inhabitants in the New World tropics. The ants depend on the acacia tree for food and a place to live, and the acacia depends on the ants for protection from herbivores and neighboring plants. Not all New World acacias (*Acacia* spp.), of which there are approximately 700 species, depend on the ants, and not all the acacia ants (*Pseudomyrmex* spp.), of which there are over 150 species, depend completely on acacias. In a few cases, a high degree of mutualism has developed, which was first described in detail by Daniel Janzen in 1966. Daniel Janzen is a conservation biologist who works in Central America and is one of the world's leading tropical ecologists. Janzen found that some of the species of ants that inhabit acacia thorns are obligate acacia ants—which means that they live nowhere else.

Swollen-thorn acacias have large, hollow thorns in which the ants live (Figure 9.2). The ants feed on modified leaflet tips called Beltian bodies, which are the primary source of protein and oil for the ants, and also on enlarged extrafloral nectaries, which supply sugars.

Daniel H. Janzen (1939–) Professor of Biology, University of Pennsylvania

Swollen-thorn acacias maintain year-round leaf production, even in the dry season, providing food for the ants. The acacia ants continually patrol the leaves and branches of the acacia tree and immediately attack any herbivore that attempts to eat acacia leaves or bark. The ants also bite and sting any foreign vegetation that touches an acacia, and they clear all the vegetation from the ground beneath their tree. Thus the swollen-thorn acacia often grows in a cylinder of space virtually free of all foreign vegetation (Figure 9.2). If all the ants are removed from swollen-thorn acacias, the trees are quickly destroyed by herbivores and crowded out by other plants. Janzen showed experimentally that acacias without ants grew more slowly and were often killed:

	Acacias with ants removed	Acacias with ants present
Survival rate over 10 months (%)	43	72
Growth increment		
May 25–June 16 (cm)	6.2	31.0
June 16–August 3 (cm)	10.2	72.9

Swollen-thorn acacias have apparently lost (or never had) the chemical defenses against herbivores found in other trees in the tropics.

- **Would you expect every species of acacia tree to evolve a similar mutualism with ants?**

Figure 9.2 Mutualism of ants and acacias. (a) *Acacia collinsii* growing in open pasture in Guatemala. This tree had a colony of about 15,000 worker ants and was about 4 m tall. (b) Swollen thorn of *Acacia collinsii* on a lateral branch. Each thorn is occupied by 20–40 immature ants and 10–15 worker ants. All the thorns on the tree are occupied by one colony. One worker ant is visible on the thorn. (c) Clearing maintained by the ants around the base of *Acacia collinsii*. (Photos courtesy of: (a) Dr Alex Wild, (b) and (c) Dr Dan Perlman, Brandeis University, Ecolibrary.)

Figure 9.3 Coral reefs are a classic illustration of foundation species in shallow tropical waters. This photo shows a red-throat emperor (*Lethrinus miniatus*) at Heron Island, on the southern Great Barrier Reef of Australia. Coral reefs provide the habitat for many invertebrates and fishes, and they ameliorate wave action and currents. The damage and loss of corals from coral bleaching (*Essay 9.1*) results in a collapse in the biodiversity of the entire community. (Photo courtesy of Ove Hoegh-Guldberg, Centre for Marine Studies, University of Queensland.)

Thus the ant–acacia system is a model system of the coevolution of two species in an association of mutual benefit. The acacias benefit from harbouring the ants, which reduce herbivore browsing and greatly reduce competition from adjacent plants—the costs to the plants of provisioning the ants are greatly exceeded by the benefits. The ants benefit because the acacias provide food and shelter.

■ 9.3 FOUNDATION SPECIES PROVIDE SHELTER FOR OTHER SPECIES

Shelter is a universal need for all organisms—for some species, shelter is provided by other organisms. **Foun-**

dation species are dominant species that form the habitat that shelters many other species. Dominant trees are the most common trees in a forest, which provide shade for other plants, and reduce temperature, wind and moisture stress for the understory. In a similar manner, coral reefs (Figure 9.3) provide shelter from waves and currents for a variety of fishes and invertebrates. Mangroves in tropical coastal zones serve as nursery habitat for many juvenile tropical fishes. Forest trees, corals and mangroves are all habitat-forming species, which are so common that we tend to overlook the shelter that they create for other plants and animals.

Salt marshes provide many examples of positive interactions between species. Salt marshes are stressful environments for plants to live in because of high

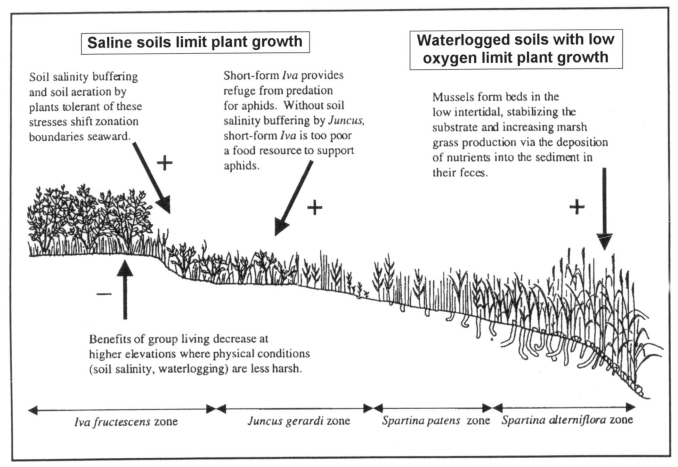

Figure 9.4 Positive interactions between plants and animals in a New England salt marsh community. The sea is to the right; the land to the left. *Spartina alterniflora* is an intertidal cord grass with a high salt tolerance that can tolerate waterlogged soils and occasional submergence by high tides, *Juncus* is a sedge called black rush that dominates in the mid-marsh area, and *Iva fructescens* (marsh elder) is a bush that lives at the upper edge of the marsh and has a relatively low salt tolerance. (Modified from Bertness and Leonard 1997.)

soil salinity levels and tidal inundation. On the east coast of North America the cord grass *Spartina alterniflora* is an important plant that stabilizes the outer part of the marsh, preventing soil erosion (Figure 9.4). Mussels in the lower part of the intertidal zone facilitate *Spartina* colonization and growth by feeding on plankton and detritus and depositing nutrients in their feces. Fiddler crabs burrow in the lower part of the marsh, and this helps to aerate the waterlogged soil, which often has little oxygen. In the upper part of the marsh, the rush *Juncus gerardi* facilitates the colonization of the shrub marsh elder (*Iva fructescens*) by reducing soil salinity. If *Juncus* is removed from this zone, soil salinity rises and *Iva* dies because it cannot tolerate high salinity soils. This means that much of the typical

structure in salt marshes is the result of positive interactions between species.

Balancing Positive and Negative Interactions

Both competition and facilitation act in plant and animal communities and the exact balance of these positive and negative interactions will depend on the harshness of the environment (Figure 9.5).

In more severe environments, such as deserts, we should expect to find more positive interactions but this is not always the case. Desert plants illustrate both positive and negative interactions well because of the overwhelming importance of water as a limiting factor in these ecosystems. In an analysis of the interactions between desert shrubs and annual plants

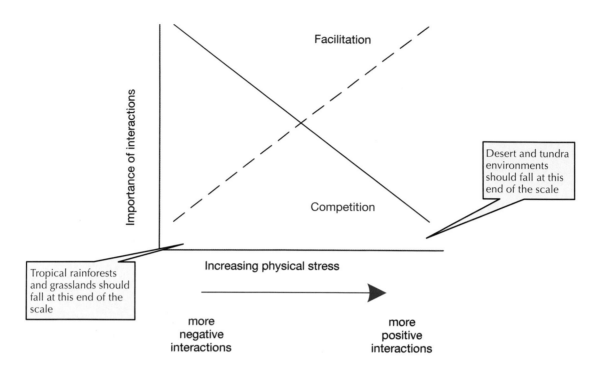

Figure 9.5 The intensity of the positive interactions of facilitation and the negative interactions of competition in relation to the harshness of the environment. As physical stress increases in desert environments (water limitation) or in polar environments (temperature limitation), facilitation should be more common. In regions with lower amounts of physical stress, such as tropical rainforests and temperate grasslands, competition should be more common. (Modified from Bertness and Leonard 1997).

in relation to water use in the Mojave Desert, plant ecologists experimentally measured the net effect of positive and negative interactions by comparing the performance of two treatments in relation to control (unmanipulated) plots:

1. Annual plants were removed completely from plots by clipping.
2. Shrubs were completely removed from a plot by clipping.

Figure 9.6 shows the experimental design, which allowed ecologists to calculate the net effect of the inter-actions between annuals and desert shrubs. Figure 9.7 gives the results of these experimental manipulations. Annuals benefit from the shelter provided by desert shrubs, but the shrubs are negatively affected by the competition for water when annuals are present. In most plant communities both positive and negative interactions will occur and experiments like this are a precise way of measuring the overall impacts. The

important principle illustrated by this study of desert plants is that interactions between species are rarely entirely negative or positive, but a mixture of costs and benefits. In this example, there are no positive benefits for the shrubs, but there are for the annual plants, while there are costs for both plant groups due to competition for the limited water resource.

Nurse Plants

In desert environments the establishment of many species of cacti depend on the presence of nurse plants that provide shade and thus improve water balance for the small cacti. Seedlings of the giant saguaro *Carnegiea gigantea* in the Sonoran Desert in Arizona were found only in the shade of nurse plants. Nurse plants provide shade and thus reduce temperature and water loss for the small seedlings. It is also possible that soil nutrient levels are higher under nurse plants than in open areas. The net result is that the cactus seedlings benefit from this association, but whether the nurse

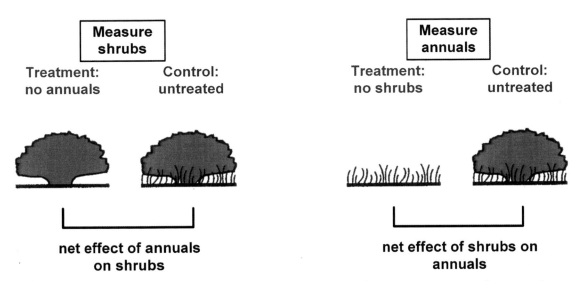

Figure 9.6 Experimental design for measuring the net effect of positive and negative interactions between desert shrubs and annual grasses and herbs in the Mojave Desert. The shrub is burrowed (*Ambrosia dumosa*). Two treatments are illustrated. By removing all grasses and herbs, the effect of annuals on shrubs can be estimated (left side of diagram). By removing shrubs, the effect of shrubs on annual plants can be estimated (right side of diagram). The net effect combines the positive and negative effects to determine which is stronger in affecting plant growth. The net effects that were measured are plotted in Figure 9.7. (Modified from Holzapfel and Mahall 1999.)

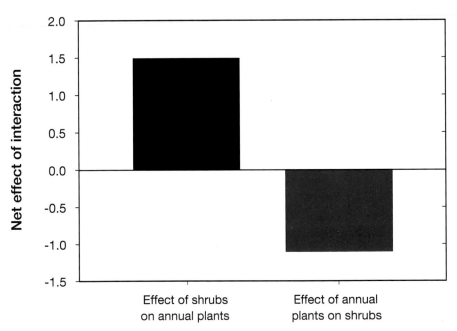

Figure 9.7 Positive effects of the desert shrub burrowed (*Ambrosia dumosa*) on annual plants in the Mojave Desert of California, and negative effects of annual plants on the shrubs. Burrowed provides shade that reduces temperature and improves water balance in annual plants, so their growth performance is enhanced. Annuals, however, compete for scarce water and nutrients with the burrowed, to the detriment of the shrubs (competition). Many interactions between species may be asymmetrical like this one. (Data from Holzapfel and Mahall 1999.)

Giant cactus (*Carnegiea gigantea*) near Tucson, Arizona with Chien Hsun Lai. (Photo by C. Krebs.)

plant is negatively affected probably depends on the size of the seedling cacti and their survival rates. As the cacti seedlings grow, they often compete for water with, and eventually replace, the nurse plant, which later dies to be replaced by the adult cactus plant. Consequently the interaction between the nurse plants and the cacti has a time and space dimension that begins as a commensal (+/0) interaction and moves into a +/− interaction as the cactus grows larger and the nurse plant dies. The key point is that the nature of interactions between plants can change with the size and age of the individuals involved, so that interactions cannot always be described simply as mutualism or commensalisms.

■ 9.4 PLANT–ANIMAL INTERACTIONS CAN BE A COST OR A BENEFIT TO PLANTS

Our analysis of plant–plant and plant–animal interactions has shown that we must carefully measure the

costs and benefits involved in these associations. Interactions that seem to be mutualistic or commensal may become competitive, as we have seen with the nurse plant–cactus interaction. This understanding has caused ecologists to re-examine some of the most obvious interactions that have always been assumed to be exploitation, such as grazing, to see if they might involve some positive interactions.

Overcompensation Hypothesis for Grazing

Herbivores eat parts of plants and, at first sight, this would appear to be detrimental to the individual plant. But, could grazing or browsing in fact be beneficial to a plant so that grazing can be considered as a mutualism in which both the plants and the herbivores profit? On a more practical level of public policy, should public grazing land be protected from sheep and cattle grazing, or should we encourage cattle and sheep production? If cattle and sheep grazing is good for plants, we would have a clear ecological reason to support current land management policies in the western United States. What is the ecological evidence that grazing might be mutualistic rather than a harmful negative interaction?

The idea that properly managed grazing is good for individual grasses, good for cattle, good for plant communities (to slow down succession to shrubs) and good for ecosystems (to speed up decomposition) has been promoted by some range managers and ecologists. This idea postulates that grazers and grasses are in a mutualistic relationship in which both gain. In order to test this interesting idea, we must be specific about the details of the interaction. It is clear to everyone that excessive grazing is detrimental to plants. The important question is whether or not some moderate level of grazing will stimulate plants to produce more biomass—a proposal called the **overcompensation hypothesis** (Figure 9.8).

One way to test the overcompensation hypothesis is to subject plants to controlled grazing in an experimental garden. In one experiment plant ecologists grew eight different populations of sagebrush (*Artemesia tridentata*) in a common garden in Utah for two years, and subjected plants to winter browsing at two intensities (Figure 9.9). Moderately-browsed plants had one-third of their shoots cut off in winter to simulate browsing by elk and deer. Severely-browsed plants

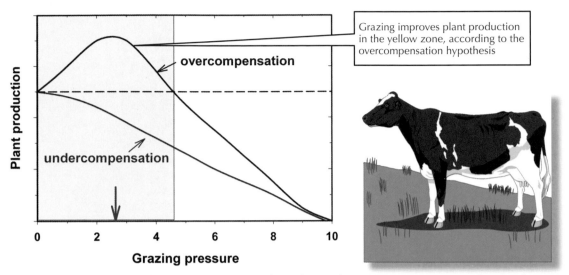

Figure 9.8 Schematic illustration of the overcompensation hypothesis of grazing. Grazing is postulated to be more and more favorable for plant production up to some optimum point of grazing pressure (indicated by the green arrow). This hypothesis predicts that plant production will be less in the absence of grazing, and that moderate grazing (the yellow zone) will increase plant growth and plant fitness. The classical view of grazing is shown by the red line, in which plants cannot completely replace grazed tissues, so that grazing is always a negative interaction. The horizontal black dashed line indicates exact compensation of lost tissues by the plant. (Modified after Belsky 1986.)

had three-quarters of shoots clipped off in winter. Total productivity was measured by the change in above-ground biomass of the plants after two years (final biomass + biomass removed by clipping – original biomass). Moderate browsing reduced total productivity only 4% while severe browsing reduced it 35%. There was no evidence of overcompensation in these sagebrush plants.

Figure 9.9 Average productivity of sagebrush subjected to moderate browsing (one-third of shoots removed) and severe browsing (three-quarters of shoots removed). Averages of eight populations of sagebrush from Colorado, Utah and Wyoming grown in a common garden. Moderate browsing reduced productivity by only 4% but severe browsing reduced productivity by an average of 35%. No evidence of overcompensation was found. (Data from Messina *et al.* 2002, photo courtesy of Markku Savela.)

Q Can the overcompensation hypothesis be adequately tested in field plots or must tests be done in experimental gardens?

To evaluate the idea that grazing could improve plant production, it is important to consider both above-ground and below-ground biomass. But there is little evidence that grazing increases plant production or improves the fitness of plants as suggested by the overcompensation hypothesis. Plants respond to grazing by regrowth, but they never recover completely from the losses caused by grazing, as illustrated for sagebrush in Figure 9.9. The prevailing view of the plant–herbivore interaction for grazing systems is that it is not a mutualistic, win–win situation but rather a negative interaction in which the herbivore gains but the plant loses. If grazing systems are not mutualistic, they may resemble more the desert shrub–annual plant interaction illustrated in Figure 9.7.

Seed Dispersal and Defense of Fruits

Many plant–animal interactions are beneficial to plants—one good example of mutualism involves plant seed dispersal by animals. Many species of plants depend on birds and mammals for the dispersal of their seeds. Positive interactions leading to coevolution between plants and their seed predators may help to explain some features of plant reproductive biology and animal feeding habits. Coevolution may be seen particularly clearly in the evolution of fruits.

Fruits are effectively discrete packages of seeds associated with a certain amount of nutritive material. Vertebrates eat these fruits and digest part of the material, but many of the seeds pass through the digestive system unharmed. Plants advertise fruits by a ripening process in which the fruits change color, taste and odor. Once fruits are ripe, they can be attacked by other damaging agents, such as fungi and bacteria, or by destructive feeders, who will not disperse the seeds.

There are four general ways for a plant to defend its fruit against damage by herbivores (Table 9.1).

The cheapest defense is to fruit during the season of the year when there are fewest pests. For example, in the temperate zone, insect damage to fruits will be minimal in the autumn and winter. Secondly, plants can reduce their exposure to fruit damage by ripening fruit more slowly. This strategy, however, will also reduce the availability of ripe fruit to dispersers and may be counterproductive to the plant. Thirdly, by making the fruit very unbalanced nutritionally, a plant may discourage insect and fungal attack. Fruit pulp is high in carbohydrates, low in lipids and extremely low in protein. Fruits are among the poorest protein sources in nature, and this may be a general pest-defense mechanism.

Table 9.1 The main defense methods employed by plants to protect fruits from destruction by pests and to facilitate seed dispersal by mutualists. The plant will benefit from having its seeds dispersed, but pays a price by producing edible fruits that can also be taken by pests that eat the fruit, but do not disperse seeds.

Defense method	Cost to the plant in energy and nutrients	Effect on non-mutualists (pests that do not disperse seeds)	Effect on mutualists (seed dispersers)
Fruiting during the time of lowest pest pressure	None	Negative	Usually positive
Reducing exposure time by increasing ripening rate	None	Negative	Often negative[a]
Unbalanced or poor nutrition in the flesh of the fruit	None	Negative	Negative
Chemical defense of fruit	Some	Negative	Negative

[a] Depending on disperser abundance and attractiveness.
Source: Modified after Herrera (1982).

Finally, plants may adopt a chemical or mechanical defense of their fruits, although this is the most expensive method. Because chemical defenses in fruits reduce herbivore attack, and thus seed dispersal, it has commonly been assumed that as fruits ripen, chemical defenses would be eliminated. For example, the alkaloid tomatine in green tomatoes is degraded by a new enzyme system that is activated in ripening red tomatoes. But the assumption that ripe fruits will not be chemically defended is probably a biased extrapolation based on cultivated fruits that have been selected for centuries. Of all the wild European plants that produce fleshy fruits, at least one-third have fruits toxic to humans. Toxic fruits are avoided by many birds and mammals, and so such fruits are dispersed only by a selected subset of vertebrates which can detoxify the specific chemicals. Toxic fruits are thus one way for a plant to 'choose' its dispersal agents.

Whitebark Pine and Clark's Nutcracker

The interaction of seed predation and seed dispersal has been analyzed particularly thoroughly in the case of Clark's nutcracker (*Nucifraga columbiana*) and the whitebark pine (*Pinus albicaulis*). Several species of pines in North America have large, wingless seeds that are not dispersed by wind. These seeds are frequently removed from their cones by jays and nutcrackers, transported some distance and cached in the soil as a

Figure 9.10 Whitebark pine trees in western USA and Clark's nutcracker, which harvests seeds from the pine and caches some of them, thereby serving as a dispersal agent for the tree. (Photos courtesy of University of Georgia and the USDA Forest Service, and D. McShaffrey, Marietta College.)

future food source. When the birds cache more seeds than they can subsequently eat, the surplus is available for germination. This interaction can be considered a case of mutualism only if both species increase their fitness because of the association. The birds gain food supplies, while the tree gains by having its seed dispersed to new, fertile sites.

Does the whitebark pine benefit from seed dispersal from Clark's nutcracker? On average, nutcrackers cache three to five seeds in each store, bury them 2 cm, avoid damp sites and place many of their caches in microenvironments favorable for subsequent tree growth. Each nutcracker stores about 32,000 whitebark pine seeds each year, which represents three to five times the energy required by the nutcracker. Thus many caches are not used, and the survival of pine seedlings arising from unused caches is very high. Nutcrackers also disperse pine seeds to new areas within the subalpine forest zone, so that they increase the local distribution of whitebark pine. Many other species of birds feed on whitebark pine seeds, as do mice, chipmunks and squirrels, but none of these species caches pine seeds in ways that favor germination. We conclude that the Clark's nutcracker–whitebark pine system is a coevolved mutualism in which both species profit—the nutcracker by obtaining food and the pine by achieving seed dispersal.

The great variety of seed dispersal systems used by plants has important consequences for seed-eating animals, and much more exploratory work will be needed before we know how many of these systems are mutualistic and how many are exploitative.

SUMMARY

Not all interactions between species are negative. Positive interactions are encounters between organisms that benefit one or both of the participants and cause no harm to either species. In physically stressful environments, such as the intertidal zone, the arctic or desert areas, neighbors may buffer one another by providing shelter from physical factors such as wave action or desiccation.

Mutualisms form the basis of many plant–animal interactions in natural communities. Pollination and seed dispersal are two examples of processes that benefit both plant and animal species. Pollination is the single most important mutualistic relationship between plants and animals, and is critical for both agricultural crops and native plants. Ants may form mutualistic relationships with plants, particularly in tropical areas. Many terrestrial plants acquire nutrients via positive interactions between plant roots and soil fungi and bacteria. Mycorrhizal fungi can take up nutrients at low concentrations in the soil and pass these nutrients via the roots to plants, while the plants pass carbohydrates via the roots to the fungi.

SUGGESTED FURTHER READINGS

Arsenault R and Owen-Smith N (2002). Facilitation versus competition in grazing herbivore assemblages. *Oikos* **97**, 313–318. (A review of whether different herbivore species in grazing ecosystems help or hinder one another.)

Côté IM (2000). Evolution and ecology of cleaning symbiosis in the sea. *Oceanography and Marine Biology Annual Review* **38**, 311–355. (A review of the symbiotic interaction of cleaner fish and their clients in coral reef communities.)

Olesen JM and Jordano P (2002). Geographic patterns in plant–pollinator mutualistic networks. *Ecology* **83**, 2416–2424. (What affects how specialized pollinators are in natural ecosystems?)

Roubik DW (2002). Tropical agriculture: the value of bees to the coffee harvest. *Nature* **417**, 708. (The first evidence that bee pollination increases the coffee harvest.)

Stachowicz JJ (2001). Mutualism, facilitation, and the structure of ecological communities. *BioScience* **51**, 235–246. (A good overview of how positive species interactions reduce physical stresses and increase ecosystem richness.)

van der Heijden MGA and Sanders IR (2002). *Mycorrhizal Ecology*. Springer, New York. (A book outlining the great significance of mycorrhizal fungi to human life.)

West SA, Kiers ET, Simms EL and Denison RF (2002). Sanctions and mutualism stability: why do rhizobia fix nitrogen? *Proceedings of the Royal Society of London, Series B* **269**, 685–694. (An evolutionary discussion of how natural selection maintains one of the most important mutualisms adding nitrogen to soils.)

QUESTIONS

1. Mutualisms are usually thought to be more common in tropical areas than in temperate or polar regions. Discuss why this might be a common belief, and then evaluate the evidence for this generalization. Stachowicz (2001) and Bertness and Leonard (1997) discuss this issue.

2. How does herbivory on seeds and fruits differ from herbivory on leaves and stems of plants? Can both of these types of herbivory result in a positive interaction between the plant and its herbivore?

3. Pollinators may be generalists or highly specific to one or two species of plants. It is often suggested that pollinators are more specialized in the tropics and more generalists at higher latitudes. How would you test this idea and what evidence exists on this idea? Olesen and Jordano (2002) discuss this issue.

4. Mangroves serve as nursery areas for a variety of tropical fish species. Do mangroves gain anything from these associations? How would you decide if this interaction is a +/0 (beneficial for the fish, neutral for the mangrove) or a +/+ (beneficial for both species) interaction?

5. Discuss possible reasons for the observation that ecologists have described and studied many more negative interactions between species than positive interactions. Does this imply that negative interactions are more common and more important in ecological communities than positive interactions? Keddy (1990) and Bertness and Callaway (1994) discuss this issue.

Chapter 10

POPULATION REGULATION AND THE BALANCE OF NATURE

IN THE NEWS

The mountain pine beetle—an insect the size of a rice grain—is one of the most important insect pests of pine forests in western North America. From 1996 to 2001 mountain pine beetles killed more than half a million ponderosa pine trees in the Black Hills of South Dakota. From 1997 to 2004 these beetles have infested about 300,000 hectares (about 1200 square miles) in central British Columbia, killing about 80 million lodgepole pine trees worth US$4 billion.

The mountain pine beetle is native to North America and occurs from Mexico to northern Canada. Female beetles bore into large pine trees and excavate galleries under the bark in which they lay their eggs. The larvae spend about 10 months under the bark, feeding on the phloem of the tree. Adults emerge the following year and within a day or two of leaving one tree they attack other trees. Trees are killed by the

Geographical distribution of mountain pine beetle

*Mountain pine beetle (*Dendroctonus ponderosae*)*

combined action of the beetles feeding and by a blue-staining fungus that the beetles carry as spores on their body. The fungus develops in the sapwood of the pine tree and cuts off the flow of water to the needles, so the tree eventually dies. The first visible sign of beetle attacks is discolored foliage: yellow and brown needles appear on the trees.

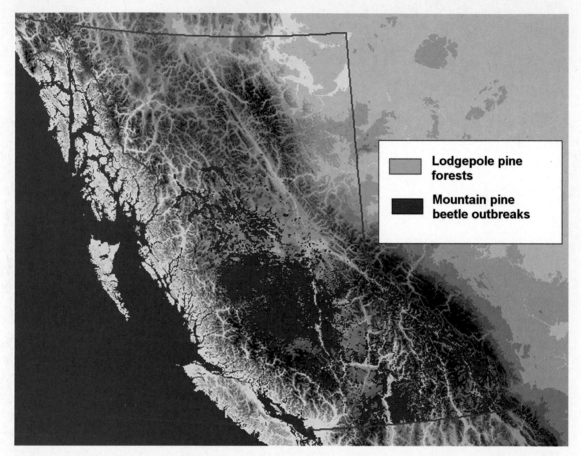

Destruction of lodgepole pine forests in British Columbia

Forest ecologists want to know what triggers mountain pine beetle populations to begin growing rapidly and what stops these population outbreaks. Beetles need large trees for good larval survival because large trees have a thick phloem layer on which the larvae feed. Cold winter temperatures (below −40°C) can cause high larval mortality. Woodpeckers feed heavily on larvae under the bark, but there are too few woodpeckers to stop a massive outbreak. Outbreaks stop when most of the large trees have been killed—the beetle then becomes rare for about 10–50 years.

Humans cannot control mountain pine beetle outbreaks. Trees can be harvested, but usually not rapidly enough once the outbreak has begun. Insecticides are not cost effective and have little effect because the spatial scale of the outbreaks is too large. The only long-term solution is to prevent

The lodgepole pine trees with red needles are dying

uniform-aged forest stands from developing over large areas by harvesting in patches. Large stands of uniform-aged pine trees have typically arisen as a result of large forest fires in the past, so that there is an interaction between fire and beetle attacks.

■ 10.1 POPULATION DYNAMICS ANALYZES POPULATION GROWTH AND AVERAGE ABUNDANCE

In discussing the effects that predation, disease, competition and facilitation have on individuals and populations, the bottom line is how much these factors affect the population dynamics of a particular animal or plant. If a predator kills a prey individual, does that automatically reduce the population level of the prey? If we kill pests with insecticides, will they necessarily become less abundant? The answer to these questions is *no*, and in this chapter we will explore why simple concepts of population arithmetic can be misleading. These questions are at the core of the problems of conservation, land management, fisheries and pest control issues that occupy our news media daily. For that reason it is important that we get our understanding of **population regulation** right.

Population regulation is defined as the maintenance of a population within a restricted zone of abundance so that the species neither becomes extinct nor increases without limits. Ecologists ask what ecological factors produce and maintain this regulation.

We can make two fundamental observations about populations of any plant or animal. The first is that abundance varies from place to place. There are some 'good' habitats, where the species is generally common, and some 'poor' habitats, where it is generally rare. The second observation is that no population goes on increasing without limit, and the problem is to find out what stops the population growing. This second problem can also be restated as the problem of explaining fluctuations in numbers. The distinction physicists make between 'statics' and 'dynamics' is exactly the same as in these two ecological problems. Figure 10.1 illustrates these two problems, which are often confused in discussions of population regulation.

Prolonged controversies have arisen over the problems of the regulation of populations. Before 1900, many authors, including Malthus and Darwin, had noted that no population goes on increasing without limit and that there are many agents of destruction that reduce the population. Darwin (1859), for example, wrote:

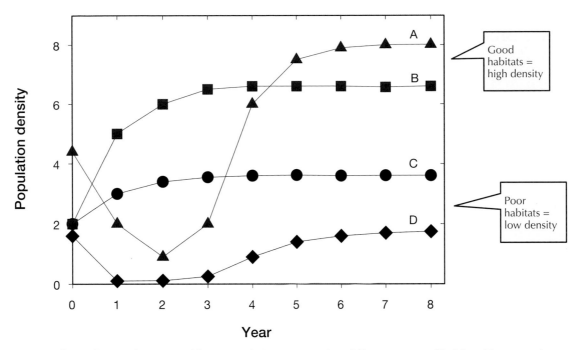

Figure 10.1 Hypothetical annual census of four populations occupying different types of habitat. Two questions may be asked about these populations: (1) Why do all populations fail to go on increasing indefinitely? (2) Why are there more organisms on the average in the good (red) habitats A and B compared with the poor (blue) habitats C and D?

'The causes which check the natural tendency of each species to increase are most obscure. Look at the most vigorous species; by as much as it swarms in numbers, by so much will it tend to increase still further. We know not exactly what the checks are even in a single instance.' (Chapter 3, page 55)

It was not, however, until the twentieth century that an attempt was made to analyze these facts more formally. The stimulus for this came primarily from eco-

Charles Darwin in 1877 (1809–1882)

nomic entomologists, who had to deal with both introduced and native insect pests. The basic principles of population regulation can be derived from a simple model, taken from the models of population growth presented in Chapter 6.

Temporal Variation in Abundance

If populations do not increase without limit, what stops them? We can answer this question with a simple graphical model. A population in a closed system[1] will increase until it reaches an equilibrium point at which

Birth rate = death rate

Figure 10.2 illustrates three possible ways in which this equilibrium may be achieved. As population density goes up, birth rates[2] may fall, death rates may

---

1 A population is closed if there are no immigrants and no emigrants, so the dynamics are driven solely by births and deaths.

2 In all discussions of population regulation, birth rates always mean per capita birth rates, and death rates always mean per capita death rates (*Essay 10.1*).

Figure 10.2 Simple graphical model to illustrate how equilibrium population density may be determined. Population density comes to an equilibrium only when the birth rate equals the death rate, and this is possible only if birth or death rates change with population density. Three ways in which this can occur are illustrated here. (Modified from Enright 1976.)

rise, or both changes may occur. To determine the equilibrium population size for any field population, we need to determine only the lines shown in Figure 10.2.

We now introduce a few terms to describe the concepts shown in Figure 10.2. The death rate per capita is defined as **density-dependent** if it increases as density increases (Figure 10.2a and 10.2c). Similarly, the birth rate per capita is defined as density-dependent if it falls as density rises (Figure 10.2a and 10.2b). Another possibility is that the birth or death rates do not change as density rises; such rates are defined as **density-independent** rates.

Note that Figure 10.2 does not include all logical possibilities. Birth rates might, in fact, *increase* as population density rises, or death rates might *decrease*. Such rates are called **inversely density-dependent** because they are the opposite of directly density-dependent

rates.[1] Inversely density-dependent rates can never lead to an equilibrium density. Figure 10.2 can be formalized into the first principle of population regulation: *No closed population stops increasing unless either the birth rate or death rate changes with density.* This principle gives guidance to any attempt to understand 'dynamics'— why all populations stop increasing sooner or later.

Spatial Variation in Abundance

To understand 'statics', or why two populations differ in equilibrium density, we can extend this simple model to answer the question of why abundance varies from place to place: why we have good and bad habitats (Figure 10.3). Consider the simple case of populations with a constant (density-independent)

---

1 Inverse density dependence = positive density dependence and direct density dependence = negative density dependence.

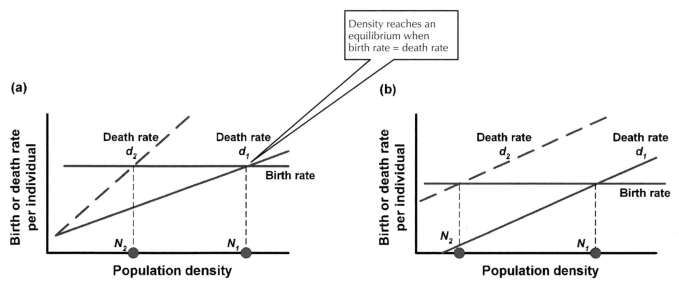

Figure 10.3 Simple graphic model to illustrate how two populations may differ in average abundance. In this example the birth rate is density independent and the death rate is density dependent. In (a) the two populations differ in the amount of density-dependent mortality because the slope of the lines differ. If the death rate goes up only slowly with population density, the equilibrium density (N_1) will be larger than if the death rate rises rapidly with density (N_2). In (b) the populations differ in the amount of density-independent mortality (the slope of the lines does not differ). Dotted lines mark the equilibrium population densities. (Modified from Enright 1976.)

birth rate. Equilibrium densities vary for two reasons: (1) Either the slope of the mortality curve changes (Figure 10.3a), or (2) the general position of the mortality curve is raised or lowered (Figure 10.3b). In case 1, the density-dependent rate is changed because the slopes of the lines differ, but in case 2, only the density-independent mortality rate is changed. From this graphic model we can arrive at the second principle of natural regulation: *Differences between two populations in equilibrium density can be caused by variation in either density-dependent or density-independent rates of birth and death.* This principle seems simple: it states that anything that alters birth or death rates can affect equilibrium density. This principle gives us a way to study differences in average abundance between good and poor habitats.

■ 10.2 POPULATIONS ARE REGULATED BY DENSITY RELATED CHANGES IN BIRTHS, DEATHS, OR MOVEMENTS

The **balance of nature** has been a background assumption in natural history since the time of the early Greeks and underlies much of the thinking about natural regulation. The simple idea of early naturalists was that the numbers of plants and animals were fixed and in equilibrium, and observed deviations from equilibrium, such as the locust plagues described in the Bible, were the result of a punishment sent by divine powers. Only after Darwin's time did biologists try to specify how a balance of nature was achieved and how it might be restored in areas where it was upset.

The balance of nature has changed dramatically from the early concepts of a nature that was fixed and in equilibrium to a modern concept of a balance of nature. Our more modern view of the balance of nature is more Darwinian in outlook—focusing on changes in populations—and it implies a more dynamic view of a shifting balance between populations that are affected by predators, competitors, diseases and mutualists, which are all influenced by changing weather and human activities.

The modern view of population regulation begins by recognizing that population events are determined by the interplay of two sets of factors: extrinsic and intrinsic (Figure 10.4). We cannot assume that the individuals that make up the population are all

ESSAY 10.1   HOW DO WE MEASURE BIRTH RATES AND DEATH RATES?

Birth rates and death rates are commonly used in population studies and we need to define clearly what they mean. Because they are rates, the first point to note is that they must always specify some **time interval**, such as one year or one month. The second point is that they must specify a base population. Two bases are commonly used:

1. Per capita or per single individual: if the birth rate per capita is used, it is equivalent to the number of births per individual per unit of time. So if a bird lays 5 eggs once a year, the birth rate is 5 eggs per female bird per year.

2. Per 1000 population: human demographers often use birth and death rates per 1000 or per 100,000 population to make them easier for people to understand. For example, in 2005 the birth rate in Australia for women aged 25–29 was 103 per 1000 women per year. In 2006, the death rate in 2006 for the USA population was 8 per 1000 (http://www.prb.org/).

These two ways of expressing birth rates and death rates are mathematically equivalent, so the birth rate for Australian women aged 25–29 could be expressed as 0.103 per capita per year, and the death rate in the USA in 2006 could be given as 0.008 per capita per year.

Birth and death rates apply to populations, even though they are often stated as per individual. Individual plants and animals either live or die, but, averaged over a whole population, we can express these vital events as population averages to tell us how fast a population is growing or declining.

identical, like atoms or marbles. Individuals may differ in **phenotype** and in **genotype**, and not all individuals are equally affected by disease or bad weather. Of course, no matter what mechanisms are operating in populations, evolution is continually occurring, and so theories of population regulation become concerned with evolutionary arguments.

The modern synthesis of population regulation has explored the practical issues involved in applying the two central principles illustrated in Figures 10.2 and 10.3 to real world populations. Let us start with a clear definition of two confusing terms:

*   **limitation**: a factor is defined to be a limiting factor if a change in the factor produces a change in average or equilibrium density. For example, a disease may be a limiting factor for a deer population if deer abundance is higher when the disease is absent.

*   **regulation**: a factor is defined to be a potential regulating factor if the percentage mortality caused by the factor increases with population density.[1] For example, a disease may be a potential regulating factor only if it causes a higher proportion of losses as deer density goes up. Regulating factors must always be density-dependent.

There is an important distinction between a potential regulating factor and an actual regulating factor. Unless the change in mortality is large enough, a regulating factor will not stop population growth. A spider that eats 1 of 1000 grasshoppers and 6 of 2000 grasshoppers is causing density-dependent mortality in the grasshopper population, but this impact is so tiny that it will not stop population growth.

---

1   Or alternatively, if the reproductive rate is reduced as the population rises.

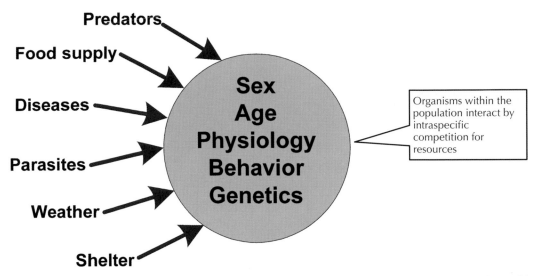

Figure 10.4 The modern view of population regulation processes. The population being studied is indicated by the blue circle. Extrinsic processes like predators and disease interact with the variable properties of individuals that make up the population (intrinsic processes), so that population regulation emerges as a result of interplay between these two kinds of processes: extrinsic and intrinsic. We cannot assume that extrinsic agents like diseases or predators act on individuals of different age or sex in the same way, so we must take into account the composition of the population when we study population regulation.

Q Are populations that fluctuate widely in abundance still regulated?

The simple model of population regulation shown in Figure 10.2 is focused on the concept of equilibrium, and we must begin by asking whether natural populations can be in equilibrium. Recent work on ecological stability has given us a more comprehensive view of the factors that affect the tendency of populations to reach a stable equilibrium (Figure 10.5). There is no reason to expect all populations to show stable equilibria. There are two sources of instability in populations. Strong environmental fluctuations in weather can produce instability, but strong biotic interactions may also promote instability. Time lags also can affect population stability. We should expect real world populations to cover the whole spectrum from those showing stable, equilibrium dynamics to those showing unstable, non-equilibrium dynamics. The simple model shown in Figure 10.2 will be difficult to detect in a real population that shows unstable dynamics over time.

Populations and Metapopulations

A second consideration in population regulation is the recognition of habitat variation in space. It is important in studying population dynamics to consider the spatial scale of the study because it can be important in measuring stability. If you study a very small population on a small area, it may fluctuate widely and even become extinct. A large population on a large study area may, at the same time, have a stable density. The important concept here is that local populations may be linked together through dispersal into **metapopulations** (Figure 10.6). To study population regulation, you must know if a population is subdivided, and how the patches are linked. Ensembles of randomly fluctuating subpopulations, loosely linked by dispersal, will persist if irruptions at some sites occur at the same time as extinctions at other sites. The result can be that the population appears stable at a regional level, while the individual subpopulations fluctuate wildly.

Butterflies on islands are a good example of metapopulations. To show that a set of local populations is a metapopulation, you must show that some populations

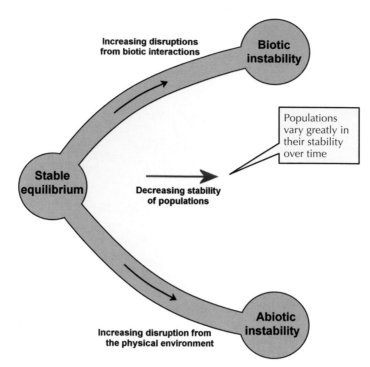

Figure 10.5 Schematic representation of ecological populations on a scale from stable equilibrium systems on the left to unstable systems on the right. Both biotic instability (caused by factors such as excessive predation or epidemic disease), shown at the top, and environmental instability (caused by strong extrinsic fluctuations, such as weather), shown at the bottom, can produce unstable populations. (Modified from DeAngelis and Waterhouse 1987.)

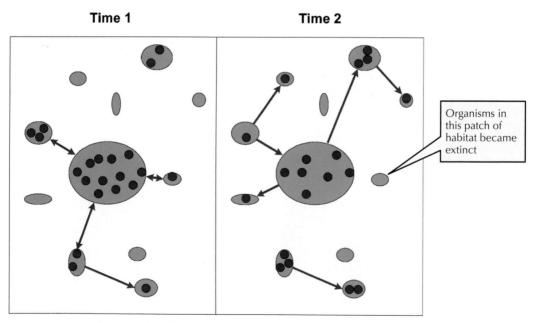

Figure 10.6 Hypothetical metapopulation dynamics. Closed green circles represent habitat patches; red dots represent individual plants or animals. Arrows indicate dispersal movements between patches. Over time the regional metapopulation changes less than each local population. Some local populations become extinct, and others are recolonized by dispersing individuals. The key to understanding metapopulation dynamics is to understand movements of individuals.

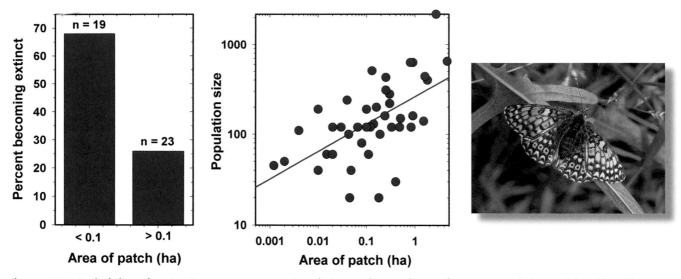

Figure 10.7 Probability of extinction over two years in relation to the patch area for metapopulations of the Glanville fritillary butterfly in the Åland archipelago, Finland. Small patches are much more likely to become extinct, and small patches tend to have smaller populations of this endangered butterfly. (Data from Hanski *et al.* 1994 and 1995, photo courtesy of Simon Coombes, Captain's European Butterfly Guide UK.)

become extinct in ecological time, and that these can be recolonized by dispersing individuals from nearby populations. Extinction and recolonization are thus essential features of a metapopulation. Ilkka Hanski and his colleagues have studied 1502 small populations of the Glanville fritillary butterfly (*Melitaea cinxia*) on islands in the Åland archipelago between Finland and Sweden. This butterfly is an endangered species that has recently become extinct on mainland Finland and now exists only on islands in the Åland archipelago. Larval caterpillars feed on two host plants and spin a web, which is easy to detect in field surveys. These butterfly populations range in size from one to 65 larval groups per meadow, but most populations are small, averaging four larval groups per patch (corresponding to about 5–50 butterflies). From 1991 to 1993 an average of 45% of these local populations became extinct. Smaller patches supported smaller populations and had a higher chance of becoming extinct (Figure 10.7). There were two reasons for small populations becoming extinct more often. Firstly, male and female butterflies tend to leave small patches in which there is a reduced chance of mating. Figure 10.8 shows the residence time for female butterflies in populations of different sizes, and the fraction of mated females.

Allee Effects

Small butterfly populations suffered reduced population growth rates—an effect called the **Allee effect**—which is the exact opposite of what is predicted by the simple density-dependent model. Allee effects produce instability in populations and may contribute to local extinctions.

Allee effects are now a focus of great interest in conservation biology. Allee effects are defined as inverse density dependence at low density (Figure 10.9) and were first described by W.C. Allee in 1931, who pointed out that 'undercrowding' could be as harmful to social species as overcrowding. If species become too rare, mates may become difficult to locate or group defenses against predators may become ineffective. The key point for populations is that there is a critical threshold density below which a social group or an entire population may become extinct.

Q **Could Allee effects ever occur in human populations?**

Analyzing Population Dynamics

If a population does not continue to increase, it is axiomatic that births, deaths or movements must change

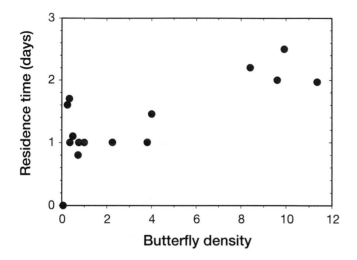

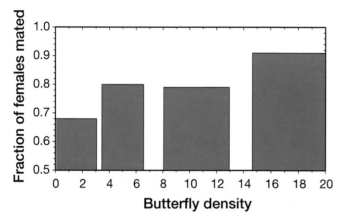

Warder Clyde Allee (1885–1955) Professor of Zoology, University of Chicago

Figure 10.8 Effect of population density on the residence time of female Glanville fritillary butterflies in the Åland archipelago, Finland. Butterflies spend less time in low density patches, and the result is that the fraction of mated females and the birth rate goes down at low density. Small populations suffer decreased growth rates. (Data from Kuussaari *et al.* 1998.)

at high density. The first step is to ask which of these parameters changes with population density. Does reproductive rate decline at high density, or does mortality increase (or both)? If mortality increases, does this have a disproportionate effect on younger or older animals, on males or females? These patterns of changing reproduction and mortality with population density can then be analyzed to see if they occur in a variety of populations. Particular attention should be given to what happens to reproductive rates and mortality rates at low densities to determine if an Allee effect is possible. This description of patterns of change

in reproduction and mortality should be the first step to understanding population regulation in animals.

The second step is to determine the reason for the changes in reproduction or mortality. Determining the cause of death of plants or animals in natural population is not always simple. If a fox or a bat has rabies—a fatal disease—the cause of death is clear. If a caterpillar has a tachinid parasite, it is certain to die from this parasitization. But as you examine more complex cases, decisions about causes of death are not clear. If a moose has inadequate winter food and the snow is deep, it may be killed by wolves. Is predation the cause of death? 'Yes,' you would answer, but only in the immediate sense. Malnutrition and deep snow have increased the probability of being killed. Because many components of the environment can affect one another, and not be independent, mortality can be compensatory. The idea of **compensatory mortality** is one of the most important concepts needed to understand population regulation. At the two extremes, mortality may be **additive** or it may be completely **compensatory**. How can we distinguish between these?

Additive mortality assumes the agriculture model of population arithmetic (*Essay 10.2*). If a farmer keeps sheep, and a coyote kills one sheep, the farmer's flock is smaller by one. Deaths are additive and to measure their total effect on a population, you simply add them

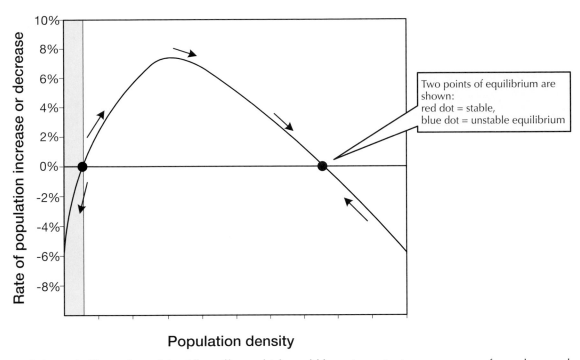

Figure 10.9 Schematic illustration of the Allee effect, which could have important consequences for endangered species driven to low population densities. The standard population regulation model (Figure 10.2) assumes that things get better for populations as density falls to a low level—the death rate goes down and the birth rate goes up. But, if an Allee effect (or negative density dependence) occurs, things get worse for a population as density falls. In the yellow zone the rate of population change is negative, and if a population falls below the density indicated by the blue dot, extinction is inevitable. The horizontal red line indicates a stable population (zero population growth).

up. But in natural populations, where there are several causes of death, the arithmetic is not so simple. Consider, for analogy, a sheep population in which winter food is limiting so that starvation will kill many individuals by the end of winter. If a coyote kills one of these sheep, it may be doomed to die anyway from starvation and, because it is gone, there is now more food for the remaining sheep. The number of sheep left at the end of winter will be the same, regardless of whether predation occurs or not (in this hypothetical scenario). In this case, predation mortality is not additive, but is compensatory, and simple arithmetic does not work. Figure 10.10 illustrates how additive and compensatory effects can be recognized. Consider for example what would happen if coyote mortality on sheep lambs increases from 10% per year to 20% per year. If coyote mortality is additive, total mortality will increase from 45% to 55% per year (in this hypothetical example). If coyote mortality is compensatory, total lamb mortality will remain unchanged at 45% per

year. Clearly if coyote mortality is very high, compensation is not possible, as shown on the right side of Figure 10.10 and the total mortality rate increases.

Compensatory mortality is the reason behind many ecological anomalies that puzzle the average person. If you kill pests, they will not necessarily become less abundant—we will discuss pest control further in Chapter 19. If you shoot grouse or catch fish, their numbers may not necessarily fall, as we will see in Chapter 18. Compensatory mortality, when it occurs, has practical consequences.

In natural populations mortality agents will rarely be completely additive or completely compensatory. We can determine if a particular cause of mortality is compensatory only by doing an experiment in which total losses are measured with and without the particular cause of death. Few of these experiments have been done, and there is a tendency to assume the agricultural model of population arithmetic that all losses are additive in natural populations.

ESSAY 10.2   WHY IS POPULATION ARITHMETIC NOT SIMPLE?

We typically approach population dynamics with a simple agricultural model of population change. Population arithmetic is simple in an agricultural model:

$$\text{Population change} = \text{births} + \text{immigrants} - \text{deaths} - \text{emigrants}$$

If a farmer owns 10 cows and one cow gets sick and dies, the farmer now has 9 cows—the arithmetic is simple. If we apply this model to wolves eating elk in Yellowstone National Park, it is certainly correct that if wolves kill 23 elk on a Monday, the elk population will be lower by 23 individuals on Tuesday. But we cannot conclude from data of this type that wolves limit the abundance of elk so that next year there will be 23 fewer elk. The reason for this complication is that these four variables on the right side of the equation are not independent, but affect one another. More births may lead to more emigrants, or more deaths may lead to more births, so that these four variables can compensate for one another. Several scenarios are possible for these elk over a time frame of several months to years:

- The elk that died from wolf predation may have been dying from a disease or may have died from starvation or severe winter weather during the next few months. The fact that the wolves killed the elk may be irrelevant to population changes.

- The elk that died from wolf predation may have been dying from a disease that they would have transmitted to other healthy elk, so that by killing diseased individuals, wolves may actually increase elk numbers by preventing an epidemic.

- By killing elk in early winter, wolves reduce the amount of elk browsing on winter food plants, and this might increase reproductive rates in elk, so that more elk are born the following spring, increasing the elk population.

By considering what happens to the population, rather than what happens to individuals, we can see a variety of compensatory mechanisms that make it impossible to project the outcome of individual deaths on to what will happen to population numbers. A predation death might *increase* population size if the individual was carrying a serious disease that could have spread to other animals. Or, in social animals, a single death might endanger a whole social group—for example in lions where the loss of a dominant male could cause the loss of a cohort of juveniles. Animal or plant populations may be limited by the number of safe sites or territories, so that only a fixed number of individuals will be able to survive, no matter how many individuals are available.

The failure of simple population arithmetic was demonstrated many years ago by bounty programs. Coyotes have been controlled in western North America for over 150 years, and bounties are still paid in 7 states. Hunters have been paid to kill coyotes for many years without any evidence that coyote abundance has been reduced or that livestock depredation has been stopped. In Utah, about 10,000 coyotes were killed in 2001, and about 1000 of these were taken for a bounty of about $20 each. There is no evidence that this kill reduced coyote numbers in Utah, and the estimated total kill was about

Coyote (Canis latrans)

10% of the coyote population of the state. Simple calculations suggest that coyote populations can recover within one year from the removal of up to 60% of the starting population. Once coyote numbers are reduced, their reproductive rate increases because of larger litters and increased survival of juveniles. These kinds of compensatory responses explain why population arithmetic is quite complicated.

If birth rates change with population density, it is important to pin down the factors that cause reproduction to change. Food supply is usually the first hypothesis to be tested for animals, or nutrient availability for plants. But other factors may cause birth rates to change as well. Social interactions can inhibit reproduction in vertebrates, and predation-risk can change the behavior of animals such that they can gather less energy and thus produce fewer offspring. These factors can most easily be identified experimentally by manipulations of field populations, or by careful descriptive studies of processes in unmanipulated populations.

Q **Would you expect to find the same causes of population regulation in all parts of a species range?**

The bottom line is that inferences about population limitation and regulation are important, but not easy to come by.

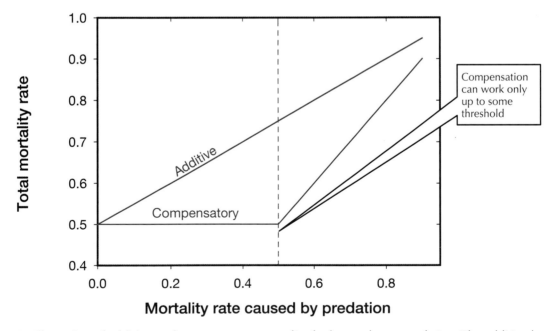

Figure 10.10 Illustration of additive and compensatory mortality for losses due to predation. The additive hypothesis (blue) predicts that for any increase in predation mortality, total mortality increases by a constant amount. The compensatory hypothesis (red) predicts that, below threshold mortality rate, any change in predation losses has no effect on total mortality. This model can be applied to any mortality agent—predation, disease, starvation or hunting. (Modified after Nichols *et al.* 1984.)

Plant Population Regulation

Because most plants are highly variable in size, population regulation in plants must be discussed as the regulation of biomass rather than numbers. Plant ecologists have not usually addressed the problem of population regulation in the same way as animal ecologists have done, but the same principles can be applied. As a plant population increases in numbers and biomass, either reproduction or survival will be reduced by a shortage of nutrients, water or light, by herbivore damage, by parasites and diseases, or by a shortage of space. Because plants are typically fixed in one location, competition for light or nutrients is often implicated in population regulation. This competition has been described by the **self-thinning rule**—an important generalization in plant population ecology.

The self-thinning rule describes the relationship between individual plant size and density in uniform-aged populations of a single species. There can be many small plants at a high density or fewer large plants at a low density, and this trade-off is known as the self-thinning rule. Mortality, or 'thinning', from competition within the population is postulated to fit a theoretical line with a slope of –4/3: This line has been suggested as an ecological law that applies both within a plant species and between different plant species. Figure 10.11 illustrates the self-thinning rule schematically. The self-thinning rule highlights the trade-offs that can occur in organisms with highly flexible growth, so that the size of an individual can become smaller as density increases.

The principle of a strict, mathematical trade-off between average plant size and total plant population density, however, is supported by all plant studies. The self-thinning rule has been replaced by a more general '–4/3 boundary rule' (Figure 10.12), which postulates the self-thinning line as an upper limit for the relationship between plant size and density in monocultures. The self-thinning rule expresses competition between plants for essential resources, and gives us insight into species differences under strong competition for light and nutrients.

The self-thinning rule has been applied to animals as well as plants. Animals with plastic growth rates, such as fish, can respond to population density by changing growth rates and body size. Animals with larger bodies use more energy and, when populations are food-limited or space-limited, there can be a trade-off between average size and population density. Salmon and trout fingerlings living in streams are a good example. For these kinds of animals with flexible growth, the self-thinning rule is a useful empirical description of these trade-offs between body size and population density.

Q Do you think the self-thinning rule could apply to mammals and birds?

■ 10.3 POPULATIONS MAY ACT AS SOURCE POPULATIONS OR AS SINK POPULATIONS

The balance of nature is not always very clearly visible when we look at local populations[1] that are the convenient units for humans to study. Local populations can be classified as **source populations**, in which there is a net excess of reproduction over mortality, and **sink populations**, in which there is a net excess of mortality over reproduction. Left to themselves, source populations would grow to infinity, and sink populations would become extinct. But, in practice, source populations do not increase forever, but are regulated. This regulation may involve a net export of animals via dispersal, so that in a source population emigration exceeds immigration. Sink populations may indeed become extinct, so we would not necessarily know about them, but more typically they persist because immigration adds sufficient individuals to counteract losses. Sink populations thus continue to exist only if they attract immigrants for nearby source populations. All the local populations in a region could be source populations, but not all could be sink populations.

Fragmentation of Source Populations

Sources and sinks have become more common in human-impacted landscapes in which large continuous areas of forest or grassland have been dissected by

---

1 Local populations typically refer to plant or animal populations on the scale of a few square meters to a few hectares in area for smaller organisms, and a scale up to a few square kilometers for larger animals.

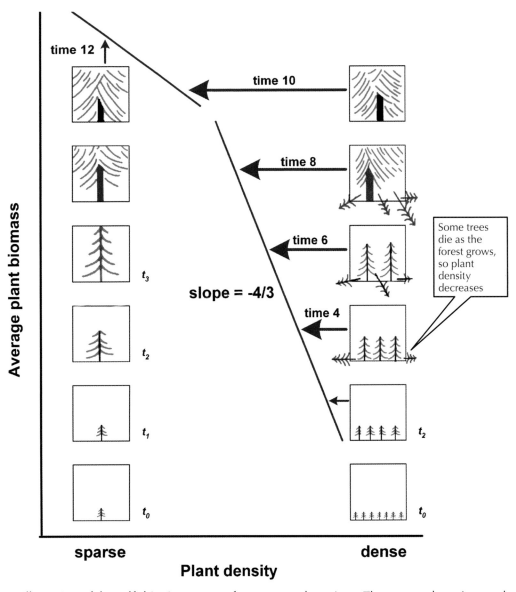

Figure 10.11 An illustration of the self-thinning process for two tree plantations. The sparse plantation on the left shows no mortality and the trees simply grow to maximal size. The dense plantation on the right shows continuous mortality (which thins the density) along with growth (which increases individual tree size). *t* indicates time steps. (Modified after Silvertown and Charlesworth 2001.)

modern agriculture into a series of small fragments. Source and sink dynamics are thus often an integral part of habitat fragmentation. Forests in agricultural landscapes have been particularly fragmented, and there is much concern that fragmentation can turn source populations into sink populations.

Delimiting Sources and Sinks

To identify source and sink populations, we need to measure reproduction, mortality and movements between a whole set of local populations. Much of the concern about source and sink populations has concerned migratory birds in North America. For the simplest model of population change for birds, we can estimate the rate of population change from three parameters: adult survival rate, juvenile survival rate and the number of juveniles produced per adult by the end of the breeding season. If we can measure these three parameters for any population, we can determine if that population is a source or a sink.

Figure 10.12 The self-thinning line for the weed lambsquarters (*Chenopodium album*). There is a direct trade-off between population density (no. per m²) and the size of individual plants (grams). As plant density goes up, crowding rapidly reduces the size of the average plant. The slope of this line is –1.33, as predicted by the self-thinning rule (–4/3). Experimental populations planted at densities on either the left side of this line or the right side would be expected to move to the line and then reach equilibrium somewhere along the line. (Data from Yoda *et al.* 1963, photo courtesy of Oregon State University, Department of Crop and Soil Science.)

Q Could all local populations of a species be source populations?

Source–sink dynamics may be characteristic of particular habitat or metapopulation, or they may be a product of variation in weather from year to year. A local population may thus vary from being a source in some years and a sink in other years. Bird ecologists have studied house sparrows on four islands off Norway to measure variation in population growth rate between islands and over time. Figure 10.13 shows that some islands on average were much more productive than others, but that all islands become sink populations in particularly severe years. Populations on each of the four islands remained nearly constant from 1993 to 1996, with immigration boosting the sinks and emigration balancing the source populations. The dynamics of source–sink populations provide a graphic illustration of how immigration and emigra-

tion can be equally important as agents of population changes as reproduction and mortality. They illustrate the dynamic changes that are part of the new concept of a balance of nature.

■ 10.4 EVOLUTIONARY CHANGES IN POPULATIONS CAN AFFECT THE INTERACTIONS THAT LIMIT ABUNDANCE

How are systems of population regulation affected by evolutionary changes? In many ecological interactions, evolutionary changes operate very slowly and are difficult to detect and so most ecologists separate the ecological time scale of a few years to tens of years, from the evolutionary time scale of hundreds of years to thousands of years. But recent work in ecological genetics has shown that evolutionary changes may occur very rapidly, so that the evolutionary time scale

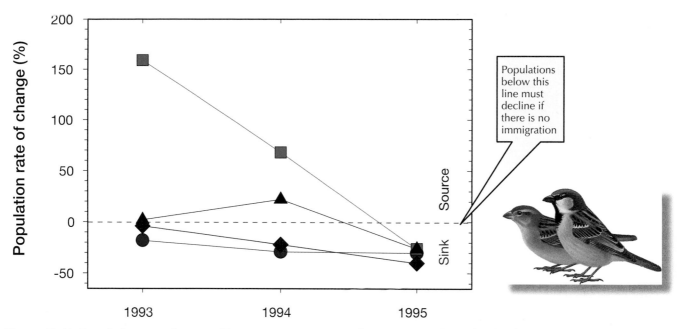

Figure 10.13 Population growth rates of house sparrows (*Passer domesticus*) on four islands of northern Norway from 1993 to 1995. All populations were sinks in 1995, and one island (Indre Kvarøy, green) was particularly productive on average, but was a sink in one year. Two islands were always sink populations because of poor juvenile survival. (Data from Sæther *et al.* 1999.)

approaches the ecological time scale in many cases. We need to ask whether natural selection may impinge upon population regulation in some organisms.

Genetic Changes Affecting Abundance

Many changes in average abundance can be attributed to changes in extrinsic factors, such as weather, disease or predation. But some changes in abundance are the result of changes in the genetic properties of the organisms in a population. Such evolutionary changes can be produced by natural selection. A good example is the myxomatosis–rabbit interaction in Australia, discussed in Chapter 8, in which there were evolutionary changes in both the disease virus and in the rabbit that occurred in just a few years and affected population abundance.

Genetic changes in populations can thus affect interspecific interactions that limit abundance. The coevolution of interacting populations of predator and prey, disease and host, food plant and herbivore all may have implications for population dynamics. The important point to remember is that we should not assume that the ecological traits of species are constant and unchanging in ecological time.

Intrinsic Population Regulation

Animal populations in which intrinsic processes are involved in regulation present yet another problem in evolutionary ecology. Under what conditions should we expect a population to be regulated by intrinsic processes? Intrinsic mechanisms of regulation involve spacing behavior in the broad sense—including territoriality, dispersal and reproductive inhibition. We might expect that vertebrates with their relatively complex behavior would be the most obvious species to show intrinsic regulation. Jerry Wolff has suggested a conceptual model that predicts which vertebrates have the potential for intrinsic regulation. He has discussed mammals in particular, but similar arguments could be made for birds and other vertebrates. The key to Wolff's model is that territoriality in female mammals has evolved as a counter-strategy to infanticide committed by strange females. Infanticide is a mechanism of competition by which intruders usurp the breeding space of residents and increase their fitness by killing the offspring of the resident female. Infanticide has in the past been thought to be a peculiarly human behavioral disorder, and thus rare or absent in natural populations. But detailed field studies on species ranging from lions in Africa, gorillas in Africa, polar bears in the Arctic, rodents in North America, and dolphins in the North Atlantic have shown that it is common in many mammals and birds. How might the possibility of infanticide affect population regulation?

Female mammals should evolve territorial behavior to defend their young from infanticide only if young are not mobile at birth. Females with precocial young (that have their eyes open and can move very soon after birth) will not be susceptible to infanticide and will not defend territories. These predictions from Wolff's model are consistent with most of what is known about mammalian social systems. For example, hares have precocial young while rabbits have altricial[1] young. There are no known cases of infanticide in hares, while there are many cases in rabbits. Many carnivores, such as lions, have altricial young, are subject to infanticide, and are territorial. By contrast kangaroos have altricial young, but carry them about in a pouch so that the young are not vulnerable to infanticide. No kangaroos are territorial.

Q Would Wolff's model predict the possibility of infanticide in humans?

A second feature of self-regulation in mammals is reproductive suppression of juveniles. If juveniles do not disperse from their natal area, they risk the possibility of breeding with close relatives. Selection against inbreeding has molded the dispersal pattern of mammals, so that male juveniles will emigrate, while female offspring remain near their natal site. But high density may make dispersal costly so that all juveniles stay near the birth place. At high density, adults may suppress sexual maturation of their offspring in order to prevent inbreeding, especially if space for breeding is limited. The result can be that a large fraction of the population is non breeding. This reproductive suppression of juveniles at high density will act as a density-dependent factor to potentially regulate the population.

Figure 10.14 summarizes Wolff's model for the evolution of intrinsic regulation in mammals. Many mammal species and many other vertebrates will not

be subject to potential infanticide, and these species would be expected to be subject to extrinsic regulation by predators, food shortage, disease or weather. Note that intrinsic regulation is not in itself an evolved strategy. What evolves are behavioral strategies like territoriality, dispersal and reproductive inhibition, and these individual strategies can result in population regulation at the level of the population. In most cases, evolution works at the level of the individual, not at the population level, but the effects of evolutionary shifts can be seen in population dynamics.

SUMMARY

Populations of plants and animals do not increase without limits, but show more or less restricted fluctuations. Two general questions may be raised for all populations: (1) What stops population growth? (2) What determines average abundance?

Populations stop growing because births decrease at high density, deaths increase or movements occur—one or more of these density-dependent changes must occur to cause populations to stop growing. Average abundance can be affected by any of the factors that affect births, deaths and movements—predation, disease, parasites, nutrients and food supplies, shelter and weather, as well as the interactions that occur between individuals within the population.

Population regulation theory focuses on stability and equilibrium conditions, but many ecologists now emphasize non-equilibrium concepts and ask what factors increase or reduce stability for populations. The spatial scale of a study affects conclusions about stability—if a population is subdivided into local populations, stability may be increased for the entire population. Metapopulations, or clusters of local populations, are critical foci for conservation efforts as habitats become broken up into small, isolated blocks.

The mortality agents affecting a particular population may cause additive or compensatory losses. Additive losses allow simple population arithmetic and additive mortality factors may limit or regulate population density. Compensatory mortality occurs when the causes of death are replaceable, so that if one thing does not kill the plant or animal, something else

---

1 Altricial young are typically blind at birth, naked, and cannot move around.

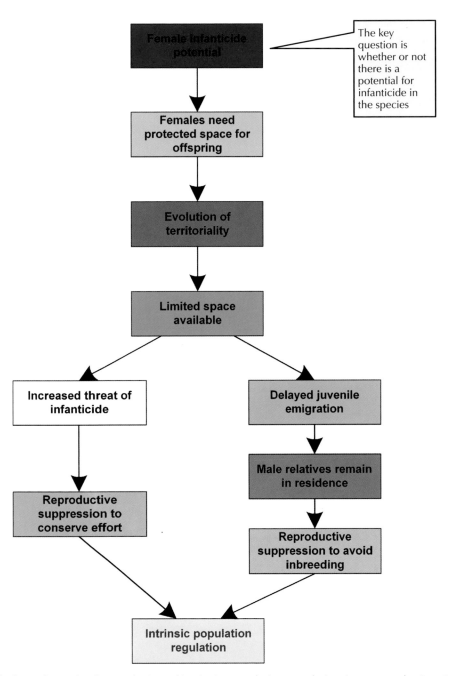

Figure 10.14. Wolff's hypothesis for the evolution of intrinsic population regulation in mammals. Spacing behavior must be the key evolutionary mechanism to evolve in species that compete for space free from infanticidal individuals. The demographic attributes that contribute to population regulation are shown in blue boxes. (Modified from Wolff 1997.)

will. Identifying the immediate causes of death may not increase understanding of population changes if compensatory mortality is occurring.

The theories of population regulation can be tested only by detailed field studies of animal and plant populations, combined with experimental manipulations of the critical processes. The limitation and regulation of populations are critical areas of theoretical ecology because they are central to many questions of community ecology and because they have enormous practical consequences, which we shall explore in the next eleven chapters.

SUGGESTED FURTHER READINGS

Hixon MA (1998). Population dynamics of coral-reef fishes: controversial concepts and hypotheses. *Australian Journal of Ecology* **23**, 192–201. (How are populations of these colorful fishes regulated?)

Murdoch WW (1994). Population regulation in theory and practice. *Ecology* **75**, 271–287. (An overview of population studies by one of our leading ecologists.)

Peterson RO, Thomas NJ, Thurber JM, Vucetich JA and Waite T (1998). Population limitation and the wolves of Isle Royale. *Journal of Mammology* **79**, 828–841. (Wolves and moose on an island provide a model system for analyzing predator–prey dynamics.)

Pierson EA and Turner RM (1998). An 85-year study of saguaro (*Carnegiea gigantea*) demography. *Ecology* **79**, 2676–2693. (The ups and downs of a long-lived plant on an area protected from grazing since 1907.)

Porter WF and Underwood HB (1999). Of elephants and blind men: deer management in the U.S. National Parks. *Ecological Applications* **9**, 3–9. (Overabundant deer in the eastern USA raise critical questions about population limitation.)

Singer FJ, Swift DM, Coughenour MB and Varley JD (1998). Thunder on the Yellowstone revisited: an assessment of management of native ungulates by natural regulation. *Wildlife Society Bulletin* **26**, 375–390. (The controversy over how an important national park should be managed.)

Wolff JO (1997). Population regulation in mammals: an evolutionary perspective. *Journal of Animal Ecology* **66**, 1–13. (A careful evaluation of how mammal populations might be regulated from the principles of evolutionary theory.)

QUESTIONS

1. The identification of the critical factors regulating mortality in animals and plants often depends on identifying a single agent, such a predator or a disease, as a cause of death. Using your knowledge of human mortality factors, discuss the idea that a single cause of death can be readily specified for animals and plants.

2. Density-dependent relationships can be looked for by studying different local populations living in different patches at the same time (spatial density dependence) or by following one local population over several years (temporal density dependence). Discuss the interpretation of density-dependence in these two types of data with regard to the problem of regulation. Is population regulation only about temporal changes in populations?

3. Suppose that you studied a plant population for many years and then made a graph of the reproductive rate (number of seeds produced per plant) for each year against population density and the mortality rate (proportion of plants dying) for each year against population density and found that none of the three density-dependent relationships shown in Figure 10.2 applied. What would you conclude about population regulation in this plant?

4. Darwin (1859) wrote in *The Origin of Species* (chap. 2): 'Rarity is the attribute of a vast number of species of all classes, in all countries.' Discuss the possible effects of rarity and commonness on population-regulation mechanisms.

5. The saguaro is a prominent columnar cactus of the Sonoran Desert of Arizona and northern Mexico. Saguaro are long-lived perennials—individuals may reach 150–200 years of age. Pierson and Turner (1998) reported these data from a long-term study of four populations in an ungrazed desert preserve:

Plot	Census			
	1964	1970	1987	1993
North	284	265	232	221
South	1308	1316	–	1087
East	1367	1394	–	1277
West	603	586	–	459

What would you conclude about the population dynamics of these cacti from these data? What additional data would you like to have to predict future population trends?

6. You have been given an assignment to study the population dynamics of a butterfly species. What information would you need to know to determine whether this species was distributed as a meta-population or not?

Chapter 11
COMMUNITY DYNAMICS—SUCCESSION

IN THE NEWS

Logging of old-growth forests around the world, from Tasmania to British Columbia, has pitted forestry workers and mill owners against environmentalists and conservationists. There are two major ecological questions that underlie the logging of old-growth forests:

1. Will the logged forest recover to become similar old-growth forest in the future?

2. What is the time frame for this recovery?

Both these questions are ecological questions about **succession**, or the recovery of communities of plants and animals from disturbance. Logging is a major disturbance in forests and it destroys stands

An intact old-growth Douglas fir forest, Umpqua National Forest, Oregon. (*Photo by Garth Lenz (2002).*)

A clear-cut old-growth forest in coastal British Columbia.

Logging of an old-growth Douglas fir in coastal British Columbia in the 1920s.

of trees in the same way as severe forest fires and windstorms. Why is it important to know about forest succession? One practical reason is that valuable forest stands may not be easily replaced because the desirable species from a forestry perspective are replaced by weedy tree species. Many ecological reasons exist for such a poor outcome from a forester's point of view. Seeds of the desirable species may not be present if the large trees are removed, or fungal diseases introduced on logging equipment might be lethal to new seedlings. Herbivores such as deer might eat all the seedlings of the desirable species, but leave the weeds.

From a more practical point of view of economics, old-growth forests might replace themselves eventually, but on a time scale of centuries rather than decades. It is therefore highly unlikely that the logging enterprises would still be around to ensure old-growth replacement 100–500 years later. Many of our desirable old-growth trees are not able to re-establish immediately after logging because their seedlings need shade and soil moisture that are no longer present in clear-cut stands. In some forests, we can shorten the time sequence of succession by planting nursery-grown seedlings of desirable species, but these healthy, well-fertilized seedlings may soon be attacked by diseases, insects and mammals so that few of them survive.

Whether or not we should be harvesting old-growth forests is a complex social issue, but it is important to have correct ecological knowledge of how recovery can be achieved and how long it will take, so that the social decisions are not taken based on exaggerated promises of recovery that our children and grandchildren will never witness.

CHAPTER 11 OUTLINE

■ 11.1 COMMUNITIES DO NOT REMAIN CONSTANT BUT CHANGE SLOWLY OR RAPIDLY

One of the most important features of communities is change. In this chapter we focus on the factors causing communities to change on an ecological time scale. If you sat in a prairie for 10 years, or in a forest for 30 years, you would see the surrounding community change. These changes can be slow or they can be dramatic. The consequences for land and water management and conservation can be severe. If, for example, you designate a tract of prairie grassland as a protected area and keep grazing animals and fire out of the site, the prairie will turn into shrubland and finally forest, and you will have lost the grassland community you set out to protect. Knowing how these changes operate is critical for land management.

Succession is the universal process of change in communities. Ecologists want to know what species arrive first in succession and what species replace them in what order. They also would like to define a timeframe for succession, so that predictions of the stages of succession can be made. Finally, ecologists would like to know what environmental factors cause succession to proceed, and what disturbances can disrupt successional trends in communities. Knowing the mechanisms responsible for succession can lead to better land management.

Types of Community Change

Community change has important repercussions both for conservation and for all forms of water and land management for agriculture, forestry or recreation. Much of the discussion of succession focuses on plant communities, because animal communities rely upon them. There are two main types of changes in plant communities: directional changes and cyclical changes. Directional changes are the normal focus of succession on disturbed landscapes, such as when trees replace grasses and shrubs after a forest fire. Cyclical changes occur on a small scale of a few square meters and center on the death of a plant or a few individual plants. We shall begin with a graphic example of succession and then focus on three questions:

1. What is the time scale of succession?
2. What factors cause succession?
3. How predictable are successional changes?

When stripped of its original vegetation by fire, flood, glaciation or volcanic activity, an area of bare ground does not remain devoid of plants and animals for long. The area is rapidly colonized by a variety of pioneer species that subsequently modify one or more environmental factors. This modification of the environment may in turn allow additional species to become established. This development of the community by the action of vegetation on the environment leading to the establishment of new species is termed succession. Succession is the universal process of directional change in vegetation during ecological time. It can be recognized by a progressive change in the species composition of the community. The sequence of vegetation stages through which succession moves over time are called **seral stages**.

Most observed successions are called **secondary succession**, and involve the recovery of disturbed sites. A few successions are called **primary succession** because they occur on a new sterile area, such as that uncovered by a retreating glacier or covered by an erupting volcano. We will now look at one example of primary succession before we discuss the theory of succession.

Primary Succession on Mount St. Helens

Mount St. Helens in south-west Washington State erupted catastrophically on 18 May 1980. About 400 m was blown off the cone of this volcano and the blast from the eruption devastated a wide arc extending up to 18 km north of the crater. The eruption produced a landscape with low nutrient availability, intense drought, frequent surface movements and erosion—a great variety of conditions for vegetative recolonization. Three main areas affected by the eruption:

- the blast zone in which trees and vegetation were blown down, but not eliminated
- the pyroclastic flows (a hot mixture of volcanic gas, pumice and ash) to the north of the crater
- the extensive mudflows and ash deposits away from the crater toward the south, east and west.

In addition the eruption spewed tephra (ash) over thousands of square kilometers (see Figure 3.4, page 50).

Primary succession following volcanic eruptions has been studied less than other forms of succession, and Mount St. Helens provided a good opportunity to study the mechanisms determining the rate of primary succession. Permanent plots have been established at several sites above the tree line around the crater, and succession has been described by Roger del Moral and his students from the University of Washington over the past 25 years (Figure 11.1).

Colonization of habitats above the tree line on Mount St. Helens has been slow. Figure 11.2 shows a time sequence from 1984–2003 of a small mudflow south-west of the crater at Studebaker Ridge. In this site there were no surviving plants in 1980, and by 1984 only one species (*Lupinus lepidus*) had colonized the area (Figure 11.3). Only two additional species had invaded by 1989, but by 1997 there were 17 species occupying this site. Plant cover has increased slowly so that only the lupine species is common 23 years after the eruption.

Figure 11.1 Plant ecologist Roger del Moral from the University of Washington on the Pumice Plains of Mount St. Helens 9 years after the volcanic eruption. The golf balls were set in seed traps to catch the colonizing seed rain as a standardized method to measure the potential plant colonizers of the barren pumice surface. (Photo courtesy of Roger del Moral.)

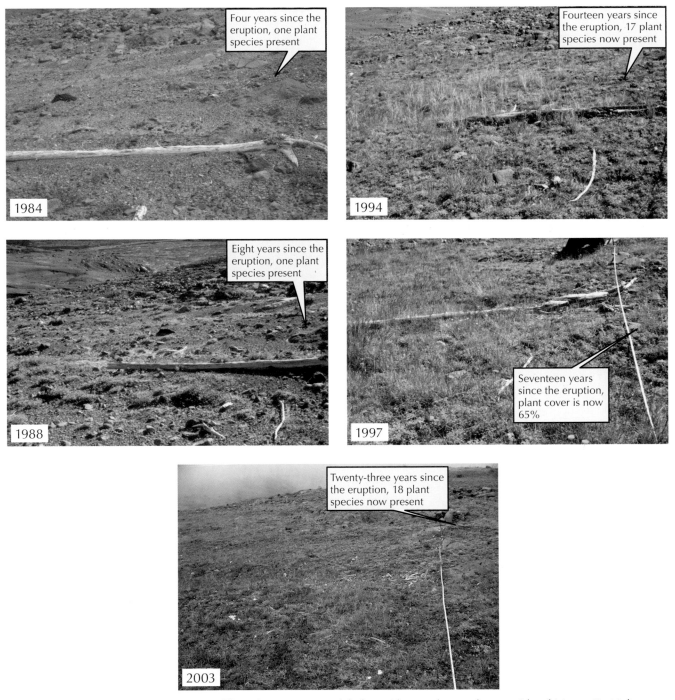

Figure 11.2 Primary succession on mudflow deposits on Studebaker Ridge on the south-west side of Mount St. Helens following the eruption of May 1980. Already in 1981 plants had colonized this devastated area and plant cover has been slowly increasing as the site undergoes primary succession. Figure 11.3 shows data from these plots. (Photos courtesy of Roger del Moral, 2004.)

Early primary succession on volcanic substrates rarely produces plant densities sufficient to inhibit the colonization of new species. Neither space nor light is a limiting resource for plants in this environment. Nurse plants facilitate the establishment of other species. Lupines (*Lupinus lepidus*) have heavy seeds and are

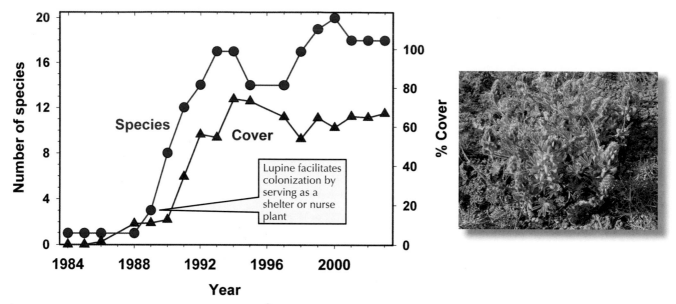

Figure 11.3 Number of species occurring in 1 m² quadrats in the Studebaker Ridge area on the south cone of Mount St. Helens, Washington. The number of species reached a plateau around 17–20 within 10 years of the irruption and the total plant cover on this area has increased only slowly because of the harsh conditions on these volcanic deposits. Photos of these plots are shown in Figure 11.2. (Data courtesy of R. del Moral, 2004, photo courtesy of Thayne Tuason.)

poorly dispersed, but have become locally common on mudflows and pyroclastic surfaces. Before lupines become very common, other wind-dispersed plants, such as *Aster ledophyllus* and *Epilobium augustifolium*, become established in lupine clumps and survive better in the shelter of these nurse plants. Lupines die after 5–6 years and, because they fix nitrogen, they contribute to increased soil nitrogen levels on a local scale.

Chance events strongly affected primary succession on Mount St. Helens. Biological mechanisms are initially very weak in the severe environments produced by volcanic flows. The ability to become established in these severe environments is directly related to seed size. But dispersal ability is related inversely to seed size. Consequently subalpine areas on Mount St. Helens receive many wind-blown seeds, but almost none of these small seeds germinated and colonized under the stressful soil and drought conditions. When plants with large seeds, such as lupine, colonize by chance, they become a focus of further community development. If by chance a single individual plant survives in the devastated landscape, it quickly becomes a center for seed dispersal to adjacent areas, so that there is a positive feedback during the early

years of succession. Primary succession has been very slow because of erosion, low nutrient soils and chronic drought stress, coupled with limited dispersal of larger seeds from plants in distant areas with undisturbed vegetation.

Q **Why is plant succession on Mount St. Helens not called secondary succession?**

Mount St. Helens provides us with a graphic example of plant succession after an extreme disturbance. Measuring the speed of change on the mudflow areas allows us to estimate that it will require more than 100 years for this landscape to return to a stable plant community. Understanding succession also requires understanding the mechanisms that drive changes in communities, and one focus has been on the effects that early successional species have on later successional species. Early species can help, hinder or not affect the establishment of later species. Competition between individual plants for resources such as water, light or nitrogen may drive succession. On Mount St. Helens we can see these processes in action and thus analyze them more easily. Understanding how naturally disturbed landscapes renew themselves can help us to predict how landscapes disturbed by humans might respond. We now turn to these broader issues, and review the theory of succession and the mechanisms involved.

■ 11.2 THREE MODELS OF SUCCESSION DEPEND ON WHETHER THE INITIAL SPECIES *HELP, HINDER* OR *IGNORE* SUBSEQUENT COLONIZERS

The idea of succession was largely developed in the 1890s by the botanists J.E.B. Warming in Denmark and Henry C. Cowles at Chicago. Warming's book, published in 1896, greatly influenced Cowles, who studied the stages of sand-dune development at the southern edge of Lake Michigan. Henry Cowles was one of the most influential plant ecologists in the United States in the early years of the 20th century, and the students he taught at Chicago became a who's who of American plant ecology. Successional studies pioneered by Warming and Cowles have led to three major hypotheses of succession.

The Facilitation Model

The first model is the classical theory of succession, called the **facilitation model**, which postulates an orderly hierarchical system of change in the community from pioneer species toward climax species

(Figure 11.4 a). The classical theory of succession was elaborated by Frederic E. Clements, who developed a complete theory of plant succession that included the **monoclimax hypothesis**. The biotic community, according to Clements, is a highly integrated superorganism. It shows development through a process of succession to a single end point in any given area—the **climatic climax**. The development of the community is gradual and progressive, from simple pioneer communities to the ultimate, or climax, stage. This succession is due to biotic reactions only—the plants and animals of the pioneer stages alter the environment in ways that favor a new set of species, and this cycle recurs until the climax is reached. According to Clements' view, development through succession in a community is therefore analogous to development in an individual organism. Secondary succession differs from primary succession in having a seed bank from plants that occur later in succession, so that late succession species are already present in the early stages of secondary succession (Figure 11.4 a, b). Secondary successions should also lead over time to the climatic climax.

The key assumption of the classical theory of succession is that species replace one another because at each stage the dominant species modify the environment to make it more suitable for other species. Thus species replacement is orderly and predictable and provides directionality for succession, leading finally to the **climax community**. These characteristics led Joe Connell and Ralph Slatyer to call this the **facilitation**

Henry C. Cowles (1869–1939)
Professor of Plant Ecology, University of Chicago

Frederic Clements (1874–1945) Professor of Plant Ecology, University of Nebraska

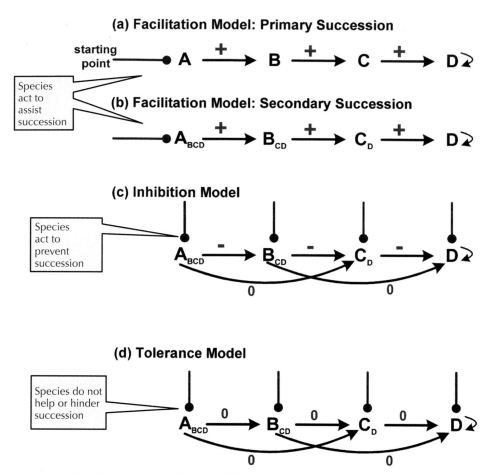

Figure 11.4 Conceptual models of succession. The capital letters A–D represent hypothetical vegetation types or dominant species; subscript letters indicate that species are present as minor components or as propagules. Black arrows represent vegetation sequences in time; blue lines represent alternative starting points for succession after disturbance. Circular arrows indicate that the species replaces itself. + = facilitation, – = inhibition, 0 = no effect. In these simplified sequences, species D would be the climax species. (Modified after Noble 1981.)

model of succession—the early species in succession facilitate the arrival of the later species.

In Clements' view, the climax community in any region is determined by climate. Other communities may result from particular soil types, fire or grazing, but these **subclimaxes** are understandable only with reference to the end point of the climatic climax. Therefore, the classification of communities must be based on the climatic climax, which represents the state of equilibrium for the area.

The Inhibition Model

A second major hypothesis of succession was proposed by Frank Egler in 1954, who called it **initial floristic**

composition. In this view, succession is very heterogeneous because the development at any one site depends on who gets there first. Species replacement is not necessarily orderly because each species excludes or suppresses any new colonists (Figure 11.4 c, d). Thus succession becomes more individualistic and less predictable because communities are not always converging toward the climatic climax, in contrast to what the classical theory of succession predicts.

Egler's hypothesis of initial floristics actually contained two ideas. Part of his hypothesis has been called the **inhibition model** of succession. The species present early in succession inhibit the establishment of the later species (Figure 11.4 c). No species in this model is competitively superior to another. Whichever plant colonizes the site first holds it against all comers until it dies. Succession in this model proceeds from short-lived species to long-lived species and is not an orderly replacement because chance always determines which species gets there first. This model has been called the **preemptive initial floristics model** to emphasize that the first species at a site pre-empt the course of succession.

The Tolerance Model

Frank Egler's initial floristics model can also describe the third major model of succession proposed by Joe Connell and Ralph Slatyer, who called it the **tolerance model**. This model is intermediate between the facilitation model and the inhibition model. In the tolerance model, the presence of early successional species is not essential—any species can start the succession (Figure 11.4 d). Some species are competitively superior, however, and these eventually come to predominate in the climax community. Species are replaced by other species that are more tolerant of limiting resources. Succession proceeds either by the invasion of later species or by a thinning out of the initial colonists, depending on the starting conditions. The tolerance model includes Egler's emphasis on the initial plant colonizers as a major influence on how succession proceeds.

The three hypotheses of succession all predict that many of the pioneer species in a succession will appear first because these species have evolved colonizing characteristics, such as rapid growth, abundant seed production and high dispersal powers (Table 11.1). The critical feature of the life-history traits given in Table 11.1 is that there is a trade-off, or inverse correlation, between traits that promote success in early succession and traits that are advantageous in late succession.

The critical distinction among the three hypotheses is in the mechanisms that determine subsequent establishment. In the classical facilitation model, species replacement is *facilitated* by the previous stages. In the inhibition model, species replacement is *inhibited* by the present residents until they are damaged or killed. In the tolerance model, species replacement is *not affected* by the present residents.

The utility of these three models of succession is that they immediately suggest experimental manipulations to test them. Removing or excluding early colonizers, transplanting seeds or seedlings of late succession species into earlier stages, and other experiments can shed light on the mechanisms involved in succession.

Q **Would animal ecologists have come up with the same three models of succession for animal communities?**

To explain a successional sequence, we must add to these idealized models of succession some additional information on seed availability, insect and mammal herbivory, mycorrhizal fungi and plant pathogens. In the same successional sequence some species may facilitate others while other species inhibit further colonization. The primary processes underlying successional changes could be competition between plant species or mutualistic interactions between plants that facilitate succession. But these plant–plant interactions are affected by other community interactions—animal grazing, plant diseases, seed dispersal and seed storage in the soil. The resulting successional sequences are thus complex and do not always proceed in a single direction to a fixed end point.

■ 11.3 THE MAJOR ECOLOGICAL MECHANISM DRIVING SUCCESSION IS COMPETITION FOR LIMITING RESOURCES

The simplest models of plant succession include only shade tolerance and seed availability for each species.

A Simple Mechanistic Model of Succession

A major mechanism driving succession is competition for limiting resources. One fundamental resource for plants is the light needed for photosynthesis. We can model succession mechanistically using an individual-based plant model that explicitly incorporates light as the limiting resource. Each individual plant is given species-specific traits of maximum size and age, maximum growth rate and tolerance to shading. Most of these models have been used for trees to model forest succession, but they could be applied to other kinds of plants as well. The key variable in these models is

Table 11.1 Physiological and life history characteristics of early- and late-successional plants.

Characteristic	Early succession	Late succession
Photosynthesis		
Light-saturation intensity	high	low
Light-compensation point	high	low
Efficiency at low light	low	high
Photosynthetic rate	high	low
Respiration rate	high	low
Water-use efficiency		
Transpiration rate	high	low
Mesophyll resistance	low	high
Seeds		
Number	many	few
Size	small	large
Dispersal distance	large	small
Dispersal mechanism	wind, birds, bats	gravity, mammals
Viability	long	short
Induced dormancy	common	uncommon?
Resource-acquisition rate	high	low?
Recovery from nutrient stress	fast	slow
Root-to-shoot ratio	low	high
Mature size	small	large
Structural strength	low	high
Growth rate	rapid	slow
Maximum lifespan	short	long

Source: From Huston and Smith (1987).

light availability—each individual plant is analyzed to see how much shading its neighbors produce and how, if light is limited, growth rates and survival rates are reduced accordingly. This type of simple mechanistic model can produce successional sequences of tree species that resemble natural succession. Figure 11.5 illustrates two scenarios with species of different life-history traits. In both cases the species that is most shade-tolerant in regeneration and grows to the largest size wins out in succession to form the climax species. The seral stages vary greatly depending on which trees are present.

The additional effects of competition for soil nitrogen can be added to these simple models, so that both light and nitrogen become the limiting resources. Mechanisms of nutrient limitation make these models more realistic, but also more complex to evaluate. The most successful models of succession are for forests.

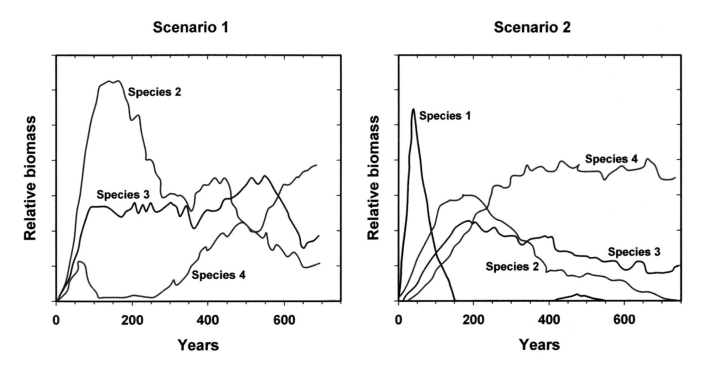

Species	Characteristics	Relative growth rate	Relative seed production	Diameter (cm)	Height (m)	Maximum age (years)	Shade tolerance
1	Fast growth, short life	2.17	5	50	15	50	intolerant
2	Moderate growth, longer life	1.43	1	100	30	300	tolerant
3	Slow growth, longer life	1.09	1	100	35	400	tolerant
4	Very slow growth, very long life	1.00	1	150	35	650	tolerant

Figure 11.5 A simple model of succession for trees. Species biomass dynamics and community biomass for hypothetical successional sequences with three and four idealized species in which competition for light is the driving variable. In scenario 1, all three species are shade-tolerant and thus have late-successional characteristics, but differ sufficiently in relative competitive abilities (growth rate) to produce a 'typical' successional replacement. In scenario 2, an early-successional species with a rapid growth rate and shade-intolerance is added to the three species in scenario 1. (Modified from Huston and Smith 1987.)

Because of their economic importance a wealth of detail is known about the life-history traits of individual tree species.

How well do natural communities fit the three hypotheses of succession? Does succession in a region converge to a single end point, or are there multiple stable end points, many climax vegetation communities? How predictable are successional sequences? Is the climax community in equilibrium or might it collapse? For Mount St. Helens, we have seen that the facilitation model of primary succession is a good description of the early stages of succession. As vegetation cover on the mudflows became more complete, inhibition began to operate. Let us look at some additional examples of succession to evaluate these models. Numerous studies of succession have been made, but there are few cases in which the succession can be related to a time scale. Some examples have been investigated in detail where the time scale is known, and we will look at three of these briefly.

Glacial Moraine Succession in South-eastern Alaska

During the past 200 years there has been a generalized retreat of glaciers in the Northern Hemisphere. As the glaciers retreat, they leave moraines, whose age can be determined by the age of the new trees growing on them or, in the last 120 years, by direct observation. The most intensive work on moraine succession has been done at Glacier Bay in south-eastern Alaska, where the glaciers have retreated about 98 kilometers since about 1750—an extraordinary rate of retreat (Figure 11.6).

The pattern of primary succession in this area proceeds as follows. The exposed glacial till is colonized first by mosses, fireweed, *Dryas*, willows and cottonwood. The willows begin as prostrate plants, but later grow into erect shrubs. Very quickly the area is invaded by alder (*Alnus*), which eventually forms dense pure thickets up to 9 meters tall. This requires about 50 years. These alder stands are invaded by Sitka spruce, which, after another 120 years, forms a dense forest. Western hemlock and mountain hemlock invade the spruce stands and, after another 80 years, the situation stabilizes with a climax spruce–hemlock forest. This forest, however, remains on well-drained slopes only. In areas of poor drainage, the forest floor of this spruce-

Needles of a Sitka spruce—a climax tree in SE Alaska

hemlock forest is invaded by *Sphagnum* mosses, which hold large amounts of water and increase the acidity of the soil greatly. As the *Sphagnum* spreads, the trees die out because the soil becomes waterlogged and too oxygen-deficient for tree roots, and the area becomes a **Sphagnum bog**, or **muskeg** (*Essay 11.1*). The climax vegetation is thus muskeg on the poorly drained areas and spruce–hemlock forest on the well-drained areas.

The bare soil exposed as the glacier retreated is quite basic, with a pH of 8.0 to 8.4, because of the carbonates contained in the parent rocks. The soil pH falls rapidly with the advent of plant colonisation. There is almost no change in the pH due to leaching in bare soil, which remains at pH 8. The most striking change

Dryas spp., a pioneering plant that fixes nitrogen

Leaves of alder—a pioneering shrub that fixes nitrogen

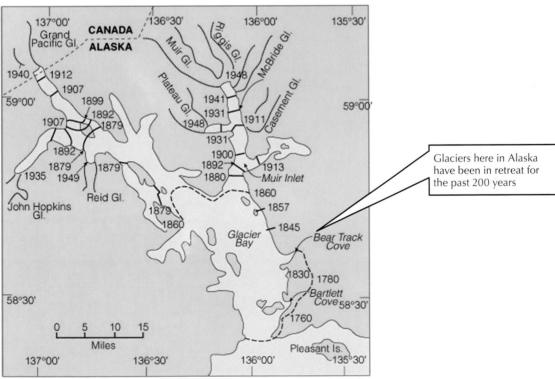

Figure 11.6 Glacier Bay fjord complex of south-eastern Alaska showing the rate of ice recession since 1760. As the ice retreats it leaves moraines along the edge of the bay on which primary succession occurs. The dashed lines on the map show the approximate edge of the ice in 1760 and 1860 from historical descriptions. (After Crocker and Major 1955, photo courtesy of US Geological Survey.)

225

ESSAY 11.1   WHY IS SPHAGNUM MOSS SO COMMON?

There may be more carbon locked up in *Sphagnum* bogs than in any other plant genus. Peat is largely poorly decomposed *Sphagnum,* and the surface of many bogs is dominated by living *Sphagnum* moss. Peat bogs are a prominent feature of many north temperate and south temperate areas, and the dominance of *Sphagnum* in these bogs makes them a good example of plant succession to a simple climax community. In cool climates, peat bogs have been an important source of fuel for humans—both Ireland and Scotland relied on peat fuel before the era of oil and gas.

Edge of a Sphagnum *bog in Minnesota*   *Individual* Sphagnum *plants*

Bogs are permanently wet areas that are fed by rainwater. The few nutrients in bogs typically come only from rainwater, so there is a shortage of essential plant nutrients for growth. In particular, all vascular plants growing in bogs must be able to tolerate very low levels of nitrogen and phosphorus. *Sphagnum* bogs have a 10–40 cm surface layer where ground water fluctuates depending on rainfall. *Sphagnum* is a moss that lacks roots and has no internal water-conducting tissues, so that it is very susceptible to desiccation.

Sphagnum is successful in bogs because it creates an environment in which few other plants can live. 'Nothing eats *Sphagnum*' is a general rule of bog ecology. Fresh *Sphagnum* has no lignin, which is the main constituent of cell walls in other plants. *Sphagnum* consists mostly of polysaccharides made up of glucose and galacturonic acids, which make it highly acidic. *Sphagnum* is also rich in phenols—defensive chemicals against herbivores. *Sphagnum* decomposes so slowly that anatomical details can be seen even after 70,000 years. Almost every other plant decomposes more rapidly than *Sphagnum* so that the fraction of *Sphagnum* material increases with depth in a bog. *Sphagnum* peat is famous for its excellent preservation of human and animal bodies and of other organic artifacts. This preservation is due to 5-keto-D-mannuronic acid, which suppresses microbial activity and aids in a tanning-like process as long as the peat is waterlogged.

Sphagnum is possibly the climax vegetation for many areas of pine and spruce forest on relatively flat areas in the north temperate zone. Stunted trees and shrubs are common on bogs, and poor growth is caused by adverse conditions below ground—low nutrients, high acidity, low oxygen and excess water. *Sphagnum* also conducts heat very poorly, so that soil temperatures remain low, giving a short growing season to vascular plants. *Sphagnum* has a negative impact on vascular plants, but a positive impact on its own growth. It has very low nutrient requirements, and it competes successfully for light, not by growing tall but by attacking the roots of its possible competitors. Because *Sphagnum* bogs contain about 20–30% of the world's carbon locked up in soils, their fate could affect atmospheric CO_2 levels, and ultimately climate change, in the coming centuries.

is caused by alder, which reduces the pH from 8.0 to 5.0 in 30–50 years. The leaves of alder are slightly acid, and as they decompose they become more acidic. As the spruce begins to take over from the alder, the pH stabilizes at about 5.0, and stays at that level for the next 150 years.

The organic carbon and total nitrogen concentrations in the soil also show marked changes with time. Figure 11.7 shows the changes in nitrogen levels. One of the characteristic features of the bare soil is its low nitrogen content. Almost all the pioneer species begin the succession with very poor growth and yellow leaves due to inadequate nitrogen supply. The exceptions to this are *Dryas* and, particularly, alder—these species have bacteria that fix atmospheric nitrogen.

The rapid increase in soil nitrogen in the alder stage is caused by the presence of nodules on the alder roots that contain microorganisms that actively fix nitrogen from the air. Spruce trees have no such adaptations, so the soil nitrogen level falls as alders are eliminated. The spruce forest develops by using the nitrogen accumulated by the alder.

One way to test experimentally for the impacts of species on each other is to seed or plant the later successional species into the earlier stages of succession. Terry Chapin and his students introduced seeds of alder and Sitka spruce into all the early successional stages at Glacier Bay to see if there was facilitation or inhibition. During the early successional stages inhibition of germination and survival was the dominant

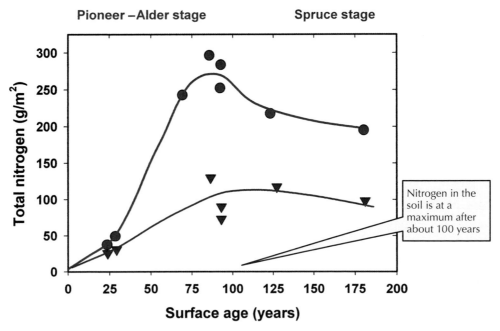

Figure 11.7 Total nitrogen content of soils recently uncovered by glacial retreat at Glacier Bay, Alaska. Nitrogen is typically a limiting nutrient for plant growth, and alders fix nitrogen in the soil. Plant successional stages are shown along the top, culminating in the climax spruce forest. (Data from Crocker and Major 1955.)

process. In later stages of succession Sitka spruce is facilitated by alder because of the nitrogen added to the soil.

The important point to notice here is the reciprocal interrelations of the vegetation and the soil. The pioneer plants alter the soil properties, which, in turn, permit new species to grow, and these species, in turn, alter the environment in different ways, bringing about succession. Succession on the moraines at Glacier Bay contains elements of both the classical facilitation theory (Figure 11.4 a) and the inhibition model (Figure 11.4 c). Life-history traits also affect succession because all the early pioneers have very light, wind-dispersed seeds. *Dryas* seeds weigh about 10 milligrams and can be blown easily by the wind. Spruce seeds, by contrast, weigh 270 milligrams and cannot be blown very far from the parent tree. No single model of succession accounts for primary succession at Glacier Bay—both facilitation and inhibition are involved.

Lake Michigan Sand-Dune Succession

Henry Cowles, from the University of Chicago, worked on the sand-dune vegetation of Lake Michigan during the 1890s and made a classic contribution to our understanding of plant succession. The sand dunes around the southern edge of Lake Michigan have been a model system for examining the theories of succession in a dynamic landscape. Cowles did not have radiocarbon dating methods available to him, but now ecologists have re-examined the successional stages in this area in relation to an absolute time scale determined by radiocarbon dates.

During and after the retreat of the glaciers from the Great Lakes area, the resulting fall in lake level left several distinct 'raised beaches' and their associated dune systems. These systems, which run roughly parallel to the present shoreline of Lake Michigan, are about 7, 12 and 17 meters above the present lake level (Figure 11.8). Ecologists can date the older dunes by radiocarbon techniques and the younger areas by tree-ring counts and recorded historical changes since 1893.

The dunes, like glacial moraines, offer a near-ideal system for studying plant succession because many of the complicating variables are absent. The initial substrate for all the area is dune sand, the climate for the whole area is similar, the relief is similar and the available flora and fauna are the same. So the differences between the dunes should be due only to **time**, **biological processes of succession** and **chance events** associated with dispersal and colonization.

Two processes produce bare sand surfaces ready for colonization. One is the slow process of a fall in lake level; the other is a rapid process—the **blow-out** of an established dune. These blow-outs result from the strong winds that come off the lake. This wind erosion sets up a moving dune that is gradually stabilized after migrating inland. The dunes are stabilized only by vegetation.

The bare sand surface is colonized first by dune-building grasses, of which the most important is marram grass *(Ammophila breviligulata)*. Marram grass usually propagates by rhizome migration, only rarely by seed. It spreads very quickly and a single plant can stabilize a bare area of 100 m^2 in 6 years. After the sand is stabilized, marram grass declines in vigor and dies out. The reason for this is not known, but the grass disappears from the stable dune areas after about 20 years.

Marram grass stabilizing beach dunes

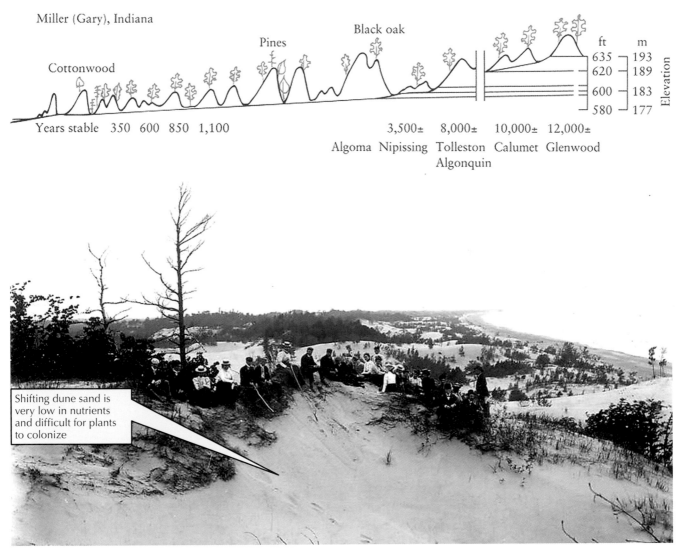

Figure 11.8 Diagrammatic profiles across Indiana sand dunes at the southern end of Lake Michigan. Successively older dune systems originated along earlier and higher beaches. The photo shows a botany field class with Henry Cowles from the University of Chicago on the Indiana Dunes about 1902, with Lake Michigan to the right. (After Olson 1958, photo courtesy University of Chicago.)

Two other grasses are important in dune formation and stabilization: sand reed grass (*Calamovilfa longifolia*) and the little bluestem (*Andropogon scoparius*). The sand cherry (*Prunus pumila*) and willows (*Salix* spp.) also play a role in dune stabilization. The first tree to appear in the young dune is usually the cottonwood (*Populus deltoides*), which may also help to stabilize the sand.

Once the dune is stabilized, it may be invaded very quickly by jack pine and white pine if seed is available; pines are usually found after 50 to 100 years of development. Under normal conditions, black oak replaces the pines, entering the succession at about 100 to 150 years. A whole group of shrubs that require considerable light invade the early pine and oak stands, but they are replaced by more shade-tolerant shrubs as the forest of black oak becomes denser.

Cowles believed that this succession to black oak might be part of the succession sequence, which would then proceed to a white oak–red oak–hickory forest and finally to the 'climate climax,' beech–maple forest. But more recent work has questioned whether this could ever occur. The oldest dunes (12,000 years) still had black oak associations, and there appears to be no tendency for any further succession. Moreover, the black oak community was very heterogeneous: Olson recognized four different types of understory communities that could occur under black oak. Successional patterns on the dunes do not lead to a single climax vegetation in ecological time.

Most of the soil improvements of the original barren dune sand occur within about 1000 years of stabilization. As a result of these trends in the soil, Olson pointed out, the nutritional conditions for succession toward beech and maple probably become *less* favorable with time (these trees require more calcium, near neutral pH and larger amounts of water). It appears improbable that this succession will move beyond the black oak stage, contrary to what Cowles had suggested. Beech and maple associations in this area are found only in favorable situations, such as moist lowlands, where the soil characteristics differ from those of the dry dunes. The low fertility of the dune soils favors vegetation, such as the black oak, that has limited nutrient and water requirements. But this sort of vegetation is ineffective in returning nutrients to the dune surface in its litter, which continues the cycle of low fertility.

As far as the dunes are concerned, probably the most striking changes in vegetation occur in the first 100 years and then the system seems to become stabilized with the black oak association. It is a mistake to distort the different dune successions into a single linear sequence leading from pine to black oak to oak–hickory to beech–maple. Successions in the dunes go off in different directions with different destinations, depending on the various soil, water and biotic factors involved at a particular site. Olson suggested that instead of convergence to a single climax, we may get *divergence* of different communities on different sites, or multiple stable end points. Furthermore, human use of the beach areas along the dunes tends to disturb the sand and slow down the whole successional sequence.

Thus dune succession around Lake Michigan begins as the classical facilitation model suggests, but then changes to the inhibition model at the black oak stage and culminates in several different stable communities.

Abandoned Farmland in North Carolina

When upland farm fields are abandoned in the Piedmont area of North Carolina, a succession of plant species colonizes the area. The sequence is as follows:

Years after last cultivation	Dominant plant	Other common species
0 (autumn)	Crabgrass ↓	
1	Horseweed ↓	Ragweed
2	Aster ↓	Ragweed
3	Broomsedge ↓	
5–15	Shortleaf pine ↓	Loblolly pine
50–150	Hardwoods (oaks)	Hickories

This is initially a striking sequence of rapid replacements of herbaceous species, and Catherine Keever attempted to find out why the initial species die out and why the later colonizers are delayed entry to the succession.

The sequence of this succession is dictated by the life history of the dominant plants. Horseweed (*Erigeron canadensis*) seeds will mature as early as August and germinate immediately. It overwinters as a rosette plant, which is drought-resistant and grows, blooms and dies the following summer (it is a biennial). The second generation of horseweed plants is stunted and does not grow well in the second year of the succession. Decaying horseweed roots inhibit the growth of horseweed seedlings. The density of horseweed individuals is much greater in second-year fields, but these small individuals do not do well with increased competition from aster. The result is a great reduction of horseweed dominance in second-year fields.

Aster (*Aster pilosus*) seeds mature in the fall—too late to germinate in the year that plowing ceases. Seeds germinate the following spring and seedlings grow slowly during their first year to reach 5 to 8 cm

Horseweed (*Erigeron canadensis*). (Photo courtesy of Dr Kenneth R. Robertson.)

Broomsedge (*Andropogon virginicus*)

(2–3 inches) by autumn. This slow growth is caused partly by shading by horseweed and also by stunted root growth caused by decaying horseweed roots. So, horseweeds do not facilitate asters; if anything, they inhibit them. Asters enter the succession in spite of horseweeds, not because of them.

Aster (a perennial) blooms in its second year, after horseweed declines, but is not drought-resistant. Seedlings of aster are present in large numbers in third-year fields, but succumb to competition for moisture with the drought-resistant broomsedge (*Andropogon virginicus*). In fields with more water available, aster is able to last into the third year, but eventually broomsedge overwhelms it.

Broomsedge seeds will not germinate without a period of cold dormancy. A few broomsedge plants are found in first-year fields, but they do not drop seed until the fall of the second year. Broomsedge is a very drought-resistant perennial that competes very well for soil moisture. There are few broomsedge plants in first- and second-year fields because few seeds are present. Once a few plants begin seeding, broomsedge rapidly increases in numbers (third year). Broomsedge grows better in soil with organic matter, especially in soil with aster roots. It grows very poorly in the shade.

Thus early succession in Piedmont old fields is governed more by competition than by cooperation between plants. The early pioneers do *not* make the environment more suitable for later species, and the later species achieve dominance in spite of the changes caused by the early species, rather than because of them. If seeds were available, broomsedge could colonize an abandoned field immediately, rather than following horseweed and aster. Old-field succession is not described well by the facilitation or tolerance models, but seems to be an excellent illustration of the inhibition model.

After this succession by herbs and grasses, the abandoned farmland is invaded in great numbers by shortleaf pine (*Pinus echinata*). Pine seeds are light and are blown by the wind, so they effectively disperse into abandoned fields if there are large pines anywhere nearby. Pine seeds can germinate only on mineral soil and are able to become established in shady, but bare, sites among the herbs and grasses. Pine seedlings are very effective competitors for soil water and,

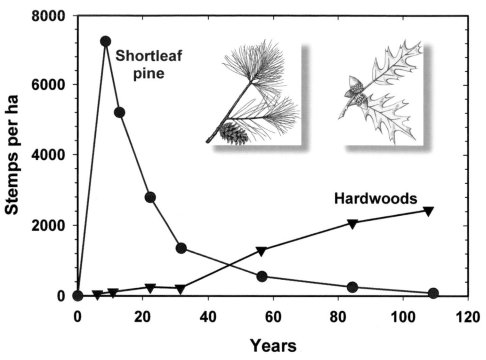

Figure 11.9 Decline in the abundance of shortleaf pine—the pioneer tree species—and increase in the density of hardwood tree seedlings during succession on abandoned farmland in the Piedmont area of North Carolina. (After Billings 1938.)

as they grow, they shade the herbs and grasses in the understory. The density of pines becomes very high, but falls rapidly as the pines lose their dominance to hardwoods (Figure 11.9). After approximately 50 years several species of oaks become important trees in the understory, and the hardwoods gradually take over the community. New seedlings of shortleaf pine are virtually absent after about 20 years because there is no bare soil for seed germination, and pine seedlings cannot live in shade. The networks of pine roots in the soil become closed very quickly, and the accumulation of litter under the pines causes the old-field herbs to die out. Oak seedlings first appear after 20 years, when enough litter has accumulated to protect the acorns from desiccation and the soil is able to retain more moisture. Hardwood seedlings persist in the understory because they develop a root system deep enough to exploit soil water, they are shade tolerant and they compete more effectively than pines for water and nutrients.

Soil properties change dramatically along with this plant succession on Piedmont soils. Organic matter accumulates in the surface layers of the soil and permeates to the deeper soil layers, which increases the moisture-holding capacity of the soil.

Thus shortleaf pine is independent of the early succession in that it requires only bare soil for germination. If all herbaceous species could be eliminated from the early succession, this would not affect colonization by pines, which fits the tolerance model of Figure 11.4d. Pines can invade old fields as soon as seeds become available. Oaks and other hardwoods, by contrast, depend on the soil changes caused by pine litter, so oak seedlings could not become established without the environmental changes produced by pines. Thus the latter part of this succession seems to fit the classical facilitation model.

The examples we have just discussed do not always fit any single model of succession, as some replacements are facilitated while others are inhibited. Most forest succession does not conform to the classical model. If the classical model is not always correct, we must reconsider the nature of the climax state—the 'end point' of succession.

■ 11.4 SUCCESSION PROCEEDS TO A CLIMAX STAGE, WHICH IS RELATIVELY STABLE OVER ECOLOGICAL TIME

In the examples of succession described in the preceding section, the vegetation has developed to a certain stage of equilibrium. This final stage of succession is called the **climax**. Numerous definitions of the climax have been made. *A climax is the final or stable community in a successional series. It is self-perpetuating and in equilibrium with the physical and biotic environment.* There are three schools of thought about the climax state: the monoclimax school, the polyclimax school and the climax-pattern view.

Monoclimax Theory

The *monoclimax* theory was an invention of American Frederic Clements. According to the monoclimax theory, every region has only one climax community towards which all communities are developing. This is the fundamental assumption of Clements—that given time and freedom from interference, a climax vegetation of the same general type will be produced and stabilized, irrespective of earlier site conditions. Climate, Clements believed, was the determining factor for vegetation, and the climax of any area was solely a function of its climate.

However, it was clear in the field that in any given area there were communities that were not climax communities, but that appeared to be stable. For example, tongues of tall-grass prairie extended into Indiana from the west, and isolated stands of hemlock occurred in what is supposed to be deciduous forest. In other words, we observe communities in nature that are non-climax according to Clements, but seem to be in equilibrium nevertheless. These communities are determined by topographic, edaphic (soil) or biotic factors.

Polyclimax Theory

The *polyclimax* theory arose as the obvious reaction to Clements' monolithic system. The British plant ecologist Arthur G. Tansley was one of the early proponents of the polyclimax idea—that many different climax communities may be recognized in a given area, such as climaxes controlled by soil moisture, soil nutrients, activity of animals and other factors. The American plant ecologist Rex Daubenmire also suggested in 1966 that there may be several stable communities in a given area, so there was no one single climax for a region.

The real difference between these two schools of thought lies in the time factor of measuring relative stability. Given enough time, say monoclimax supporters, a single climax community would develop, eventually overcoming the edaphic climaxes. The problem is, should we consider time on a geological scale or on an ecological scale? The important point, as we know now, is that climate fluctuates and is never constant. We see this vividly in the Pleistocene glaciations and more recently in the advances and retreats of mountain glaciers in the last 300 years (Figure 11.6). *So the condition of equilibrium can never be reached because the vegetation is subject not to a constant climate, but to a variable one.* Climate varies on an ecological time scale as well as on a geological time scale (*Essay 11.2*). Succession then in a sense is continuous because we have variable vegetation interacting with a variable climate.

Climax-pattern Hypothesis

In 1953, the Cornell plant ecologist Robert Whittaker proposed a variation of the polyclimax idea—the **climax-pattern hypothesis**. He emphasized that a natural community is adapted to the whole pattern of environmental factors in which it exists—climate, soil, fire and biotic factors Whereas the monoclimax theory allows for only one climatic climax in a region, and the polyclimax theory allows for several climaxes, the climax-pattern hypothesis allows for a continuity of climax types, varying gradually along environmental gradients and not neatly separable into discrete climax types. Thus the climax-pattern hypothesis is an extension of the idea that vegetation changes gradually and continuously in response to environmental gradients. The climax is recognized as a steady-state community with its constituent populations in dynamic balance with environmental gradients. We do not speak of a climatic climax, but of prevailing climaxes that are the end result of climate, soil, topography and biotic factors, as well as fire, wind, salt spray and other influences, including chance. The usefulness of the climax as an operational concept is that similar sites in a

ESSAY 11.2 HOW PALEOECOLOGY CAN ADD A LONG-TERM PERSPECTIVE TO UNDERSTANDING SUCCESSION

Ecologists are limited to the past 100–200 years for direct observations of plant succession. To obtain a longer timeframe of changes, paleoecologists have developed a clever array of techniques for analyzing fossil shells, pollen and bones. Depending on the methods used, it is possible to look back 20,000 to 30,000 years, and ask how communities have changed during that time. The most spectacular changes have occurred in areas that were strongly affected by the last Ice Age, and a great deal of information is now available to help us reconstruct these past environments.

The most common method used in paleoecology is pollen analysis. Researchers take cores in lake beds or peat bogs and extract pollen grains from the sediment. Pollen grains from different plant species have different shapes, which makes it relatively easy to tell what species were present in the vegetation when the pollen was deposited. By counting the relative numbers of different pollen grains in a sediment sample, it is possible to judge the composition of the vegetation at the time. This information, added to radiocarbon dates obtained from the same level in the core, allow paleoecologists to reconstruct the vegetation changes that have occurred in the vicinity of the lake or peat bog. By studying many such cores, it is possible to reconstruct vegetation changes on a continental scale over thousands of years.

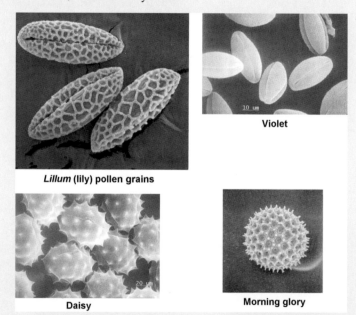

Pollen grains found in lake sediments in central Canada

An example of a pollen profile from the subarctic of Canada illustrates the vegetation changes in this area since the melting of the ice sheets 10,000 years ago. The lake sediments examined were in central Canada in an area that is now shrub tundra. The core containing sediments with pollen was about 80 cm long, and from radiocarbon dating spanned deposition in the lake over the last 8500 years.

Map showing location of lake

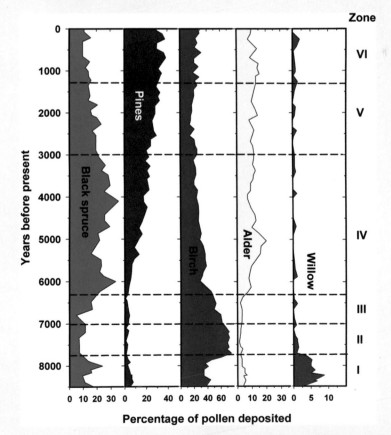

Percentage of pollen of various species deposited in a lake core over the past 8500 years

235

In the oldest part of the lake core—about 8500 years ago (Zone I in the diagram)—there was relatively little pollen deposited, and this possibly represents a shallow lake with some black spruce, willow and birch around the edges in a barren landscape. An abrupt increase in birch pollen occurred about 7700 years ago (Zone II), and the vegetation in the area became a peat bog with birch tundra. About 7000 years ago, black spruce pollen began to increase, and by 6300 years ago, shrubs and trees began to increase dramatically (Zone IV). Pines and alder, along with spruce, became common in the area, and the vegetation around the lake was a forest. About 3000 years ago, black spruce began to decrease, while pines continued to increase in abundance (Zone V). More recent vegetation was present from about 1300 years ago, characteristic of the forest tundra transition zone that presently surrounds the lake (Zone VI). The broad picture is of forest advance in this subarctic region beginning about 6000 years ago, followed by a forest retreat beginning about 3000 years ago. The inference is that the climate was warmer during the time of forest advance, and then became colder, with a shorter growing season, about 3000 years ago. These kinds of conclusions can be cross-checked with data from other pollen cores in bogs and lakes across a broad region to provide a way of estimating how general the climatic trends were during the last 10,000 years in this part of North America.

The general message that has come from this research in paleoecology is that the vegetation around the globe has changed dramatically on a time scale of hundreds of years, reflecting the changes in global climate caused by ice ages and other large-scale disturbances. A stable equilibrium in vegetation has not been observed, even in tropical areas.

region should produce similar climax stands. This stand-to-stand regularity should allow prediction for new sites of known environment—we can say that a particular site should develop in, say, 100 years to a stand of sugar maple and beech of specified density.

There is a tendency for plant ecologists to assume that vegetation succession is driven by climate, soils and plant–plant interactions, so that animals are left out of the picture. A good example of how animals may affect plant succession is found in the uplands of north-west Scotland. Sheep grazing selects against trees and favors grassland. Several vegetation types occur in the Scottish uplands. Under low grazing pressure and without fire, Scot's pine and birch woodlands tend to be favored. Under high grazing pressure and frequent fires, only grassland is self-sustaining, and when grazing is moderate and fires are occasional both heather moor (Calluna vulgaris) and bracken fern (Pteriduium aquilinum) communities are favored. There is no single climatic climax on these Scottish uplands—the plant communities are a mosaic, which changes with

the level of sheep grazing and fire frequency. These communities are best described by Whittaker's climax-pattern hypothesis.

How can we recognize climax communities? The operational criterion is the attainment of a steady state over time. Because the time scale involved is very long, observations are lacking for most presumed successional sequences. We assume, for example, that we can determine the time course of succession for a spatial study of younger and older dune systems around Lake Michigan (Figure 11.8), but this translation of space and time may not be valid. In forests, we can use the understory of young trees to look for changes in species composition, because the large trees must reproduce themselves on a one-for-one basis if steady state has been achieved. Forest changes may be very slow. Ken Lertzman studied a subalpine forest stand at 1100 meters elevation near Vancouver, Canada. This site had been undisturbed by fire for almost 2000 years and still was not in equilibrium because the dominant western hemlocks were not replacing themselves, while Amabilis fir seedlings were invading the understory in large numbers. Climax vegetation on this site had not been reached after two millennia.

Some communities may appear to be stable in time and yet may not be in equilibrium with climatic and soil factors because of grazing effects. A striking example of this occurred after the outbreak of the disease myxomatosis in the European rabbit in Britain. Before 1954, rabbits were common in many grassland areas. Myxomatosis devastated the rabbit population in 1954 and the consequent release of grazing pressure caused dramatic changes in grassland communities. The most obvious change was an increase in the abundance of flowers. Species that had not been seen for many years suddenly re-appeared in large numbers. There was also an increase in woody plants, including tree seedlings that were commonly grazed by rabbits. No one anticipated these effects. The vegetation would reach two quite different states: with and without rabbit grazing.

Q Would the three climax vegetation hypotheses lead to different vegetation management policies of you were in charge of a national park?

We conclude from this discussion that climax vegetation is an abstract ideal that is seldom reached, owing to the continuous fluctuations of climate. The climate of an area has a clear overall major influence on the vegetation, but within each of the broad climatic zones are many other variables, such as soil, topography and animals, which lead to many climax situations. The rate of change in a community is rapid in early succession, but becomes very slow as it approaches the potential climax community. But, like Nirvana, it may never quite reach this state.

■ 11.5 SMALL PATCHES MAY BE CHANGING IN A REGENERATION CYCLE WITHIN A CLIMAX LANDSCAPE

Communities are dynamic and are changing continually. In 1947 A.S. Watt—a British plant ecologist—first drew attention to cyclical events in communities that occur in patches on small spatial scales and are repeated over and over in the whole of the community.

A plant community over a region may be moving slowly towards a climax state, while on a local scale the internal dynamics of the community are producing cyclical changes that are more rapid. These cyclical changes occur in patches, or gaps, in the community, and the study of gap dynamics has shed interesting light on the overall processes of community change. We will discuss three examples of **patch dynamics.**

Cyclical Vegetation Changes

Alex Watt studied several examples of cyclical vegetation changes in British vegetation. One of these was the *Calluna* heath that covers large areas in Scotland and has made heather almost a synonym for Scotland. The dominant shrub in this community is heather *(Calluna),* which loses its vigor as it ages and is invaded by the lichen *Cladonia.* In time the lichen mat dies to leave bare ground. This bare area is invaded by bearberry *(Arctostaphylos),* which in turn is invaded by *Calluna.*

Heather *(Calluna)* is the dominant plant, and *Arctostaphylos* and *Cladonia* occupy the area that is temporarily vacated by *Calluna.*

The cycle of change can be divided into four phases (Figure 11.10):

- *Pioneer*: establishment and early growth in *Calluna*—open patches, with many plant species (years 6 to 10).
- *Building*: maximum cover of *Calluna* with vigorous flowering—few associated plants (years 7 to 15).
- *Mature*: gap begins in *Calluna* canopy and more species invade the area (years 14 to 25).
- *Degenerate*: central branches of *Calluna* die—lichens and bryophytes become very common (years 20 to 30).

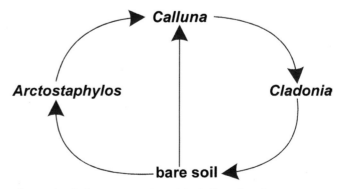

The cyclical change on a Scottish *Calluna* heath

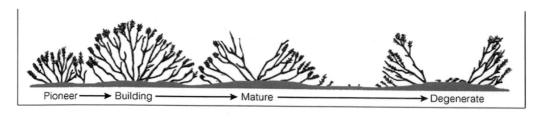

Figure 11.10 Patch dynamics in a plant community. Profile of the four phases in the heather (*Calluna*) cycle in Britain. Like many perennial plants, heather loses vigor with age so older plants die back naturally. These patches occupy only a few square meters. (Modified after Watt 1955, photo courtesy of Paul Busselen, Catholic University of Louven, Belgium.)

Plant ecologists have described this sequence in detail from maps of permanent quadrats in Scotland. The life history of the dominant plant *Calluna* controls the sequence.

Bracken fern *(Pteridium aquilinum)* is a cosmopolitan species that lives in a great variety of soil types and climatic conditions. It forms rhizomes in the soil and, once established, it spreads vegetatively. Bracken is fire-resistant because of its underground rhizomes and, when it invades grassland, there is a vigorous 'front' of invading bracken, but reduced vigor in older fronds (Figure 11.11). The obvious explanation for this marginal effect is that a particular soil nutrient is depleted by the advancing fern and is in short supply in the older stands. However, there is no accompanying soil change to account for the reduced vigor, and the addition of fertilizers to sample plots in the older stands produced no effect. The significant variable seems to be rhizome age. Younger rhizomes produce more vigorous fronds. The explanation for this is not known.

Watt divided all these cycles of change into an **upgrade series** and a **downgrade series** and pointed out that the total productivity of the series increases to the mature phase and then decreases. What initiates the downgrade phase? A possible explanation lies in the relationship between the general vigor of a perennial plant and its age. There seems to be a general relationship between age and performance in most perennial plants and, consequently, between age and

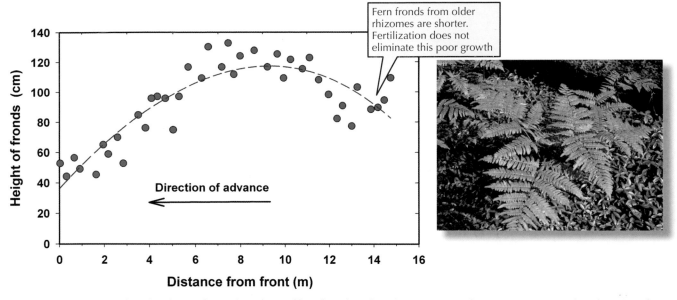

Figure 11.11 Average height above the soil surface of bracken fern fronds across an advancing margin as bracken invades a grassland in England. Bracken fern is widely distributed around the world (the photo was taken in northern Michigan). (After Watt 1940, photo courtesy of Dr Robert Beall.)

competitive ability. Several studies on the relation of leaf diameter to age also support this idea.

For this reason, a stable community will be in a constant state of phasic fluctuation—with one species becoming locally more abundant as another species reaches its degenerate phase. These dynamic interrelationships in natural communities tend to operate on small spatial scales and may not be conspicuous without detailed measurement.

Gap Dynamics in Forests

Patch dynamics will result in a single dominant species—a monoculture—if one species is able to replace itself exclusively and also replace all the other species, so that it is the best competitor in the system. As early as 1905, French foresters suggested that in virgin forests, individual trees tended to be succeeded in time by those of another species. This is called **reciprocal replacement**. Reciprocal replacement occurs at the individual tree level and could explain why old-growth forests have a mixture of tree species rather than a single dominant species. If, for example, species A seedlings were found predominately under large species B trees, and species B seedlings were found predominately under large species A trees, we

would have reciprocal replacement at the individual tree level.

One striking example of codominance occurs in the eastern deciduous forests of North America, where American beech and sugar maple form a climax forest with both species being abundant. Why do these forests not succeed towards a monoculture of the most competitive, shade-tolerant tree species? Good examples of the beech–sugar maple forest are found in old-growth forests in southern Michigan, and reciprocal replacement at the individual tree level has been suggested as one possible mechanism of codominance. But reciprocal replacement does not occur at the individual tree level in this forest stand. Beech seedlings were not concentrated under sugar maple trees, and sugar maple seedlings were not concentrated under beech trees. Codominance is caused by differing growth habits of seedlings in gaps in the forest. As older trees die and fall down, they create gaps in the forest canopy. Large gaps produce a large area of sunlight that favors sugar maples, which grow quickly, while beech seedlings cannot take advantage of large gaps because they grow slowly (Figure 11.12 top). When the forest is like this, large gaps are rare and beech outnumber sugar maples by 7 to 1. Small beech trees spread their leaves laterally

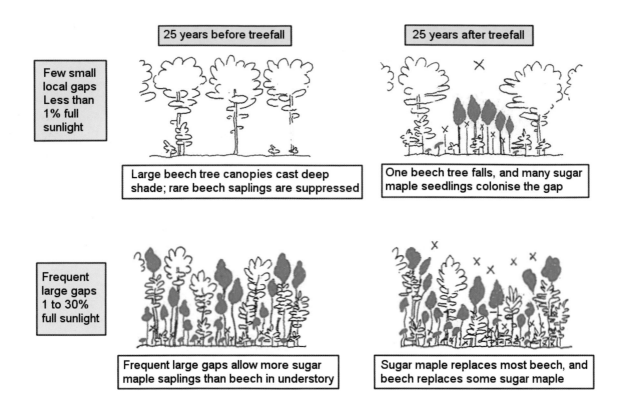

Figure 11.12 Conceptual model of how the American beech–sugar maple forests of Michigan maintain codominance of two climax tree species instead of succeeding to a single dominant climax species. The left column is before tree-fall; the right column is after tree-fall. Infrequent falls of large beech trees every 100+ years (top panel) favor sugar maple (green tree silhouettes) while frequent gaps every 10–40 years (bottom panel) favor a mixture of sugar maple and beech (open tree silhouettes). X marks the former positions of the crowns of fallen trees in the right column. Maples from the understory were suppressed on average 20 years before being released to grow in a gap, while beech trees from the understory were suppressed 121 years on average. Intermediate conditions between these two extreme scenarios will often favor beech over sugar maple. This type of succession mechanism will maintain a climax forest containing both beech and sugar maple. (Modified from Poulson and Platt 1996.)

and capture light in flecks in the understory. Beech trees thus thrive better than maples in the understory of these forests with little light. If there were no gaps produced by tree falls, beech trees would eventually come to dominate the forest. But, when old trees fall more frequently and many gaps are produced (Figure 11.12 bottom), sugar maples increase in relative abundance in the gaps. Codominance of these two tree species is stable because of continual tree falls producing a mixture of gaps of different sizes that are created over time spans of hundreds of years.

Gap dynamics play an important role in the regeneration of forest communities, and other communities in which space is a key resource. Space in forests, and in many other plant communities, is equivalent to light, so competition for space is usually thought to be equivalent to competition for light. But light is not the only resource for which plants compete—nitrogen levels, other soil nutrients and water may modify competition between plants. Thus communities are dynamic and they change because of the interactions between the life-history traits of the dominant species and patterns of physical disturbances. An apparently stable community at a regional level may be a mosaic of patches undergoing changes at a local level. Succession is change and we will now consider how biological communities are organized and what mechanisms control their structure.

SUMMARY

Communities change over time and the study of succession has been an important focus of community ecologists for a century. Succession is the process of directional change in communities—the process of recovery from disturbances. Most of the work on succession has been done on plants, but the same principles apply to animal communities.

Three conceptual models of succession have been proposed to explain the direction of vegetation changes. All models agree that pioneer species in a succession are usually fugitive or opportunistic species with a high dispersal rate and rapid growth. These pioneer species are replaced in three possible ways. The classical model states that species replacements in later stages of succession are facilitated by organisms present in earlier stages. At the other extreme, the inhibition model suggests that species replacements are inhibited by earlier colonizers and that successional sequences are controlled by the species that gets there first. The tolerance model suggests that species replacements are not affected by earlier colonizers and that later species in succession are able to tolerate lower levels of resources than earlier species. No single model explains an entire successional sequence. Ecologists now try to analyze succession as a dynamic process resulting from a balance between the colonizing ability of some species and the competitive ability of others. Succession does not always involve progressive changes from simple to complex communities.

Succession proceeds through a series of seral stages from the pioneer stage to the climax stage. The monoclimax hypothesis suggested that there was only a single predictable end point for whole climatic regions and that, given time, all communities would converge to a single climatic climax community. This hypothesis has been superseded by the climax pattern hypothesis, which suggests a variety of different climaxes occur along environmental gradients, controlled by soil moisture, nutrients, herbivores, fires or other factors.

A stable community contains small patches undergoing cyclical changes that are repeated over and over as part of the internal dynamics of the community. The life cycle of the dominant organisms dictates the cyclical changes, many of which are caused by the decline in vigor of perennial plants with age. In many forests, tree-fall gaps create a mosaic of patches undergoing cyclical changes within a relatively stable climax community.

Communities are not stable for long periods in nature because of disturbances caused by short-term changes in climate, fires, windstorms, diseases or other environmental factors. For most communities, we can observe changes over time, but we need to determine the mechanisms that cause the changes. Unless we understand the mechanisms behind succession, we will be unable to suggest manipulations to alleviate undesirable successional trends caused by human activities such as logging.

SUGGESTED FURTHER READINGS

del Moral R and Jones C (2002). Vegetation development on pumice at Mount St. Helens, USA. *Plant Ecology* **162**, 9–22. (Twenty years of succession on the Mount St. Helens volcano.)

Finegan B (1996). Pattern and process in neotropical secondary rain forests: the first 100 years of succession. *Trends in Ecology and Evolution* **11**, 119–124. (What happens when agricultural land is abandoned in the wet tropics, and how long will tropical forests take to recover?)

Henry, HAL and Aarssen LW (1997). On the relationship between shade tolerance and shade avoidance strategies in woodland plants. *Oikos* **80**, 575–582. (A good discussion of how light availability drives succession in forests, and some of the physiological mechanisms involved.)

McCook LJ (1994). Understanding ecological community succession: causal models and theories, a review. *Vegetatio* **110**, 115–147. (A critical review of succession theory and an outline for further synthesis.)

Torti SD, Coley PD and Kursar TA (2001). Causes and consequences of monodominance in tropical lowland forests. *American Naturalist* **157**, 141–153. (How do some species-rich tropical rainforests come to be dominated by a single tree species?)

Walker LR and del Moral R (2003). *Primary Succession and Ecosystem Rehabilitation.* Cambridge University Press, Cambridge. (A comprehensive summary of how succession proceeds on devastated landscapes, and how to use this knowledge to rehabilitate damaged land.)

QUESTIONS

1. Discuss whether you think the succession concept can be applied to communities in the sea. What evidence would you like to have to decide this question?

2. In discussing forest succession as a plant-by-plant replacement process, Horn (1975, p. 210) states: 'Copious self-replacement does not guarantee a species' abundance or even its persistence in late stages of succession.' How can this be true?

3. In the primeval forest landscape, where were the plants that are abundant today in old fields (page 230)? Discuss how the evolution of colonizing ability might have occurred in plants that evolved in temporary forest openings. What plant traits might be needed for success? Compare your discussion with that of Marks (1983).

4. How can species that facilitate other species in a successional sequence evolve? For example, why should species that fix nitrogen from the air leak this nutrient into the soil to assist their competitors that will replace them in the successional sequence? Is this an example of altruistic behavior?

5. Discuss how much the present state of a plant community, such as a forest, depends on past history. If you were suddenly put in charge of managing a forest undergoing succession, would you need to know the history of the site to be able to predict how it would develop in the future?

6. Do communities always grow more complex and species rich as they move through succession? In your local vegetation area, can you find any examples of succession leading to a simpler community with fewer species?

7. What ecological principles would you use to make a decision about whether or not to allow logging on a stand of old-growth forest? Discuss this question from the perspective of a politician and from the perspective of a forest ecologist.

Chapter 12

COMMUNITY DYNAMICS—BIODIVERSITY

IN THE NEWS

Over the last 15 years, many species of amphibians (frogs, toads, salamanders and newts) throughout the world have declined markedly in numbers. A few species have disappeared completely from areas in Australia and Central America where they were previously common. The map below shows the global nature of amphibian declines.

In many instances, these declines in biodiversity can be attributed to local factors, such as deforestation, draining of wetlands or pollution. In 1988, however, amphibian biologists from many parts of the world began to report declines in amphibian species that lived in protected, apparently pristine habitats, such as national parks, mountain streams and nature reserves, where no local disturbances could be implicated. Might there be a global factor that is acting on a variety of populations behind many of these declines? Possible candidates for global influences are climatic and atmospheric changes, such as increased ultraviolet radiation, widespread pollution from acid rain, and disease.

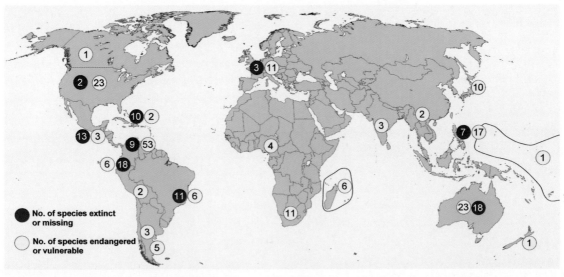

Map showing global decline in amphibians

The picture is not yet clear about which factors are causing the loss of amphibian biodiversity. Increased temperatures and more intense ultraviolet light may stress amphibians and increase their susceptibility to fungal diseases. Newly emerging diseases caused by a chytrid fungus and by pathogenic iridoviruses are now under study. Concern about continuing losses of frogs, toads and salamanders has stimulated many biologists to begin studies of how climate change might interact with disease agents and pollution to affect populations worldwide. The answers to amphibian declines are not simple, and more research is being done to test whether combinations of detrimental changes could be lethal.

*The Corroboree frog (*Pseudophryne corroboree*) from alpine areas of south-eastern Australia (threatened). (Photo courtesy of David Hunter.)*

*The golden toad (*Bufo periglenes*) from Costa Rica, last seen in 1989*

*The yellow-spotted tree frog (*Litoria castanea*) from eastern Australia, last seen in 1973. (Photo courtesy of Jean Hero.)*

CHAPTER 12 OUTLINE

■ 12.1 BIODIVERSITY DESCRIBES THE VARIETY AND NUMBER OF SPECIES IN COMMUNITIES

Populations of many species combine to make an ecological community. Ecological communities do not all contain the same number of species. This variety of species in ecological communities is called **biodiversity**, and the study of biodiversity is one of the hot topics in ecology today. Over a hundred years ago, Alfred Wallace recognized that animal and plant life was on the whole more abundant and varied in the tropics than in other parts of the globe. Other patterns of variation have long been known—small or remote islands have fewer species than large islands or those nearer continents. Ecologists try to describe and explain these trends in biodiversity as a first step in understanding ecological communities.

Biodiversity is now an everyday term in newspapers and television reports, and the preservation of biodiversity has become a public goal in many countries. Biodiversity measurement is important because without an inventory of species we cannot decide on conservation goals. While conservation biologists often worry about one particular species, such as the giant panda, community ecologists tend to group the species and condense information into counts of number of species. This community-based approach looks for large patterns in groups of species and tries to understand what has caused them. To do this, we first need to know how to identify species of plants and animals and then how to measure biodiversity.

Measurement of Biodiversity

The concept of biodiversity can be applied at several ecological levels. We can speak of genetic diversity within a species—this is one form of biodiversity that is important because of local adaptation of species to their environment. At the next level, species diversity is the more usual concept of biodiversity, and refers to the variety of species that are found in a community. Community and ecosystem biodiversity can also be measured as the diversity of ecological communities and ecosystems in a landscape. We shall confine our discussion of biodiversity here to the diversity of species in a community.

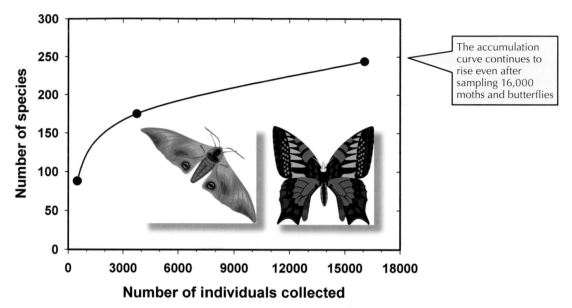

The accumulation curve continues to rise even after sampling 16,000 moths and butterflies

Figure 12.1 Species richness of Lepidoptera collections (moths and butterflies) at Rothamsted Experimental Station in England in relation to sample size. More and more species accumulate in these light-trap samples as the collection of individuals grows larger, making it difficult to answer the simple question of how many moth species live in this agricultural landscape. (Data from Williams 1964.)

The simplest measure of biodiversity is to count the **number of species** or **species richness**. In such a count, we should include only resident species, not accidental or temporary immigrants. The number of species is the first, and oldest, concept of biodiversity. A community that has 12 species of frogs in it has higher biodiversity than a community with only four species of frogs.

But, from an ecological viewpoint, species richness is not a complete measure of biodiversity. A second concept of species biodiversity is needed, that of **evenness**. One problem with counting the number of species as a measure of diversity is that it treats rare species and common species as equals. A hypothetical community with two species might be divided in two extreme ways:

	Community 1	**Community 2**
Species A	99 individuals	50 individuals
Species B	1 individual	50 individuals

Both communities have two species and both contain 100 individuals. The first community is almost a monoculture. The second community would seem intuitively to be more diverse than the first, yet the two communities have the same species richness. Ecologists have combined the concepts of number of species and relative abundance into a single concept of **heterogeneity**. Heterogeneity is higher in a community when there are more species and when the species are equally abundant.

At first glance, we might think that it is easy to count the number of species in a community, but this can be difficult for many groups. *Species counts depend on sample size.* Figure 12.1 gives one example of how counts of the species richness of moths and butterflies changes with sample size. Adequate sampling can usually get around this difficulty, particularly with birds or other vertebrate species, but not so easily with insects and other arthropods, in which species are not all counted because hundreds or thousands of species may live in a single community. In the samples shown in Figure 12.1, new species of moths and butterflies were still being found after 16,000 individuals had been collected. The point to remember is that determining the biodiversity of a community is not simple unless one deals only with the larger vertebrate or

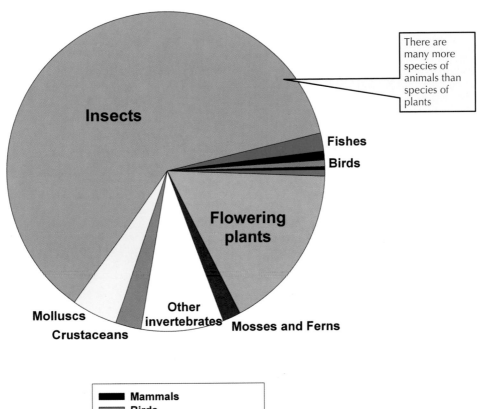

There are many more species of animals than species of plants

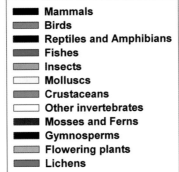

Legend	
■	Mammals
▦	Birds
■	Reptiles and Amphibians
▦	Fishes
▦	Insects
▢	Molluscs
▦	Crustaceans
▢	Other invertebrates
▦	Mosses and Ferns
■	Gymnosperms
▦	Flowering plants
▦	Lichens

Figure 12.2 The fraction of species that have been described for the major groups of plants and animals. The tiny black pie slice is the mammals and the tiny pink slice above it is the birds. Most known animals are insects. (Data from Larsen 2004.)

plant species, and for many groups we can only obtain an index of biodiversity.

How Many Species Exist at Present?

The current interest in biodiversity would suggest that biologists have a good inventory of the species that occupy the Earth. Nothing could be further from the truth. Figure 12.2 shows the current breakdown of the numbers of species that have been described by scientists for the whole Earth. This picture of the distribution of biodiversity among different taxonomic groups

would be a good reflection of what is on the Earth at present if all the taxonomic groups were equally studied, but this is not the case. About 13% of the species on Earth have been formally described and named by taxonomists—biologists who describe, name and classify organisms. Most of the vertebrates have been described, but only 10–12% of the insects and 1% of the viruses (Figure 12.3). Partly because of a shortage of taxonomists to describe species, we have only a crude estimate of the total number of species on Earth—13 to 14 million is an average guess, and much

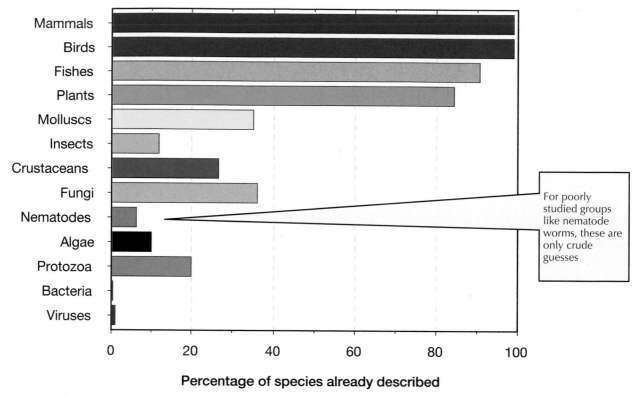

Figure 12.3 The percentage of species that have already been described and named for various groups. Most plants and higher vertebrates have been described, but species in most other groups, and particularly the smaller forms, are only poorly known to science. (Data from Hawksworth and Kalin-Arroyo 1995.)

work remains to be done just to catalog this diversity (*Essay 12.1*). About 1.75 million species have been described so far, but a shortage of funding and of scientists who are trained to identify organisms is holding back progress.

Q What economic arguments might be made to encourage more funding for cataloging biodiversity on Earth?

■ 12.2 THE MAJOR GLOBAL PATTERN FOR BIODIVERSITY IS A GRADIENT FROM THE TROPICS TO THE POLAR REGIONS

Where do all these species live? Ecologists try to understand the Earth's biodiversity first by describing where species occur. Once the patterns are well described, we need to determine what ecological factors are associated with the observed patterns. We begin with the most well-known pattern in biodiversity—from the species-rich tropics to the species-poor polar regions.

Tropical Biodiversity

Tropical habitats support large numbers of species of plants and animals, and this diversity of life contrasts starkly with the impoverished faunas of temperate and polar areas. A good example of this global gradient can be seen in trees, most species of which have been described and named. Tropical rainforest contains many more tree species than temperate or subarctic forests. A 50-hectare plot of tropical rainforest in Malaysia contained 830 species of trees, and a 6.6 hectare area in Sarawak contained 711 tree species. A deciduous forest in Michigan contains 10 to 15 species on a plot of 2 hectares, and the part of Europe north of the Alps has 50 tree species. The Yukon Territory in northern Canada has two tree species on a plot of 2 hectares.

There are 620 native tree species in North America. Figure 12.4 shows the details of the polar to temperate

ESSAY 12.1   BIODIVERSITY: A BRIEF HISTORY

There are somewhere between 5 million and 30 million species of animals and plants on the Earth. About 1.7 million of these have been described by taxonomists—perhaps 13% of life. This situation is a scandal that no one outside of biology seems to recognize. If only 13% of the companies being traded on Wall Street were known, or if the catalog of the Louvre Museum included only 13% of its paintings, right-thinking people would be outraged. Not so with biodiversity.

Our taxonomists are the heroes of biodiversity, and without them working quietly in the background we would not know even the 13% we do, and our appreciation of community organization and dynamics would be much less. Fortunately, a few taxonomists rise to public recognition, and Edward O. Wilson of Harvard University is one. Wilson is an ant taxonomist by training, and a naturalist by nature. While working on ant distributions on islands, he met and joined forces in 1961 with Robert MacArthur

Edward O. Wilson (1929–): Entomologist, ecologist and conservation biologist, Harvard University

from the University of Pennsylvania to produce one of the most famous books of community ecology, *The Theory of Island Biogeography* (1967). Wilson has become a champion spokesman for biodiversity through his books on ants and, more recently, through a series of popular books on biodiversity and its conservation. He is one of the few ecologists to have an autobiography (*Naturalist*, Island Press, Washington D.C., 1994).

Many other ecologists have cooperated to bring biodiversity into the public arena at the start of the 21st century. Some of them, like Ed Wilson, Paul Ehrlich and David Suzuki are well known. But many others work to bring biodiversity to the fore in biological research agendas and in political and social action. The exploding human population and its increasing impact on the globe is combining with our undescribed biodiversity to produce extinctions of species we will never have named or described—a loss that we should not bequeath to our children and grandchildren.

gradient in tree species. Superimposed on an overall tropical to polar gradient in the number of tree species are other patterns that need explanation. For example, more species occur in south-eastern US forests than occur in western forests, so that there is another gradient across the United States from west to east. Very few tree species occur in the rain shadows just east of the Rocky Mountains and the Sierra Nevada.

Ants are another well-studied group that are much more diverse in the tropics:

	No. of ant species	Latitude
Brazil	222	10°S
Trinidad	134	11°N
Cuba	101	21°N
Utah	63	39°N
Iowa	73	42°N
Alaska	7	61°N
Arctic Alaska	3	70°N

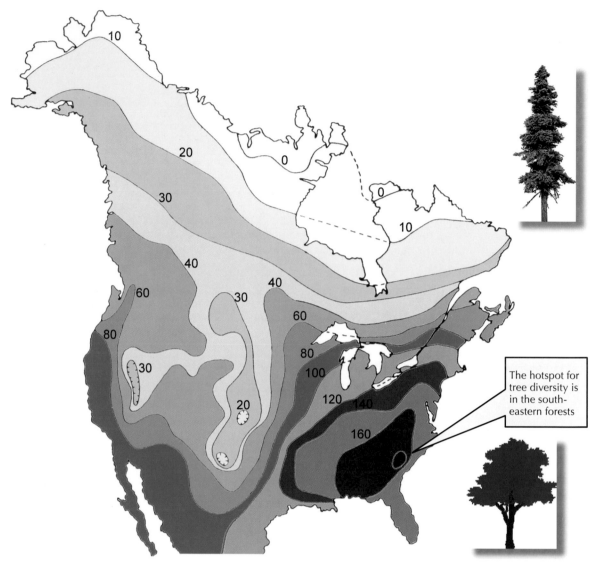

Figure 12.4 Tree species richness (number of species) in Canada and the United States. Contours connect points with the same number of species. (Modified from Currie and Paquin 1987.)

Other vertebrate groups follow the tropical–polar gradient. There are 293 species of snakes in Mexico, 126 in the United States and 22 in Canada. Bird species are much more numerous in tropical countries. Figure 12.5 shows the number of breeding land-bird species in different parts of North America. There are more than 600 bird species in Central America and only one-third this number in southern Canada.

Freshwater fishes are much more diverse in tropical rivers and lakes. Lakes Victoria, Tanganyika and Malawi in East Africa contain about 1450 species of freshwater fish. Over 1000 species of fishes have already been found in the Amazon River in South America, but exploration is still incomplete in this region. By contrast, Central America has 456 fish species, and the Great Lakes of North America have 173 species. Lake Baikal in Asia has 39 fish species and Great Bear Lake in north-western Canada has 14 species of fish.

There are exceptions to this general rule—not all animals and plants show a smooth trend of biodiversity from the poles to the tropics. Figure 12.6 shows the diversity of Alcid seabirds and of seals and sea lions. Species diversity is greatest in both these groups in the high latitudes—few species occur in tropical regions.

Figure 12.5 Geographic pattern of species richness in the land birds of North and Central America. Contour lines give the numbers of species present. (From Cook 1969.)

There are few seaweed species in the polar regions, but seaweeds are equally diverse in the temperate regions and in the tropics.

Q **Would you expect cold-blooded animals to show an exception to the tropical–polar biodiversity gradient?**

This brief look at species diversity gradients can assist us in looking at some factors that may affect latitudinal gradients in species diversity. The overall pattern of increase in biodiversity from the poles toward the tropics is clearly only a general trend—the exceptions to this rule are useful because they permit us to untangle some of the ecological factors that influence biodiversity.

Hotspots of Biodiversity

The global pattern of tropical-to-polar gradients in biodiversity interacts with evolutionary history to

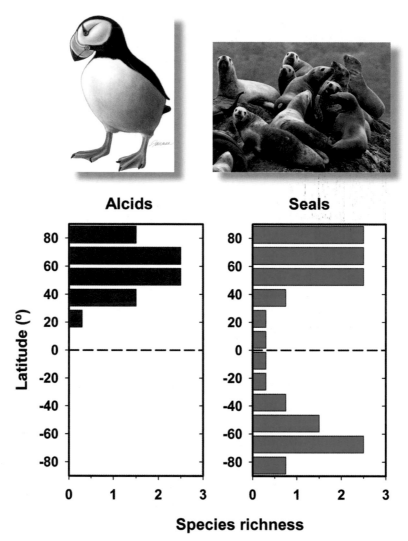

Figure 12.6 Species diversity of Alcid seabirds and of seals and sea lions (Pinnipedia) in relation to latitude. Southern latitudes are given as a minus sign. In neither of these marine groups is there a tropical to polar gradient in biodiversity. (Data from Proches 2001; drawing of horned puffin courtesy of Fairman Studios; photo of sea lions courtesy of Andrew Trites.)

produce an unequal distribution of species around the Earth. Species arise primarily by geographic isolation, and the patterns of isolation that have arisen from continental drift have resulted in some areas being much more species-rich than others. Many, but not all, of these areas of high biodiversity occur in the tropics. Identification of these hotspots of biodiversity has become important in recent years because humans have cleared more and more areas for agriculture and forestry, thereby endangering many species. **Hotspots** are defined in several different ways, but in general the measure used to define a hotspot is the number of

endemic species that it contains. **Endemic species** are those that occur in only one relatively small geographic area. The Hawaiian goose, for example, is an endemic bird found only on the islands of Hawaii and Maui.

There are 34 hotspots of biodiversity around the globe (Figure 12.7). Hotspots are defined as areas containing at least 1500 endemic plant species. One surprising feature of this map is that not all the hotspots are in tropical countries. Many hotspots are tropical, but the Cape Floristic Province of South Africa and New Zealand are two examples of temperate hot spots. Polar regions contain no hotspots. Table 12.1 lists the

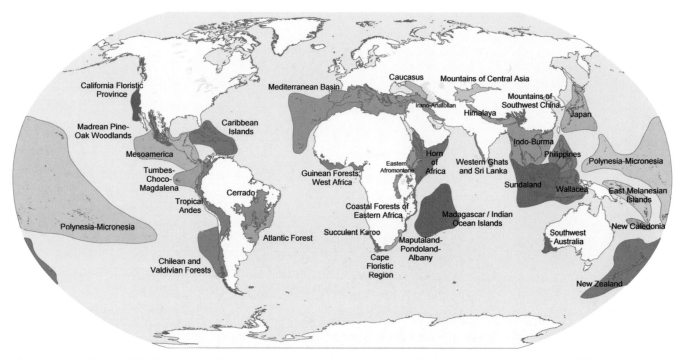

Figure 12.7 The 34 global hotspots of biodiversity, defined on the basis of high plant species richness. Table 12.1 lists the number of endemic plant species in the top 25 global hotspot areas. (From Myers *et al.* 2000; map modified from Conservation International 2006.)

size of the most important hotspots and the number of plant species and vertebrate species they contain. These data are incomplete and err in the direction of minimal species counts. For example, Brazil has the world's richest flora—and probably has at least 50,000 species of plants—but there is no up-to-date list of Brazil's plant species.

The 34 hotspots mapped in Figure 12.7 contain a minimum of 50% of the world's terrestrial plant species and 77% of the world's vertebrate species, all within 2.3% of the land surface of the Earth. The implication of this concentration of biodiversity is that these hotspots should be a focus of our conservation efforts at the present time.

The hotspot concept has an underlying assumption that hotspots occur in the same geographic region for all the different plant and animal groups. This general idea is the concept of **umbrella species**—that one species, or a group of species, will serve as a guide to many other groups of species that are less well known or less studied. For example, butterflies could serve as umbrella species for the community of all insects and

plants in a region. Consequently, before designing a conservation action plan, it is important to check how much overlap there is among hotspots for different taxonomic groups. Hotspots for one group of species, such as plants, might not coincide with hotspots for other groups, such as butterflies. Andy Dobson and his colleagues mapped the geographical distribution of endangered plants, birds, fish and mollusks in the United States (Figure 12.8). They found that the hotspots for one taxonomic group did not coincide with hotspots for other groups. This means that recovery plans for species will have to be area-specific and species-specific. For any particular group of species, such as the molluscs, the hotspots for that group can be targeted for conservation.

Q Should global conservation efforts be directed predominately to hotspots?

The global distribution of hotspots provides another dimension to the overall trend of a drop in biodiversity as we move from the tropics toward the poles. These

Table 12.1 Characteristics of 25 of the highest ranked biodiversity hotspots. There are approximately 300,000 described plant species on Earth, and approximately 28,595 described vertebrate species (excluding fish, which are not included in the vertebrate tally). The eight hottest hotspots are shown in bold. Figure 12.7 shows a map of these regions. (From Myers *et al.* 2000, and Conservation International 2006.)

Hotspot	Original extent of vegetation (km²)	Per cent remaining original vegetation	No. of plant species	No. of endemic plant species	No. of vertebrate species	No. of endemic vertebrate species
Tropical Andes	1,258,000	25.0	30,000	15,000	3389	1567
Mesoamerica	1,155,000	20.0	17,000	2941	2859	1159
Caribbean	263,500	11.3	13,000	6550	1518	779
Brazil's Atlantic forest	1,227,600	7.5	20,000	8000	1361	567
Turnbes/Choco/Western Ecuador	260,600	24.2	11,000	2750	1625	418
Brazil's Cerrado	1,783,200	20.0	22,000	10,000	1268	117
Chile/Valdivian forest	300,000	30.0	3892	1957	335	61
California	324,000	24.7	3488	2124	584	71
Madagascar	594,150	9.9	13,000	11,600	987	771
Eastern Afromontane and Coastal Forests of East Africa	30,000	6.7	11,598	4106	1019	121
Guinean West African forests	1,265,000	10.0	9000	1800	1320	270
Cape Floristic Province	74,000	24.3	9000	6210	562	53
Succulent Karoo	112,000	26.8	6356	2439	472	45
Mediterranean Basin	2,362,000	4.7	22,500	11,700	770	235
Caucasus	500,000	10.0	6400	1600	632	59
Sundaland	1,600,000	7.8	25,000	15,000	1800	701
Wallacea	347,000	15.0	10,000	1500	1142	529
Philippines	300,800	3.0	9253	6091	1093	518
Indo-Burma and Himalaya	2,060,000	4.9	23,500	10,160	2185	528
South-west China	800,000	8.0	12,000	3500	1141	178
Western Ghats/Sri Lanka	182,500	6.8	5916	3049	1073	355
South-west Australia	309,850	10.8	5571	2948	456	100
New Caledonia	18,600	28.0	3270	2432	190	84
New Zealand	270,500	22.0	2300	1865	217	136
Polynesia/Micronesia	46,000	21.8	5330	3074	342	223

patterns raise the question of what environmental factors cause these large differences in species diversity.

■ 12.3 DIFFERENCES IN BIODIVERSITY MAY BE PRODUCED BY SIX CAUSAL FACTORS

Differences in species richness may be produced by six causal factors that are difficult to untangle (Table 12.2). Many causes have interacted over evolutionary and ecological time to produce the assemblages of species we see today, so that no single cause will explain all the patterns we have described. For any particular diversity gradient, we can ask which of these six factors are involved and which are most important.

Evolutionary Speed Hypothesis

This idea is a historical hypothesis with two main variants. The hypothesis postulates that the tropics contain more species because there has been more speciation there, either because speciation rates are faster in the tropics or because speciation has gone on longer in these regions. The first mechanism assumes that temperature increases the rate of speciation, so that plants and animals in the warm, humid tropics evolve and diversify more rapidly than those in the temperate and polar regions. A constant favorable environment causes this rapid diversification because generation times are shorter, mutation rates are higher

and natural selection acts more quickly in tropical regions. The second mechanism proposes that species diversity is a product of evolution and therefore is dependent on the length of time through which the ecological community has developed in an uninterrupted fashion. Tropical biotas are examples of mature biotic evolution over a long timeframe, whereas temperate and polar biotas are immature communities, continuously interrupted by glaciation and severe climate shifts. So, even if evolutionary rates are the same everywhere, more species will evolve in tropical communities. In short, all communities diversify in time, and older communities consequently have more species than younger ones.

The time factor in evolution need not be confined to tropical regions. The key point is that to create biodiversity an ecological community must have a long, uninterrupted evolutionary history. Lake Baikal in the former Soviet Union is a particularly striking illustration of the role of time in generating species diversity. Baikal is one of the oldest lakes in the world—about 4 million years old—yet it is situated in the temperate zone. Baikal contains a very diverse fauna. For example, there are 580 species of benthic invertebrates in the deep waters of Lake Baikal. A comparable, but much younger, lake of the same area in glaciated northern Canada—Great Slave Lake—contains only four species in this same zone. Time has allowed the evolution of many species in ancient lakes.

Table 12.2 Ecological and evolutionary factors that can have an influence on biodiversity. More than one of these factors can operate in any particular ecological community or in any particular taxonomic group. Some factors operate at a local scale and others at regional scales.

Factor	Rationale
1. Evolutionary speed	More time and more rapid evolution permits the evolution of new species
2. Geographical area	Larger areas and physically or biologically complex habitats provide more niches
3. Interspecific interactions	Competition affects niche partitioning and predation retards competitive exclusion
4. Ambient energy	Fewer species can tolerate climatically unfavorable conditions
5. Productivity	Richness is limited by the partitioning of production or energy among species
6. Disturbance	Moderate disturbance retards competitive exclusion

(Modified after Currie 1991, and Willig *et al.* 2003).

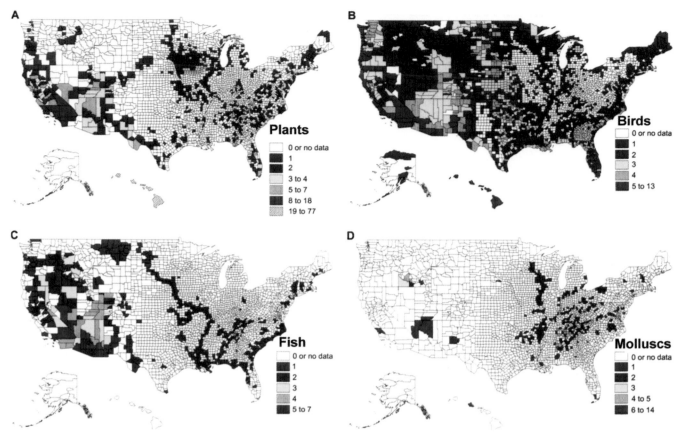

Figure 12.8 The geographic distribution of endangered species in the United States. (A) Plants, (B) Birds, (C) Fish, (D) Molluscs. Areas colored red have more endangered species. While there is some overlap in hotspots among these taxonomic groups, there are large differences as well, so that protecting all the areas with endangered plants will not also protect all endangered birds. (Modified from Dobson *et al*. 1997.)

Fossil data support the assumption that species richness increases over geological time, further supporting the view that more time equals more species. The number of species of terrestrial plants found as fossils appears to have increased in two waves during the last 450 million years (Figure 12.9). Episodes of extinction have occurred, but the overall pattern is a gradual increase. No plateau in biodiversity has yet been reached for terrestrial plants, and evolution has not stopped in the 21st century. For fossil mammals, by contrast, species richness seems to have reached a plateau during the last 30 million years (Figure 12.10).

Note that the species diversity of a community is a function not only of the rate of addition of species through evolution, but also of the rate of loss of species through extinction or emigration. Compared with polar communities, the tropics could have a more rapid rate

of evolution or simply a lower rate of extinction—these two rates act together to determine species diversity.

A major assumption of the first version of the evolutionary speed hypothesis is that favorable climatic factors permit a more rapid rate of evolution. In birds there is no molecular evidence that evolution is occurring more rapidly in tropical species. Further tests using molecular techniques will permit a more critical analysis of this important assumption, but at present this hypothesis must be questioned as an adequate explanation of tropical–polar gradients in species richness. The second version of the evolutionary hypothesis awaits more data on rates of extinction and speciation in tropical and temperate communities.

Geographical Area Hypothesis

This hypothesis begins with the assumption that larger areas support more species than smaller areas, which

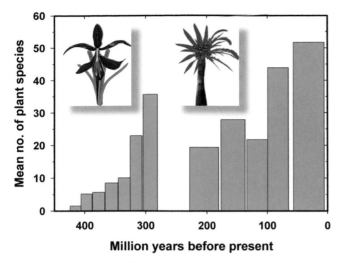

Figure 12.9 Pattern of increase in the number of terrestrial plant species over evolutionary time. These trends are consistent with the prediction that more evolutionary time will produce more biodiversity. Extinction episodes (like that in the Permian, 250 million years ago) can reduce diversity, but then recovery begins again. Data are derived from fossils. (Data from Nicklas *et al.* 1980.)

seems to be universally true. Given this assumption, the postulate is that the tropics support more species than the temperate zone because they have a larger land area. Larger areas contain more habitats and more individuals, reducing the risk of extinction. If there are more habitats in a region, we would expect there to be more species. There could be a general increase in the number of habitats per square km as one proceeds towards the tropics, and the more complex the plant and animal communities, the higher will be the species diversity.

Q **What ecological mechanisms might make it possible for larger areas to support more species than smaller areas?**

But geographical area does not explain the detailed distribution of some species groups. Topographic relief—mountains and hills—may have a strong effect on species diversity in some groups of organisms. The highest diversities of mammals in the United States occur in the mountain areas of the western states. The explanation seems simple: areas of high topographic relief contain many different habitats and hence more species (*Essay 12.2*). Also, mountainous areas tend to isolate populations geographically and so may promote

speciation. But this conclusion does not fit all taxonomic groups. The trees of North America are most diverse in the south-eastern states (Figure 12.4) and less diverse in the mountainous western states. The land birds of North America, by contrast, are most diverse in the western states (Figure 12.5). These differences reflect a principle that is important for understanding biodiversity—not all taxonomic groups are affected in the same way by the same factors.

Tropical to polar gradients in the oceans seem unlikely to be explained by geographical area. The oceans are not uniform water masses, yet the extent of coastal and deep waters does not show any clear trend with latitude. Benthic marine invertebrates become more diverse as you move from shallow waters on the continental shelf to deeper waters at the edge of the shelf. There is no obvious change in the extent of bottom sediments to explain this increase in biodiversity, so the idea that larger areas harbour more biodiversity is contradicted.

Spatial heterogeneity does not explain the observed tropical–polar diversity gradient. It cannot be a general explanation because many aquatic habitats, such as shallow mudflats, in the ocean show these gradients in the absence of any change in spatial area. The explanation of higher diversity in tropical areas must lie in factors other than geographical area.

Interspecific Interactions Hypothesis

Several hypotheses suggest that high tropical diversity is associated with greater interspecific competition and higher predation rates. How might interactions between species affect the latitudinal gradient in species diversity?

This hypothesis assumes that natural selection in the temperate and polar zones is controlled mainly by the physical factors of these severe environments, whereas biological competition becomes a more important part of evolution in the tropics, where physical conditions are more favorable for life. Competition between animals and plants for resources may thus be more intense in tropical areas. As a result of this intensive competition, natural selection should produce animals and plants that are more specialized in their habitat requirements in the tropics. Animals in the tropics should also have more restricted diets in each

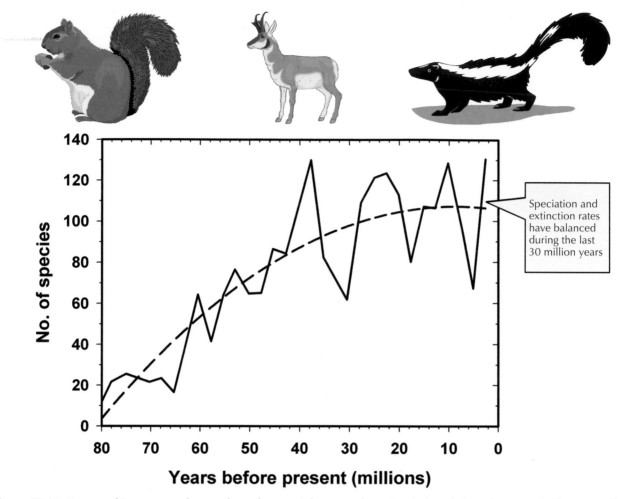

Figure 12.10 Pattern of increase in the number of terrestrial mammal species in North America over the last 80 million years. Rapid evolution occurred around 50–60 million years ago and then slowed significantly, so that mammal species richness reached a fluctuating asymptote around 30–40 million years ago. Data are derived from fossils. (Data from Alroy 1999.)

habitat. If competition is keener in the tropics, niches should be smaller. Tropical species are more highly evolved and possess more specialised adaptations than do temperate species. Consequently, more species can be fitted into a given habitat in the tropics.

The possible effect of competition on species richness can be visualized by considering the niche relations of species in a community. Consider the case of one resource, such as soil water for plants or food item size for animals (Figure 12.11 a). Two niche measurements are critical: **niche breadth** and **niche overlap**. We can recognize two different cases. If there is no niche overlap between the species, the wider the average niche breadth, the fewer the number of species in

the community (Figure 12.11 b). At the other extreme, if niche breadth is constant, the lower the niche overlap, the fewer the species in the community (Figure 12.11 c). In this hypothetical analysis, tropical animal communities might have more species because tropical species have smaller niche breadths or higher niche overlaps. We assume here that competition between species will not permit total niche overlap.

In relatively few cases have the detailed measurements been made to test the schematic model of Figure 12.11. The best example is the work undertaken in the Caribbean on *Anolis* lizards. Lizards of the genus *Anolis* are small, insect-eating iguanid lizards that forage during the day. Most species perch on tree trunks or

ESSAY 12.2   HOW SPATIAL SCALE AFFECTS BIODIVERSITY MEASUREMENTS

Counts of the numbers of species will always be affected by the spatial scale of the measurements. The larger the area sampled, the greater the number of species that will be counted, and this makes it difficult to compare the data obtained from different studies. A simple example will illustrate this problem. Consider the herbaceous plant species in a region that has three different types of communities—a grassland, a dry forest and a wet riverine forest (species are indicated by a letter):

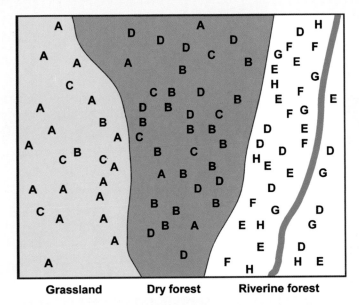

Grassland   Dry forest   Riverine forest

Hypothetical distribution of species in three habitats within a region

The grassland (yellow) has three species, the dry forest (pink) has four species, and the riverine forest (white) has five species. There are two simple ways to describe the diversity of this artificial landscape. If we stay within one community, we can measure **within-habitat diversity** (also called α–**diversity**), but it we sample two of the communities together, we can measure **between-habitat diversity** (also called β–**diversity**).

Habitats sampled	Total number of species	No. of species shared in the combined communities	No. of new species added by combining
Grassland	3	–	–
Dry forest	4	–	–
Riverine forest	5	–	–
Grassland + dry forest	4	3	1
Dry forest + riverine forest	8	1	4

For this hypothetical example, between-habitat diversity is low if we compare the grassland and the dry forest, because we add only one more species by combining these two communities. But if we consider the two forest types, between-habitat diversity is high, because four more species are added.

The key point is that the species diversity of a region is not simply the number of species in each community (α-diversity) added together. The number of shared species can be used to measure a second component of biodiversity, between-habitat or β-diversity.

bushes. They are sit-and-wait predators, and food size is a critical niche dimension. *Anolis* lizards are a dominant component of the vertebrate community on islands in the Caribbean. Niche breadth in *Anolis* lizards is smaller when more species occur together on an island, exactly as predicted in Figure 12.11. The results support the suggestion that niche breadth is reduced in species-rich communities. Enclosure experiments have confirmed that *Anolis* lizards compete fiercely when their diets are similar. Competition for food is a major factor determining the species diversity of these lizards.

Another factor acting on species richness could be predator impacts. Bob Paine has argued that there are more predators and parasites in the tropics than elsewhere and that these hold down their prey populations to such low levels that competition among prey organisms is reduced. This reduced competition allows the addition of more prey species, which in turn support new predators. Thus, in contrast to the competition proposal—in this hypothesis, there should be *less* competition among prey animals in the tropics.

Paine supported his ideas with some experimental manipulations of rocky intertidal invertebrates of the Washington coast. The food web of these areas on the Pacific coast is remarkably constant with about 15 species being relatively common (Figure 12.12). The starfish *Pisaster* is one of the common predators on rocky shores, which feeds preferentially on mussels and barnacles. Paine removed the predatory starfish *Pisaster* from a section of the shore and observed a decrease in diversity from a 15-species system to a two-species system five years later. A bivalve, *Mytilus*, took over the rock area, crowding out the other species, so the community became simpler. By continual predation,

starfish prevent barnacles and bivalves from monopolizing space so there is room for other species. Thus local species diversity in intertidal rocky zones appears to be directly related to predation intensity. Paine called the starfish a **keystone species** in this community.

The prediction from Paine's experimental work in the intertidal zone is that increased predation will lead to higher diversity of prey species, but this prediction depends on the ability of one prey species to be competitively dominant. For the predation hypothesis to operate on a broad scale, the predators involved must be very efficient at regulating the abundance of their prey species. In terrestrial food webs, predators are usually specialized and in some cases do not seem to regulate prey abundance. The predation hypothesis can explain tropical species diversity if it can be applied to all trophic levels. If the species diversity of the herbivore trophic level is determined by the predators, we are left with explaining the diversity of the plants. Keystone species, such as the starfish *Pisaster*, might be more common in tropical communities, but at present there is no evidence that this is correct.

The predation factor can be extended to the primary-producer level. Tropical lowland forests contain many species of trees, and one of the great challenges of ecology has been to explain the high diversity of tropical rainforests. Most large trees of a given species in tropical forests are spread out in a regular pattern at very low densities so that distances between trees of the same species are relatively large. Dan Janzen and Joe Connell suggested independently that these characteristics of tropical trees might be explained by the predation hypothesis—the species that eat seeds or destroy seedlings being analogous to the predators

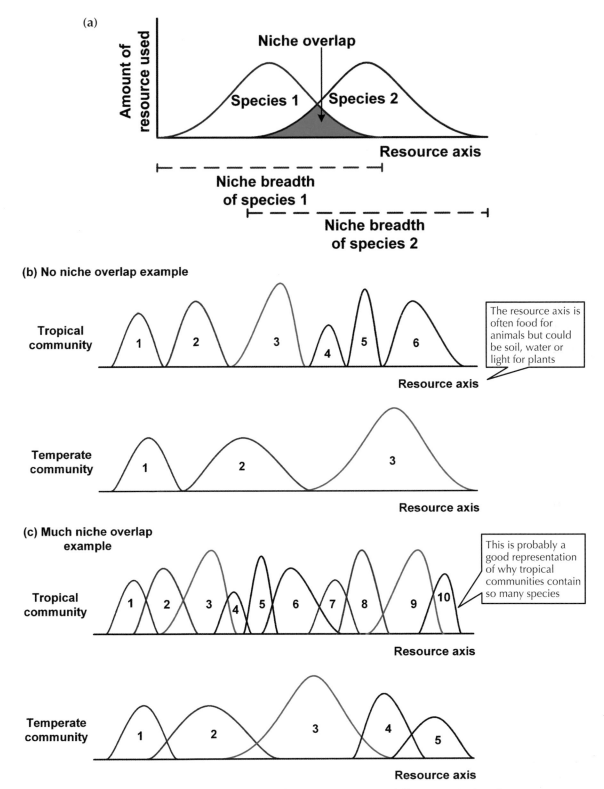

Figure 12.11 Two hypothetical extreme cases of how niche parameters may differ in tropical and temperate communities. Both niche breadth and niche overlap are determined by competition within the communities. If there is no niche overlap (b), the number of species that can be fitted into a community is determined by niche breadth. If there is variable niche overlap (c), the number of species is determined by the amount of niche overlap.

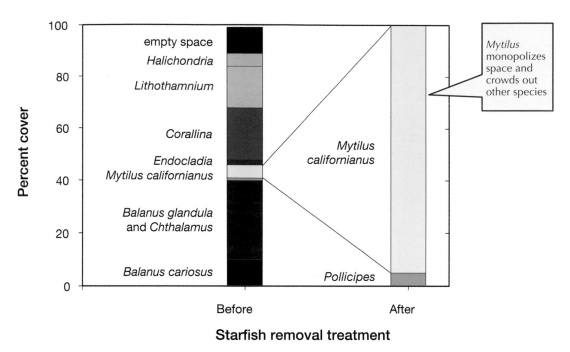

Figure 12.12 Keystone predator effect in the rocky intertidal zone. Bob Paine removed predatory starfish from rocky intertidal sites in Washington State and observed a collapse of the community to a near-monoculture of California mussels (*Mytilus californianus*) over 5 years. (Data from Paine 1974.)

just discussed. The Janzen–Connell model for the maintenance of tropical tree species diversity is shown schematically in Figure 12.13. This model predicts that tree seedlings will do poorly if they are close to a large tree of the same species. To find out if this is correct, Steve Hubbell and Robin Foster in 1981 established a 50 ha forest plot on Barro Colorado Island in Panama. Censuses were carried out in this plot in 1981–83, 1985 and 1990–91, and 244,000 stems of 303 species were measured and geographically mapped in 1990–91. Many of the species in this rainforest plot show few seedlings near adult trees of the same species, exactly as predicted in Figure 12.13. Most of the negative interactions in tropical rainforests seem to occur within tree species rather than between different tree species, so that competition for resources does not seem important in maintaining diversity. Pests and pathogens may be the key to understanding Figure 12.13. Each adult tree casts a 'seed shadow,' in which survival of its own kind is reduced. As one moves from the lowland tropical forests to temperate forests the seed and seedling herbivores and diseases are less efficient at preventing establishment of seedlings close to the parent tree. But this seed shadow effect may not be the whole story for lowland rainforests. Data from tropical rainforests have supported the Janzen-Connell model in some, but not all, cases.

Predation and competition are important interactions in tropical and temperate communities, but it is unlikely that they may be responsible for the tropical–polar trends in species richness. The problem is that these interactions between species are more likely to be the *result* of tropical species richness than the *cause* of the high richness.

Ambient Energy Hypothesis

This hypothesis states that energy availability generates and maintains species richness gradients. Climate determines energy availability, and the key variables for terrestrial plants and animals are solar radiation, temperature and water. More stable and favorable climates cause higher productivity, and these factors work together to support more species.

The **ambient energy hypothesis** is one of the simplest and most elegant of the climatic explanations for the tropical–polar gradient in terrestrial biodiversity.

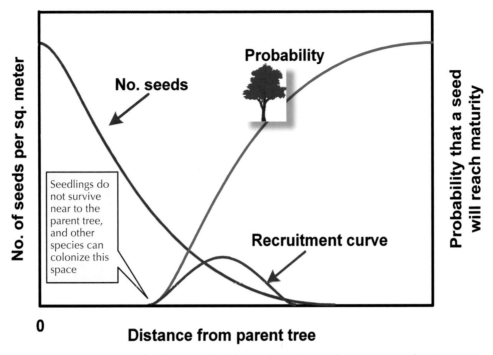

Figure 12.13 The Janzen–Connell model for high tropical-forest diversity. Each tree species has its own specific pests and diseases. Most seeds fall near the parent tree, and these seeds and seedlings are killed by herbivores and diseases living on the parent tree. The product of the number of seeds falling in an area and their probability of survival defines a recruitment curve with a peak at a distance from the parent tree where a new adult tree is most likely to appear. The empty space near the parent tree can be occupied by another tree species, maintaining high diversity. (Modified after Janzen 1970.)

A wealth of data has now been presented in support of this hypothesis. This simple hypothesis fits data for trees, British birds, and vertebrates and butterflies from North America and Europe. Figure 12.14 illustrates the relationship between biodiversity and available energy for trees and vertebrates. Available energy can be measured by annual evapotranspiration, which measures the energy balance at a site and can be calculated from solar radiation and temperature. Vertebrate biodiversity in North America below about 45°–48°N (approximately the Canadian border) is not correlated with ambient energy, but is more directly correlated with water availability. There seems to be a general threshold in mid latitudes of the Northern Hemisphere, where, in all terrestrial groups, water becomes the major factor predicting species richness (Figure 12.15).

How might we test the energy hypothesis? The energy hypothesis makes specific predictions about seasonal bird migrants, so this is one test of the idea.

In temperate areas, energy levels in summer should control the diversity of summer birds, while energy levels in winter should control the numbers of resident birds in winter. In Britain, the biodiversity of summer birds is correlated with summer temperature and the diversity of winter birds is correlated with winter temperature at 75 localities, exactly as the energy hypothesis predicts.

A second test of the energy hypothesis can be made with coral reefs. The coral reefs in tropical waters are species-rich and the number of taxa falls off rapidly as you move from warm tropical seas to cooler temperate waters (Figure 12.16). High species diversity in corals has usually been attributed to the historical factor associated with high rates of speciation in the Indo-Pacific region. The best predictors of coral species diversity are ocean temperature and coral biomass, so that energy-rich areas have more coral genera and species. Historical, evolutionary factors are responsible for the major differences between the number of coral

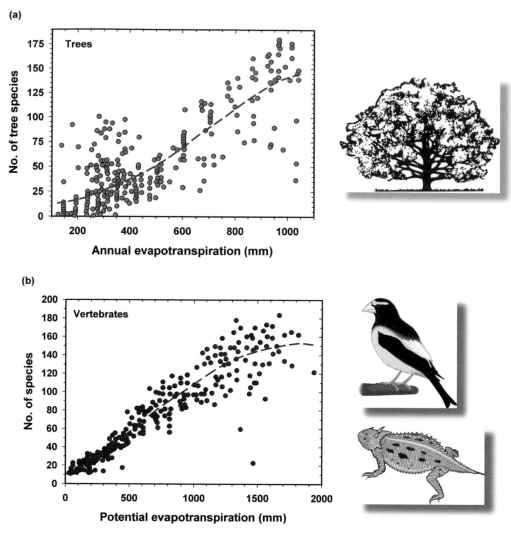

Figure 12.14 The species richness-energy hypothesis for biodiversity. Species richness of (a) trees and (b) vertebrates from North America are related to annual available energy at each site, measured by evapotranspiration (which combines solar radiation and temperature). (Data from Currie 1991.)

species in the Atlantic and the Indo-Pacific regions, so that both history and available energy are important overall. Caribbean coral reefs are less than 10,000 years old and were affected by glaciation in the Northern Hemisphere, while Pacific coral reefs are up to 60 million years old and have been less affected by climatic oscillations. Caribbean reefs contain only 10–20% of the number of coral species found on Pacific reefs.

Favorable climates on a broad geographic scale thus support high biodiversity. This idea explains a large proportion of the global tropical–polar diversity gradient—on average about 63% of the variation. Ambient energy theory works well for large-scale patterns in global diversity. It will not explain local habitat-scale variations in species diversity, such as why a grassland has more herb species than an adjacent forest. No single factor will explain all gradients in biodiversity from the local scale to the regional, but ambient energy and water provide a good explanation for the regional patterns.

Productivity Hypothesis

Tropical habitats are generally more productive than temperate and polar habitats, and this might suggest that productivity is a key to biodiversity and a good explanation of the tropical–polar trend in species richness. The productivity hypothesis in its pure

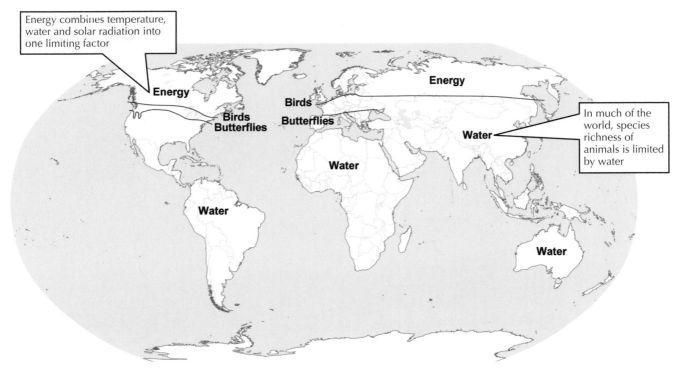

Figure 12.15 The ambient energy hypothesis to explain the tropical–polar gradient in species richness of terrestrial animals. Across the globe, animal species richness is constrained by the interaction of energy and water. The bold red lines show the geographical position of thresholds—north of these lines for birds and butterflies energy is the limiting component, whereas south of these lines water is the key limiting component. Energy is less relevant in the Southern Hemisphere because landmasses do not extend so far south to extreme latitudes. (Modified from Hawkins *et al.* 2003.)

form states that greater production results in greater biodiversity—everything else being equal. The data available do not support this view, and this is a classic example of an appealing, intuitive idea that is wrong. David Tilman has described several examples in which plant biodiversity is maximal in resource-poor habitats of low productivity. Two of the world's most diverse plant communities are the fynbos in the Cape Floristic Province of South Africa and the heath scrublands of south-western Australia (see Figure 12.7 and Table 12.1). Both of these communities occur on nutrient-poor soils, and in both cases adjacent areas with better soils and more productive vegetation have fewer species. Productivity in plant communities seems to lead to reduced biodiversity on a local scale, exactly the opposite of what one might expect.

Q **What ecological mechanisms might lead to reduced biodiversity when productivity is high?**

Productivity could be more important for animal communities on a global scale, but again the available data do not agree with this conclusion. There is no relationship between productivity and vertebrate biodiversity in North America. Productivity by itself does not seem to be the key to understanding diversity gradients on a global scale.

Intermediate Disturbance Hypothesis

If ambient energy can explain much of the global pattern of species richness, local patterns on much smaller scales must involve other mechanisms that affect diversity. Disturbance is one such local factor. If natural communities exist at equilibrium, and the world is spatially uniform, competitive exclusion ought to be the rule, and each community should come to be dominated by a few species—the best competitors. But if communities exist in a non-equilibrium state, competitive equilibrium is prevented. A whole range of factors can prevent equilibrium—predation, herbivory,

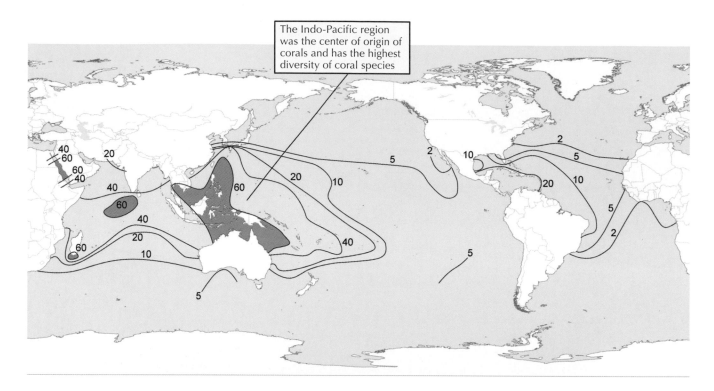

The Indo-Pacific region was the center of origin of corals and has the highest diversity of coral species

Figure 12.16 Biodiversity of corals in the tropical and subtropical areas of the world. Contour lines connect areas with the same number of genera. The Indo-Pacific region is much richer in corals than the eastern Pacific or the Atlantic region. Within these three regions, coral diversity can be predicted from the available energy, as measured by ocean temperature. The coral photo shows the Great Barrier Reef of Australia. (Modified after Fraser and Currie 1996, coral photo courtesy of Julian Caley, Australian Institute of Marine Sciences.)

fluctuations in physical factors and catastrophes, such as fires, and we group these together as 'disturbance.'

When disturbances occur too often, species become extinct if they have low rates of increase. When disturbances are rare, the system moves to a competitive equilibrium and species of low competitive ability are lost. The idea that in between these extremes is a level of disturbance that maximizes biodiversity (Figure 12.17) is called the **intermediate disturbance hypothesis**. If population growth rates are low for all members of a community, the competitive equilibrium is approached so slowly that it is never reached. Thus species diversity is maintained by periodic disturbance or by environmental fluctuations. If this model is correct, the worst thing land managers can do to a natural community is to prevent disturbances such as fire. There is a lesson here for supervisors of national parks.

Disturbance can operate on a local scale of a few square meters to produce patches that undergo succession. Within each patch in a local area, the species composition may be changing, but on a larger spatial scale of tens to hundreds of meters the species composition may be constant and include all the pioneer species, as well as the climax species. Defining the geographic scale is important for specifying the mechanisms affecting biodiversity.

The disturbance dimension of the intermediate disturbance hypothesis is directly related to succession (Figure 12.17). When disturbances are frequent, the community is kept in the pioneer stage of succession. When disturbances are reduced, succession proceeds and species richness reaches a peak in the middle of succession—a prediction of this hypothesis that can be checked with data on successional sequences.

Q **Would you expect a disturbance such as fire to affect all the species in the community equally strongly?**

Disturbance does not always produce the pattern predicted by the intermediate disturbance hypothesis, but it does so in the vast majority of cases that have

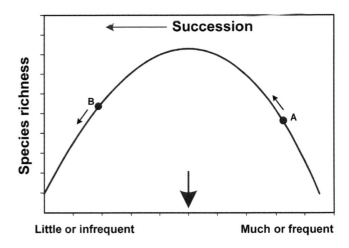

Amount of disturbance

Figure 12.17 The intermediate disturbance hypothesis of species diversity. This model predicts that at some intermediate level of disturbance (red arrow) biodiversity will be at a maximum. Succession is assumed to be proceeding in the direction of the blue arrow. Note that if a community is at point A, a reduction in disturbances would result in an increase in species richness. At point B, a reduction in disturbances would result in a decrease in species richness, if this model is correct. (Modified from Huston 1979.)

been analyzed. One example illustrates that different patterns can be found within a single community. On rocky shores in Massachusetts the periwinkle snail *Littorina littorea* is the most common herbivore. In tide pools, *Littorina* feeds on the alga that is competitively dominant. Moderate grazing by the snails permits many competitively inferior algae to survive (Figure 12.18a), as predicted by the intermediate disturbance hypothesis. But on emergent rocks, perennial brown and red algae are competitively superior and the snails do not eat them, but feed on the competitively inferior algae. Consequently, on emergent rocks *Littorina* grazing reduced algal diversity (Figure 12.18b). The critical factors are the food preferences of the grazer and the competitive abilities of the algae.

The intermediate disturbance hypothesis is an attractive hypothesis for the maintenance of high species diversity in local communities, but it does not apply to all communities and cannot apply on a global scale. In particular, land managers should not assume

(a) Tide pools

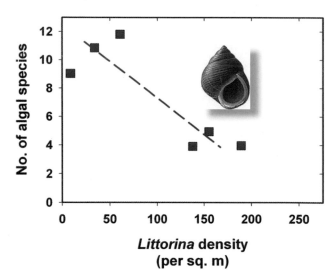

(b) Emergent substrates

Figure 12.18 The effect of periwinkle snail grazing on the diversity of algae in (a) high-tide pools and (b) on emergent rocks in the low intertidal zone in Massachusetts and Maine. The intermediate disturbance hypothesis applies only in the tide pools, but not on the emergent rocks. (Modified from Lubchenco 1978.)

that the intermediate disturbance hypothesis operates when making management plans for national parks or other protected areas.

■ 12.4 COMMUNITIES COULD BE SATURATED WITH SPECIES THROUGH EVOLUTION BUT THEY APPEAR UNSATURATED

Biodiversity in local habitats could be limited by either evolutionary or ecological causes. As we have just seen, the mixing of evolutionary processes on a long time scale and ecological processes on a short time scale has made it difficult to untangle the reasons for the latitudinal change in species diversity. One way to separate out evolutionary and ecological causes is to ask whether or not each community is saturated with species. A **saturated community** would mean that no more species can be squeezed into it, that all the niches are filled, and that competition is the main factor limiting diversity. An unsaturated community would be open to invasion and not all niches would be occupied.

Regional Species Saturation

We can test for saturation by plotting *local* species diversity against *regional* species diversity. To do this we need to define what is local and what is regional. Local diversity is measured on a scale in which all the species in the community could interact with each other in ecological time—typically a generation in length. For example, fish species in a lake or stream, herb species in a meadow or bird species in woodlots would all be examples of local diversity. Local diversity would typically be measured on a scale of a few hectares to a square kilometer or two. Regional diversity on the other hand refers to a larger spatial scale—typically 100 or more times the size of the local scale. Within the region, species could disperse to and colonize a local patch through dispersal over tens of generations. Examples of regional diversity would be the fish species of the Great Barrier Reef, the grass species of Britain or the bird species of the boreal forest of Canada and Alaska. Regional species richness can be specified only if the flora and fauna of a region are well known. For this reason, studies of local and regional diversity have concentrated on the better-known taxonomic groups, such as birds, butterflies and trees.

Local Community Saturation

Communities would be saturated if there were intense competition among the existing species. Figure 12.19 illustrates the idea of testing for local community saturation, and the key is that we expect a linear relationship if communities are unsaturated and a curvilinear relationship if communities are saturated. The best data for comparisons are from a single defined habitat sampled in several geographically distinct regions. Figure 12.20 shows a comparison of local and regional diversity at the continental scale for a range of taxa from amphibians to trees. In a broad, global sense there is no evidence of local community saturation, which implies that biodiversity at the local level is not constrained by intensive competition and communities are not closed to new invaders.

Q **How would you find and identify an empty niche in a community?**

On a smaller spatial scale David Pearson studied the bird communities of six undisturbed lowland tropical forest sites in the Amazon, Borneo, New Guinea and West Africa. Animal communities in rich tropical forests might be expected to be saturated with species. He spent from 200 to 700 hours in each local plot of about 15 ha taking a census of birds. Local and regional richness was linearly related, suggesting that bird communities in these tropical forests were not saturated with species—a surprising conclusion, since backed up by many other studies.

The majority of communities studied so far are unsaturated. At present, the assumption that ecological communities are saturated with species does not appear to be correct for many species groups. Biodiversity does not appear to have any fixed upper limit in natural communities.

SUMMARY

Biodiversity can be measured at the genetic level, at the species level or at the ecosystem level. Most often species are the focus of study, and by counting all the

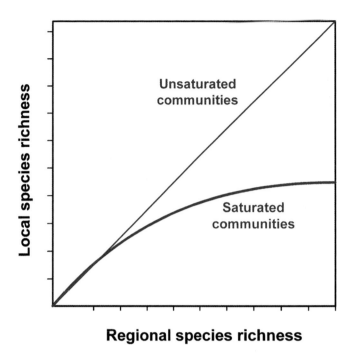

Figure 12.19 Local and regional biodiversity plot. If local communities are unsaturated, the diversity of communities will go on increasing with regional diversity in a linear manner (blue line). Richer regions will have richer local communities. On the other hand, if local communities are saturated with species they will reach an asymptote or maximum species richness (red line) set by competition for resources and niche overlap. To test for saturation several communities in several different regions are needed.

species in an area we measure species richness as an index of biodiversity. Communities vary greatly in the numbers of species they contain, and this chapter describes the patterns of biodiversity over the globe and then relates these patterns to ecological mechanisms that can affect biodiversity.

Many species remain to be described, and only some groups, such as birds, mammals and butterflies, are named and well described. Insects comprise more than half of the total biodiversity of the Earth's 15–30 million species. These species are not evenly spread over the Earth. Hotspots of biodiversity are regions with large numbers of rare species, and they occur largely in tropical areas and in a few temperate regions, such as California. Hotspots contain nearly 50% of all species on only about 1–2% of the Earth's land surface.

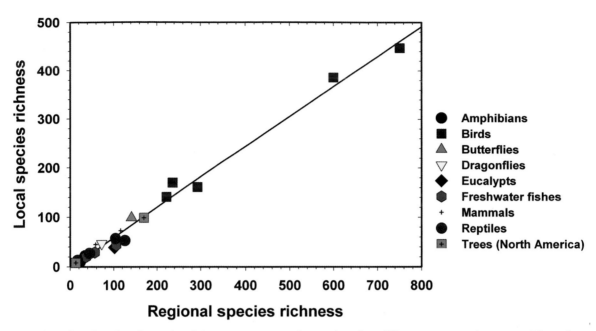

Figure 12.20 Local and regional species richness across continents for nine different taxonomic groups. There is no indication of saturation in this relationship and remarkably all the different groups on different continents appear to follow the same linear regression. (Data from Caley and Schluter 1997.)

Tropical environments support more animal and plant species than temperate and polar areas. The latitudinal gradient in biodiversity from the tropics to the polar regions is one of the most obvious, global patterns found in community ecology. The trees and birds of North America illustrate the complexity of species diversity gradients, which are not always smooth, steady trends from the equator to the poles.

Many different hypotheses have been proposed to explain the observed patterns of species richness. The focus has been on the increase in biodiversity from the polar regions to the tropics. Six major hypotheses are described—evolutionary speed, geographical area, interspecific interactions, ambient energy, productivity and disturbances. On a regional scale, ambient energy—involving temperature, water and solar energy—is the best predictor of species richness in terrestrial plants and animals. Evolutionary speed, which summarizes the evolutionary history of a region, is the

most difficult factor to evaluate and is potentially important on a regional scale. On a local scale of a few tens of square meters, disturbance, competition and predation interact to affect species richness.

There is no single, general answer to the question 'What controls biodiversity?' As we have found in most ecological systems, the answer depends on the scale of analysis. At the global level, the ambient energy or climate model seems to predominate as the best explanation of biodiversity. Some factors such as climate may provide insight on a global scale, while others, such as predation, help to explain biodiversity in local habitats.

At the regional scale, communities do not appear to be saturated with species, because local community diversity rises in step with regional diversity. If communities are not saturated, they may be open to invasion, and we are led to the broader question of what controls community organization.

SUGGESTED FURTHER READINGS

Hawkins BA, Field R, Cornell HV, Currie DJ, Guégan J-F, Kaufman DM, Kerr JT *et al.* (2003). Energy, water and broad-scale geographic patterns of species richness. *Ecology* **84**, 3105–3117. (An overview of the role of climate in controlling biodiversity.)

James CD and Shine R (2000). Why are there so many coexisting species of lizards in Australian deserts? *Oecologia* **125**, 127–141. (Why are Australian desert lizards more diverse than lizard communities on all other continents?)

Levin SA (Ed.) (2001). *Encyclopedia of Biodiversity*. Academic Press, San Diego. (The bible of biodiversity, with short articles on every critical issue.)

Myers N, Mittermeier RA, Mittermeier CG, da Fonseca GAB and Kent J (2000). Biodiversity hotspots for conservation priorities. *Nature*, **403**, 853–858. (Where are the richest areas on Earth for species, and how can we conserve these areas?)

Possingham HP (2001). The business of biodiversity: applying decision theory principles to nature conservation. *Environment, Economy and Society* **9**, 1–37. (How decision theory can help us to conserve biodiversity.)

Wilson EO (2001). *The Diversity of Life*. 2nd edn. Penguin, London, UK. (A plea for the conservation of biodiversity by one of our greatest living biologists.)

QUESTIONS

1. Would you expect to have latitudinal gradients in the species richness of parasites of mammals and birds? What factors might control species richness in parasites? Rohde (1999) discusses this problem.

2. The tree flora of Europe is very poor compared with that of eastern North America or eastern Asia (Grubb 1987). Why should this be? Compare your explanations with those of Grubb (1987) and of Currie and Paquin (1987).

3. Grazing by cattle is a disturbance that has a long history of conflict with conservation biologists. Is there any evidence that grazing increases plant species diversity, at least up to intermediate levels of grazing? Stohlgren *et al.* (1999) report data from grazing exclosures in place for up to 60 years, and discuss this issue.

4. The longest experiment in ecology is the Park Grass Experiment begun in 1856 at Rothamsted, England. A mowed pasture was divided into 20 plots, and a series of plots were fertilized annually with a variety of nutrients, including nitrogen. Discuss the general predictions you would make regarding biodiversity on fertilized and unfertilized plots for this experiment, using the six hypotheses discussed above. Tilman (1986, pp. 62–63) shows the observed results

5. In Antarctica species richness in soft-bottom invertebrates (sponges, bryozoans, polychaetes and amphipods) is higher than that of almost all other tropical and temperate zone soft-bottom communities (Clarke 1990). What observations or experiments would you perform to find out why this high biodiversity occurs in Antarctica?

6. Suppose you were put in charge of finding out why a group of frogs were declining in abundance and becoming threatened with extinction. What general information would you like to know to determine the cause of this biodiversity loss?

7. Hydrothermal vents are like geysers in the deep ocean on the sea floor. They continuously spew super-hot, mineral-rich water that helps support a diverse community of organisms. Vents are typically 2–100 km apart on the ocean floor and, in the Pacific, exist for 10–100 years before they collapse. How could you test the intermediate disturbance hypothesis for these vents? Tsurumi (2003) discusses diversity at hydrothermal vents in the Pacific Ocean.

Chapter 13
COMMUNITY DYNAMICS—FOOD WEBS

IN THE NEWS

On 24 March 1989 the oil tanker *Exxon Valdez* grounded on a reef in Prince William Sound, Alaska and spilled 42 million liters of crude oil. At least 1900 km of shorelines in the area were contaminated. Marine mammals and seabirds that routinely use the sea surface were immediately affected. Mass mortality of 1000–3000 sea otters and an estimated 250,000 seabirds occurred in the first few days after the spill. Mass mortality also occurred among marine algae and invertebrates caused by both the oil itself and chemicals used in the immediate clean-up operation.

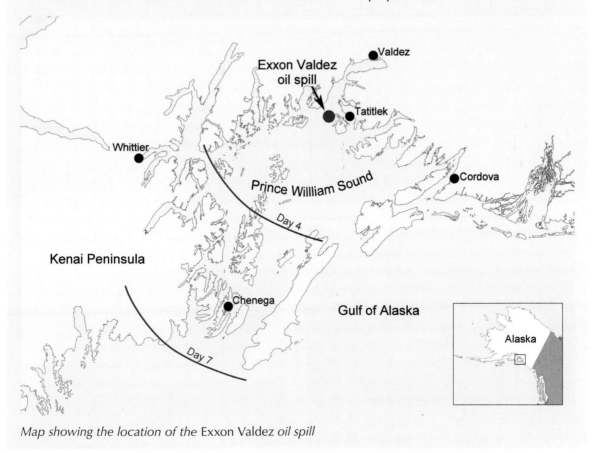

Map showing the location of the Exxon Valdez *oil spill*

To determine the impact of the oil spill, ecologists first constructed a web of all the feeding interactions between the major groups of animals and plants in this marine system, by asking the simple question who eats whom (shown in red arrows in the diagram). To complete the interaction picture, we add the competitive interactions in the community (who competes for space with whom, blue arrows) and, finally, who provides shelter to whom (green arrows). The resulting interaction web is quite complex:

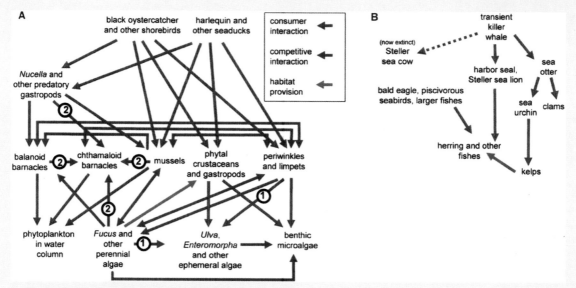

Web of interactions in Prince William Sound

The impact of the oil spill spread through the marine community in Prince William Sound, depending on the species interactions shown in these webs. The immediate destruction of marine algae and invertebrates affected both the intertidal and subtidal invertebrates, as well as the birds feeding on these organisms. After the immediate acute mortality in the few weeks after the spill, the conventional belief was that the marine community of Prince William Sound would recover quickly to its former abundance. This has not happened, and we can now view this ecological disaster from a perspective of 15 years to see what lessons it can tell us about restoring damaged communities.

Harlequin duck

Prince William Sound, Alaska

Sea otters have recovered since 1989 at 4% per year—much less than the 10% per year that was expected. At heavily oiled Knight Island, no recovery of sea otters has occurred at all in 15 years. Chronic exposure to oil and its breakdown products has increased juvenile mortality. Harlequin ducks—a threatened species—suffered high losses of about 1000 birds after the oil spill, and they also have not recovered since the spill. Continued exposure to residual oil has prevented successful reproduction. Harlequin ducks feed on intertidal invertebrates, and were showing physiological signs of oil exposure from their food 9 years after the oil spill.

The ecological messages that have come from the *Exxon Valdez* oil spill have been strong. The original assumption that all the damage to the ecosystem would be short term—because the oil would be dispersed and degraded by microbial action—have turned out to be wrong. Long-term toxicity to fish, birds and mammals has now been documented, and the focus has shifted from the catastrophic immediate impacts to considering the sub-lethal, long-term health problems that have slowed recovery and continue even after 15 years.

■ 13.1 COMMUNITIES ARE ORGANIZED BY A NETWORK OF INTERACTIONS THAT INVOLVE COMPETITION, PREDATION, AND MUTUALISM

Ecologists wish to understand the organization of ecological communities in the same way that sociologists wish to understand the organization of human communities. Both kinds of scientists proceed by finding the components of the community and the links that tie them together. We need to understand community organization in order to manage communities—including those impacted by human agriculture and forestry and those communities protected in national parks and reserves. Undesirable changes in communities can be avoided only if we understand how communities are organized.

Community Organization

By **community organization**, ecologists mean what species are present, how abundant they are, which species eat which, and how the species interact positively or negatively by competition or mutualism. Four biological processes—competition, predation, herbivory and mutualism—could organize ecological communities. Competition among plants, herbivores and carnivores could control the diversity and abundance of the species in a community. Predation and herbivory (who-eats-whom) could organize the community so that the framework of community organization is set by the animals. Mutualism is an important process that links species and could serve to increase community organization by linking species to the benefit of all. Physical processes set limits to these four biological processes—variation in temperature, salinity and other physical factors have potential implications for the species in a community. To study community organization, we need to look at the component species and the processes that tie them together. We do not expect to find that only one process operates to control community organization, but the focus is on the relative importance of these four processes. In an experimental sense, or a management sense, we would like to know which process to change to make a desired change in the community.

When we speak of community organization, we imply that there is some regularity in the biomass or the numbers of the species that make up the community. Naturalists looking for particular birds, butterflies or flowers have an implicit model of community organization in their heads, so they know that pine forests have a certain group of species in them, and oak–hickory forests have a different set of species. Conservation biologists have an implied model of community organization when they discuss the preservation of the Everglades of Florida or other natural landscapes. The first question we have to ask is whether or not community composition is strictly defined. Natural communities could either be very loosely organized or very tightly organized. How can we determine this for any particular community?

Communities contain so many different species that we cannot study all the individual species separately. We need ways of reducing the complexity of many species to fewer units. Ecologists have developed three different ways of measuring and studying community complexity. The first way to do this is to measure the **biodiversity** of the community, as we saw in the last chapter. If we measure the species richness of a community, we implicitly assume that each species is equal to every other species in the community. A second way to group species is to define feeding roles in the community and to group species according to their roles. We can group species into **trophic levels** (such as herbivores or carnivores) or at a finer level into **feeding guilds**. A third way is to look at particular types of species and to ask 'Are all species of equal importance in a community?' This question is purposely vague because we must define **importance**, and we can do this in several ways. We could consider a species important if, when we remove it, the diversity or abundance of other species in the community changes. Such species are known as **keystone species**. Alternatively, we could determine which species are most common in the community—the **dominant species**. Dominant species could be major players in defining the organization of the community. In this chapter, we will discuss each of these three approaches to understanding community organization.

Equilibrium Communities

We will begin this analysis with the classical assumption that communities are in equilibrium. In **equilibrium communities** species abundances remain constant over time—nature is in a state of balance. In most cases equilibrium means **stable equilibrium** (Figure 13.1a). In different habitats the equilibrium point may differ, so that there is spatial variation in species abundances, but the key point is that at each spatial location the community is in equilibrium and remains relatively constant. This equilibrium will usually be stable within a specified environmental range. In some cases, the equilibrium can be stable over a broad range, so that over all environmental conditions the system will return to the equilibrium point following any disturbance (Figure 13.1b).

The classical equilibrium assumption of community ecology is an abstraction, like the frictionless pendulum of physics, and will not be found in its pure state in natural communities. Real communities will be spread along a spectrum from equilibrium to non-equilibrium. Equilibrium communities are claimed to show stability, and stability can be measured in several different ways. The mathematician's idea of **local stability** (points A_1 and A_2 in Figure 13.1a) is the simplest meaning. Stability can be measured by the *time* it takes for a community to recover from disturbance. Stable communities recover quickly from disturbances. Stability can also be measured as the **variability** of a community over time, so that if the populations that make up the community fluctuate in size dramatically from year to year, the community would be judged to be unstable. This is the most common meaning ecologists attach to the word **stability**. Stability can also be measured as the **persistence** of a community over time. An ideal equilibrium community would score high on all these measures of stability. Such a community would have many biotic interactions involving competition and predation. Equilibrium communities would also be saturated with species, so that species invasions would be rare. Weather catastrophes would rarely occur and the community would form a tightly coupled biotic unit—an interlocking web of life.

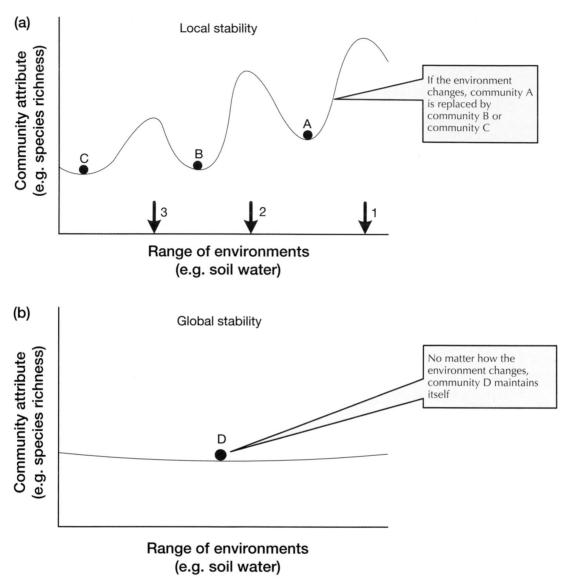

Figure 13.1 Schematic view of local and global community stability: (a) Three locally stable communities A, B, and C are shown. Red arrows mark the boundaries of stability between the communities. The community at A is locally stable between the environmental range from arrow 1 to arrow 2, and community B is locally stable between arrows 2 and 3. (b) A globally stable community at D will remain in the same state no matter how much the environment changes. The environmental variable might be soil water or soil pH for a plant community. Community attributes could be any measure of species composition or abundances. Most real-world communities are only locally stable as in (a), and only a few would be globally stable like (b). (Modified from DeAngelis and Waterhouse 1987.)

Non-equilibrium Communities

By contrast, **non-equilibrium communities** would score low on all these measures of stability. Species would operate individualistically and climatic catastrophes would occur frequently. Species would come and go regularly, so the composition of the community would be highly variable over time. We will discuss non-equilibrium community dynamics in the next chapter, and try to determine where most natural communities fall on this continuum from equilibrium to non-equilibrium. For the moment, let's assume that all communities are equilibrium communities.

■ 13.2 WHO-EATS-WHOM DETERMINES THE BASIC STRUCTURE OF A COMMUNITY

One of the first questions an ecologist asks to determine community organization is 'who-eats-whom?' The transfer of food energy from its source in plants through herbivores to carnivores is referred to as the **food chain**. Charles Elton was one of the first to apply this idea to ecology and to analyze its consequences. He recognized in 1927 that the length of these food chains was limited to four or five links. Thus we may have a pine tree–aphids–spiders–warblers–hawk food chain. Elton recognized that these food chains were not isolated units, but were hooked together into **food webs**—interconnected food chains in a community. Let's look at a few examples of food chains.

Food Chains and Food Webs

The Antarctic pelagic food chain is a good example of a food chain found in seasonally productive oceans. The dominant herbivores are euphausids (krill) and copepods. These zooplankton species are fed upon by an array of carnivores from penguins to seals, fish and baleen whales (Figure 13.2). Squid, which are carnivores feeding on fish as well as zooplankton, are another important component of this food chain because they in turn provide food for seals and the toothed whales. During the whaling years, humans became the top predator of this food chain. Having reduced the whales to low numbers, humans are now fishing for krill (page 440).

Terrestrial food webs in tropical rainforest can be complex like some marine ecosystems (*Essay 13.1*). Polar ecosystems are somewhat simpler than tropical ones, and are thus easier to analyse. In the boreal forests of north-western Canada and Alaska snowshoe hares are the dominant herbivores, along with red squirrels. An array of mammalian and avian predators feed on the herbivores in the forest, and the herbivores feed on an array of trees, shrubs and herbs (Figure 13.3). However, this food web has a few twists. Red squirrels and arctic ground squirrels are usually herbivores, but they kill and eat baby snowshoe hares when they are available—so they are mostly herbivores, but occasionally act as predators. Also, predators will kill other predators when they are hungry and have the chance. Coyotes will kill lynx, lynx will kill foxes and greathorned owls will kill a variety of other birds-of-prey. The important point is that not all the links in food webs are linear from predators to herbivores to plants. Predators may eat other predators, and herbivores may become predators when the opportunity arises.

In many cases ecologists simplify food webs. Two approaches are taken. Firstly, some taxonomic groups are lumped into **trophic species**. Trophic species are a set of organisms with identical prey species and identical predator species. Species are typically grouped depending on the focus of the study. In terrestrial studies, often all the vertebrate species are identified individually, but grasses, other plants or insects are lumped together, depending on the exact focus of the food web. The boreal forest food web in Figure 13.3 illustrates this approach. Secondly, instead of mapping the entire food web, only a part of the whole food web is isolated for analysis to keep things relatively simple. Figure 13.4 shows a partial food web for the red squirrel in the boreal forest—abstracted from Figure 13.3. The red squirrel has only a few food sources and a relatively small suite of predators, so that this partial web is much less complex than the full food web shown in Figure 13.3. Partial food webs or compartments can sometimes be used to simplify large food webs.

Trophic Levels

Within food webs we can recognize several different **trophic levels**:

Producers	=	Green plants	=	First trophic level
Primary consumers	=	Herbivores	=	Second trophic level
Secondary consumers	=	Carnivores, insect predators	=	Third trophic level
Tertiary consumers	=	Higher carnivores, insect hyperparasites[1]	=	Fourth trophic level

[1] Insect predators that feed on herbivores are also called insect parasitoids, and the insect predators that eat other insect predators are called insect hyperparasites.

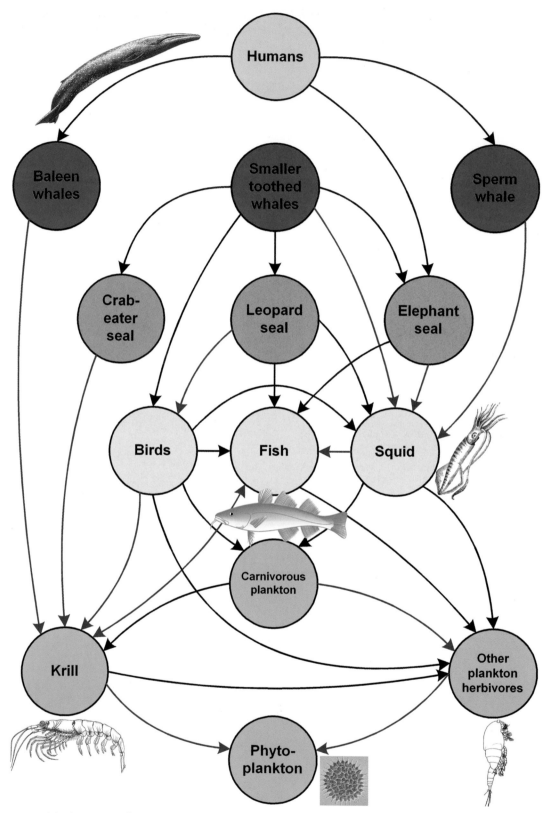

Figure 13.2 A simplified version of the Antarctic's marine food chain. Blue arrows indicate the major trophic interactions before whaling by humans began. (Modified after Knox 1970.)

The marine continental shelf of the New England states and eastern Canada has one of the most complex food webs yet deciphered for a marine environment. It illustrates the complexity that natural communities display and that ecologists must try to understand. Link (2002) illustrates this food web and describes it as follows:

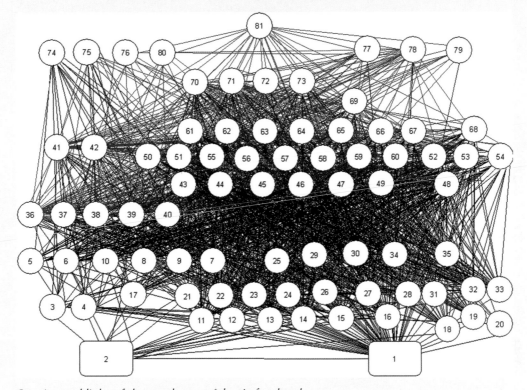

Species and links of the north-west Atlantic food web.

This tangled 'bird's nest' represents interactions at the approximate trophic level of each species, with increasing trophic levels towards the top of the web. The left side of the web typifies pelagic (open water) organisms, and the middle and right represent more bottom-dwelling species. Red lines indicate predation on fish. The species and species groups are identified by numbers as follows:

1 = detritus, 2 = phytoplankton, 3 = *Calanus* sp., 4 = other copepods, 5 = ctenophores, 6= chaetognatha (arrow worms), 7 = jellyfish, 8 = euphausiids, 9 = Crangon sp., 10 = mysids, 11 = pandalids, 12 = other decapods, 13 = gammarids, 14 = hyperiids, 15 = other crabs, 16 = isopods, 17 = pteropods, 18 = cumaceans, 19 = mantis shrimps, 20 = tunicates, 21 = porifera, 22 = cancer crabs, 23 = other crabs, 24 = lobster, 25 = hydroids, 26 = corals and anemones, 27 = polychaetes, 28 = other worms, 29 = starfish, 30 = brittle stars, 31 = sea cucumbers, 32 = scallops, 33 = clams and mussels, 34 = snails, 35 = urchins, 36 = sand lance, 37 = Atlantic herring, 38 = alewife, 39 = Atlantic mackerel, 40 = butterflyfish, 41 = loligo, 42 = ilex, 43 = pollock, 44 = silver hake, 45 = spotted hake, 46 = white hake,

47 = red hake, 48 = Atlantic cod, 49 = haddock, 50 = sea raven, 51 = longhorn sculpin, 52 = little skate, 53 = winter skate, 54 = thorny skate, 55 = ocean pout, 56 = cusk, 57 = wolf-fish, 58 = cunner, 59 = sea robins, 60 = redfish, 61 = yellowtail flounder, 62 = window-pane flounder, 63 = summer flounder, 64 = witch flounder, 65 = four-spot flounder, 66 = winter flounder, 67 = American plaice, 68 = American halibut, 69 = smooth dogfish, 70 = spiny dogfish, 71 = goosefish, 72 = weakfish, 73 = bluefish, 74 = baleen whales, 75 = toothed whales and porpoises, 76 = seals, 77 = migratory scombrids, 78 = migratory sharks, 79 = migratory billfish, 80 = birds, 81 = humans.

Two important observations flow from this complex food web. Firstly, the web has a high number of connections—about 31%—which is much higher on average than other food webs analyzed. This might be typical of marine food webs. Secondly, even at this level of complexity, the food web is highly aggregated. Although the fish are well documented, the 33 invertebrate groups in the lower trophic levels are groups of species, as are the birds and the whales. The detailed web would be even more complex than this 'bird's nest'.

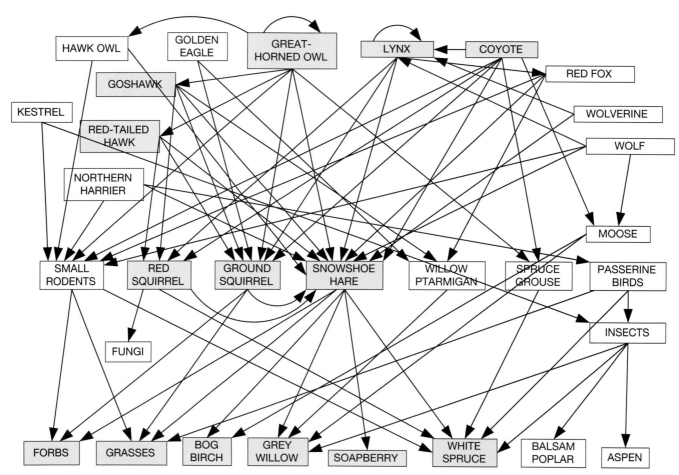

Figure 13.3 Feeding relationships of the snowshoe hare-dominated food web in the boreal forests of north-western Canada. The dominant species in this community are shown in yellow. Arrows go from the herbivore to the plant or from the predator to the prey. (After Krebs *et al.* 2001.)

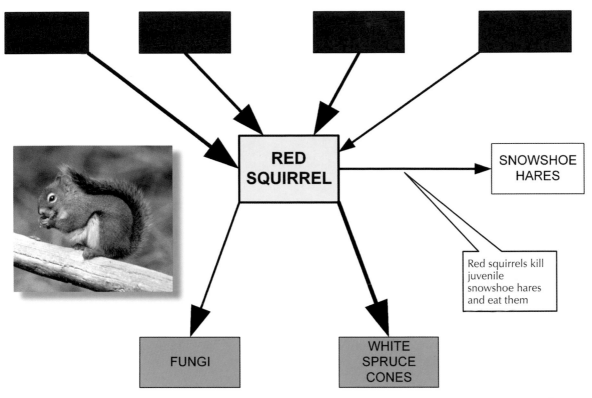

Figure 13.4 Partial food web for the red squirrel in the boreal forests of Canada. This partial web has been abstracted from the more complete web shown in Figure 13.3. Partial webs are one way of simplifying complex community food webs by concentrating on only a few of the interactions. (Data from Boutin *et al.* 1995.)

For long food chains, there are possible fifth and higher trophic levels. Trophic levels are the most aggregated classification that can be applied to food webs—the great complexity of a food web can be collapsed into four or five trophic levels for further analysis.

The classification of organisms by trophic levels is one of function and not of species. A given species may occupy more than one trophic level. For example, male horseflies feed on nectar and plant juices, whereas the females are blood-sucking ectoparasites. Sea nettles in Chesapeake Bay are secondary consumers of zooplankton and tertiary consumers of fish larvae, which themselves are secondary consumers of zooplankton.

Size has a great effect on the organization of food chains. Animals of successive trophic levels in a food chain tend to be larger (with the exception of parasites). There are, of course, definite upper and lower limits to the size of food a carnivorous animal can eat. The structure of an animal puts some limit on the size of food it can take into its mouth. Except in a few cases, large carnivores cannot survive on very small food items because they cannot catch enough of them in a given time to provide for their metabolic needs. One exception is the baleen whales that feed on tiny zooplankton in the world's oceans. The other obvious exception is the omnivore *Homo sapiens*, and part of the reason for our biological success is that we can prey upon almost any level of the food chain and can eat any size of prey.

One question we can ask about food webs is whether there are limits to their complexity. As more and more species are involved in a food web, two possibilities exist. Firstly, each species may interact with a constant number of other species, no matter whether the food web is highly diverse (as in the tropics) or of low diversity (as in polar regions). Alternatively, each species may interact with more species in diverse food webs, so that, for example, predators in tropical food webs would prey on more species than predators in polar food webs. Figure 13.5 shows that the connectivity of

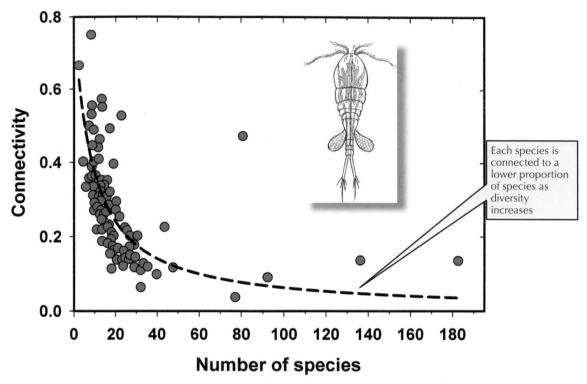

Figure 13.5 Relationship between connectivity and the size of the food web for the 89 food webs from marine, freshwater, and terrestrial communities. Connectivity is the number of linkages as a proportion of the number of potential linkages in a food web. Once a food web exceeds about 40 species, the connectivity is approximately 10%. (Modified after Link 2002.)

food webs falls with higher diversity, and this loose connectivity should promote community stability. In aquatic food webs each species interacts with about two to three other species no matter what the diversity. This principle seems to apply to marine, freshwater and terrestrial food webs.

Length of Food Chains

The length of a food chain is defined as the number of links running from a top predator to a basal species. A second generalization about food webs is that food chains tend to be short. If we count for each food web all the possible routes from a basal species to a top predator, we can get a set of chain lengths and a maximum chain length for the food web. The Ythan Estuary in north-east Scotland has been studied in great detail, covering 95 species in this community. The most common food chain length is five links, and the range was one to nine links.

Why should food chains be relatively short like this? There are several hypotheses. The **energetic hypothesis** is the most popular explanation for food chain length. It suggests that the length of food chains is limited by the inefficiency of energy transfer along the chain. This is the classical hypothesis articulated by Charles Elton in 1924. If this idea is correct, food chains should be longer in habitats of higher productivity—a clear prediction that can be tested.

Another explanation for short food chains emphasizes the stability of the food chain interactions. The **stability hypothesis** explains short food chains by suggesting that longer food chains are not stable, so that fluctuations at lower levels are magnified at higher levels and top predators become extinct. In a variable environment, top predators must be able to recover from catastrophes. The longer the food chain, the slower the recovery rate from catastrophes for top predators. If catastrophes occur too often, the top

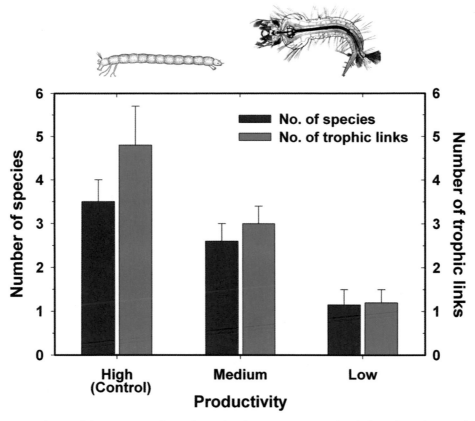

Figure 13.6 Experimental test of the energetic hypothesis for the restriction on food chain length. Tree-hole communities in Queensland were manipulated with litter input at three levels: high litter input = natural (control) rate of litter fall, medium = 1/10, and low = 1/100 natural rate. Reducing energy input reduced food chain length, in agreement with the hypothesis. The tree-hole community consists of microbes that break down leaf litter, mosquito larvae that feed on these microbes and predatory midges and other insects that feed directly on leaf litter. (From Jenkins *et al.* 1992.)

predators will become extinct. This hypothesis predicts shorter food chains in unpredictable environments—this prediction can also be tested.

These two hypotheses to explain food chain length are difficult to test in large communities, but can be tested in smaller communities. The organisms that inhabit natural, water-filled tree holes in the subtropical rainforest of Queensland have been used to test these ideas in a small system. Tree holes were standardized and replicated with 1-liter plastic containers to which the ecologists added leaf litter in varying amounts. By reducing leaf litter input to 1/10 and 1/100 the natural rate over one year, they found that both the number of species supported and the number of trophic links were reduced as leaf-litter input was reduced (Figure 13.6). These results sup-

port the prediction of the energetic hypothesis that reduced energy input will result in reduced food chain lengths.

A third generalization about food webs is that there is a nearly constant proportion of species that are top predators, intermediate species and producers, regardless of the size of the food web. For example, there is an approximately constant ratio of 2 to 3 prey species for every predator species in food webs, regardless of the total number of species in the web.

It is important to understand food webs because the structure of food webs has implications for community persistence. Some food webs can support additional species without suffering any losses, but other food webs are unstable and thus subject to species losses if another species invades the community. If we can understand the structure of food webs better, we should be able to design better management strategies for conservation.

■ 13.3 FUNCTIONAL ROLES AND GUILDS HELP DEFINE COMMUNITY ORGANIZATION

Trophic levels provide a good description of a community, but, by themselves, may be too coarse a classification for defining community organization. A better approach is to use the food web to subdivide each trophic level into **guilds**, which are groups of species exploiting a common resource base in a similar fashion. For example, nectar-feeding insects and hummingbirds form a guild exploiting flowering plants. The seed-eating guild may contain ants, birds and mammals, as well as other invertebrates. We expect competitive interactions to be potentially strong between the members of a guild. By grouping species from a particular trophic level into guilds, we may also identify the basic **functional roles** played in the community.

Guilds and Functional Groups

There are three advantages to using guilds in the study of community organization:

1. Guilds focus attention on all the potentially competing species, regardless of their taxonomic relationship.
2. Guilds allow us to compare communities by concentrating on specific functional groups. We do not need to study the entire community, but can concentrate on a manageable unit.
3. Guilds might represent the basic building blocks of communities and thus help us to analyze community organization.

A community can be viewed as a complex assembly of component guilds or **functional groups**, each containing one or more species. Guilds may interact with one another within the community and thus provide the organization we see. No one has yet been able to analyze all the guilds in a community, and at present we can deal only with a few guilds that make up part of a whole community. Two examples of the

organization of guilds show how this concept can be applied to communities.

The eastern deciduous forest of the United States contains 62 different tree species that can be grouped into 9 guilds based on their regeneration ecology. Three broad groups were defined as **pioneer guilds** that regenerate quickly in highly disturbed areas, **opportunistic guilds** that regenerate in a wide variety of conditions, and **persistent guilds** that regenerate in shade and are long-lived. These broad groups each contained three guilds, based on the tolerance of species to dry or wet soils and seed dispersal characteristics. For example, red maple (*Acer rubrum*) is a member of the pioneer guild that tolerates moist sites, and sugar maple (*Acer saccharum*) is a member of the persistent guild that contains slow-growing shade-tolerant species. The nine tree guilds are functional groups that simplify our task of understanding how tree communities will respond to large-scale disturbances such as climate change.

Agricultural ecosystems occupy much of the Earth today, and agricultural production depends on a food web of other plants, predators and parasites that are often not recognized as relevant to the farm's operation. If we aggregate the species in an agricultural ecosystem into guilds or functional groups, the food web is relatively simple. Figure 13.7 illustrates one agricultural food web for an organic farm in the Netherlands. Different farm management systems will alter the relative abundances of many of these guilds, including both specialized and generalized insect predators that can hold crop pest species in check. Constructing a food web like Figure 13.7 for agricultural systems is a useful approach that can tie together agricultural production science and ecological community organization. How these guilds on farms change will determine both the conservation value of the farm to wildlife and the amounts of herbicides and pesticides that are needed to control crop pests.

The guild or role concept of community organization is an important idea, and suggests four generalizations that require further testing in natural communities:

1. Many species form interchangeable members of a guild from the point of view of the rest of the community. These species are functional equivalents.

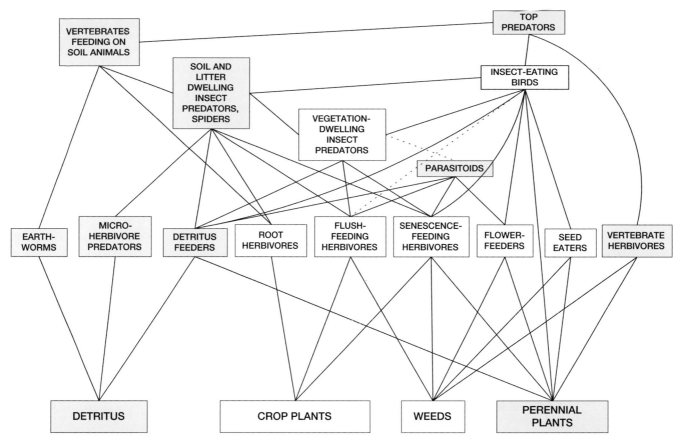

Figure 13.7 Food web for a generalized organic farm in the Netherlands. Boxes in yellow indicate a relatively high quantitative value, and thick lines indicate strong interactions. Dotted lines refer to minor interactions. The detritus food web on the left side is typically very large on farms and interacts with the herbivores and predators of the crops. By maintaining perennial plants in hedgerows, the ecological infrastructure of the farm can be improved to assist in crop pest control. The top predators in this farming system are carnivores such as the barn owl and the weasel. Vertebrates that feed on soil animals include birds, such as starlings, and mammals, such as shrews and moles. (Modified from Smeding and de Snoo 2003.)

2. The number of functional roles within a community is small in relation to the number of species and might be constant among different communities.
3. There may be a limit on the number of species that can simultaneously fill a given functional role. A community always has a set of roles, but the guilds may be packed with different numbers of species.
4. Species within guilds fluctuate in abundance in such a way that the total biomass or density of the guild remains stable.

The usefulness of the guild concept is that it reduces the number of components in a community and should help us to study how communities are put together. It also emphasizes that ecological units are not taxonomic units. Ants, rodents and birds can all eat seeds in desert habitats and can thus form a single guild of great taxonomic diversity.

Keystone Species

A functional role may be occupied by a single species, and that role may be critical to the community. Such important species are called **keystone species** because their activities determine community structure. Bob Paine was the first ecologist to recognize keystone species from his research in the rocky intertidal zone of Washington State (discussed in Chapter 12, page 262). Keystone species are typically not the most common species in a community, and their impact is much larger than would be predicted from their relative abundance. One way to recognize keystone species is by removal experiments.

The starfish *Pisaster ochraceous* is a classic keystone species in rocky intertidal communities of western North America and it is instructive to examine the ecological details of how this interaction was discovered by Bob Paine. The rocky intertidal zone of the West Coast includes a large number of species. Typically starfish are present, but do not appear to be common, along with many species of limpets, chitons, mussels and barnacles with no single species predominating. Starfish (Figure 13.8) were known to eat mussels, as well as barnacles and other invertebrates, in the intertidal zone, but they were relatively uncommon so no

one thought them to be very important in the community. But when Bob Paine manually removed the starfish *Pisaster* from intertidal areas, the mussel *Mytilus californianus* increased to monopolize all the space on the rocks, thus preventing other invertebrates and algae from being able to attach.

The mussel *M. californianus* is an ecological dominant species that is able to compete for space effectively in the intertidal zone. Predation by *Pisaster* reduces the abundance of the mussel and opens up space on the rocks, allowing other species to colonize and persist. *Pisaster* is not able to eliminate mussels because *Mytilus* can grow too large to be eaten by starfish, so the large adults are safe from predation. Size-limited predation

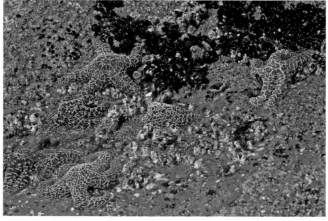

Robert T. Paine (1933–) Professor of Zoology, University of Washington, collecting starfish in the rocky intertidal

Figure 13.8 The starfish *Pisaster ochraceous* from the Pacific Coast of North America. These starfish occur in an array of colors from purple and brown to orange, but no one knows the ecological significance of this color variation. (Photos from Dave McShaffrey, Marietta College.)

provides a refuge for the prey species, and these large mussels are able to produce large numbers of fertilized eggs. Mussels and starfish interact strongly in the rocky intertidal zone, and the result of this predator-prey interaction is that many other species are able to occupy the area. Predation can prevent a species that would otherwise become a monoculture from taking over completely.

Sea otters are a keystone predator in the North Pacific. Once extremely abundant, they were reduced by the fur trade during the 19th century to near extinction by 1900. Since they were protected by international treaty during this century, sea otters began to increase and by 1970 had recovered in most areas to near maximum densities. Sea otters feed on sea urchins, which in turn feed largely on macroalgae (kelp). Early natural history observations showed that in areas where sea otters were abundant, sea urchins were rare and kelp forests were well developed. Similarly, where sea otters were rare, sea urchins were common, and kelp forests were non-existent. This condition is referred to as 'sea urchin barrens' because there are many sea urchins, but no visible kelp, and the urchins have to feed on microscopic surface algae on the bottom rocks. Sea otters are thus a good example of a keystone species in a marine subtidal community. During the last 15 years sea otters have declined precipitously in large areas of western Alaska (Figure 13.9), often at rates of 25% per year. The loss of this keystone species has allowed sea urchins to increase with the attendant destruction of kelp forests. Killer whales are the suspected cause of the sea otter decline. Killer whales have begun to attack sea otters in the last 15 years because their prey base (seals and sea lions) has

Sea otter (*Enhydra lutris*). (Photo courtesy of Christian Handl.)

declined along with the fishes on which they feed. Fish have probably declined owing to human over-harvesting in the North Pacific, illustrating that the interactions in food webs can propagate from top predators to basal plant species in unexpected ways.

A third example of a keystone species is the African elephant. The African elephant is a relatively unspecialized herbivore, but relies on a diet of browse supplemented by grass. By their feeding activities, elephants destroy shrubs and small trees and steer woodland habitats towards open grassland (Figure 13.10). Elephants feeding on the bark can destroy large mature trees. As more grasses invade the woodland habitats, the frequency of fires increases, which accelerates the conversion to grassland. This conversion works to the elephants' disadvantage, however, because grass is not a sufficient diet for elephants, and they begin to starve as woody species are eliminated. Other ungulates that graze the grasses benefit from the elephants' activities.

The critical impact of keystone predators is that they can reverse the outcome of competitive interactions. The impact of keystone predators is seen very clearly in aquatic communities. Amphibians are a major component of temporary ponds. In the coastal plain of North Carolina, a single pond can support five species of salamanders and 16 species of frogs and toads. Salamanders are the major predators in these temporary ponds, and the broken-striped newt *Notophthalmus viridescens* acts as a keystone predator. It selectively preys on the dominant competitors *Rana utricularia* and *Bufo americanus* and allows less competitive frogs, such as the cricket frog (*Hyla crucifer*), to survive. When the keystone predator is absent in a pond, there are fewer frog species present. Predation prevents competitive exclusion by dominant species.

Keystone species may be rare in natural communities, or they may be common but not recognized. At present, few terrestrial communities are believed to be organized by keystone species, but in aquatic communities keystone species may be more common. There seems to be no simple way of recognizing keystone species in food webs without doing detailed experimental study. The important message is that some species of low abundance can have strong impacts on

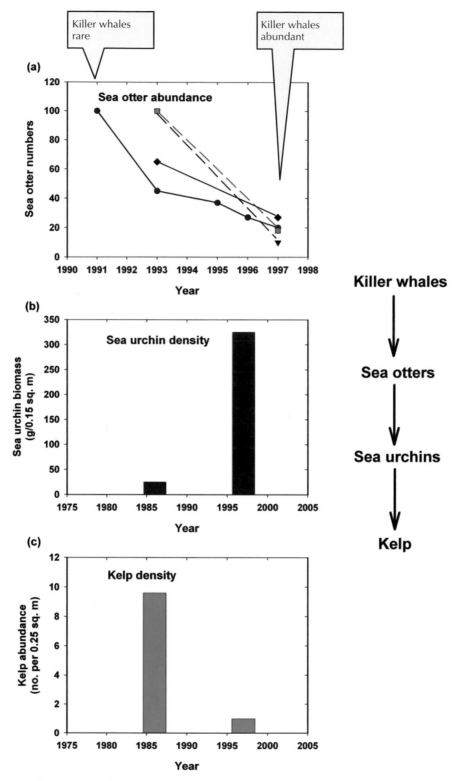

Figure 13.9 Sea otters as keystone predators in the North Pacific. (a) Changes in sea otter abundance over time at several Aleutian islands and concurrent changes in (b) sea urchin biomass, and (c) kelp density. The food chain is shown on the right. During the 1990s the number of killer whales—the top predator— increased, resulting in lower numbers of sea otters, higher numbers of sea urchins and much less kelp. (Modified from Estes, *et al.* 1998.)

Figure 13.10 African elephant browsing in open woodland, Tarangire National Park, Tanzania. Elephants are keystone herbivores because of their impact on tree regeneration. They feed on small trees—converting woodland habitats into grasslands. (Photo courtesy of Dan Perlman, EcoLibrary.)

community structure, so that land managers and conservationists should be concerned about both common and uncommon species in communities. We cannot determine which species might be keystone species unless we understand the detailed structure of the food web and analyze it experimentally.

Dominant Species

Some species in a community may exert a powerful control over the occurrence of other species, and the concept of dominance has long been engrained in community ecology. **Dominant species** are recognized by their numerical abundance or biomass and are usually defined separately for each trophic level. For example, the sugar maple is the dominant tree species in part of the climax forest in eastern North America and, by its abundance, determines in part the physical conditions of the forest community. Domi-

nance usually means numerical dominance—keystone species are not usually the dominant species in a community (Figure 13.11).

Dominant species are usually assumed to achieve their dominance by competitive exclusion. Competitive exclusion is considered to follow directly from linear competition (Figure 13.12). The simplest illustration of competition is a linear hierarchy (species A out-competes B and B out-competes C). In this case, competitive exclusion can occur, so only one species (the best competitor) will win and the all the others will become extinct. But competition between species can take on other forms as well. A more complex case of competition is circular competition, which occurs when no single species can be called dominant. In circular networks of spatial competition, species A out-competes species B, B out-competes C, but C, in turn, is able to out-compete A. This type of competitive

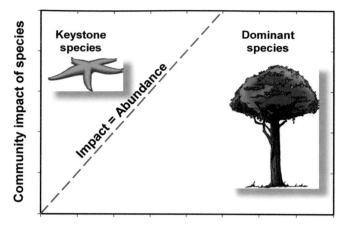

Figure 13.11 Schematic illustration of the difference between dominant species and keystone species. Species whose total impact is exactly proportional to their abundance would fall on the line. Both the dominant and keystone species in a community are assumed to have a high community impact, but keystone species have low biomass yet high impact. Trees, giant kelp and corals are examples of dominant species in their particular communities. (Modified after Power *et al.* 1996.)

interaction has no end point. Competitive exclusion does not occur in circular networks. If circular competition is the rule in natural communities, species diversity during succession would not decline because of competitive exclusion.

Competitive dominance in linear hierarchies is not the only explanation for a species becoming dominant. Australian mangrove forests show a complex zonation pattern across the intertidal zone. This zonation has usually been explained by mechanisms of physiological tolerance to seawater inundation or by tidal sorting of seeds by size. But in mangrove forests seed predation by small grapsid crabs is very high, and dominance in four of five mangrove species is correlated with the amount of seed predation. Predation may override competition in some communities, and dominance may result from a species being resistant to predation. A similar result arises in human societies because of resistance to disease.

Communities that develop under similar ecological conditions in a particular geographic region are expected to go through the process of succession and

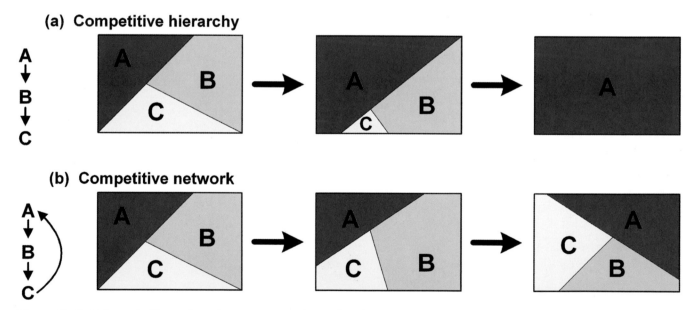

Figure 13.12 Schematic illustration of competitive relationships between three species competing for space. Each rectangle represents a plot of ground or a rock surface in the intertidal zone. (a) The conventional competitive hierarchy showing linear competition in which species A out-competes species B, which in turn out-competes species C. As time progresses, competitive exclusion will occur and only species A will be left in the community. (b) A competitive network showing circular competition in which species A out-competes B and B out-competes C, but C is able to out-compete A. This system changes in time but does not move toward competitive exclusion; so all three species remain in the community in spite of competition between them. (Modified after Buss and Jackson 1979.)

to become dominated by the same species in the climax state. For example, a deciduous forest in Ohio is expected to have beech and sugar maple as the dominant trees—botanists would be surprised if rare species, such as black walnut or white ash, became dominant. Ecologists would like to know if the same pattern occurs in aquatic communities, particularly those of the open ocean. The Central Gyre of the North Pacific Ocean has a rich diversity of phytoplankton, zooplankton and fish species. Because the waters of the ocean mix and the gyre is so large, the dominant species might be expected to vary from place to place within this large area. This does not seem to occur. In the Central Gyre, about 30 species of copepods were abundant and the dominance structure of this community has remained the same in samples collected up to 16 years apart. The same stability of community structure was evident in the phytoplankton community and the fish community of the Central Gyre. These oceanic communities appear to be as constant in their dominant species as are temperate zone forests—a result that has surprised ecologists.

The removal of a dominant species in a community has occurred frequently because of the impact of humans, but, unfortunately, few of these removals have been studied in detail. The American chestnut was a dominant tree in the eastern deciduous forests of North America before 1910—making up more than 40% of the overstory trees. This species has now been eliminated as a canopy tree by a disease called chestnut blight. The impact of this removal has been negligible as far as anyone can tell, and various oaks, hickories, beech and red maple have replaced the chestnut. Fifty-six species of Lepidoptera fed on the American chestnut. Of these seven species became extinct, but the other 49 species did not rely solely on the chestnut for food and they still survive.

Size-Efficiency Hypothesis

Dominance has been studied in freshwater communities in considerable detail. The zooplankton community of many lakes in the temperate zone is dominated by large species when fish are absent and by small species when fish are present. John Brooks and Stanley Dodson first observed this change in Crystal Lake,

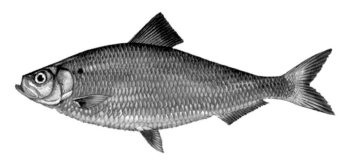

Alewife (*Alosa pseudoharengus*)

Connecticut, after the introduction of a herring-like fish—the alewife (Figure 13.13). They proposed the **size-efficiency hypothesis** as a wide-ranging explanation of the observed shift in dominance in the zooplankton community. The size-efficiency hypothesis is based on two assumptions:

1. planktonic herbivores (zooplankton) all compete for small algal cells (1–15 μm) in the open water
2. larger zooplankton feed more efficiently on small algae than do smaller zooplankton—large animals are able to eat larger algal particles that small zooplankton cannot eat.

Given these two assumptions, the size-efficiency hypothesis makes three predictions:

1. When predation on zooplankton is at low intensity or absent, the small zooplankton herbivores will be completely eliminated by large forms (dominance of large cladocera and calanoid copepods).
2. When predation is at high intensity, predators will eliminate the large zooplankton and allow the small zooplankton (rotifers, small cladocera and small copepods) to become dominant.
3. When predation is at moderate intensity, predators will reduce the abundance of the large zooplankton so that the small zooplankton species are not eliminated by competition.

Thus competition forces communities towards larger-bodied zooplankton, while fish predation forces them toward smaller-bodied species. These three predictions of the size-efficiency hypothesis are consistent with the keystone-species idea discussed in the preceding section.

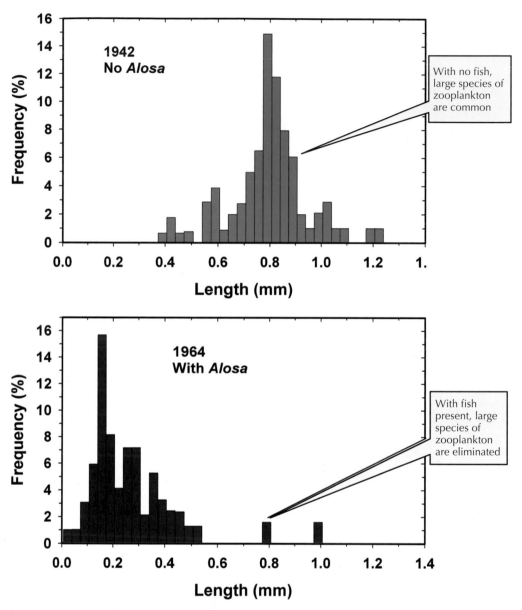

Figure 13.13 The composition of the crustacean zooplankton of Crystal Lake, Connecticut, before (1942) and after (1964) the introduction of the alewife—a plankton-feeding fish. The composition of the zooplankton shifted dramatically to smaller organisms after fish were introduced. Dominance in aquatic communities depends on the predators that are present. (Modified after Brooks and Dodson 1965.)

The second and third predictions of the size-efficiency hypothesis have been tested in several lakes and the predictions seem to describe adequately the zooplankton distributions in many lakes. Fish do tend to prefer the larger zooplankton species, as predicted by the size-efficiency hypothesis. But invertebrate predators in the plankton prey more heavily on the smaller zooplankton species, and, in some lakes, efficient invertebrate predators reduce the smaller zooplankton. The result is that large zooplankton may predominate in lakes with no fish, either because they are superior competitors (as the size-efficiency hypothesis predicts) or because small zooplankton are selectively removed by invertebrate predators. The recognition of the importance of invertebrate predators has added another dimension to the search for

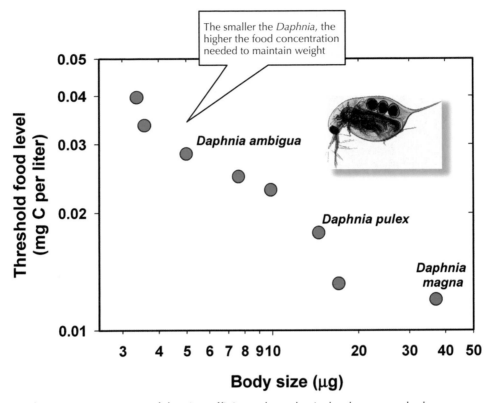

Figure 13.14 A test of a major assumption of the size-efficiency hypothesis that large zooplankton are competitively superior to small-sized zooplankton in their feeding ability. Eight *Daphnia* species of different sizes were raised in the laboratory on a constant food source. Larger species of *Daphnia* can maintain weight on a lower food concentration, and are thus competitively dominant as predicted. (Modified after Gliwicz 1990.)

explanations of the size distributions of zooplankton in freshwater lakes.

The second assumption of the size-efficiency hypothesis is that large zooplankton are superior competitors that are better able to harvest food resources. In a laboratory microcosm, eight species of *Daphnia* were fed constant food levels and the concentration of algae needed to maintain body weight was measured. The more efficient filter feeders ought to be able to graze the algae down to a lower level and still obtain enough food to maintain weight. Figure 13.14 shows that large copepods can indeed subsist on much lower algal concentrations than small copepods, so that large species are superior competitors, as assumed by the size-efficiency hypothesis.

The importance of fish predation in structuring zooplankton communities is now well established, but the competitive nature of feeding relationships in the zooplankton may not always favor large-sized species because of fluctuating food conditions in lakes and ponds. There are ponds and lakes where, in the absence of fish predators, small zooplankton predominate, and further research on the feeding strategies of zooplankton of different sizes is required to explain these anomalies. In lakes with abundant green algae, it may be possible to manipulate the abundance of large zooplankton species by restricting fish predation, resulting in improved water quality.

Dominance is an important component of community organization, although it is poorly understood. Dominant species may be the focal point of interactions that structure many of the other species in a community. Dominant plants often ameliorate the physical conditions of a community. The characteristics of dominant species may affect the stability of the community as well as its organization.

■ 13.4 STABILITY IS A CRITICAL PROPERTY OF ECOLOGICAL COMMUNITIES

Stability is a dynamic concept that refers to the ability of a system to bounce back from disturbances. If a brick is raised slightly from the floor and then released, it will fall back to its original position. This is the physicists' concept of **neighborhood stability** or local stability. The system will respond to temporary slight disturbances by returning to its original position. Thus, for example, a rabbit population would show neighborhood stability to moderate hunting pressure if it returned to its normal density after hunting is prohibited.

Physicists discuss stability in terms of small perturbations, but ecological systems are subject to large disturbances. To deal with these, we must consider a second type of stability—**global stability**. A system that has local stability shows global stability only if the system returns to the same point after large disturbances. Our brick, for example, shows both local and global stability because if we raise it 10 mm or 10 meters from the floor and release it, it will fall back to the floor. Ecological communities are not passive objects like bricks, and global stability is probably rare. One of the problems of ecology is to map out the limits of stability for various communities.

Biodiversity and Stability

One of the classical tenets of community ecology, and a hallowed tenet of conservation biology, has been that **biodiversity causes stability**. Charles Elton in 1958 suggested several lines of circumstantial evidence that support this conclusion:

- Mathematical models of simple systems show how difficult it is to achieve numerical stability.
- Many laboratory experiments on simple organisms, such as protozoa, confirm the difficulty of achieving numerical stability in simple systems.
- Small islands are more vulnerable to invading species than are continents.
- Outbreaks of pests are often found in simple communities on cultivated land or land disturbed by humans.
- Tropical rain forests do not usually have insect outbreaks like those common to temperate forests.

- Pesticides have caused outbreaks of insect pests by the elimination of predators and parasites from the insect community of crop plants.

We now know that many of these statements are only partly correct, but the simple, intuitive and appealing notion that biodiversity leads to stability has become an important cornerstone in conservation appeals.

Field ecologists have supported Elton in believing that complex communities are indeed more stable than simple ones. A few direct experiments have been done on the relationship between stability and diversity, and these support the idea that in many systems higher diversity does lead to higher stability. All the studies to date have been carried out on plant communities.

In 1982 David Tilman set up a long-term study of four Minnesota grasslands in order to measure the relationships between species diversity in plants and community functions. By measuring the variability in the biomass of plants at the end of the growing season for 11 years on 207 plots, Tilman could directly assess the diversity–stability hypothesis. Figure 13.15 shows the data from one field with 54 plots. Plots with higher plant diversity showed less fluctuation in yield than plots with low diversity, in keeping with the expectations of the diversity–stability hypothesis of Elton.

Stability is usually measured as variability in numbers or biomass, but it can also be measured as **resistance to change**. More stable communities will change

David Tilman (1949–) Professor of Plant Ecology, University of Minnesota

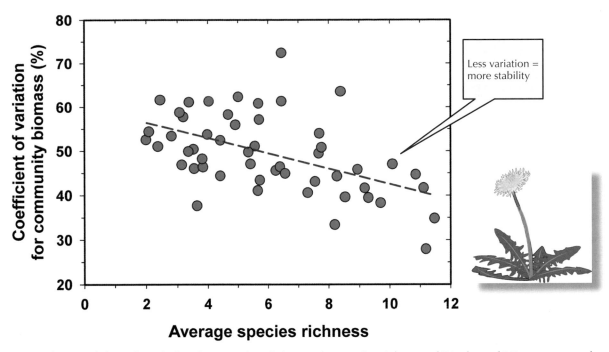

Figure 13.15 The variability of total plant biomass in relation to the species richness of 54 plots of Minnesota grasslands, averaged over 11 years. Each point represents one plot, 4 × 4 m. At the peak of the growing season each summer, the above-ground living biomass of all plant species was clipped and dried to estimate the community biomass. There is a significant negative relationship between variability in biomass and species diversity, as predicted by the diversity–stability hypothesis. (Data from Tilman 1996.)

less if an external stress is imposed on them. Drought is a major stress for temperate plant communities and plant ecologists have used droughts to test the diversity–stability hypothesis. In Yellowstone National Park a severe drought in 1988 allowed ecologists an opportunity to study the impact of diversity on stability in grassland communities. Species-rich communities in Yellowstone Park had higher resistance to change, as predicted by the diversity–stability hypothesis.

Resilience of Communities

While much ecological interest has focused on the stability of communities, scientists have approached community organization from another direction by focusing on **resilience**. Resilience is the capacity of a community to absorb disturbance, undergo change and still retain essentially the same function, structure, identity and feedbacks. Communities vary greatly in resilience—while some are fragile others are remarkably robust.

Variability and flexibility are both needed to maintain the resilience of communities and ecosystems.

Attempting to stabilize such systems in some perceived optimal state—whether for conservation or production—tends to reduce resilience and often results in the system moving close to a critical threshold.

There are three aspects of the resilience of a community:

1. **Latitude:** The maximum amount of change a community can experience before losing its ability to recover, or before crossing a threshold which, if breached, makes recovery difficult or impossible

2. **Resistance:** The ease or difficulty of changing the community

3. **Precariousness:** How close the community currently is to a limit or threshold.

A good example to illustrate the ideas of resilience is coral bleaching (Chapter 9, page 174)—a serious ecological problem in all of the world's coral reefs over the last 10 years. If sea surface temperatures rise above a threshold of 30°C, coral polyps expel the symbiotic algae that provide both essential nutrients and the rich coloration of the coral, leaving the corals with a bleached appearance. If the sea temperature does not drop below 30°C within about 3 days, the coral polyps die and the coral reef community changes dramatically in organization. As global temperatures increase, the coral reef community moves closer to a precarious threshold of change—while it is resistant to change, the resistance is limited to the sea surface temperature range of 24–30°C.

The conjecture of Charles Elton in 1958 that species diversity imparts stability to ecological communities has been supported for many, but not all, ecological systems, and we now know that species diversity by itself does not make a community resilient or stable. The practical application of these ideas to our human-degraded landscapes is the focus of the applied area of restoration ecology. We can attempt to restore species diversity to degraded landscapes, not only to improve the environment but also to restore stability to communities. What progress have we made so far in restoring damaged ecological systems?

■ 13.5 RESTORATION ECOLOGY APPLIES ECOLOGICAL KNOWLEDGE TO REPAIR DAMAGED COMMUNITIES

Communities recover from disturbances through a whole series of biological restoration mechanisms. Succession is a major pathway in restoration ecology, which aims to harness natural processes to restore systems adversely impacted by humans. There is a key starting principle in restoration ecology that the spatial scale of the impact and the recovery time are related, so that the larger the scale of the disturbance, the longer the timeframe for restoration (Figure 13.16). There appears to be no difference in this relationship for human-caused versus natural disturbances. If we can identify the processes that limit the speed of recovery we can alter this curve to reduce the impacts of human disturbances.

The first principle of restoration ecology is that environmental damage is not irreversible. This optimistic principle must be tempered by the second principle of restoration ecology, that communities are not infinitely

resilient to damage. Figure 13.1a illustrates these ideas schematically—there is only a finite range of environments over which a community is locally stable. To illustrate the first principle of restoration ecology, let's look at an example of lake restoration that has been successful. Aquatic communities have been disturbed by pollution caused by humans for many years, and the stability of aquatic systems under pollution stress is a critical focus of restoration ecology today. Several large-scale uncontrolled experiments have been performed by the diversion of sewage into large lakes near cities. Lake Washington is one such instance.

Lake Washington is a large, formerly unproductive, lake in Seattle, Washington that was used for sewage disposal until the late 1960s. In the early phases of development, Lake Washington was used for raw sewage disposal, but this practice was stopped between 1926 and 1936. However, with additional human population pressure, a number of sewage-treatment plants built between 1941 and 1959 began discharging increasing amounts of treated sewage into the lake. By 1955 it was apparent that the sewage was destroying the clear-water lake, and a plan to divert sewage from the lake was voted into action. More and more sewage was diverted to the ocean from 1963 to 1968, and almost all was diverted from March 1967 onwards. Thus, the recent history of Lake Washington consists of two pulses of nutrient additions, followed by a complete diversion.

Since the diversion of sewage began in 1963, ecologists from the University of Washington have recorded the changes in Lake Washington in detail. Figure 13.17 shows the rapid drop in phosphorus in the surface waters and the closely associated drop in the standing crop of phytoplankton (algae). The nitrogen content of the water has dropped very little, which suggests that phosphorus is a limiting nutrient to phytoplankton growth. The water of the lake has become noticeably clearer since the sewage diversion. The phosphorus tied up in the lake sediments is apparently released back into the water column rather slowly.

In 1976 Lake Washington suddenly became much clearer than had been recorded previously. This clearing of the water coincided with the colonization of Lake Washington by *Daphnia* sp.—a small crustacean.

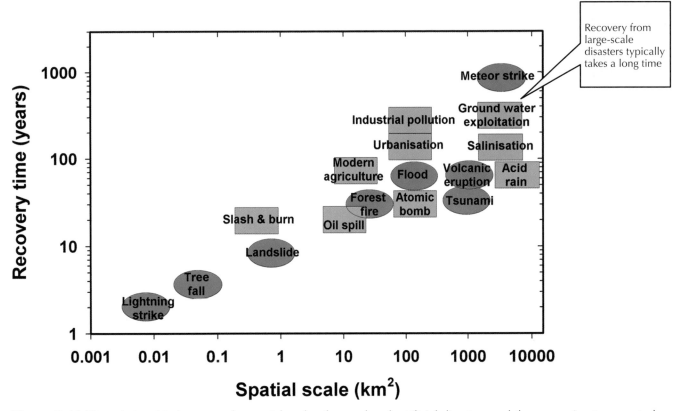

Figure 13.16 The relationship between the spatial scale of natural and artificial disasters and the approximate expected time to recovery. Natural disasters are depicted as red ellipses and human-caused changes are represented by black rectangles. The aim of restoration ecology is to reduce the recovery time by manipulating ecological factors restricting the time sequence of recovery. (Modified from Dobson *et al.* 1997.)

Daphnia are filter feeders that sweep small green algae out of the water and, in the process, make lake water clearer. *Daphnia* have always been present in lakes around Lake Washington so their sudden appearance in 1976 was not a simple matter of colonization and dispersal. Two factors combined to hold *Daphnia* numbers low before 1976. The blue-green alga *Oscillatoria* was common in Lake Washington during its polluted phase, but gradually declined after 1968 when sewage was diverted. *Oscillatoria*'s long filaments clog the filter-feeding apparatus of *Daphnia*, which could not feed properly until *Oscillatoria* became scarce. *Daphnia* are also a major prey item for the shrimp *Neomysis mercedis*. The numbers of *Neomysis* decreased during the 1960s because of fish predation by increased populations of longfin smelt in the lake. Numbers of smelt increased during the 1960s when their spawning grounds in the Cedar River were protected from

dredging and construction in the river. The aquatic community of Lake Washington shows a complex web of predator–prey interactions that have profound impacts on community structure.

The Lake Washington experiment is of considerable interest because it suggests that detrimental changes in lakes may be *stopped and reversed* if the input of nutrients can be stopped. The restoration of Lake Washington shows that this community displays a considerable amount of global stability and resilience, and is a good example of an equilibrium community.

Restoration ecology can depend on the natural time scale of succession, as happened in Lake Washington, or it can speed processes of recovery by adding nutrients when they are deficient, seeding areas with a shortage of available colonists, and adding microbes to break down organic compounds such as oil. This area of applied ecology will assume more importance in the

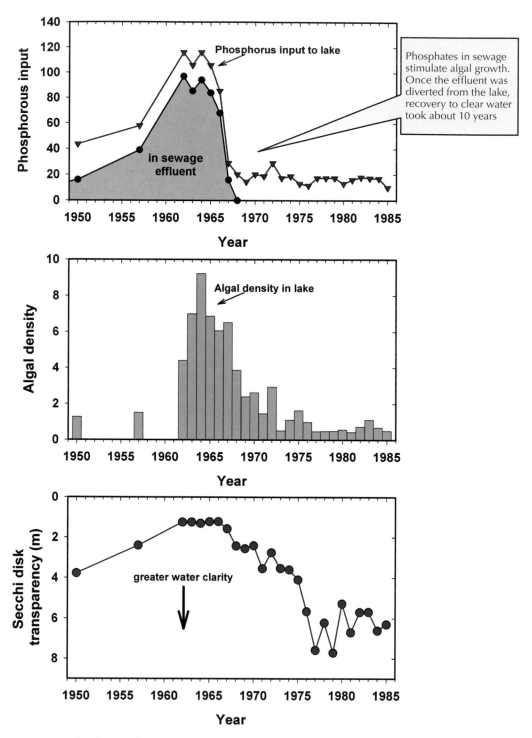

Figure 13.17 Recovery of Lake Washington in Seattle, 1950–1984. Treated sewage flowed into the lake in increasing amounts during the 1950s. Sewage was diverted from the lake gradually from 1963 to 1968. The phosphorus content of the lake decreased rapidly after sewage was diverted. Algal density dropped in parallel with phosphorus levels because phosphate is the nutrient that limits algal growth in this freshwater lake. In 1976 the small crustacean *Daphnia* increased greatly in abundance and, because *Daphnia* eats small algae, the algal abundance fell further, so the lake water became clearer. Transparency is measured by the depth at which a standard white plate can be seen in the water during midsummer. (Data courtesy of W.T. Edmondson.)

years to come as we try to speed the recovery of degraded landscapes. But we should remember the third principle of restoration ecology—it is always simpler, cheaper and better to preserve ecosystems than to destroy and then restore them.

SUMMARY

Communities can be organized by competition, predation and mutualism working within a framework set by the physical factors of the environment. Of these three processes, most emphasis has been placed on the roles of competition and predation in organizing communities—the role of mutualism has been largely unstudied.

Two broad views of communities are postulated as explanations of community organization. The classical model is that communities are in equilibrium—their species composition and relative abundance are controlled by biotic interactions. According to the equilibrium model, interspecific competition, predation and spatial heterogeneity are the major processes controlling organization. The non-equilibrium model does not assume stable equilibria, but that communities are always recovering from disturbances, as we will discuss in the next chapter.

Species in a community can be organized into food webs based on who-eats-whom. Trophic levels may be recognized in all communities from the level of the producers (often green plants) to the higher carnivores

and hyperparasites. Within a trophic level, we can recognize guilds of species exploiting a common resource base. Guilds may serve to pinpoint the functional roles species play in a community—species within guilds may be interchangeable in some communities.

Keystone species are relatively uncommon, but highly influential, species. They determine community structure almost single-handedly and can be recognized by performing removal experiments. Dominant species are the species of highest abundance or biomass in a community. Dominance is often achieved by competitive superiority, but some dominant species can be removed from the community and replaced with subdominants with little effect on community organization.

The characteristics of dominant species may affect community stability, in the sense of the ability of the system to return to its original configuration after disturbance. The ecological generalization that diversity causes stability is supported by field data. The attributes of individual species and compartments in food webs may be significant in determining the amount of community stability.

Restoration ecology tries to apply ecological knowledge of community dynamics to restore damaged landscapes. Community succession can heal damaged landscapes, but may take more time than we might like. We can try to speed recovery by knowing how succession operates in a community and what limits its rate of progress. Communities can recover from disasters, but there is a limit to their resilience.

SUGGESTED FURTHER READINGS

Beisner BE, Haydon DT and Cuddington K (2003). Alternative stable states in ecology. *Frontiers in Ecology and the Environment* **1**, 376–382. (A discussion of local and global stability in communities.)

Chase JM (2000). Are there real differences among aquatic and terrestrial food webs? *Trends in Ecology and Evolution* **15**, 408–412. (An evaluation of whether we need different hypotheses and explanations for terrestrial and aquatic food webs.)

Estes JA, Tinker MT, Williams TM and Doak DF (1998). Killer whale predation on sea otters linking oceanic and nearshore ecosystems. *Science* **282**, 473–476. (How apex predators like killer whales can affect community structure.)

Link J (2002). Does food web theory work for marine ecosystems? *Marine Ecology Progress Series* **230**, 1–9. (Marine food webs may be more connected than terrestrial and freshwater webs.)

Power ME, Tilman D, Estes JA, Menge BA, Bond WJ, Mills LS *et al.* (1996). Challenges in the quest for keystones. *BioScience* **46**, 609–620. (A discussion of the role of keystone species in communities.)

Tilman D (1996). Biodiversity: population versus ecosystem stability. *Ecology* **77**, 350–363. (An evaluation of the diversity-stability problem by a leading ecologist.)

Worm B and Duffy JE (2003). Biodiversity, productivity and stability in real food webs. *Trends in Ecology and Evolution* **18**, 628–632. (How the biodiversity crisis might affect whether we can achieve conservation goals.)

QUESTIONS

1. Bracken fern (*Pteridium aquilinum*) often invades lowland heaths in Britain and develops a dense, uniform stand with a much reduced flora and fauna. To reverse this habitat deterioration, various chemical and physical control methods were carried out for 18 years, but the objective of this restoration scheme (to restore heather heathland) was not achieved. List some reasons why such a failure might occur. Read Marrs *et al.* (1998) and discuss the reasons he gives for the failure of this restoration program.

2. Over what time period does a community need to be observed before its food web is complete? Discuss the implications of constructing a time-specific food web versus a cumulative food web over a long period. Compare your analysis with that of Schoenly and Cohen (1991).

3. How might you define functional groups of plant species? Would the same criteria work for animals? Gitay and Noble (1997) discuss this problem.

4. Elton (1958, p. 147) claims that natural habitats on small islands are much more vulnerable to invading species than natural habitats on continents. Find evidence that is relevant to this assertion, and evaluate its importance for the question of community stability.

5. In a few tropical rainforests, 50–100% of the canopy trees are one species (Connell and Lowman 1989). List several possible mechanisms by which a single tree species can maintain its dominance in a species-rich forest system. Which of these mechanisms could not operate in species-poor temperate forests? Compare your list with that of Connell and Lowman (1989, p. 97).

6. The *Exxon Valdez* oil spill in Alaska was the classic oil spill of recent years. How much recovery has occurred to the marine and intertidal communities since this spill, and what is the prognosis for full recovery in these northern ecosystems? Garshelis and Johnson (2001) and Peterson (2001) provide background data on the spill.

Chapter 14

COMMUNITY DYNAMICS—DISTURBANCE ECOLOGY

IN THE NEWS

Forest fires are large, infrequent disturbances to ecological communities around the world. Large fires, such as those in Yellowstone National Park in the summer of 1988 and in south-eastern Australia in 2006, cause social and political pressures for ecologists to do something about the desolation of the landscape, the loss of vegetation and animal life, and the loss of human life and property.

In August and September 1988 a large fire in the border area of Montana and Wyoming burned about 500 km^2 of Yellowstone National Park, sparking criticism of Park management.

Lodgepole pine stands in Yellowstone National Park during the fires of 1988.

Elk standing in lodgepole pine stand just after the 1988 Yellowstone fires.

For almost 75 years, the response to forest fires in many parts of the world was to put them out immediately. However, forest ecologists recognized in the 1950s and 1960s that fires are a natural disturbance in many ecosystems—this has caused a policy shift in government agencies to permit natural fires to burn unimpeded, unless humans and property are at risk. Many of our vegetation communities are fire-prone, and preventing fires in these communities can change the community structure and composition, delay the inevitable and cause even worse ecological damage when fire does arrive. Most severe fires, such as the Yellowstone fires of 1988 and the Australian fires of 2006, are weather-driven and, in spite of massive human intervention, cannot be stopped except by changes in the weather. A total of $120 million was spent fighting the Yellowstone fires of 1988, but the fires only began to dwindle in mid-September when it began to snow.

The desolation left after the Yellowstone fires quickly disappeared as the vegetation recovered. Perennial plants, grasses and shrubs were not killed, but re-sprouted in 1989 and flowered profusely in 1990 and 1991. The dominant trees in Yellowstone are lodgepole pines, which are killed by fire, but many lodgepole pines produce seeds in serotinous cones that do not open unless they are burned in a fire. Lodgepole pines may retain their cones on the tree for 40 years or more, so millions of seeds are released after a fire. Their seeds germinate best in severely burnt soil surfaces, and lodgepole seedlings often form very dense stands. The lodgepole pine forests of Yellowstone will grow to maturity over the next 125 years, assuming there is not another fire disturbance.

Lodgepole pine seedlings ten years after the 1988 Yellowstone fires.

Aspen is another common tree in Yellowstone and their seedlings also established strongly after the fires of 1988. Aspen are a preferred food for elk, and severe browsing of small aspen trees has retarded their growth and reduced their ability to colonize some areas. Elk do not eat lodgepole pine seedlings, so elk browsing favors the recovery of pine forests.

The Yellowstone fires of 1988 illustrate the major changes that fire can impose on an ecosystem and also the mechanisms by which recovery occurs in a landscape over many years. The story of Yellowstone is the story of disturbance ecology.

■ 14.1 COMMUNITIES ARE NOT IN EQUILIBRIUM IF THEY ARE CONTINUALLY SUBJECT TO DISTURBANCES

Communities have traditionally been thought to be in equilibrium with their environment, so that when they are subjected to a disturbance, such as fire or logging, they undergo succession back to their original state. The equilibrium model for communities is focused on stability, and this is a good description of how some communities operate. But in many other communities of a few hectares or more, *change* seems to be the rule rather than stability—this has caused many ecologists to search for a broader model for communities, which includes the ecology of disturbances. Ecologists are now forging a new model of community organization called the **non-equilibrium model** to replace the classical equilibrium model. The non-equilibrium model focuses on a small spatial scale, and has two central ideas to aid understanding of how communities operate—patches and disturbance. A **patch** is any discrete area and a **disturbance** is any discrete event that disrupts a community. In this chapter, we explore some of the new concepts the non-equilibrium model has stimulated.

Patches and Disturbance

Disturbances can affect ecological communities in many different ways, depending on their strength and frequency of occurrence. In general, ecologists have considered that communities subject to disturbances recover slowly back to their original state through a process of succession (Figure 14.1). But, if several disturbances hit a community at the same time, or in rapid succession, the community may not be able to recover, and will be pushed into an altered state. These impacts of disturbances can be particularly severe on communities that are already stressed by human impacts, such as pollution or climate change. In this chapter we will try to understand what determines how a community responds to disturbances.

Communities often are heterogeneous—they exhibit patchiness rather than uniformity—and the impacts of disturbances may vary in different patches. The patchiness of different communities varies, and

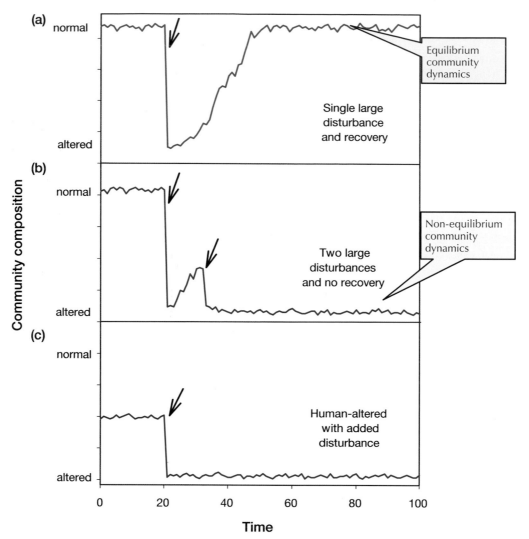

Figure 14.1 Schematic illustration of the effects of disturbances (arrows) on ecological communities. (a) A community is subjected to a single large disturbance, such as a fire, at time 20 and then recovers through a process of succession to its original state. (b) A community is subjected to two disturbances (arrows) at times 20 and 35, and the combined effects lead to a change in community composition and no recovery. (c) A community altered by human activities, such as farming or forestry, is then subjected to a disturbance by fire or flooding, and the combination of stresses changes community composition and prevents it from recovering in a short time. 'Normal' is used here as a shorthand for 'previous community state'. (Modified from Paine *et al.* 1998.)

the patches may differ in size. We can recognize five spatial scales at which ecologists work:

1. **individual space:** occupied by one plant or sessile animal, or the home range of one individual animal (typically a few square meters to a few hundred square meters)
2. **local patch:** occupied by many individual plants or animals (a few to many hectares)
3. **region:** occupied by many local patches or local populations linked by dispersal (a few square kilometers)
4. **landscape:** a region large enough to be closed to immigration or emigration (tens to hundreds of square kilometers)
5. **biogeographical zone:** including zones of different climate and different communities (hundreds to thousands of square kilometers).

Understanding community dynamics at small spatial scales of a few hectares, and aggregating the resulting dynamics into a regional or landscape scale, is one important approach needed to predict large-scale community dynamics in response to disturbances.

Most field studies of communities, and virtually all experimental manipulations of communities, are carried out at the local patch scale, but large fires, such as those of Yellowstone in 1988, affect whole landscapes. There is no general definition of a 'patch' that will be appropriate for all ecological communities, but, in general, a patch will cover a few square meters for small plants and animals to several hectares for larger plants and animals. Patches do not need to be completely homogeneous like a concrete floor—they will be somewhat variable internally.

A patch is similar to what many naturalists would call a habitat. A habitat is a place where a species normally lives, and so the definition of a habitat is species-specific. A bird may occupy one habitat for nesting, a second habitat for feeding and migrate in winter to a third habitat for overwintering. A butterfly will occupy one habitat containing the favored food plants for the larval stage, a second habitat for the adults that feed on nectar in flowers and a third habitat for roosting at night.

Landscape Patchiness

Habitats and patches are typically mapped as vegetation communities for larger plants and animals. These maps can be constructed from ground surveys, aerial photos or satellite images. Figure 14.2 illustrates a map

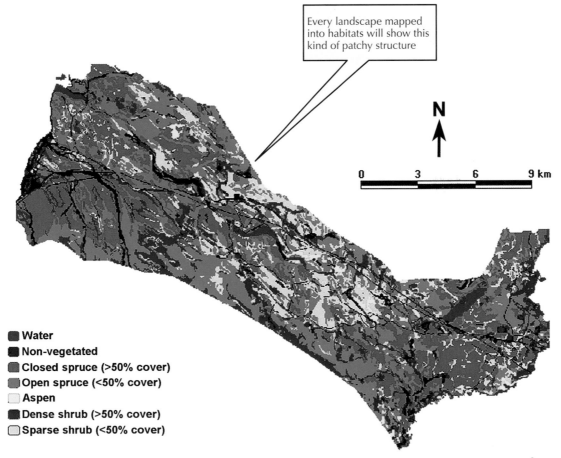

Every landscape mapped into habitats will show this kind of patchy structure

N

0   3   6   9 km

- ■ Water
- ■ Non-vegetated
- ■ Closed spruce (>50% cover)
- ■ Open spruce (<50% cover)
- □ Aspen
- ■ Dense shrub (>50% cover)
- ▨ Sparse shrub (<50% cover)

Figure 14.2 Landscapes are patchy, not homogeneous. This is an example of a vegetation map of a 350 km² section of the boreal forest near Kluane Lake in the south-western Yukon, Canada, based on Landsat satellite images. Five broad vegetation types are identified in this habitat map. The habitat patchiness shown at this scale is a product of forest fires and soils. (Modified from Krebs and Boonstra, 2001.)

of vegetation habitats in a valley of the Yukon boreal forest, and illustrates the heterogeneity that is typically seen in habitat maps. Patches of some vegetation communities are small and widely spaced in this landscape, while others occupy larger patches that are highly connected. The scale at which these habitats are used by animals will depend on the size and mobility of the particular animal species concerned.

Landscape patchiness is a joint product of disturbance, soil type and topography. A disturbance disrupts community structure and changes available resources, substrate availability or the physical environment. Disturbances can be destructive events, such as fires or landslides, or an environmental fluctuation, such as a severe frost. Disturbances such as fire operate as a **pulse**—the disturbance is over and finished quickly. Other disturbances, such as a prolonged drought, operate as a **continuous press**, slowly pushing in one direction. The notion of what is 'normal' for the community is excluded from the ecologists' view of disturbance (in contrast to the everyday use of the word 'disturbance'). This is an important change of focus that has implications for conservation and management. We cannot assume under the non-equilibrium model that communities in the 'good old days' were 'normal' and had no disturbances, and that the job of conservationists or land managers is to get back to what the community was like before human disturbances. Disturbances in this new model are part of community organization, and are always occurring. For some communities, disturbances are frequent, but in others disturbances are rare (*Essay 14.1*).

Disturbances can be measured in a variety of ways to provide a time perspective and a spatial perspective (Table 14.1). Disturbances may be classified via their source—either coming from physical factors (such as fire) or from biological interactions (such as an introduced predator). These two classes are the end points of a continuum of types of disturbances, and many communities are affected by a combination of physical and biological disturbances. The ecological challenge is to measure the various disturbances in the field and to see how they affect particular communities.

Table 14.1 Definitions of measures of disturbance.

Measure	Definition
Spatial	
Distribution	Spatial distribution, including relationship to geographic, topographic, environmental and community gradients
Area or size	Area disturbed. This can be expressed as area per event or area per time period
Temporal	
Frequency	Mean number of events per time period
Return interval, or turnover time	The inverse of frequency: mean time between disturbances
Rotation period	Mean time needed to disturb an area equivalent to the study area (the study area must be defined)
Predictability	Low if the return interval is highly variable
Magnitude	
Intensity	Physical force of the event per area per time (e.g. wind speed for hurricanes)
Severity	Impact on the community (e.g. basal area removed)
Synergism	How one disturbance affects another disturbance (e.g. drought increases fire intensity, or insect damage increases susceptibility to windstorm)

(Modified from Pickett and White 1985.)

ESSAY 14.1   AUSTRALIA – THE FIRE-PRONE CONTINENT

Of all the continents, Australia is the most fire prone and this has produced an ever-changing illustration of our ecological and social views about the role of fire in natural ecosystems. A highly variable climate with recurrent droughts has made Australia a showcase of disturbance ecology and non-equilibrium community dynamics. Fires can be natural (caused by lightning) or human induced—either by arson or by deliberate prescribed burns.

In northern Australia large areas of open savannah woodland are burnt every year—or sometimes twice a year—with prescribed burns. In 2001, 10% of the entire continent was burnt, and much of the fire was concentrated in tropical savannas. In south-eastern Australia wildfires caused by lightning or arson in years of drought have brought a large human toll and produced a bitter debate about how fire-prone communities should be managed. In 2003 in Canberra 500 houses were destroyed by a fire started by lightning and fanned by 43°C temperatures and high winds.

A few of the houses destroyed by the Canberra fires in January 2003

Many of Australia's plant communities are fire-adapted and there has been considerable controversy over the 'natural' fire regime. Before Europeans settled Australia, the Aboriginal people used fire extensively both to obtain immediate food and to open up the bush. In northern Australia fires are set regularly as an important component of land management. For example, almost half the area of Kakadu National Park in the Northern Territory has been burnt each year.

Controlled burns are used in many parts of Australia to protect property, and there is controversy over whether enough controlled burning is being done. But land management issues emerge as to what kind of fires should be set (spring or early dry season, autumn or late dry season?) and how frequently.

The ecological community in Australia has been divided over the question of how much of the continent was burned regularly by the Aboriginal people. This uncertainty

A fire front in Kakadu National Park near Darwin, Australia

means that we cannot determine whether the continent's vegetation, as first seen by white explorers, was in equilibrium or a non-equilibrium condition. This question is easily confused with the more practical management question of how much controlled burning should be carried out each year to maintain desirable plant and animal communities. The ecologically desirable objective to maximize biodiversity often conflicts with the economic and social objective to protect livestock, farms and urban communities from runaway wildfires.

In large parts of the Red Centre and northern Australia that are sparsely settled, fire management has had both positive and negative consequences for biodiversity.

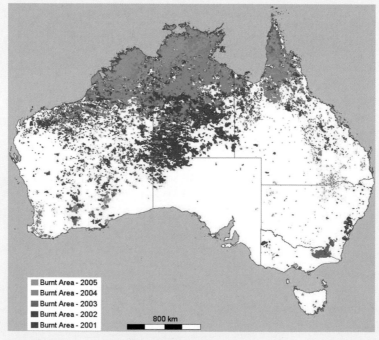

Burnt Area - 2005
Burnt Area - 2004
Burnt Area - 2003
Burnt Area - 2002
Burnt Area - 2001

800 km

Map of Australia showing the areas burnt since 2001

But under this human-determined fire regime across a broad region of northern Australia, many small mammal species are declining in abundance for reasons that are not immediately obvious, but are associated with changes in the fire regime from Aboriginal management to pastoral management (Woinarski *et al.* 2001). By contrast, frequent burning in wetlands along the South Alligator River in Kakadu National Park has reduced the abundance of the grass *Hymenachne acutigluma*, which forms a dense monoculture in the absence of disturbance. This successful application of Aboriginal land management ideas has increased bird abundance at Kakadu, as shown in the following diagram:

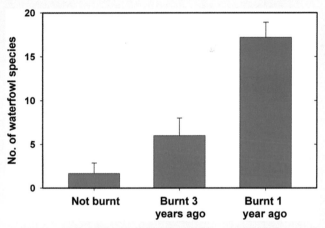

Burning of wetlands had led to an increase in the number of waterfowl species in Kakadu

A magpie goose

For any particular fire regime, some species will profit and some will lose—the ecologists' job is to determine this for each ecosystem. The first lesson learned in fire management is that there are no simple fire rules that apply to all ecosystems. Burning can help to protect biodiversity if it is undertaken with the right frequency and intensity.

■ 14.2 DISTURBANCES ARE HIGHLY SPECIFIC TO COMMUNITIES

Disturbances are highly specific to communities, and so we need to study how communities operate in the field to understand the role of particular disturbances. Many communities (such as coral reefs) are assumed to be in equilibrium, but detailed studies show they are strongly affected by disturbances and so are continuously changing. Let us look at two examples of biological communities responding to disturbance.

Coral Reef Communities

Coral reefs have been present in tropical oceans for at least 60 million years, and this long history has produced the great diversity of organisms we find on reefs today. Coral reefs have long been viewed by ecologists as the classical equilibrium community—living in tropical waters, constant in composition and relatively unaffected by disturbances. Our view of coral reefs now is very different because we have the benefit of long-term studies and of recent dramatic events affecting corals worldwide (see *Essay 9.1,* page 174). We will consider two aspects of coral reefs—the corals themselves and the coral reef fishes.

Coral reefs are subject to a variety of physical disturbances associated with tropical storms. At Heron Island reef, which is at the southern edge of the Great Barrier Reef of Australia, marine ecologists have followed changes in coral cover over a 30-year period using permanently marked quadrats. They measured the percentage of the area covered by corals to estimate abundance. They measured larval recruitment of corals by sequential sets of photographs of permanent quadrats.

Violent storms were the main source of disturbance to Heron Island reefs, and the amount of damage caused by cyclones was strongly affected by the position of the coral colonies on the reef (Figure 14.3). Five cyclones passed near to Heron Island during the 30 years of study from 1962 to 1992. Of the four study areas shown in Figure 14.3, only the protected area of the inner flat was relatively unaffected by cyclones. Virtually every cyclone caused a reduction in coral cover in the exposed pools. The 1967 cyclone completely removed coral cover on the exposed crest—the most severe disturbance observed. Recovery on the exposed crest was slow for the next 25 years. Gradual declines in coral cover on the protected sites was caused by increasing exposure to air as the corals grew upward over the 30 years of study.

Recruitment rates of larval corals were highly variable, which is typical of many marine invertebrates whose larvae drift in the plankton before settling. Figure 14.4 illustrates the differences in recruitment rates among years and among sites on the Heron Island reef. There are no particularly 'good' or 'bad' years for coral recruitment in the sense that the whole reef varied in unison. The variability in recruitment was partly associated with how much free space was available in different areas. Coral larvae cannot attach to other living coral or macroalgae and need free space to settle.

The picture that emerges from this work on the Great Barrier Reef is of a coral community that changes continually because of disturbance caused by tropical cyclones and internal processes of growth and recruitment. The coral community is not in equilibrium at the spatial scale of the reef because the frequency of disturbance is greater than the rate of recovery. Corals are therefore a good example of a non-equilibrium community.

Coral reef fishes are known to everyone and are one of the draw cards for ecotourism to coral reef areas. The diversity of coral reef fishes is astonishing. For example, at One Tree Reef—a small island at the southern edge of the Great Barrier Reef—nearly 800 species of fish have been recorded, while at the northern edge of the Great Barrier Reef, over 1500 species of fish have been recorded. What determines the community structure of these very rich coral reef fish communities? Are these fish communities stable in time and space?

There are two extreme schools of thought about what controls the organization of coral reef fish communities. The first view suggests that coral reef fish communities are equilibrium systems. This equilibrium is achieved by competition between the fish species. The equilibrium view is called the **niche-diversification hypothesis**, which suggests that coral reef fish communities are equilibrium competitive systems in which each species has evolved a very specific niche with respect to food

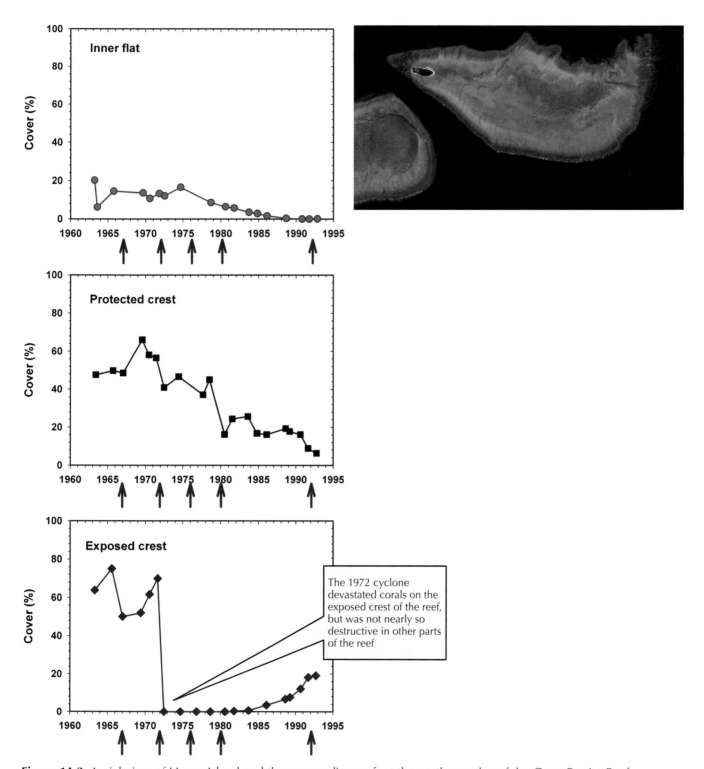

Figure 14.3 Aerial view of Heron Island and the surrounding reefs at the southern edge of the Great Barrier Reef, Australia. The island itself is small—800 m long and 300 m wide—covering about 8 hectares (left side of photo). Most of the photo shows the coral reefs around the island. The percentage cover of corals in three areas of the coral reefs surrounding Heron Island is shown from 1963 to 1992. Years when tropical cyclones affected the area are indicated by red arrows. Permanent quadrats were measured in these shallow water sites from 1963 to 1992. Damage from cyclones was highly variable depending on how much the site was protected by the island. (From Connell *et al.* 1997.)

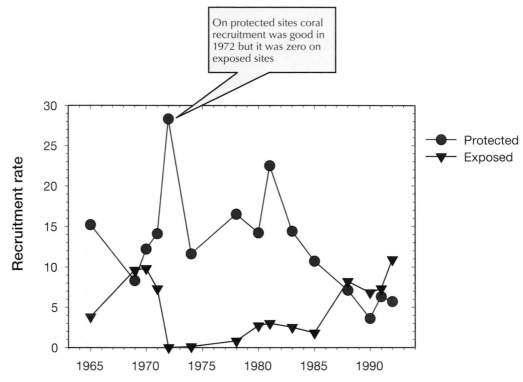

Figure 14.4 Coral larval recruitment rate from 1963 to 1992 on protected and exposed crests of the Heron Island reef in Australia. Recruitment is highly variable among sites and among years, and does not correlate among the different sites, so there are no general 'good' and 'bad' years. Recruitment was measured as number of larval recruits per m² per year on permanent plots. (After Connell *et al.* 1997.)

and microhabitat. According to this idea, competition between species is strong at the present time and maintains niche differences, species segregation and high diversity. The second school of thought supports the **variable recruitment hypothesis**, which suggests that coral reefs are non-equilibrium systems in which larval recruitment is an unpredictable lottery. Competition between species is present, but the winner in competition was unpredictable, and the local community present on a reef is a random sample from a common larval pool. Mortality after larval settlement is not dependent on density, and so local populations fluctuate under the control of unpredictable recruitment—the effect of 'who-gets-there-first'. How can we distinguish the equilibrium and non-equilibrium hypotheses of community organization for coral reef fishes?

The first question we may ask is 'how specialized are reef fishes?' Reef fish exhibit both food and habitat specialization, but there are often several species within one restricted niche, and many generalist feed-

ers are present. Many studies have now been done of feeding habits of coral reef fishes. Herbivorous fishes are more generalized feeders than predatory fishes. But, even among the specialist feeders, it is common to find two or three species with identical specialization. Feeding niches are thus not organized as tightly as the niche-diversification hypothesis would predict, and thus competition for food would not appear to be a major process of importance in organizing these fish communities. Are there other niche axes that might permit species to coexist on coral reefs?

Habitat specialization could be another way that reef fishes have evolved niche differences. Adult reef fishes are very sedentary and could therefore have very narrow habitat requirements, but this does not seem to be the case. Habitat partitioning does occur to the extent that few species range over all regions of the reef. Species tend to occur in broadly defined habitats, such as 'reef flat' or 'surge channel,' but when microhabitats are assigned more carefully, there is extensive overlap of species. There is not a high degree of habitat specialization among coral reef fishes.

Natural history information tends to go against the niche-diversification hypothesis. Can we test these two hypotheses experimentally to provide a more powerful analysis? If the variable recruitment hypothesis is correct, reef fish communities ought to be unstable in species composition and highly variable from reef to reef. Also, the species structure at a given site ought not to recover to the same configuration following artificial removals or additions of species. We should be able to predict population size on a reef from the number of recruits that arrive.

To test the first prediction, coral reef ecologists put out standard cement building blocks to create artificial reefs of constant size and shape. Forty-two fish species colonized these artificial reefs, averaging 17 species per reef. Although these reefs were set out in the same lagoon at the same time within a few meters of one another, only about 32% of the fishes colonizing them were of the same species. A survey of natural coral isolates of about the same size as the artificial reefs (0.6 m^3) showed only a 37% similarity. Moreover, on the artificial reefs, there was very high turnover of fish species from month to month. Of the species on a reef on a particular month, 20 to 40% would have disappeared by the next month and have been replaced by a new species not previously present. Clearly, reef fish communities are very unstable and highly variable from one small reef to the next.

The variable recruitment hypothesis assumes that there are no resource limitations on populations and no competitive effects. Population size is set by how many recruits arrive at a reef. One prediction from this hypothesis is that if recruits are experimentally added to a coral reef population, adult numbers will rise. In one experiment, marine ecologists transplanted juveniles of the damselfish *Pomacentrus amboinensis* for three years to small natural patch reefs approximately 8 m^2 at the southern edge of the Great Barrier Reef in Australia. Figure 14.5 shows that, at first, adult densities increased as more recruits were added. However, at high recruitment levels, density-dependent interactions between adults and potential recruits creates a ceiling on numbers—a 'carrying capacity' that is con-

trary to the predictions of the variable recruitment hypothesis. The important point illustrated by Figure 14.5 is that there is a range of recruitment that fits the variable recruitment hypothesis, and a threshold above which this hypothesis does not hold because the population has reached carrying capacity.

Discussions among coral reef ecologists about the importance of pre-recruitment and post-recruitment processes in limiting adult fish densities has remained controversial because some results favor the variable recruitment hypothesis and others do not. Both recruitment and post-recruitment processes affect the abundance of coral reef fishes. Moreover, the results from these studies may depend on the spatial scale of the fish populations. Most studies have been done on small patch reefs—small enough to be censused by one or two divers. The variable recruitment model may apply to many small, isolated patch reefs, but it cannot necessarily be extrapolated to fish communities in large sections of continuous coral reefs.

Coral reef fishes are similar to many marine organisms in having a life history that includes a larval phase that is transported with the plankton by ocean currents. This type of life cycle implies that local reproduction is not linked to local recruitment—in complete contrast to the life cycle of birds and mammals. For these marine organisms, the population or the community can never be a closed system, and the physical factors controlling recruitment may control the system. This has been called 'supply-side ecology'—an idea common to economics and ecology with a long history. The structure of an ecological community driven by supply-side ecology cannot be understood solely as a result of competition, predation and disturbance, but only by finding out what controls variable recruitment, which keeps populations under the carrying capacity.

How is coexistence of so many species permitted if the variable recruitment hypothesis is correct? Peter Sale has argued that the reef fish community is a **lottery competition**. Individuals compete for access to units of resources (space for these fish) without which they cannot join the breeding population. A lottery competition is a type of interference competition in which an individual's chances of winning or losing are

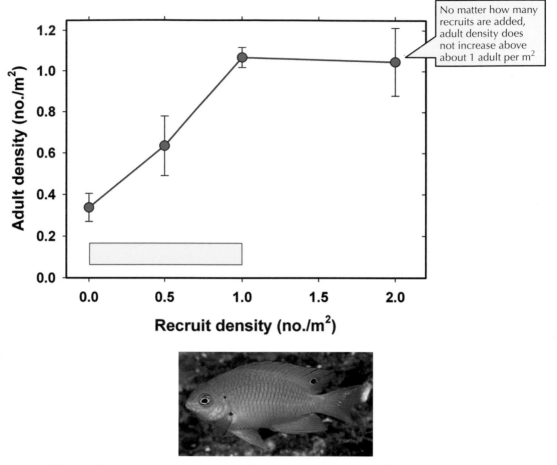

Figure 14.5 Test of the variable recruitment hypothesis for coral reef fish. Four different recruitment levels were experimentally provided for the damselfish *Pomacentrus ambionensis* for three years. Adult densities increased directly with recruitment up to 1 recruit/m^2 (yellow zone), as predicted by the variable recruitment hypothesis, but above 1 recruit/m^2 a ceiling was reached. (Data from Jones 1990, Photo courtesy of Ian Shaw, Australian Museum.)

determined by **who gets there first**. Lottery competitive systems are very unstable, but can persist if there is high environmental variability in birth rates. Because recruitment of reef fishes depends on larvae settling from the plankton, high variability is the rule, and vacated space is allocated at random to the first recruit to arrive from the larval pool. The lottery competition model can explain the coexistence of so many species of coral reef fishes.

The high diversity of coral reef fish communities is not achieved by precise niche diversification in an equilibrium community, but rather by highly variable larval recruitment, which causes a competitive lottery for vacant living spaces in which the first to arrive wins. Reef fish communities are not in equilibrium at

a local scale, but continually fluctuate in species composition, while retaining high and stable diversity over a whole region.

Rocky Intertidal Communities

The rocky intertidal zone is a stressful environment between land and sea. Disturbances in the form of waves and storms are an important feature of the physical environment. Space is the key limiting resource in the rocky intertidal because organisms must have space to be able to hold on to the rock to prevent being washed away in waves. Competition for space has been a key component of many studies of this community. Many key concepts in community ecology have their origin in the rocky intertidal—for

Jane Lubchenco (1947–) Professor of Marine Biology, Oregon State University

example, the keystone species concept and the intermediate disturbance hypothesis—and the same is true for theories of community organization. Let us look at two cases that illustrate the ways in which rocky intertidal communities can be organized.

Jane Lubchenco carried out classic work on communities of seaweeds and invertebrates on the rocky coasts of New England to measure the effects of predation and physical disturbance on seaweed abundance. Seaweeds in New England can be split into two groups:

- **ephemeral seaweeds** live for weeks or months, grow rapidly, and are eaten rapidly by herbivores like limpets.
- **perennial seaweeds** can live for many years, grow more slowly, and except in their juvenile stages are relatively inedible.

Seaweeds compete for space and for light, but the primary resource in the rocky intertidal is space on

which to attach their holdfasts so that they do not get washed away. Using wire mesh cages to exclude herbivores, Lubchenco found that there was no simple answer to the question of what controlled seaweed abundance (Figure 14.6). On protected areas in summer, limpet grazing reduced the abundance of ephemeral algae so much that there was no competition for space among the algae. On wave exposed sites, where limpets cannot easily live, seaweeds are washed away by wave action, so that there is plenty of open space on the rocks and the amount of competition is reduced. In this system, herbivores set the stage for competitive interactions, and the exact dynamics of a small patch of rock depends on the physical environment (wave action) and the season of the year.

Coralline algae are encrusting algae in the rocky intertidal that compete with each other by overgrowth. In the absence of herbivores, such as chitons and limpets, there is a clear dominance of competitive interactions—competition for space was linear and one species always won and took over the available space. But in natural communities three of the coralline algae won some encounters and lost others—grazers, such as chitons and limpets, act to slow down the rate of succession of coralline algae and induce competitive uncertainty, which acts to promote patchiness and species diversity. In the absence of disturbance from grazing, a single competitive dominant alga would monopolize space in the rocky intertidal and the equilibrium community would be very simple. The algal community of rocky shores is not an equilibrium assemblage under natural conditions because grazing changes the system from a competitive network that has a fixed stable outcome—with one species winning all the space—to an unpredictable network with no single winner—hence the coexistence of several different species undergoing continuous change.

■ 14.3 THEORETICAL MODELS SHOW WHAT ECOLOGICAL PROCESSES LEAD TO NON-EQUILIBRIUM COMMUNITIES

These two examples of community dynamics on coral reefs and the rocky intertidal zone illustrate some of

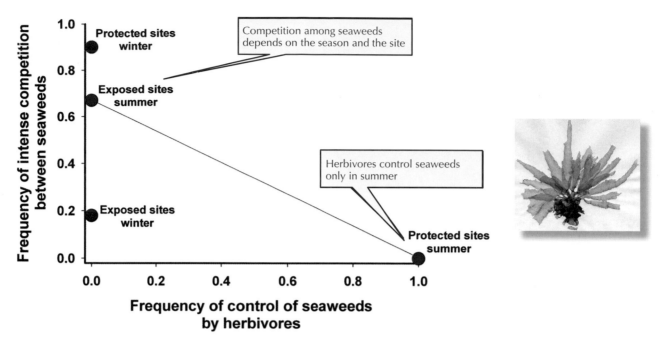

Figure 14.6 Effect of herbivores on the frequency of competition between ephemeral seaweeds (such as sea lettuce, *Ulva* sp.) in the rocky intertidal zone of New England. Season and wave exposure are indicated for each point. Red points = summer, blue points = winter. Herbivores control seaweeds only in summer on sites protected from wave action. Herbivores cannot stay attached to rocks on wave-exposed sites. The amount of competition among seaweeds for space depends both on the season and the physical environment. (Modified from Lubchenco 1986.)

the ideas that have been central to the development of non-equilibrium models of communities. Communities can be positioned along a gradient from stable, biotically interactive communities, which are equilibrium-centered, to unstable, interactive communities in which the biotic interactions do not lead to a stable equilibrium, to weakly interactive communities in which physical factors, such as temperature, salinity or fire, prevent any stable equilibrium (Figure 14.7).

Models have been developed all along this gradient to capture the ecological complexities of these systems. The key question is: what ecological processes can lead to non-equilibrium communities? There are three main types of non-equilibrium models of communities:

Fluctuating Environment Models

The simplest deviation from the classical equilibrium model of a community is a model with temporal variability. Competition is viewed in these models as the major biological interaction, but the environment changes seasonally or irregularly and the competitive rankings of species also fluctuate so that no single spe-

cies can win out. These models may include movements between patches and each patch may have a different environment.

Directional Changing Environment Models

Variable environment models usually consider environments to fluctuate about a mean value that remains constant with time. What happens to these models when the mean itself changes? Much depends on the amount of fluctuation and the speed the community reacts to change. The current concern with global climate change (see Chapter 21) makes these models very significant for the future. Unlike many community models, these models cannot ignore history—the response of a community to change, such as global warming, depends on its past history. Modern communities, these theories argue, cannot be understood only by looking at present environmental conditions.

Slow Competitive Displacement Models

If competitive abilities of species are nearly equal, the process of competitive exclusion will be very slow and

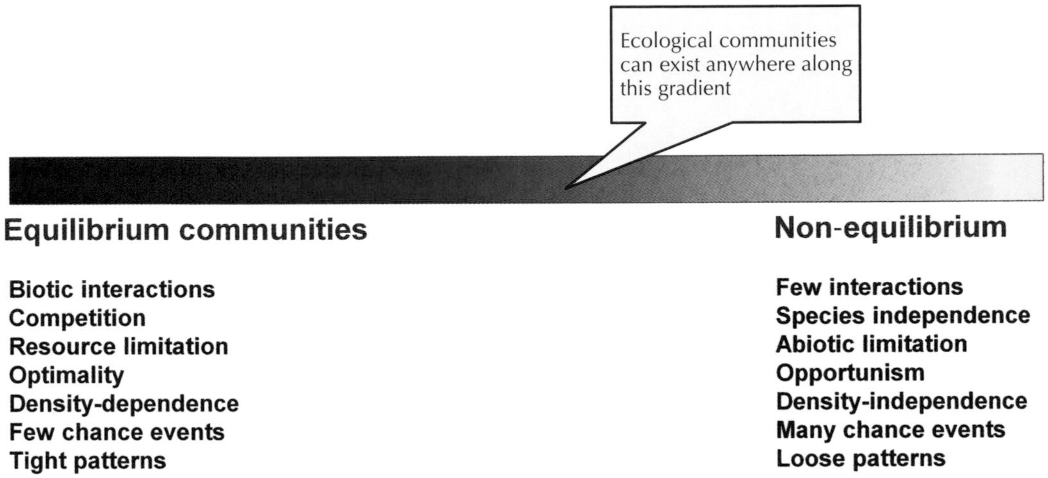

Figure 14.7 Natural communities may be arrayed along a spectrum of states from equilibrium to non-equilibrium. At either extreme, several attributes of community organization and dynamics can be anticipated. For example, density-dependent population regulation predominates in equilibrium communities, while density-independent processes predominate in non-equilibrium communities. (Modified after Wiens 1984.)

random variation in success will obscure any obvious displacements of one species by another. The most controversial application of these models has been to understanding the species richness of tropical rainforests. Steve Hubbell and his colleagues have argued that the tropical rainforest has many species that are ecologically identical. Community composition is the net consequence of a slow change in tree species densities. Competition occurs in these models all the time, but, because all species are identical in competitive abilities, there is no time trend or succession. Community structure, under these models, is strongly affected by chance and by history, and changes occur only on a geological time scale.

The purpose behind all of these theoretical models is to understand what enables a community to persist over time. If there is no stable equilibrium in community composition, species in a community should by chance disappear and drift to extinction. This does not usually happen on a regional scale because of spatial variation among local patches. Species may become extinct in local patches, but, as long as local patches are out of phase with one another, the species will persist somewhere in the landscape. In a paradoxical manner, instability at the local scale of the patch leads to stability at the regional and landscape level.

To translate these ideas on non-equilibrium community dynamics into the real world, ecologists have developed a series of conceptual models of community organization that can be tested with field studies.

■ 14.4 PHYSICAL DISTURBANCE, PREDATION AND COMPETITION ARE THE THREE ECOLOGICAL DETERMINANTS OF COMMUNITY ORGANIZATION

A series of models has been proposed by field ecologists to try to capture the interrelations between physical factors and biological interactions in organizing communities. All of these models recognize that there are many different kinds of ecological communities, that disturbances are an important part of community dynamics, and that the important processes will not be the same in all ecological systems.

Menge–Sutherland Model

The most comprehensive model of community organization has been proposed by Bruce Menge and John Sutherland. They recognized three ecological processes as the main determinants of community organization—physical disturbance, predation and competition

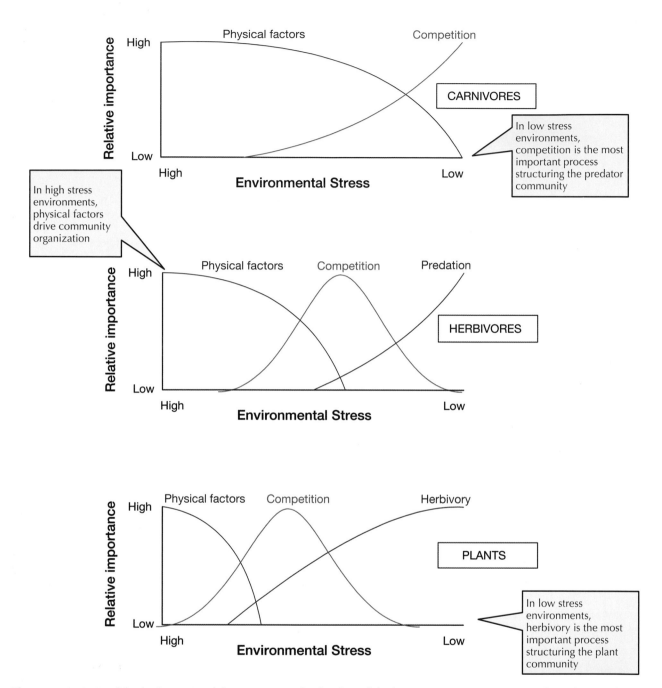

Figure 14.8 A simplified schematic of the Menge–Sutherland model of community organization. Three factors drive community organization, and the relative importance of physical factors, competition between species and predation change with trophic level. Environmental stress can be measured by disturbance levels, whether due to temperature fluctuations, rainfall, fires, wave action or salinity fluctuations. (Modified after Menge and Sutherland 1987.)

(Figure 14.8). A central assumption of this model is that food web complexity will decrease with increasing environmental stress. This model makes three predictions for communities. Firstly, in stressful environments herbivores have little effect because they are rare or absent, and plants are regulated directly by environmental stress. Neither predation nor competition is significant. An example of such a community would be the arctic tundra or a desert. Secondly, in moderately stressful environments, consumers are ineffective at controlling plants, so plants attain high densities. Competition between plants is the dominant biological interaction in these communities. Thirdly, in benign environments, consumers control plant numbers and plant competition is rare. Predation is the dominant biological interaction affecting herbivores, and herbivory is the key interaction affecting plants.

Bottom-up and Top-down Models

Freshwater ecologists have proposed several simplified models for community organization in freshwater and terrestrial ecosystems that parallel the Menge–Sutherland model. The key to these conceptual models is to consider the interactions between two adjacent trophic levels. For example, consider plants (V) and herbivores (H). There are three possible relationships:

$$V \rightarrow H \qquad V \leftarrow H \qquad V \longleftrightarrow H$$

The arrows indicate that a change in one trophic level causes a change in the other. Thus $V \rightarrow H$ means that an increase in vegetation will cause an increase in the numbers or biomass of herbivores, but not vice versa. Similarly $V \leftarrow H$ means that an increase in herbivore numbers will cause an impact on vegetation (a decrease), but not vice versa. A double arrow, $V \longleftrightarrow H$, means a **reciprocal interaction**—an increase in vegetation will increase herbivore numbers, and an increase in herbivore numbers will act to reduce the vegetation.

Given these three simple interactions between trophic levels, we can define two polar views of community organization: the **bottom-up model** and the **top-down model**. The bottom-up model postulates $V \rightarrow H$ linkages, which means that nutrients control community organization because nutrients control plant numbers, which in turn control herbivore numbers, which in turn control predator numbers. The simplified bottom-up model is thus:

$$N \rightarrow V \rightarrow H \rightarrow P$$

where N = nutrients, V = vegetation H = herbivores, and P = predators.

By contrast, the top-down model postulates that predation controls community organization, because predators control herbivores, which in turn control

plants, which in turn control nutrient levels. The simplified top-down model is thus:

$$N \leftarrow V \leftarrow H \leftarrow P$$

The top-down model has been called the **trophic cascade model** by Steve Carpenter. It predicts that there will be strong interactions between species—a series of +/- effects across all the trophic levels. Thus predators will strongly depress herbivore numbers, and depressed herbivore numbers will have only a minor impact on plant abundance, so the abundant plants will strongly depress nutrients. The trophic cascade model predicts that, for freshwater systems with four trophic levels, removing the top (secondary) carnivores[1] will increase the abundance of primary carnivores, decrease herbivores and increase phytoplankton. The effects of any manipulation will thus move down or up the trophic structure as a series of +/- effects (*Essay 14.2*).

The top-down and the bottom-up models are clearly not the only models that can be postulated for food chains. Tony Sinclair and his colleagues have derived 27 different models from various combinations of →, ← and ↔ arrows. For example, a **pure reciprocal model** would postulate two-way impacts at all trophic levels: $N \longleftrightarrow V \longleftrightarrow H \longleftrightarrow P$. The important point is to start with simple models—it is unlikely that all communities will fit only one model.

The usefulness of these simple models is that they immediately suggest experimental manipulations of communities to search for trophic level impacts and to determine which model best describes a particular ecosystem. An extensive winter kill of fish in Lake St. George, Ontario, allowed Don McQueen and his colleagues to test these models in a freshwater ecosystem. Winter kill occurs in lakes that are ice covered and the exchange of gases between the water below the ice and the air above is not sufficient to maintain sufficient oxygen levels to support fish. During the winter, oxygen normally enters the water of a frozen lake through inlet water streams, cracks in the ice and slow diffusion through the ice. A thick snow cover on a lake

---

1 In aquatic systems secondary carnivores are piscivorous (= fish-eating) fish, and primary carnivores are planktivores (= zooplankton-eating fish).

ESSAY 14.2   BIOMANIPULATION OF LAKES

Many freshwater lakes have been degraded by nutrient pollution, and one of the major thrusts of applied aquatic ecology has been to work out methods for lake recovery from such pollution.

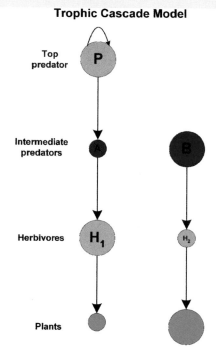

Trophic Cascade Model

The size of the circle reflects the abundance of each group of species.

Trophic cascade model

If the trophic cascade model of lake communities is correct, it immediately suggests ways of improving water quality. In lakes with four trophic levels (see diagram at top), adding top predators (P) should improve water quality by reducing algal populations. In lakes with three trophic levels, removing fish (intermediate predators, B) should improve water quality. This tool for lake restoration has been called **biomanipulation** and we can illustrate it simply:

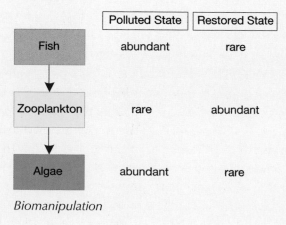

	Polluted State	Restored State
Fish	abundant	rare
Zooplankton	rare	abundant
Algae	abundant	rare

Biomanipulation

Many attempts have been made at lake restoration using biomanipulation, but the results have been mixed, possibly dependent on the depth and size of the lake. One of the largest food web manipulation trials yet carried out was carried out at Lake Vesijärvi in southern Finland. Lake Vesijärvi is a large lake (110 km²) with a mean depth of only 6 m. It was heavily polluted with city sewage and industrial wastewater until 1976 and, once these inputs were stopped, it began to recover in water quality. But by 1986, massive blooms of blue-green algae began to appear, and these algal blooms coincided with a very dense population of roach (*Rutilus rutilus*)—a planktivorous cyprinid fish that had built up during the years of nutrient input. To reverse these changes, from 1989 to 1993, fish ecologists removed 1018 tons of fish from Lake Vesijärvi, reducing roach to about 20% of their former abundance. At the same time, they began to stock the lake with pikeperch (*Sander lucioperca*)—a predatory fish that feeds on roach—thus adding a fourth trophic level to the lake.

Biomanipulation was a success in Lake Vesijärvi—the water became clear and blue-green algal blooms stopped in 1989. The lake continued to remain clear 7 years later, even though the fish removal stopped in 1993. But the mechanism was not as suggested in the diagram above, because zooplankton density in the lake did not change and the same zooplankton species were present. The reduction of algal blooms was achieved because nutrient excretion by roach was greatly reduced, and it was this nutrient regeneration from fish that was stimulating the excessive algal growth in the lake. This additional pathway for nutrients directly from fish excretion to the phytoplankton could be an important additional mechanism to consider when attempting lake restoration. Lake Vesijärvi may be an example of a lake with two alternative stable states defined by nutrient transfer from fish to algae.

Ironically, the success of the restoration of Lake Vesijärvi resulted in a significant building boom around the lake, and the added human input of nutrients into the lake since 2000 has been eroding the water quality gains achieved by the earlier fish removal.

can reduce the amount of oxygen passing through the ice. Shallow lakes are particularly susceptible to winter kill. Figure 14.9 shows the changes in community structure that occurred in the 5 years after the winter kill. The top predators (bass, pike and yellow perch) recovered in 5 years to their former levels of abundance. Planktivorous fishes like bluegill (primary carnivores) increased after the winter kill, and herbivorous zooplankton declined. Phytoplankton changes (measured by chlorophyll) occurred, but these were not correlated with zooplankton numbers. Phosphorus levels seemed to determine phytoplankton numbers. The results shown in Figure 14.9 suggest that trophic cascades damp out as they move down the food chain and, at the level of phytoplankton and zoo-

plankton, both nutrient limitation and predation could be controlling.

The results of species removals are complex because the interactions between species are complex and may be habitat-specific, as we have just seen. Food webs can also be impacted directly by disturbances. River food webs in northern California exist in two states. In regulated rivers with dams, large caddis fly larvae become abundant because they have gravel cases which make them invulnerable to fish predators, and they graze algae to low levels. In unregulated rivers, floods roll rocks that kill many caddis flies and, by reducing their numbers, allow algae to increase (Figure 14.10). The food webs of rivers with dams thus change dramatically from their original composition.

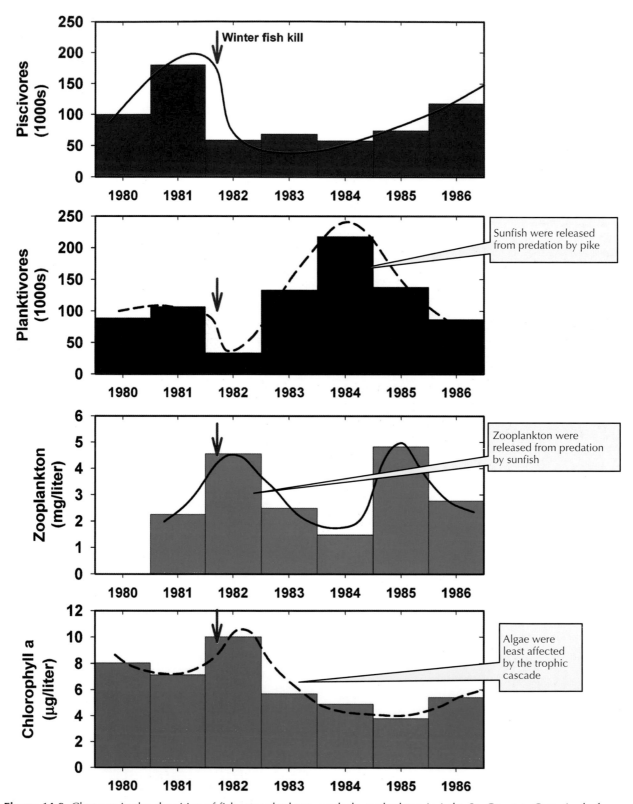

Figure 14.9 Changes in the densities of fish, zooplankton, and phytoplankton in Lake St. George, Ontario, before and after a winter kill of fish in 1982 (arrow). A trophic cascade propagated down the food chain over the next 4 years. Algal abundance is measured by the amount of chlorophyll in the water. (Modified from McQueen *et al.* 1989.)

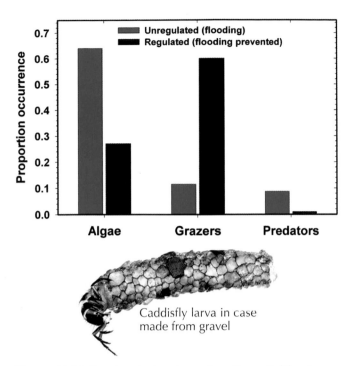

Caddisfly larva in case
made from gravel

Figure 14.10 Food web structure of northern California rivers illustrating a trophic cascade driven by flooding disturbances. In regulated (dammed) rivers with no flooding, algae are relatively scarce, grazers (caddis flies) abundant, and predators scarce during the summer growing season. Disturbance by flooding (scours) reduces the caddis flies, and releases the algae from grazing. Flooding mimics a grazer-removal experiment. (Data from Wootton *et al.* 1996.)

Disturbances can act like species removal experiments in changing the dynamics of food webs. With frequent disturbances, the community structure will differ dramatically from that expected under equilibrium conditions with no disturbances.

■ 14.5 ISLANDS HAVE HIGHLIGHTED THE ROLE OF AREA AND ISOLATION IN STRUCTURING COMMUNITIES

Islands can be thought of as special kinds of traps that catch species able to colonize and disperse there successfully. Since Darwin's visit to the Galápagos Islands, biologists have been using islands as microcosms to study evolutionary and ecological problems. Because they are bounded, islands are useful for analyzing community structure, and for determining the role of disturbances in affecting communities.

Species-area Curve

One of the oldest generalizations of ecology is that the number of species on an island is related to the area of the island. In 1807 Alexander von Humboldt wrote that larger areas harbor more species than smaller ones. This can be seen most easily in a group of islands such as the Galápagos (Figure 14.11). The relationship between species and area can be described by the simple equation called the **species-area curve**, which is illustrated in Figure 14.11.

The species-area curve relates the size of the island to the number of species, and is a straight line if plotted on logarithmic axes. A key feature of the species-area curve is the slope of the line, or how fast species accumulate as you move to larger and larger islands. For the Galápagos land plants shown in Figure 14.11, the slope (z) of the species-area curve is 0.32.

Michael Rosenzweig, an evolutionary ecologist at the University of Arizona, has championed the species-area relationship as a fundamental ecological law. The species-area curve is a useful descriptive model for both plants and animals.

What determines the number of species on a plot of ground? The number of species living in any habitat, whether an island or an area on the mainland, is a balance between immigration and extinction. If the immigration of new species exceeds the extinction of old species already present, the plot or island will gain species over ecological time. We can treat the problem

Michael L. Rosenzweig (1941–) Professor of Ecology and Evolutionary Biology, University of Arizona

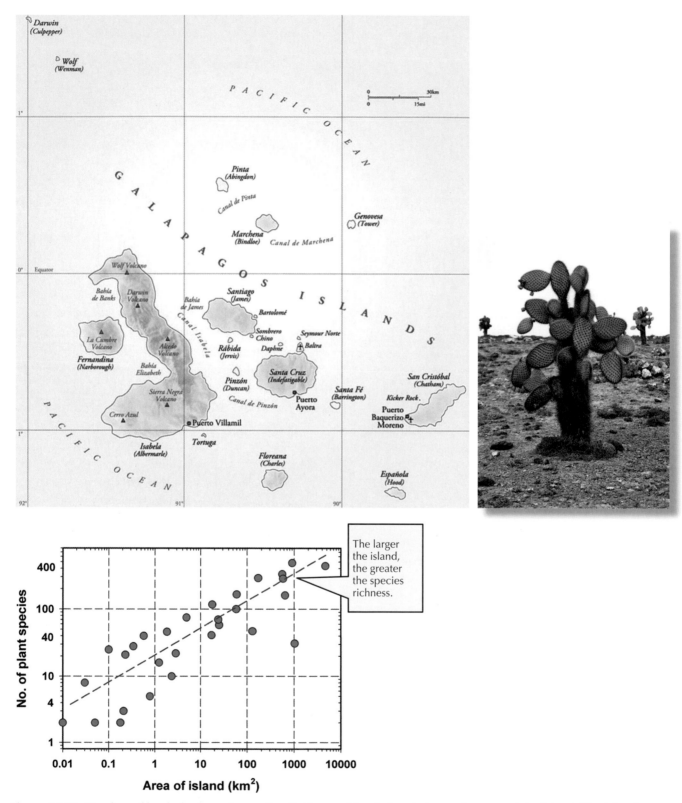

Figure 14.11 Number of land-plant species on the Galápagos Islands in relation to the area of the island. The islands range in area from 0.05–4669 km^2 and contain from 2 to 481 plant species. Photo shows a prickly pear cactus. (Data from Yeakley and Weishampel 2000, photo courtesy of Daniel Russel, Stanford University.)

of species diversity on islands by an extension of the approach used in population dynamics, in which changes in population size were produced by the balance between immigration and births on the one hand, and emigration and deaths on the other hand.

MacArthur–Wilson Theory of Island Biogeography

This insight, that the number of species on any area could be treated with the same kinds of approaches used in population dynamics, was a fundamental breakthrough in understanding island communities. Robert MacArthur and Edward Wilson developed this approach in detail in 1967, and Figure 14.12 shows their simplest model for the immigration rate and the extinction rate on islands.

The immigration rate is expressed as the number of new species gained per unit of time. This rate falls continuously because as more species become established on the island, most of the immigrants will be from species already present. The upper limit of the immigration curve is the total fauna for the region. The extinction rate (the number of species lost per unit time) rises because the chances of extinction depend on the number of species already present. The point where the immigration curve crosses the extinction curve is by definition the equilibrium point for the number of species on the island (Figure 14.12).

The shapes of the curves of immigration and extinction are critical for making predictions about island situations. Assume for the moment that distance affects only the immigration curve: nearby islands will receive more dispersing organisms per unit time through diffusion than will distant islands. Assume also that small islands will differ from large islands in their extinction rate so that the chances of becoming extinct are greater on small islands. Figure 14.13 illustrates these assumptions and shows why distant islands should have fewer species than nearby islands (if island size is constant) and why small islands should have fewer species than large islands (if distance from the source area is constant).

The MacArthur–Wilson theory was developed as an equilibrium theory, and many of the early studies of species on islands assumed that islands were equilibrium communities. More recent work has found that many islands are not in equilibrium because of disturbances caused by physical factors, such as hurricanes, and by humans. Other islands are not in equilibrium because they are now too isolated to have any immigration and yet can lose species through extinction.

The MacArthur–Wilson theory stimulated much work on island faunas. By concentrating on predictions of the *number* of species, it has ignored the more difficult questions of *which* particular species will occur where— often the question conservationists will ask. Habitat heterogeneity is a major cause of the species-area curve and detailed studies of habitats are needed to further our understanding of island communities. Individual species differ greatly in their ability to occupy islands, and, by understanding the population and community dynamics of individual species, we can improve on our understanding of island faunas and floras.

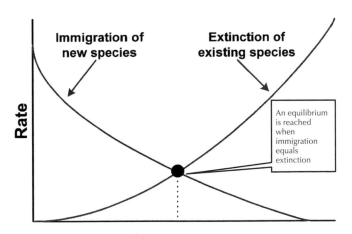

Figure 14.12 Equilibrium model of a biota of a single island. The equilibrium species number (dotted line) is reached at the intersection point between the curve of rate of immigration of new species not already on the island and the curve of extinction of species from the island. (After MacArthur and Wilson 1967.)

■ 14.6 COMMUNITIES CAN EXIST IN SEVERAL ALTERNATIVE STABLE STATES

If many communities are not equilibrium assemblages of species, their composition will change over time and we will not observe a single stable configuration. But if

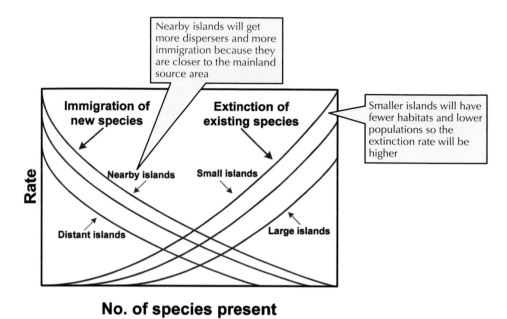

Figure 14.13. Equilibrium model of biotas of several islands of varying distances from the principal source area and of varying size. An increase in distance (nearby to distant) lowers the immigration curve; an increase in island area (small to large) lowers the extinction curve. There is an equilibrium of species richness at each crossing point of the immigration and extinction curves. (After MacArthur and Wilson 1967.)

natural communities can exist in **multiple stable states**, changes in community composition that appear to be in non-equilibrium may instead be the result of two or more alternative states for the same community. Communities with alternative stable states are in equilibrium in each state, and will not go back to the original state once the disturbance is stopped. What evidence is required to test this model? Four criteria should be satisfied to show that a community exhibits multiple stable states:

1. The community must show an equilibrium point at which it remains, or to which it returns if perturbed by a disturbing force.
2. If perturbed sufficiently, the community will move to a second equilibrium point, *at which it will remain after the disturbance has disappeared.*
3. When multiple stable states are believed to exist, the abiotic environment must be similar in the two communities.
4. The communities on both sites that are postulated to be alternative stable states must persist for more than one generation of the dominant species.

Multiple stable states have been described in some communities without satisfying all these criteria. In some cases, the alternative state persists only when artificial inputs are maintained. For example, Lake Washington (page 000) certainly can exist in two alternative states, but the enriched lake community could be maintained only by adding sewage nutrients continually to the lake. But more and more evidence of multiple stable states in communities is being found. Let us look at two examples that illustrate this point.

The woodlands and grasslands of east Africa may represent multiple stable states of a grazing system with two possible communities—woodland or grassland. Woodlands in parks and reserves over much of the savannah areas of east Africa have declined in the past 30 years, and been replaced with grasslands. Three hypotheses have been proposed to explain the woodland decline (Figure 14.14).

1. Human-induced fires have eliminated woodland. There is one stable state, and, if fires could be reduced, woodlands would return.

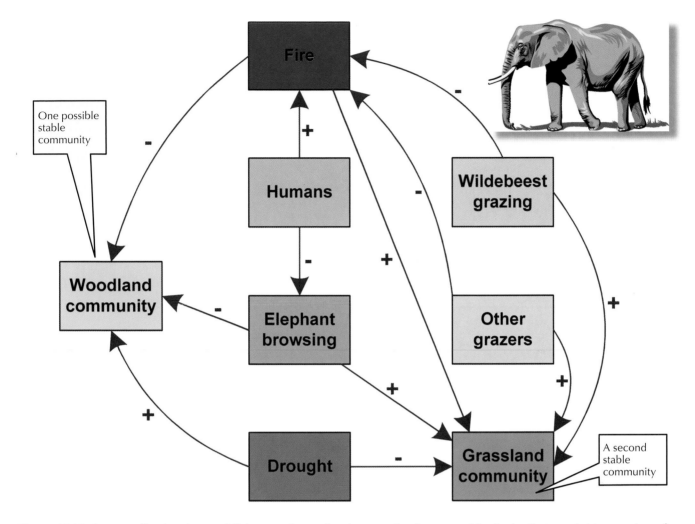

Figure 14.14 Factors affecting the establishment of woodland or grassland communities in the Serengeti–Mara region of East Africa. This community exists in two alternative stable community states, woodland or grassland. (Based on Dublin *et al*. 1990.)

2. Elephants eliminate woodland, and the resulting grassland is maintained by fire. There are two stable states. If fires were eliminated, woodlands would return to their former abundance, even if elephants remained.
3. Fire eliminated woodlands, and elephants hold tree regeneration in check by eating small trees so the woodlands can never return. Eliminating fire will not cause woodlands to return, and there are two stable states.

The evidence available supports the third hypothesis. During the 1960s, fire burned on average 62% of the Serengeti each year and, even with no elephants or other browsing or grazing mammals, tree recruitment would be too low to sustain woodlands. Elephant and wildebeest numbers increased in the parks and reserves by the 1980s. Wildebeest grazed much of the grass each dry season so the fuel load in the 1980s was reduced and fires burned only 5% of the area each year. Elephant browsing in the 1980s was severe and by itself capable of preventing woodlands from re-establishing. If elephants and wildebeest are reduced by poaching in the future, fires will increase and woodland will not return. The Serengeti–Mara ecosystem seems to be locked into a grassland state and the woodlands will not return.

A second example of multiple stable states comes from eastern United States where white-tailed deer

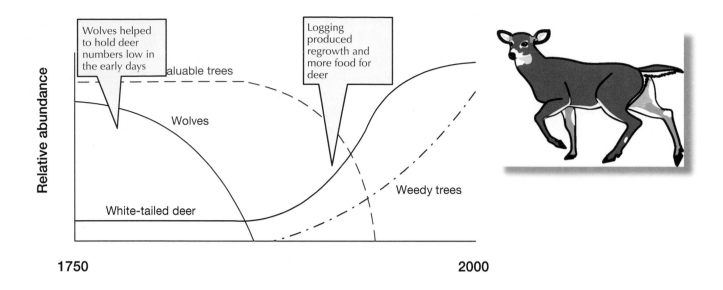

Figure 14.15 Changes in the relative abundance of white-tailed deer in the New England states after logging and wolf control during the 18th and 19th centuries. Deer are now overabundant, and suppress the seedlings of valuable tree species, so that the forest community becomes dominated by less desirable trees that the deer do not eat. (After Tilghman 1989.)

have increased greatly in recent years. By their browsing, deer may be creating alternative stable states of woody plant communities. Between 1890 and 1920 much of the hardwood forests in Pennsylvania were clear felled. These stands contained valuable trees, such as white ash, sugar maple and red maple. Deer populations increased rapidly in the regenerating stands that produced much browse (Figure 14.15). At the same time, predators such as wolves were removed from the system so deer numbers increased dramatically and deer are now considered 'overabundant'. Deer browsing has been shown to reduce hardwood regeneration, particularly of the valuable timber trees. With sufficient browsing, the seed bank of these hardwoods becomes exhausted within 3–4 years, and no regeneration is then possible. Ferns and grasses invade the forest floor and suppress regeneration of desirable hardwoods completely. The result is a community of trees dominated by black cherry and by other tree species less preferred by foresters and less browsed by deer—an alternative tree community that may be stable in 300 years or more, even if the deer are somehow removed.

Multiple stable states may occur in other communities affected by humans and may be confused with non-equilibrium communities. In some cases, the community may revert to its original configuration once human disturbance is removed, but in others the community may become locked into a changed configuration even after the disturbance is stopped. In many cases this will be because no viable seeds are available for plants, or no dispersing individuals for animals. For conservation purposes and for land management, it is important to determine which model of community organization applies to natural communities. We cannot assume that all communities subjected to human disturbance will return to their original configuration once the disturbance is ameliorated.

SUMMARY

Many community ecologists question the existence of an equilibrium state for biological communities. Patchiness is an inherent property of natural communities, and the disturbances that lead to patchiness have been the main focus of non-equilibrium theories of community organization. Non-equilibrium communities exist when disturbance intervals are shorter than recovery times, so the community never reaches equilibrium in ecological time. Disturbances may include fires or weather events, as well as biotic events such as grazing, predation or disease.

Coral reefs have been thought to be classic examples of stable, equilibrium communities, but careful long-term studies have shown that reefs vary dramatically because of disturbances caused by cyclones and oceanographic changes due to weather fluctuations. Coral reef fish communities may be driven by the lottery of variable recruitment, which can cause irregular population fluctuations and maintains high biodiversity that is not in equilibrium on a local scale.

The two extreme conceptual models of community organization are the top-down model in which changes in food webs are driven from above by predators, and the bottom-up model in which nutrients and plants control the food web. The trophic cascade model is a particular top-down model that emphasizes the alternation of +/- effects in food webs. When top predators are removed, the effects cascade as alternating + and – impacts down the trophic ladder. Trophic cascades are common in aquatic systems, but not all systems follow trophic cascades. Some communities are driven bottom-up by nutrients.

Islands are a special case of communities driven by the interaction between species colonization and extinction. The species-area curve describes how biodiversity increases with island size and is one of the grand generalizations of community ecology. Island communities will not be in equilibrium if colonization and extinction rates are not equal.

Some communities may exist in multiple stable states and these may be confused with non-equilibrium assemblages. If a community is disturbed sufficiently, it may change to a new configuration at which it will remain even when the disturbance is stopped. There is considerable controversy about how common multiple stable states are in natural ecosystems, and the answer to this question is important for conservation and land management.

SUGGESTED FURTHER READINGS

Beisner BE, Haydon DT and Cuddington K (2003). Alternative stable states in ecology. *Frontiers in Ecology and the Environment* **1**, 376–382. (An overview of the existence of alternate stable states in ecosystems.)

Bond WJ, Woodward FI and Midgley GF (2005). The global distribution of ecosystems in a world without fire. *New Phytologist* **165**, 525–538. (What would our world look like if we suppressed all fires?)

Dublin HT, Sinclair ARE and McGlade J (1990). Elephants and fire as causes of multiple stable states in the Serengeti–Mara woodlands. *Journal of Animal Ecology* **59**, 1147–1164. (One of the best examples of multiple stable states in a classic African ecosystem.)

Knowlton N (2001). The future of coral reefs. *Proceedings of the National Academy of Sciences, USA* **98**, 5419–5425. (An examination of the human disturbance issues affecting coral reefs.)

Moritz MA (1997). Analysing extreme disturbance events: fire in Los Padres National Forest. *Ecological Applications* **7**, 1252–1262. (How extreme and infrequent fire events are affected by climate and forest management.)

Paine RT, Tegner MJ and Johnson EA (1998). Compounded perturbations yield ecological surprises. *Ecosystems* **1**, 535–545. (Why understanding how a single type of disturbance affects communities is not sufficient when several disturbances occur at the same time.)

Payette S, Fortin M-J, and Gamache I (2001). The subarctic forest-tundra: the structure of a biome in a changing climate. *BioScience* **51**, 709–718. (How a climate and fire-sensitive ecosystem has responded to disturbance.)

Shears NT and Babcock RC (2002). Marine reserves demonstrate top-down control of community structure on temperate reefs. *Oecologia* **132**, 131–142. (An experimental study of predator control of abundances in a New Zealand marine ecosystem.)

QUESTIONS

1. Classify the following examples of plant–herbivore or predator–prey interactions with regard to the
   $V \rightarrow H$ $V \leftarrow H$ $V \leftrightarrow H$ models discussed on page 325.

   (a) Red squirrels feeding on the seeds in white spruce tree cones.
   (b) Eastern grey kangaroos feeding on pasture grass in semi-arid grasslands.
   (c) Blue whales feeding on krill in the Southern Ocean
   (d) Coyotes feeding on domestic sheep in Montana ranching areas
   (e) Wolves feeding on elk in Yellowstone National Park

2. The species-area curve (Figure 14.11) rises continually as area is increased, and this implies that there is no limit to the number of species in any community. Is this a correct interpretation? What hypotheses can you suggest to explain why the number of species rises as area increases?

3. Analyze the elephant/fire multiple stable state model of Dublin *et al.* (1990) (Figure 14.14) using the four criteria for the existence of multiple stable states given on page 332. Does this example satisfy all four criteria? Would the white-tailed deer example satisfy the four criteria?

4. What role does history play in community organization defined by the equilibrium model and the non-equilibrium model? Do you need to know anything about the history of a community to predict its future course? Tanner *et al.* (1996) discuss this issue for coral reefs.

5. The immigration and extinction curves in the MacArthur–Wilson theory of island biogeography are concave upwards (Figure 14.13). What difference would it make if these curves were straight lines?

6. The equilibrium–non-equilibrium distinction in community dynamics might be a reflection of the geographical scale of study with small study sites seeming to be non-equilibrium while larger areas would be considered in equilibrium. Discuss the influence of spatial scale on the concept of equilibrium.

Chapter 15

ECOSYSTEM ECOLOGY—ENERGY FLOWS AND PRODUCTION

IN THE NEWS

Production of crops is one of the most practical applications of agricultural ecosystem ecology, and the history of crop improvements gives a graphic illustration of the factors that can limit plant production. The essence of farming is to simplify the ecosystem so that the production of a crop can be maximized. Wheat is a good example because it is one of the staple crops of many countries in the world. Australia is one of the major wheat producers in the world, and the history of wheat production in Australia mirrors what has happened in North America and Europe over the last 150 years.

The average yield of wheat in Australia went through an initial decline from about 1850 to 1900 because of soil nutrient exhaustion. Input must equal output for nutrients in the soil to remain constant, and crop production continually removes nutrients in the wheat that is harvested for human use. By about 1900 wheat yield was only one-half what it had been 50 years earlier, and much early research was directed to improving wheat yields. The most important nutrient for many Australian soils was phosphorous, and around 1900 farmers began fertilizing their fields with superphosphate. At the same time new genetic varieties of wheat were developed that were more suited to the temperature

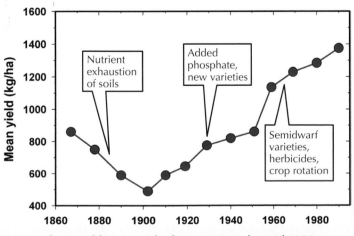

Mean wheat yield in Australia from 1860 to the mid 1990s

339

Norman Borlaug (1914–) Father of the Green Revolution

and rainfall conditions of Australia. These improvements increased wheat yields about 70% over the next 50 years.

But adding fertilizer to conventional wheat strains in the 1930s and 1940s caused the wheat plants to grow taller, making them more susceptible to wind-throw, which prevents mechanical harvesting. By the 1950s additional varieties of wheat were developed to overcome this problem by the genetic breeding of semi-dwarf wheat varieties—part of the 'Green Revolution' in agriculture. Norman Borlaug was awarded the Nobel Peace Prize in 1970 for his pioneering work in developing high-yielding, semi-dwarf wheat (http://www.nobel.se/peace/laureates/1970/borlaug-bio.html). Semi-dwarf wheat plants do not put as much energy into growing tall, and use this extra energy to produce more seed.

The introduction of improved genetic varieties of wheat accelerated yields dramatically after 1950, and genetic improvements have continued to the present time—one of the great success stories of modern agriculture. In addition to these genetic improvements in the wheat plant, farmers in the 1960s and 1970s began to use better-designed crop rotations, including legumes that fix nitrogen in the soil, and herbicides to reduce weeds. Recent selection for wheat varieties that can compete better with weeds has also improved yields. All of these factors have played an important role in the rising agricultural production of wheat throughout the world.

■ 15.1 SOLAR ENERGY FIXED IN PHOTOSYNTHESIS SUSTAINS ALL TROPHIC LEVELS

There are two broad approaches we can take to the study of communities and ecosystems of plants and animals. We can treat the species as biological entities with all the specific adaptations and interrelationships they show. This has been the approach of the last four chapters and can be considered a species-by-species-ecological approach to community and ecosystem dynamics. The second broad approach moves beyond the details of particular species and concentrates on the physics of ecosystems as energy machines and nutrient processors. Plants and animals process energy and materials, and exactly how they do this has important implications for humans. This second approach to ecosystem dynamics is the subject of the next two chapters.

Ecosystem Metabolism

The metabolism of ecosystems can be most easily understood by appreciating that it is the sum of the metabolism of individual animals and plants. Individual organisms require a continual input of new energy to balance losses from metabolism, growth and reproduction. Individuals can be viewed as complex machines that process energy and materials. There are two major ways in which organisms pick up energy and materials. **Autotrophs** pick up energy from the sun and materials from non-living sources. Green plants are autotrophs. **Heterotrophs** pick up energy and materials by eating living matter. Herbivores are heterotrophs that live by eating plants, and carnivores are heterotrophs that live by eating other animals. Communities are mixtures of autotrophs and heterotrophs. Energy and materials enter a biological community, are used by the individuals, and are transformed into biological structure only to be ultimately released again into the environment. The **ecosystem** level of integration includes both the organisms and their abiotic environment and is a comprehensive level at which to consider the movement of energy and materials.

Materials and Energy as Currency

The first step in the study of ecosystem metabolism is to determine the food web of the community (Chapter 13).

Once we know the food web, we must decide how we can judge the significance of the different species to community metabolism. Three measurements might be used to define relative importance of a species in an ecosystem:

1. **Biomass.** We could use the weight or standing crop of each species as a measure of importance. This is useful in some circumstances, such as the timber industry, but it cannot be used as a general measure. In a dynamic situation in which **yield** is important, we need to know how rapidly a community produces new biomass. When metabolic rates and reproductive rates are high, production may be very rapid, even from a low standing crop. Figure 15.1 illustrates the idea that yield need not be related to biomass.

2. **Flow of chemical materials.** We can view the ecosystem as a kind of super-organism, taking in food materials, using them and passing them out. Note that all chemical materials can be recycled many times through the community. A molecule of phosphorus may be taken up by a plant root, used in a leaf, eaten by a grasshopper that dies and released by bacterial decomposition to re-enter the soil and then go through this same loop again.

3. **Flow of energy.** We can view the ecosystem as an energy transformer that takes solar energy, fixes some of it in photosynthesis and transfers this energy from green plants through herbivores to carnivores. Note that most energy flows through an ecosystem only once, and is not recycled, but is transformed to heat and ultimately lost to the system. Only the continual input of new solar energy keeps the ecosystem operating.

To study the dynamics of ecosystem metabolism, we must decide what to use as the base variable. Most ecologists have decided to use either **carbon** or **energy.** Because most energy is not recirculated, it is one of the easier variables to measure in an ecosystem. Figure 15.2 illustrates the flows of energy and materials through the food chain.

Q **How might the analysis of ecosystem metabolism be affected in general if you decided to use calcium instead of carbon as the currency of interest?**

■ 15.2 GREEN PLANTS PROCESS THE SUN'S ENERGY UNDER LIMITATIONS IMPOSED BY TEMPERATURE, MOISTURE AND NUTRIENTS

The process of photosynthesis is the cornerstone of all life and the starting point for studies of community metabolism. Green plants are responsible for **primary production**—the energy and materials produced by plants as a result of photosynthesis. The bulk of the Earth's living biomass is green plants (99.9% by weight); only a small fraction of life consists of animals. Photosynthesis is the process of transforming solar energy into chemical energy and can be simplified as

$$12H_2O + \underset{\text{(from air)}}{6CO_2} + \text{solar energy} \xrightarrow{\text{chlorophyll +}\atop\text{enzymes}}$$

$$\underset{\text{(carbohydrate)}}{C_6H_{12}O_6} + 6O_2 + \underset{\text{(to air)}}{6H_2O}$$

If photosynthesis were the only process occurring in plants, we could measure production by the rate of accumulation of carbohydrate, but unfortunately, at the same time, plants respire, using energy for maintenance activities. Respiration is the opposite of photosynthesis, in an overall view:

$$C_6H_{12}O_6 + 6O_2 \xrightarrow{\text{metabolic}\atop\text{enzymes}}$$

$$6CO_2 + 6H_2O + \text{energy for work and maintenance}$$

At metabolic equilibrium, photosynthesis equals respiration, and this is called the **compensation point**. Photosynthesis and respiration are rate processes, and are always measured as amount of material or energy per unit of time. If plants always existed at the compensation point, they would neither grow nor reproduce. We define two terms:

Gross primary production =

energy (or carbon) fixed in photosynthesis per unit time

Net primary production =

energy (or carbon) fixed in photosynthesis −

energy (or carbon) lost by respiration per unit time

Eugene Odum from the University of Georgia was one of the key founders of production ecology. He recognized by the 1950s that the measurement of production and energy flow through communities could

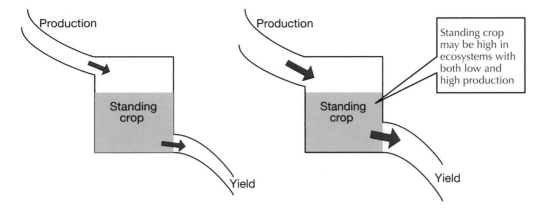

Figure 15.1 The relationship of standing crop and production in ecosystems. Illustration of two equilibrium communities (where input equals output): (a) low input, low output, slow turnover, high standing crop (b) high input, high output, rapid turnover, high standing crop. Standing crop is not related to production or yield because turnover time for all systems is not a constant. Production (input) must equal yield (output) in equilibrium communities, but many communities are not in equilibrium all the time, so standing crop may rise and fall. Note that knowing the standing crop tells you nothing about production or yield.

Eugene P. Odum (1913–2002) Professor of Ecology, University of Georgia

provide ecological insight into how ecosystems work, and in particular how humans were affecting the energy flows in natural communities.

Measuring Primary Production

How can we measure primary production in natural systems? For terrestrial plants, the direct way is to measure the change in CO_2 or O_2 concentrations in the air around plants[1]. The simplest indirect method of measuring primary production is the **harvest method**. The amount of plant material produced in a unit of time can be determined from the difference between the amounts present at the two times:

$$\Delta B = B_2 - B_1$$

where ΔB = biomass change in the community between time 1 (t_1) and time 2 (t_2)

B_1 = biomass at t_1

B_2 = biomass at t_2

Two possible losses must be recognized:

L = biomass losses by death of plants or plant parts

G = biomass losses to consumer organisms

If we know these values, we can determine net primary production in biomass:

Net primary production in biomass = NPP

$$= \Delta B + L + G$$

This may apply to the whole plant, or it may be specified as **aerial production** or **root production**. The net primary production in biomass may then be con-

---

1 Energy units have been reported in many forms in the literature, often in calories, and can be standardized to *joules* with the following conversion factors: 1 joule (J) = 0.2390 gram calorie (cal) = 0.000239 kilocalorie (kcal); conversely, 1 gram calorie = 4.184 joules and 1 kilocalorie = 4184 joules or 4.184 kilojoules (kJ).

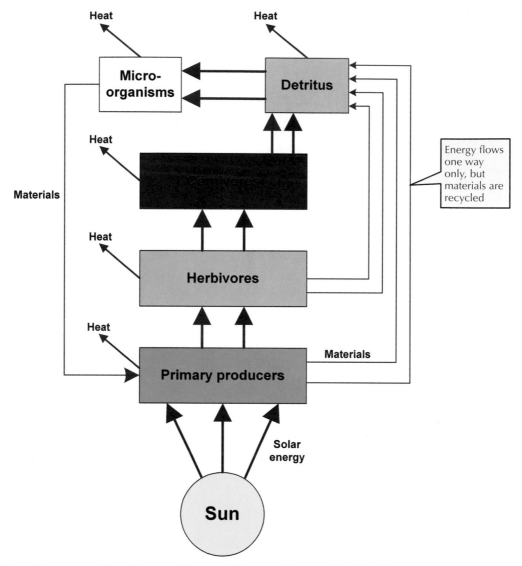

Figure 15.2 General representation of energy flows (red lines) and material cycles (blue lines) in the biosphere. The energy flows included are solar radiation, chemical energy transfers (in the ecological food web) and radiation of heat into space. Materials flow through the trophic levels to detritus and eventually back to the primary producers. (Modified from DeAngelis 1992.)

verted to energy by measuring the caloric equivalent of the material in a bomb calorimeter.

In aquatic systems, primary production can be measured in the same general way as in terrestrial systems. Gas-exchange techniques can be applied to water volumes—usually oxygen release, rather than carbon dioxide uptake, is measured. This procedure is usually repeated with a dark bottle (respiration only) and a light bottle (photosynthesis and respiration), so that both gross and net production can be measured.

Q Does Liebig's law of the minimum also apply to primary production?

How does primary production vary between the different types of vegetation on the Earth? This is the first general question we can ask about community metabolism. Figure 15.3 illustrates the yearly production for ocean and land areas of the globe. In general, primary production is highest in the tropical rain

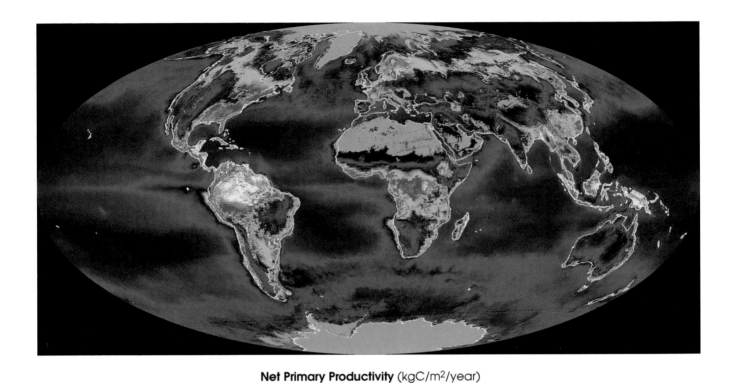

Net Primary Productivity (kgC/m²/year)

0   1   2   3

Figure 15.3 Annual net primary productivity (kilograms of carbon per m² per year) for the globe in 2002 calculated from satellite data gathered by the Moderate Resolution Imaging Spectroradiometer (MODIS). The yellow and red areas show the highest rates of net production, and the green, blue and purple areas show progressively lower productivity. Grey areas indicate no net primary production. Of total global primary production the ocean contributes 46% and the land 54%. Plant matter is about 50% carbon, so these carbon data can be readily converted to vegetation biomass by multiplying by 2. (From NASA, http://earthobservatory.nasa.gov/Newsroom/NPP/npp.html, 2004.)

forest and decreases progressively toward the poles. Productivity of the open ocean is very low, approximately the same as that of the arctic tundra. But because oceans occupy about 71% of the total surface of the Earth, total oceanic primary production adds up to about 46% of the overall production of the globe. Grassland and tundra areas are less productive than forests in the same general region.

Efficiency of Primary Production

How efficient is the vegetation of different communities as an energy converter? We can determine the efficiency of utilization of sunlight by the following ratio:

$$\text{Efficiency of gross primary production (\%)} = \frac{(100) \text{ energy fixed by gross primary production}}{\text{energy in incident sunlight}}$$

The amount of solar radiation intercepted by the Earth is 21×10^{24} J per year, or about 8.1 J per cm² per minute. Of this total solar energy, plants use only about 0.02% of this amount for photosynthesis, and most of the remaining energy is reflected back by the atmosphere or converted to heat. The energy in incident sunlight at the Earth's surface is reduced by more than half by atmospheric reflection or absorption, but still only a small amount of this incident energy is used in primary production.

Phytoplankton communities have very low efficiencies of primary production—usually less than 0.5%—although rooted aquatic plants and algae in shallow waters can have higher efficiencies. The efficiency of gross primary production is higher in forests (2.0–3.5%) than in herbaceous communities (1.0–2.0%) or in crops (less than 1.5%). Forest communities are relatively efficient at capturing solar energy, but no vegetation type captures more than 3–4% of the solar energy falling on the Earth.

Q **Do you think you could change the methods of planting agricultural row crops to make them more efficient at capturing the sun's energy?**

How much of the energy fixed by photosynthesis is subsequently lost by respiration of the plants themselves? A great deal of energy is lost in converting solar radiation to gross primary production. Net primary production, which is the useful primary production for herbivores or humans, must therefore be even less efficient. In forests, 50 to 75% of the gross primary production is lost to respiration, so that net production may be only a quarter of gross production. Forests have larger amounts of stems, branches and roots to support than do herbs, and thus less energy is lost to respiration in herbaceous and crop communities (45–50%). The result of these losses is that for a broad range of terrestrial communities, about 1% of the sun's energy during the growing season is converted into net primary production.

■ 15.3 LIGHT, TEMPERATURE, RAINFALL AND NUTRIENTS LIMIT PRIMARY PRODUCTIVITY

One important question is 'What controls the rate of primary production in natural communities?' What factors could we change to increase the rate of primary production for a given community? The control of primary production has been studied in greater detail for aquatic systems than for terrestrial systems. Let us look first at some details of production in aquatic communities.

Marine Communities

Light is the first variable one might expect to control primary production and the depth to which light penetrates in a lake or ocean is critical in defining the zone of primary production. Water absorbs solar radiation very readily. More than half of the solar radiation is absorbed in the first meter of water, including almost all the infrared energy. Even in 'clear' water, only 5 to

10% of the radiation may be present at a depth of 20 meters. Very high light levels can inhibit photosynthesis of green plants—this inhibition can be found in tropical and subtropical surface waters throughout the year. When surface radiation is excessive, the maximum in primary production will occur several meters beneath the surface of the sea.

Considerable research on ocean production has occurred in the North Pacific Central Gyre (Figure 15.4). Because the Central Gyre is large and relatively stable, biological production can be measured in the absence of external inputs. Figure 15.5 shows the vertical distribution of nutrients, temperature and primary production in the Central Gyre. Virtually all the primary production occurs above the 1% light level (above 90 m depth). Primary production is relatively low in the surface waters (due to excessive light) and is highest in the warm waters near the surface at 10–30 m depth.

If light is the primary variable limiting primary production in the ocean, there should be a gradient of productivity from the poles toward the equator. Figure 15.3 shows the global distribution of primary production in the oceans. There is no gradient of production from the poles to the equator. Large parts of the tropics and subtropics, such as the Sargasso Sea, the Indian Ocean and the Central Gyre of the North Pacific, are very unproductive. In contrast, the North Atlantic, the Gulf of Alaska and the Southern Ocean off New Zealand are quite productive. The most productive areas are coastal areas off the western side of Africa and North and South America.

Why are tropical oceans unproductive when the light regime is good all year? **Nutrients** appear to be the primary limitation on primary production in the ocean. Two elements—nitrogen and phosphorus—often limit primary production in the oceans. One of the striking generalities of many parts of the oceans is the very low concentrations of nitrogen and phosphorus in the surface layers where the phytoplankton live (Figure 15.5), whereas the deep water contains much higher concentrations of nutrients.

Nitrogen may be a limiting factor for phytoplankton in many parts of the ocean. The discovery that nitrogen limits primary production in many parts of the ocean was completely unexpected because nitrogen is

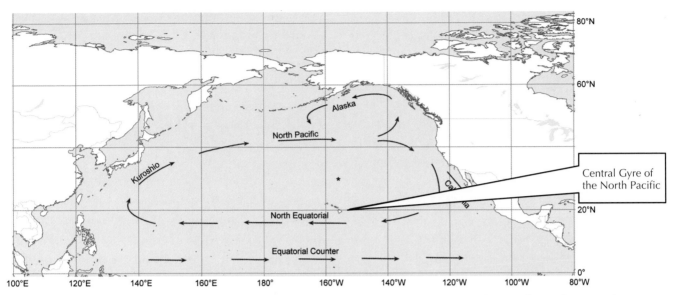

Figure 15.4 Map of the North Pacific Ocean showing the major upper ocean currents. The North Pacific Central Gyre is the large clockwise-flowing circulation system. The surface waters of the Central Gyre form nearly a closed section of the ocean with relatively little inflow and outflow at the edges. The biological structure of the Central Gyre has been most intensively studied at the location indicated by the asterisk. (Modified from Hayward 1991.)

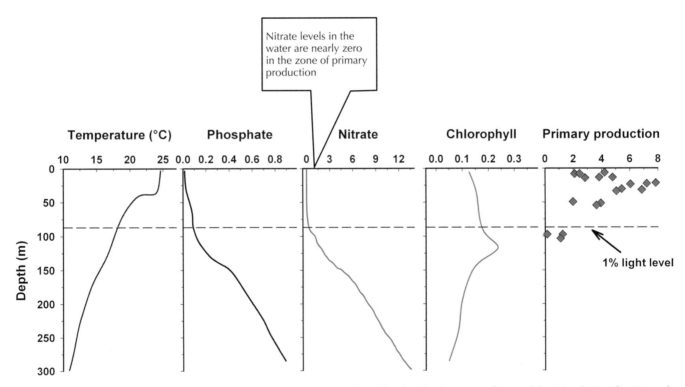

Figure 15.5 Vertical distributions of temperature, nutrients and production in the upper layer of the North Pacific Central Gyre during summer. The curves are an average of several vertical profiles made at a single location (28°N, 155°W) shown by the asterisk in Figure 15.4. The dashed line illustrates the depth of 1% of surface light. Nitrate—a critical limiting factor in the ocean—is depleted to undetectable levels above the 1% light level, and most of the primary production takes place above this depth. Maximum primary production occurs about 20 m below the surface waters. (From Hayward 1991.)

abundant in the air and can be converted into a usable form by nitrogen-fixing cyanobacteria. The expectation had been that phosphorus must be limiting productivity in the ocean because phosphorous does not occur in the air. But this has turned out to be completely wrong—a good example of why 'obvious' conclusions in science may not be correct. But the importance of nitrogen as a limiting factor raises another dilemma because several large parts of the oceans contain high amounts of nitrate and low numbers of phytoplankton. For example, the surface waters of the equatorial Pacific have both high nitrate and high phosphate concentrations but low algal biomass. One explanation for these oceanic regions is that they are communities dominated by top-down processes in which herbivores control plant biomass, and nutrients are always in excess. Alternatively, these could be bottom-up communities limited by some nutrient other than nitrogen or phosphorous.

The Sargasso Sea is an area of very low productivity in the subtropical part of the Atlantic Ocean. The sea water there is among the most transparent in the world, and the surface waters are very low in nutrients. Nitrogen and phosphorus, however, do not seem to be limiting primary production—iron seems to be critical. This was shown by a series of nutrient-enrichment experiments in which surface water from the Sargasso Sea was placed in bottles and enriched with various nutrients.

The demonstration of iron limitation in the Sargasso Sea stimulated the hypothesis that iron limitation could be responsible for the low productivity of the equatorial Pacific. Iron comes to the oceans largely as wind-blown dust from the land, and dust is particularly scarce in the Pacific Ocean and in the Southern Ocean. Iron is an essential component of the photosynthetic machinery of the cyanobacteria that fix nitrogen in the oceans. The impact of iron on primary production is mainly through its impact on nitrogen fixation, so we have a sequence of potential limitations that operate in iron-poor parts of the ocean:

$$\text{iron} \xrightarrow{+} \text{cyanobacteria} \xrightarrow{+}$$
$$\text{nitrogen fixation} \xrightarrow{+} \text{phytoplankton}$$

In most of the open oceans, light is always available for photosynthesis, but nitrogen is not.

To quantify the relative effect of different limiting nutrients in the oceans, John Downing and his students analyzed 303 comparisons between nutrient-addition and control experiments carried out over the last 30 years. They found that nitrogen addition stimulated phytoplankton growth most strongly, followed closely by iron addition (Figure 15.6). These results are consistent with the conclusion that nitrogen and iron in the oceans are key limiting resources.

Compared with the land, the ocean is very unproductive; the reason seems to be that fewer nutrients are available. Rich, fertile soil contains 5% organic matter and up to 0.5% nitrogen. One square meter of soil surface can support 50 kg dry weight of plant matter. In the ocean, by contrast, the richest water contains 0.00005% nitrogen, four orders of magnitude less than that of fertile farmland soil. One square meter of rich sea water could support no more than 5 grams dry weight of phytoplankton. In terms of standing crops, the sea is a desert compared with the land. And, although the maximal rate of primary production in the sea may be the same as that on land, these high rates in the sea can be maintained for a few days only, unless upwelling enriches the nutrient content of the water.

Areas of upwelling in the ocean are exceptions to the general rule of nutrient limitation. The largest area of upwelling occurs in the Antarctic Ocean, where cold, nutrient-rich, deep water comes to the surface along a broad zone near the Antarctic continent (Figure 15.3). Other areas of upwelling occur off the coasts of Peru and California, as well as in many coastal areas where a combination of wind and currents moves the surface water away and allows the cold, deep water to move up to the surface. In these areas of upwelling, fishing is especially good and, in general, there is a superabundance of nitrogen and phosphorus for the phytoplankton.

One of the most exciting recent developments in marine ecology is the ability to estimate primary production from satellite remote sensing data. Chlorophyll concentration in the surface water can be estimated by spectral reflectance using blue/green ratios. Figure 15.7 illustrates how a plankton bloom can be detected by chlorophyll levels in coastal waters in the Sea of Cortez off Baja California. Remote sensing allows marine ecologists to analyze large-scale production changes in the

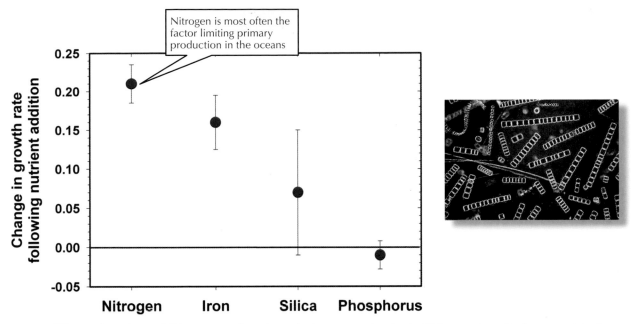

Figure 15.6 Effects of nutrient addition on marine phytoplankton growth rates in 303 experiments. Excess nutrients were added to a large water sample and followed for 2–7 days. Nitrogen and iron are clearly the main limiting factors. Silica limitation occurs when diatoms are the dominant species in the phytoplankton. The black line marks the line of zero effect. Phosphorus almost never seems to be the limiting factor. The photo shows marine diatoms from a phytoplankton bloom. (After Downing *et al.* 1999.)

ocean without being limited to a few measurements made off a ship. These techniques promise to enlarge our understanding of primary production in the oceans and how it varies in space and time.

In summary, total primary production in the ocean is rarely limited by light, but by the shortage of nutrients, particularly nitrogen and iron, which are critical for plant growth. Phosphorus limitation of primary production is very rare in oceanic ecosystems.

Freshwater Communities

In freshwater communities, the same limiting factors that operate in the ocean do not seem to operate. Solar radiation limits primary production on a day-to-day basis in lakes and, within a given lake, you can predict the daily primary productivity from the solar radiation. Temperature is closely linked with light intensity in aquatic systems and is difficult to evaluate as a separate factor. Nutrient limitations operate in freshwater lakes, and the great variety of lake ecosystems is associated with a great variety of potential limiting nutrients (*Essay 15.1*). For growth, plants require nitrogen, calcium, phosphorus, potassium,

sulfur, chlorine, sodium, magnesium, iron, manganese, copper, iodine, cobalt, zinc, boron, vanadium and molybdenum. These nutrients do not all act independently, which has made the identification of causal influences very difficult.

During the 1970s the problem of what controls primary production in freshwater lakes became acute because of increasing pollution. Nutrients added to lakes directly in sewage or indirectly as run-off had increased algal concentrations and had shifted many lakes from phytoplankton communities dominated by diatoms or green algae to those dominated by blue-green algae. This process is called **eutrophication**. Before we can control eutrophication in lakes, we have to decide which nutrients need to be controlled. Three major nutrients were suggested: nitrogen, phosphorus and carbon—much experimental work was carried out from 1960 to 1985 to determine the major limiting factor. The conclusion was simple: phosphorus is the limiting nutrient for phytoplankton production in the majority of freshwater lakes. The standing crop of phytoplankton is highly correlated with the total amount of phosphorus in the lake water (Figure 15.8).

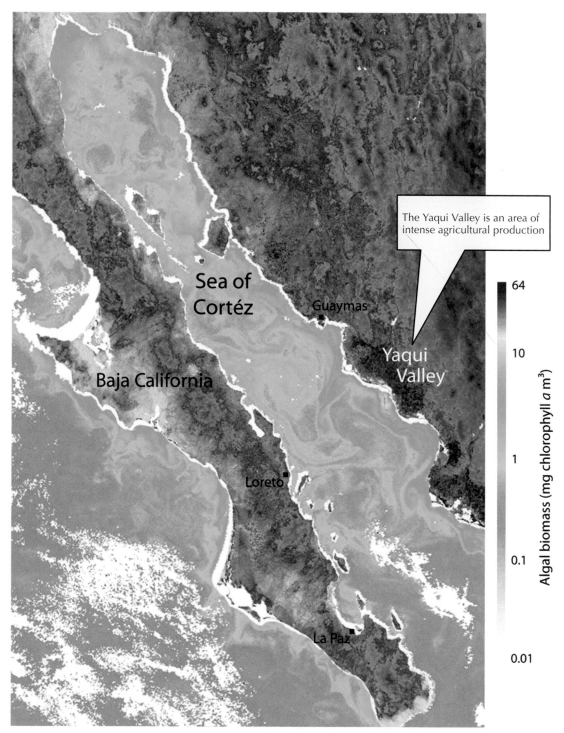

Figure 15.7 A satellite image covering the Sea of Cortez, a 700-mile-long stretch of the Pacific Ocean that separates the Mexican mainland from the Baja California Peninsula. Algal biomass in the water is colour coded from low (grey) to high (green to red). A phytoplankton bloom is shown off the mouth of the Yaqui River in the center of the image on August 12, 2004. 'Bright' water from the algal production in the bloom is colored green to red. This plankton bloom arose from agricultural run-off of nutrients from the irrigated wheat grown in the Yaqui Valley and moved over a two-week period under the influence of coastal currents and then dissipated. (Images courtesy of the NASA SeaWifs satellite and Mike Beman, Stanford University.)

ESSAY 15.1   NUTRIENT RATIOS AND PHYTOPLANKTON

Chemistry is an important aspect of ecosystem science and this is illustrated in the nutrient ratios of primary producers. In 1958 the oceanographer A.C. Redfield discovered that samples of organisms from the open ocean consistently exhibited the atomic ratio $C_{106}N_{16}P_1$, which is now referred to as the Redfield ratio in his honor. In contrast to the constant Redfield ratio found in the open ocean, the composition of phytoplankton from freshwater lakes is highly variable, suggesting that different limiting factors operate in the oceans compared with freshwaters. The ratio of C:N:P in phytoplankton varies with the ratio supplied in the water and the pH of the water. In a series of 51 lakes surveyed, the C:N ratio varied from 4 to 20, and the C:P ratio from 100 to 550, so that Redfield proportions are the exception, rather than the rule, in freshwater lakes.

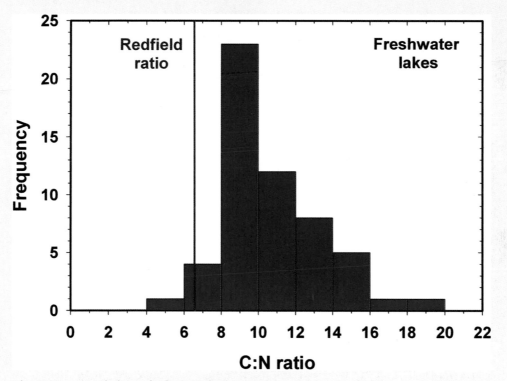

The C:N ratios of phytoplankton in freshwater lakes

In general there is much more carbon in freshwater phytoplankton relative to nitrogen and phosphorus. Why should this variation matter? Algae with high ratios of C:P are poor quality food for herbivores, such as zooplankton, but are good quality food for microbes. The ratio of C:N:P could impact on the structure of the food web. Different zooplankton species have differing ratios of C:N:P, and so survive better feeding on different algal species. In general there is much variation in C:N:P ratios in plants, less variation in bacteria, even less in zooplankton and much less in fish species. For animals, you are indeed what you eat, but for plants, you are what nutrients you are able to pick up from the environment.

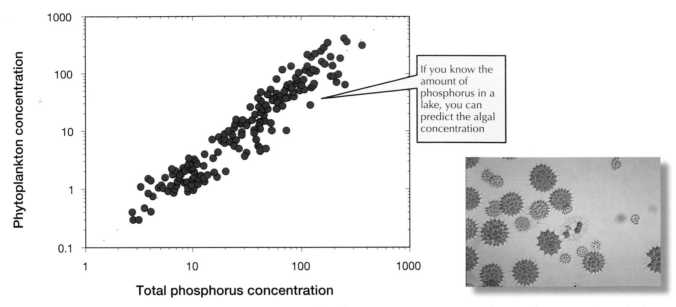

Figure 15.8. Freshwater lakes: the relationship between phosphorus concentration (g per liter) and summer phytoplankton standing crop (measured by chlorophyll as g per liter). Phosphorus concentration varies 100-fold in these lakes and algal concentration in the water varies 1000-fold. Phosphorus levels limit primary production in many freshwater ecosystems. The algae pictured are *Pediastrum boryanum*—a colonial green alga. (Data from Ahlgren *et al.* 1988.)

The practical implication from these and other experiments is to control phosphorus input to lakes and rivers as a simple means of checking eutrophication. The permissible amount of phosphorus that can be added to a lake can also be calculated so that planners can determine the desirability of human developments near a lake.

One of the changes that often accompany eutrophication in lakes is that the blue-green algae tend to replace green algae. Blue-green algae are 'nuisance algae' because they become extremely abundant when nutrients are plentiful and form floating scum. Blue-green algae become dominant in the phytoplankton for several reasons. They are not grazed heavily by zooplankton or fish, which prefer other algae. Zooplankton can often not manipulate the large colonies and filaments of blue-green algae. Some species of blue-green algae also produce secondary chemicals that are toxic to zooplankton. Blue-green algae are also poorly digested by many herbivores, so that they are low quality food. Finally, many blue-green algae can fix atmospheric nitrogen, putting them at an advantage when nitrogen is relatively scarce. In eutrophication, more and more phosphorus is contin-ually loaded into a lake so that nitrogen can become a limiting factor. The phytoplankton community in many temperate freshwater lakes therefore may have two broad configurations at which it can exist: one with low nutrient levels organized by predation and dominated by green algae and one with high nutrient levels organized by competition and dominated by blue-green algae.

Estuaries are mixtures of fresh water and salt water, and are often heavily polluted with nutrients from sewage and industrial wastes. Because they form an interface between salt water, in which nitrogen is often limiting to phytoplankton, and fresh water, in which phosphorus is typically limiting, estuaries are complex gradients of nutrient limitation in which added phosphorus and nitrogen from pollution can strongly affect primary production throughout the estuary.

To summarize, in freshwater communities, primary production is usually limited by light, temperature, and phosphorus levels.

Terrestrial Communities

In terrestrial habitats, temperature ranges are much greater than in aquatic habitats, and the great variation

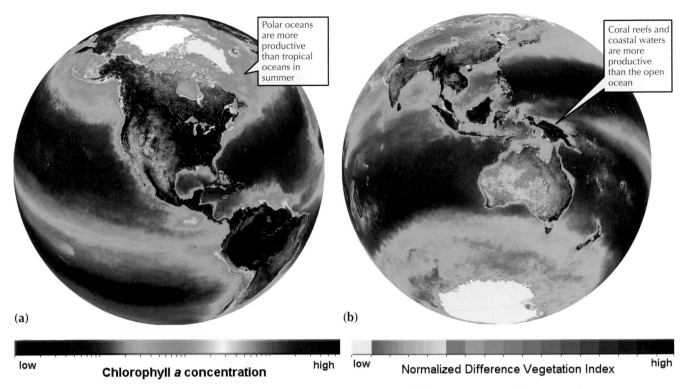

Figure 15.9 Summer productivity of the land (NDVI) and oceans (chlorophyll concentration) for (a) North America region, and (b) Australia region. On land the darker green areas are more productive and brown areas have minimal production. In the ocean the red areas are most productive. Note the high productivity of the polar oceans in summer, and coastal zones of higher production. These images were obtained with the SeaWiFS sensor on board the SeaStar spacecraft which circles 705 km above the Earth in a sun-synchronous orbit. Spatial resolution is 1.1 km. (Images provided by the SeaWiFS Project, NASA/Goddard Space Flight Center and ORBIMAGE; http://seawifs.gsfc.nasa.gov/SEAWIFS.html.)

in temperature from coastal to alpine or continental areas makes it possible to uncouple the solar radiation–temperature variable, which is so closely linked in aquatic systems. The large seasonal changes in radiation and temperature are reflected in the global patterns of primary production. Using satellite imagery, we can now look at continental and global patterns of terrestrial productivity. Satellites, such as the NOAA[1] series of meteorological satellites and the NASA Sea-Star spacecraft, have sensors on board that record spectral reflectance in the visible and infrared region of the electromagnetic spectrum. As green plants photosynthesize, they display a unique spectral reflectance pattern in the visible (0.4–0.7 µm) and the near-infrared (0.725–1.1 µm) wavelengths. Vegetation

indices that discriminate living vegetation from the surrounding rock, soil or water have been developed by combining these spectral bands. The AVHRR (advanced very high resolution radiometer) sensor on the NOAA satellite and the Sea WiFS (Sea-viewing wide field-or-view sensor) on the SeaStar spacecraft are especially useful for monitoring global vegetation because they have global coverage at a resolution of 1.1 km at least once per day in daylight hours.

The availability of satellite data to estimate primary production on a global basis has made it relatively easy to obtain measurement data of primary production. Figure 15.9 illustrates the detail that can be obtained on a daily or weekly basis. These patterns of primary production on land and in the sea set the stage for trying to determine what limits primary production. On land, production is limited by temperature, water and nutrients in the soil.

---

1 National Oceanic and Atmospheric Administration of the USA.

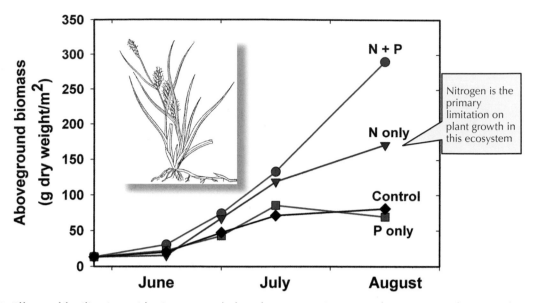

Figure 15.10 Effects of fertilization with nitrogen and phosphorus on primary production in a subarctic salt-marsh dominated by the sedge *Carex subspathacea*, southern Hudson Bay, Canada. Nitrogen and phosphorus were the primary limiting factors to growth. (From Cargill and Jefferies 1984.)

Q What are the possible limitations to the use of satellite data for measuring primary production?

Within the climatic constraints dictated by temperature and rainfall, soil nutrients limit production. Farmers and fertilizer companies have known this for many years. Ecologists wish to know exactly which nutrients are limiting and exactly how much primary production can be stimulated by adding nutrients. Nutrient-addition experiments on local sites can be used to determine how much primary production is limited by nutrients. In one experiment nitrate and phosphate were added to salt-marsh sedges and grasses in the subarctic to test for nutrient limitation. Figure 15.10 shows that, in the absence of grazing, the addition of nitrate doubled primary production of the sedges and grasses, and the joint addition of phosphate and nitrate quadrupled production. In this marsh, as in many terrestrial communities, nitrogen is the major nutrient limiting productivity, and when nitrogen is suitably increased, phosphorus becomes limiting. Figure 15.11 illustrates the idea of a sequence of limiting factors on terrestrial primary production.

In unexploited virgin grassland or forest, all nutrients that the plants take up from the soil and hold in various plant parts are ultimately returned to the soil as litter to decompose. The net flow of nutrients must

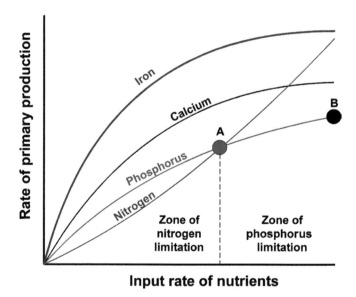

Figure 15.11 Illustration of a hypothetical sequence of limiting soil nutrients that affect terrestrial primary production. Primary production will follow the red line up to A and be limited by nitrogen in the soil. Above point A, nitrogen is superabundant and phosphorus becomes limiting between A and B. In this hypothetical soil, iron and calcium are always superabundant. In reality, many crops are nitrogen limited, as illustrated here.

354

be stabilized (input = output), or the site would deteriorate over time. But in a harvested community the situation is fundamentally different because nutrients are being continuously removed from the site. This makes it necessary to study the nutrient demands of crops so that the soils will not be progressively exhausted of their nutrients.

Terrestrial communities, especially forests, have large nutrient stores tied up in the standing crop of plants. In this way they differ from communities in the sea and in fresh water. This concentration of nutrients in the standing vegetation has important implications for nutrient cycles in forest communities. If the community is stable, the input of nutrients should equal the output, and a considerable amount of research effort is now being directed at studying nutrient cycles in terrestrial communities. We will discuss this work in Chapter 16.

From a global perspective, primary production is largely driven by the physical environment in the form of light, temperature, rainfall and nutrient supplies. Plants have adapted to these constraints from the environment to produce, through photosynthesis, the materials that drive all the subsequent biota, including ourselves.

■ 15.4 ENERGY FIXED BY GREEN PLANTS FLOWS EITHER TO HERBIVORES OR TO DETRITUS, OR IS LOST IN RESPIRATION

The biomass of plants that accumulates in an ecosystem as a result of photosynthesis can eventually go in one of two directions: to herbivores or to **detritus** feeders. Detritus is non-living, particulate organic matter and its associated microbial populations that result from decomposition. If we view animals as energy transformers, we can ask about the efficiency of energy transfer from plants down the food chain.

Efficiency of Secondary Production
Once plants have captured solar energy in primary production, this energy flows on to the rest of the food chain, leading to **secondary production**, or the aggregate of growth and reproduction in herbivores

and carnivores. In an analogous manner we can ask questions about what happens to the energy and materials fixed in primary production, and what fraction flows on to other trophic levels. One measure that we can use to describe ecosystems involves transfers between trophic levels and is called **trophic efficiency**:

$$\text{Trophic efficiency} = \frac{\text{net production at trophic level i} + 1}{\text{net production at trophic level i}}$$

This measure of energy transfer efficiency gives the fraction of production passing from one trophic level to the next. The energy not transferred is lost in respiration or to detritus. For aquatic ecosystems, trophic efficiencies vary from 2% to 24% and average 10.1% (Figure 15.12).

If we assume a trophic efficiency of 10%, we can calculate how much primary production is required to support a particular fishery. Consider the case of tuna caught in the open oceans. Tuna are top predators operating at trophic level 4[1], and, in 1990 2,975,000 tons of tuna were taken, or 0.1 g C per m^2 of open ocean per year. To support this yield of tuna to the fishery, assuming equilibrium conditions, we can calculate how much carbon is needed at each of the trophic levels (Figure 15.13). For example, to produce one 175 g can of tuna, we need to have produced 1750 g C of pelagic fishes to be eaten by the tuna, and 17,500 g C of zooplankton to be eaten by the pelagic fishes and finally 175,000 g C of phytoplankton. Furthermore, if we know the net primary production of the phytoplankton in the oceans, we can calculate what fraction of this production the tuna fishery is taking.

Using this approach, Daniel Pauly and Villy Christensen aggregated all the data for the fisheries of the world and showed that on average 8% of global aquatic primary production was being used to produce the global fisheries catch. But this average masked high variation among different fisheries (Table 15.1). Continental shelf and upwelling fisheries take a quarter to one-third of the net primary production—a fraction

---

1 Trophic level 1 = plants, 2 = herbivores, 3 = primary carnivores, 4 = secondary carnivores.

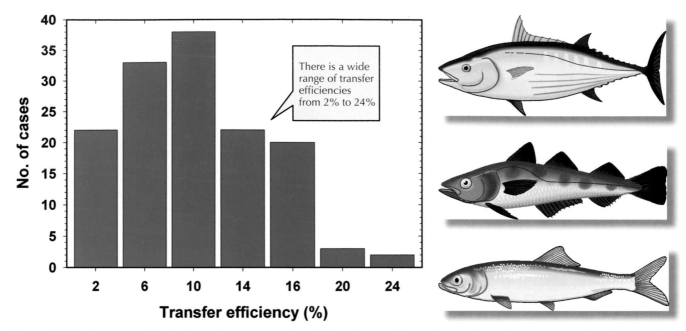

Figure 15.12 Frequency distribution of energy transfer efficiencies for 48 trophic models of freshwater and marine aquatic ecosystems. The 140 estimates express for herbivores to carnivores the fraction of production passing from one trophic level to the next. The mean transfer efficiency is 10.1%, so that on average about 90% of food energy is lost between each trophic level. (From Pauly and Christensen 1995.)

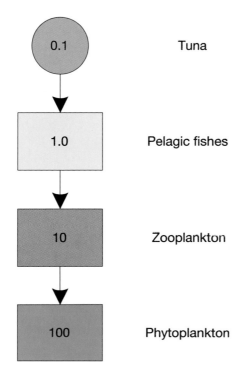

Figure 15.13 Hypothetical example of how much carbon is needed at each trophic level to support 0.1 g of tuna in the ocean, if we assume the efficiency of energy transfer is 10%. Note that these values are not standing crops, but are production or yield values expressed in units of carbon (C).

that is very high and leaves little margin for maintaining ecosystem integrity and a sustainable fishery.

This analysis for aquatic ecosystems raises the question of whether terrestrial and aquatic ecosystems operate in the same way. Many terrestrial systems are dominated by decomposers, and most of the energy in the system flows through the decomposer link in the food web. Figure 15.14 illustrates this for a temperate deciduous forest. For a typical deciduous forest, about 96% of the net primary production moves directly into dead organic matter and thence to the decomposers. This loss is greatly reduced at higher trophic levels, so that most of the production of herbivores is taken by carnivores and only 10% flows directly into the decomposer food chain.

Q At what trophic level should decomposers be placed?

The amount of herbivory varies in different ecosystems (*Essay 15.2*). Herbivores in aquatic ecosystems consume a higher fraction of the primary production than they do in terrestrial ecosystems (Figure 15.15).

Table 15.1 Global estimates of net primary production and the total catch to world fisheries (including discarded catches), and the calculated % of primary production required to support the observed fishery catches. Data from 1988–1991 were used in these estimates. Production and yield expressed in grams of carbon per square meter per year. (From Pauly and Christensen 1995.)

Ecosystem type	Area $(10^6 km^2)$	Net primary production $(gC\ m^{-2}yr^{-1})$	Fishery catch[a] $(gC\ m^{-2}yr^{-1})$	Primary production required (%)
Open ocean	332.0	103	0.012	1.8
Upwellings	0.8	973	25.560	25.1
Tropical shelves	8.6	310	2.871	24.2
Temperate shelves	18.4	310	2.306	35.3
Coastal/reef systems	2.0	890	10.510	8.3
Rivers and lakes	2.0	290	4.300	23.6
Weighted means		126	0.330	8.0

a The fishery catch includes an estimated 25% discards that are not counted in official fishery catch statistics.

Zooplankton herbivores in aquatic food webs consume an average of 79% of the net primary production of phytoplankton, while only 18% of the terrestrial primary production is eaten. Thus we can distinguish aquatic ecosystems, which are dominated by grazing, from terrestrial ecosystems, which are dominated by decomposers. Some average values are:

	Net primary production going to animal consumption (%)
Tropical rain forest	7
Temperate deciduous forest	5
Grassland	10
Open ocean	40
Oceanic upwelling zones	35

In forest ecosystems, almost all of the primary production goes into the decomposer food chain.

How do these differing consumption rates affect the standing crop of plants in different ecosystems? The standing crop of vascular plants in terrestrial, marine and freshwater communities is reduced by herbivores on average as follows:

	Reduction in standing crop of vascular plants from herbivores
Terrestrial ecosystems	26%
Marine ecosystems	65%
Freshwater ecosystems	31%

By excluding herbivores experimentally, we would expect a larger impact on the standing crop of plants in marine ecosystems compared with the results of the same experiment in terrestrial ecosystems.

Q **Why might herbivores consume more of the primary production in the ocean than they do on land?**

One consequence of low ecological efficiencies is that organisms at the base of the food web are much more abundant than those at higher trophic levels. Charles Elton recognized this in 1927, and linked it with the observation that predators are usually larger than the prey they consume. The result of these two processes is a pyramid of numbers or biomass that has been called an **Eltonian pyramid** in his honor. They illustrate graphically the rapid loss of numbers and biomass

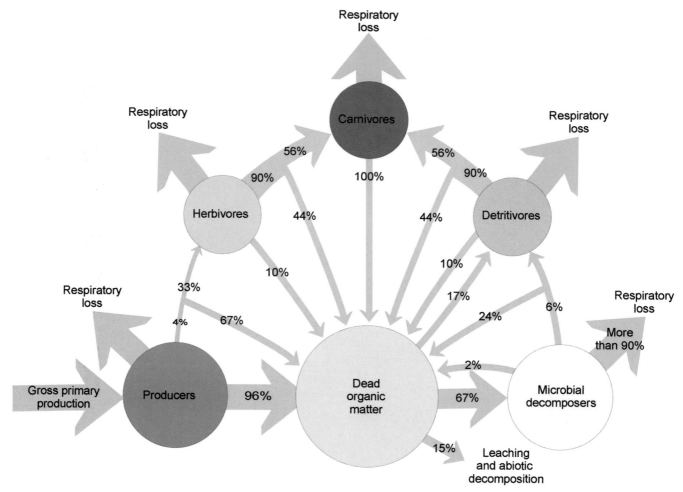

Figure 15.14 Estimates of energy flow through a temperate deciduous forest ecosystem. The energy at each trophic level splits into three parts: some energy is used for respiration (blue arrows) and the remaining energy can be lost as detritus (dead organic matter) or be taken up by the next trophic level as food. Some of this food is in turn wasted by the next trophic level and goes into the detritus pool. The vast majority of the net primary production in temperate deciduous forests goes directly into detritus and to decomposers, and about two-thirds of the total organic matter in the forest is in the detritus pool. (From Hairston and Hairston 1993.)

as one moves from smaller to larger animals in an ecosystem, a biological illustration of the second law of thermodynamics and the constraints of foraging.

Productivity of Grazing Systems

Agricultural grazing systems are a particularly good illustration of secondary production and ecological efficiency. Are human-designed grazing systems more or less efficient in using primary production than the natural grazing systems they typically replace? One practical application about grazing system efficiencies arose over the question of whether or not to encourage

game ranching in Africa. The question is this: 'are game animals more efficient producers of meat than domestic animals on the same range?' If game animals are more efficient, it would pay farmers to protect and harvest the native animals and use them for meat, rather than raising cattle or sheep.

The savannah and plains areas of Africa support a great diversity of ungulates, and the conservation of these ecosystems has been a primary concern of conservationists during the last 50 years. The history of human settlement on most continents has been a repeating sequence of the elimination of wild animals

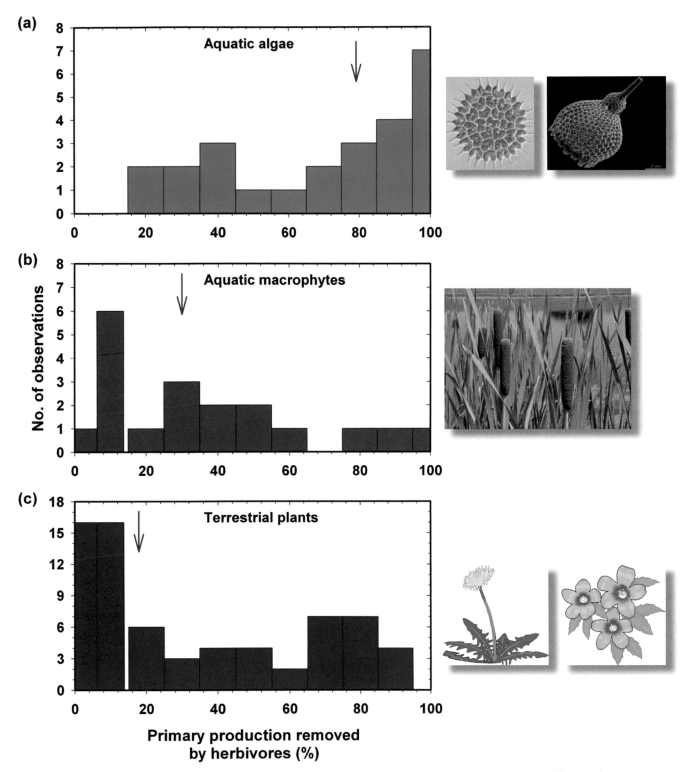

Figure 15.15 Percentage of net primary production removed by herbivores in ecosystems dominated by (a) algae (phytoplankton), (b) rooted aquatic plants, and (c) terrestrial plants. Red arrows indicate average values. Herbivores have significantly more impact on phytoplankton than on aquatic plants and even less impact on terrestrial plants. (Modified from Cyr and Pace 1993; diatom photo courtesy of Eric Condliffe, School of Earth and Environment, University of Leeds; cattail photo courtesy of Missouri Botanic Garden.)

ESSAY 15.2   WHY IS THE WORLD GREEN?

According to the **Green World Hypothesis**, the world is green because herbivores are held in check by their predators, parasites and diseases so that they cannot consume all the plant biomass. On land there is about 83×10^{10} metric tons of carbon in plant biomass, and about 5×10^{10} metric tons of plant matter are produced each year. Only 7–18% of this production on land is consumed by herbivores, so it is quite correct to say that grazing by herbivores plays a comparatively minor role for land plants on a global scale. But this global conclusion must be tempered by the observation that on occasion herbivores, such as the gypsy moth, do indeed destroy their plant resources, so that it is possible for the world not to be green. How can we reconcile these observations? Could evolution produce a super-herbivore that might eat all the green plants?

There are at least six reasons why the world is green.

1. **Plants are not passive agents, waiting to be eaten.** Plants contain much woody lignin as well as many secondary compounds that inhibit herbivores. All that is green is not edible.

2. **Nutrients limit herbivores, not energy.** Nutrients, such as nitrogen, are critical for animals and are often in short supply in plant materials (White 1993). Even in a world full of green energy, many herbivores cannot gain enough nutrients to grow and reproduce.

3. **Abiotic factors limit herbivores.** Seasonal changes in temperature, precipitation and other climatic factors depress herbivore numbers.

4. **Spatial variation reduces the availability of plants.** The world is not uniform and herbivores must search for food plants and cannot always locate them efficiently.

5. **Herbivores limit their own numbers.** Self-regulation through interference competition could limit the numbers of some herbivores through territoriality, cannibalism or other forms of interference, such as infanticide.

6. **'Enemies' limit herbivore numbers.** This limitation is the primary one suggested by the Green World Hypothesis. Enemies are effective in some communities, in which predators, parasites and diseases limit herbivores and prevent them from consuming all green plants. But not all herbivores are so limited, and the previous five mechanisms also act on predators to limit their numbers. On a global scale, enemies may not be the most important limitation on herbivores.

All six mechanisms act in varying ways to limit the off-take of herbivores, and each mechanism is important in keeping the world green. The key issue for any particular plant community is to identify the relative importance of each one of these factors. No single mechanism by itself explains why the world is green.

and replacement by domestic cattle and sheep, and during the 1950s Africa seemed next on the list. Against this background, Fraser Darling in 1960 proposed that, in many areas of Africa, game animals were more productive than domestic cattle and sheep, and hence a sustained yield of game would be more profitable than a sustained yield of cattle or sheep. If this idea is correct, local people could use wild game as a source of protein rather than turning to agricultural grazing systems with domestic cattle. Game-cropping or game-ranching schemes have been tried in many parts of Africa during the last 30 years, and these provide a practical example of the problem of what limits secondary production in ecosystems.

Game ranching has been attractive to conservationists because it seems to rest firmly on an interlocking set of theoretical ecological postulates. The argument can be summarized as follows. African wildlife has evolved within its ecosystems for millions of years, and thus is uniquely adapted to the African environment. The diversity of herbivores in Africa therefore should use the vegetation more efficiently and be more productive than a cattle or sheep monoculture would be. Thus African wildlife should attain a higher biomass than cattle or sheep on native African ranges. Furthermore, natural selection must have ensured that the different game species partition their food resources such that competition is reduced. So the diverse complex of wild species should cause less overgrazing than cattle or sheep and also be more resistant

to disease. In brief, African wildlife should provide a higher sustained yield and higher net revenue to the African people.

Although this theory seems to be theoretically sound, and eminently reasonable, attempts to demonstrate its validity since the 1960s have been unsuccessful. There does not appear to be a single case in which a sustained yield of game animals has been shown to be more valuable economically than a sustained yield of livestock in a comparable area. In all these African ecosystems, rainfall is the key limiting factor for primary production. The standing crop of herbivores on pastoral areas in Africa is always above that observed for wildlife areas, so that the original ideas about game ranching being more productive were all wrong. Figure 15.16 shows that this is also true in South America—for a given plant production, about 10 times more secondary production occurs in areas stocked with cattle compared with the production achieved by wild populations. Human-organized grazing systems are more productive than natural grazing systems because natural systems support predator populations, diseases

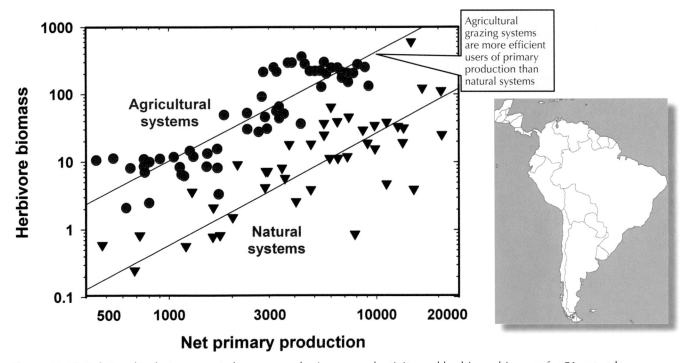

Figure 15.16 Relationship between net above-ground primary productivity and herbivore biomass for 51 natural ecosystems and 67 agricultural ecosystems for southern South America. For a given net primary production, there is about 10 times more biomass of herbivores on the agricultural ecosystems. Net primary production measured as energy by kJ per m^2 per year and herbivore biomass expressed as kJ per m^2. (From Oesterheld *et al.* 1992.)

and parasites, but cattle and sheep are protected from predators and inoculated against parasitic infections. Consequently some of the factors causing mortality in natural populations are absent in farming systems.

Game cropping is thus not an efficient way to provide cheap meat to low-income native peoples. From a conservation viewpoint, game cropping may have adverse impacts. The creation of a market for game-meat provides an outlet for illegal trade in these species. From an economic viewpoint, game farming in fenced areas can provide luxury products (meat) and services (tourism) to foreigners at a substantial economic advantage. But private game farms typically do not promote conservation of the entire community. Large predators are not welcome on game farms and grassland species are often preferred over forest species. The net result is that private game farms conserve only a few species in the ecosystem. Thus game ranching in Africa does not serve either the purposes of conservation or those of sustainable, high-yield agriculture for the African people. It is a good example of a suggested conservation strategy gone wrong.

SUMMARY

An ecological community can be viewed as a complex machine that processes energy and materials. To study community metabolism, we must determine the food web of the community and then trace the flows of chemical materials or energy through the food web. Many ecologists prefer to use energy to study community metabolism because energy is not recycled within the community.

Only about 1% of solar energy is captured by the green plants and converted into primary production. Forests are relatively efficient and aquatic communities inefficient at capturing solar energy. Primary production varies greatly over the globe; it is highest in the tropical rainforests and lowest in arctic, alpine and desert habitats. Global primary production is contributed nearly equally by the oceans and the land. The sea is less productive than the land per unit of area, except for coastal areas and upwelling zones, because of limitations imposed by nitrogen and iron. In freshwater lakes and streams, light, temperature and nutrients restrict primary production, and phosphorus is the limiting nutrient in many lakes.

Terrestrial primary productivity can be predicted from temperature and rainfall, which together determine the length of the growing season. Nutrient limitations (particularly nitrogen limitation) further restrict productivity levels set by these climatic factors, and the stimulation of plant growth achieved by fertilizing forests and crops indicates the importance of studying nutrient cycling in biological communities.

Considerable energy, often 90% or more, is lost at each step of the food chain, and thus for a given biomass of green plants, only a much smaller biomass of animals can be supported. Herbivores consume a higher fraction of the primary production in aquatic ecosystems than they do in forest or grassland ecosystems. Much of the energy flow in terrestrial systems goes directly from plants to the decomposer food chain.

Secondary production increases in step with primary production, so secondary production is broadly limited by primary production. Levels of secondary production in grazing agricultural ecosystems are up to 10 times higher than those achieved in natural grazing ecosystems. By understanding the climatic and nutrient limitations on primary and secondary production, we will be better able to predict the impact of climate change on ecosystems.

SUGGESTED FURTHER READINGS

Coale KH, Johnson KS, Chavez FP, Buesseler KO, Barber RT, Brzezinski MA, *et al.* (2004). Southern Ocean Iron Enrichment Experiment: carbon cycling in high- and low-Si waters. *Science* **304**, 408–414. (A large-scale set of experiments to test the idea that iron limits oceanic primary production in the Southern Ocean.)

Cyr H and Pace ML (1993). Magnitude and patterns of herbivory in aquatic and terrestrial ecosystems. *Nature* **361**, 148–150. (How much primary production do herbivores eat in different ecosystems?)

DeLucia EH, Hamilton JG, Naidu SL, Thomas BB, Andrews JA, Finzi A *et al.* (1999). Net primary production of a forest ecosystem with experimental CO_2 enrichment. *Science* **284**, 1177–1179. (How will added carbon dioxide affect forest primary production?)

Lizotte MP (2001). The contributions of sea ice algae to Antarctic marine primary production. *American Zoologist* **41**, 57–73. (The contribution of sea ice algae to productivity in the Southern Ocean.)

Pauly D and Christensen V (1995). Primary production required to sustain global fisheries. *Nature* **374**, 255–257. (Can we work backwards from the known fish catch to estimate how much primary production was utilized?)

Running SW, Nemani RR, Heinsch FA, Zhao M, Reeves M and Hashimoto H (2004). A continuous satellite-derived measure of global terrestrial primary production. *BioScience* **54**, 547–560. (An outstanding illustration of the power of satellite data to monitor global productivity on a weekly basis.)

Saugier B, Roy J and Mooney HA (Eds) (2000). *Terrestrial Global Productivity*. Academic Press, New York. (The definitive analysis of global patterns of productivity and the factors controlling it.)

QUESTIONS

1. 'Red tides' are spectacular dinoflagellate blooms that occur in the sea and often lead to mass mortality of marine fishes and invertebrates. Human deaths from eating shellfish poisoned with red tide algae are a worldwide problem. Review the evidence available about the origin of red tides, and discuss the implications for general ideas about what controls primary production in the sea. Anderson (1994) and Landsberg (2002) discuss this problem, and information on line is available at http://www.whoi.edu/redtide/.

2. Crop productivity has improved greatly during the last 60 years. Some of this improvement in production is due to genetic changes in the crops, and some is due to increased nutrients or water. How could you evaluate the relative contribution of these two components to increasing primary productivity in a particular crop? Boyer (1982) discusses this problem and provides some data for major U.S. crops. Calderini and Slafer (1998) discuss the historical trends in wheat yields. Recent data for the United States are available at http://www.nass.usda.gov.

3. Photosynthetic organisms produce about 300×10^{15} g of oxygen per year (Holland 1995). If this oxygen accumulated, the oxygen content of the atmosphere would double every 2000 years. Why does this not happen? Is the global system regulated? If so, how is this regulation accomplished?

4. Compile a list of the efficiency of some of our common physical machines, such as automobiles, electric lights, electric heaters, and bicycles.

5. Why should agricultural grazing systems be more efficient than natural grazing systems (Figure 15.16) at utilizing primary production? Read MacNab (1991) and Oesterheld *et al.* (1992) and discuss the problems of measuring efficiency in both kinds of ecosystems.

6. Could herbivores remove a high fraction of the net primary production in an ecosystem without depressing the standing crop of plants? How might this happen?

7. If the secondary production of herbivores increases in step 1:1 with the primary production of plants in different ecosystems, does this mean that all herbivore populations must be food-limited and unaffected by predators and diseases?

8. The discovery that iron was a primary factor limiting primary production in large areas of the ocean caused the oceanographer John Martin to say in the late 1980s that 'Give me a half tanker of iron, and I will give you an ice age.' List the causal links that could make this prediction come true, and read the evaluation by Buesseler *et al.* (2004) which suggests this prediction could not possibly be correct.

Chapter 16
ECOSYSTEM ECOLOGY—NUTRIENT RECYCLING

IN THE NEWS

The Mississippi River drains nearly one-third of North America and changes in water quality in the river over the last 50 years have caused drastic impacts in the ecosystems of the northern part of the Gulf of Mexico. The problem is nitrogen in the water, and the principal cause is a dramatic increase in fertilizer nitrogen input into the Mississippi River drainage basin between the 1950s and 1980s. Since 1980, the Mississippi River has discharged, on average, about 1.6 million metric tons

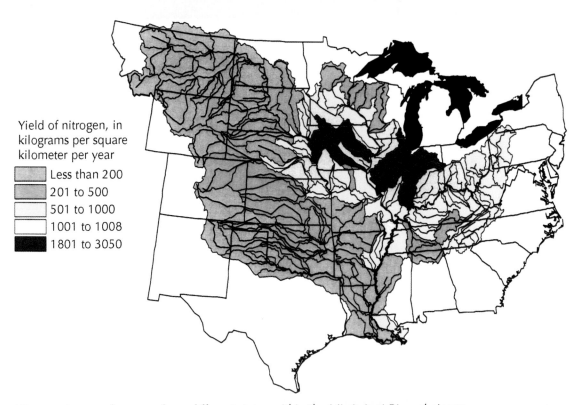

Yield of nitrogen, in kilograms per square kilometer per year

- Less than 200
- 201 to 500
- 501 to 1000
- 1001 to 1008
- 1801 to 3050

Nitrogen inputs of streams from different states within the Mississippi River drainage.

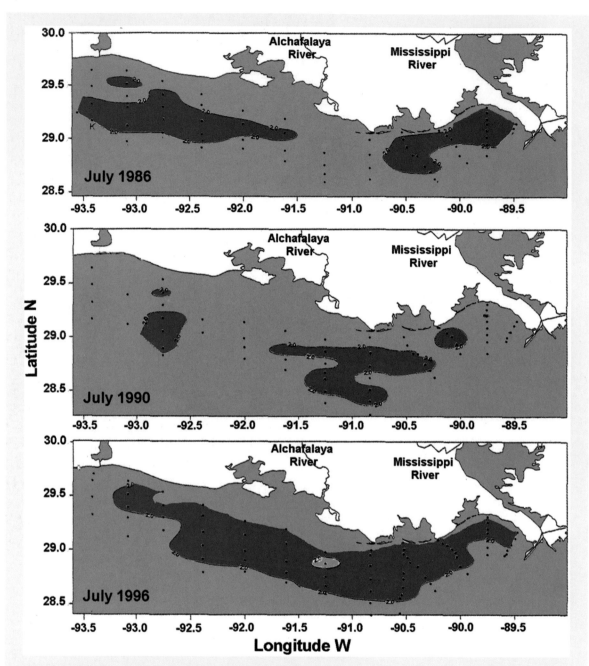

Size of the hypoxia zone (red) off the Mississippi River mouth in midsummer 1986, 1990 and 1996. The dots are the sampling stations in the Gulf of Mexico.

of total nitrogen to the Gulf each year. The most significant trend in nutrient loads has been in nitrate, which has almost tripled from 0.33 million metric tons per year during 1955–70 to 0.95 million metric tons per year during 1980–96. Other nutrients, such as phosphorus, have not increased, but may have even decreased over the last 50 years. About 90% of the nitrate in the river comes from excess fertilizer draining off agricultural land and drainage from feedlots for cattle.

The principal sources of nitrate are river basins that drain agricultural land in southern Minnesota, Iowa, Illinois, Indiana and Ohio.

Nitrogen does not seem to be a primary limiting factor for primary production in river systems—the ecological damage starts when these waters reach the coastal zone in the Gulf of Mexico. Coastal waters around the world are suffering from pollution—nutrients draining from the land are stimulating algal growth in the sea.

In coastal waters off Louisiana, the excess nitrogen stimulates algal growth and associated zooplankton growth. Fecal pellets from zooplankton and dead algal cells sink to the bottom and, as this organic matter decomposes, the bacteria use all the oxygen in the bottom layer of water. Stratification of fresh and saline waters prevents oxygen replenishment that would normally occur by the mixing of oxygen-rich surface water with oxygen-depleted bottom water. At dissolved oxygen levels of less than 2 mg/L all animals either leave or die. This shortage of oxygen in the bottom layer of coastal waters is called **hypoxia**, and these zones are called 'dead zones'.

Each summer, the Mississippi River outflow produces a hypoxic zone in the northern Gulf of Mexico along the Louisiana–Texas coast that varies in size up to 20,000 km^2, which is the size of New Jersey. The hypoxic zone is most pronounced from June to August, but can begin as early as April and last until October, when storms and winds mix up the surface and bottom water.

Spawning grounds of fish and migratory routes of commercially harvested fish species are affected by the hypoxic zones. To reduce hypoxia in the Gulf of Mexico, the most effective actions would be to reduce the amount of fertilizer used, to keep the nitrogen in the agricultural fields with alternative cropping systems and to increase the area of wetlands, which pick up nitrogen from the river water. The important message is that alleviating the problem of hypoxia in the Gulf of Mexico requires an ecosystem approach to the whole catchment of the Mississippi River. An ecological understanding is needed of how the whole catchment works and how nutrients distributed in fertilizer to grow corn in Iowa can impact algal populations thousands of kilometers away in the Gulf of Mexico. Actions taken by a single farmer or one feedlot owner will not be effective in solving these large-scale problems.

■ 16.1 NUTRIENTS CYCLE AND RECYCLE IN ECOSYSTEMS

Living organisms are constructed from chemical elements, and one way to investigate an ecosystem is to follow the transfer of chemical elements between the living and the non-living worlds. Carbon makes up most of the biomass of plants and animals, and there is widespread concern about increased CO_2 levels causing climate change. Sulfur is released when we burn coal and oil and the sulfur returns as acid rain to cause ecological problems with fish and forests. Interest in nutrient cycles goes back hundreds of years in agriculture, where organic fertilizers were used to stimulate crop production.

As we saw in the last chapter, elements such as nitrogen often set some limitation on the primary productivity of a community. Plants need a suite of nutrients for growth. Some elements such as carbon, oxygen, and hydrogen, come from the air directly. Other essential elements, such as phosphorus and potassium, come from the soil or water. These elements are classified as **macronutrients**, which are needed in large amounts, and **micronutrients**, which are needed in very small amounts. Macronutrients, in turn, can be broken into two more groups: primary and secondary macronutrients. The primary macronutrients are nitrogen, phosphorus and potassium. These major nutrients can be in short supply in the ocean, lakes and the soil because plants use large amounts for growth. The secondary macronutrients are calcium, magnesium and sulfur. These elements are not usually limiting nutrients in ecosystems. Micronutrients or **trace elements** are those elements that are essential for plant growth, but are needed in only very small quantities. The micronutrients plants need are boron, copper, iron, chlorine, manganese, zinc and molybdenum.

Ecosystems recycle all these nutrients with the net result that the world is green. To stimulate primary production, humans could add limiting nutrients. Adding nutrients in the form of fertilizers has become essential in modern agriculture and increasingly common in forestry. These additions of nutrients can also have detrimental impacts, as we have just seen for the Mississippi River Delta. In this chapter, we will

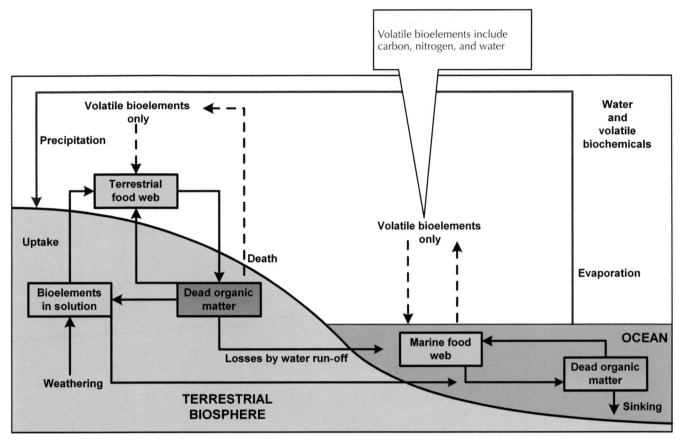

Figure 16.1 General schematic of nutrient cycling on a global scale. Movement of non-volatile elements, such as phosphorus, is largely one way—towards ocean sediments. Movement of volatile bioelements, such as nitrogen, is cyclical—from the atmosphere into plants and animals and back again. (Modified from DeAngelis 1992.)

describe how nutrients cycle and recycle in natural systems—linking together the living and the dead material in the ecosystem. When we look at nutrient cycles, we begin by considering the whole ecosystem as a machine processing nutrients, rather than looking at separate species and individuals.

Global Nutrient Cycles

We can view the biological community as a complex processor in which individuals move nutrients from one site to another within the ecosystem, which includes the abiotic environment. These biological exchanges of nutrients interact with physical exchanges and, for this reason, nutrient cycles are also called **biogeochemical cycles**. Chemical elements that cycle through living organisms are called **bioelements**. Figure 16.1 illustrates the general pattern of bioelement

or nutrient cycles on a global scale. Nutrient cycles are closed on a global scale, because none of these elements escape from the Earth, but they are open on a local scale as elements move across ecosystem boundaries or into the atmosphere. The individual atoms that make up the cycle are indestructible and can be recycled in plants and animals. Ecologists are interested in understanding and measuring global nutrient cycles because human activities can alter these cycles with possible impacts on global primary production and climate. An analysis of nutrient cycling thus ends with an assessment of human impacts on nutrient cycles and its consequences for animals and plants.

Global nutrient cycles represent the summation of local events occurring in different biotic communities and to understand global nutrient cycles we must begin at the level of the local community. All nutrients

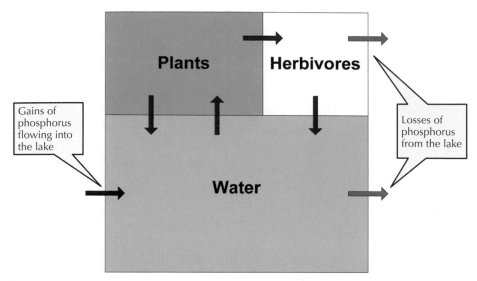

Figure 16.2 Hypothetical compartment model of a nutrient cycle for phosphorus in a simple lake ecosystem composed of three compartments—plants, herbivores and water. Arrows indicate rates of inflow or outflow of phosphorus per unit of time. Red arrows shows flows between compartments; brown arrows show flows going outside the lake ecosystem to other ecosystems; and the blue arrow indicates the input flow of phosphorus into the lake water either from ground water or a stream. Compartments are standing crops or amounts of phosphorus, and may be small or large. If the system is in equilibrium, the standing crops and the flow rates will be constant. If the system is not in equilibrium inflows may exceed outflows or vice versa.

reside in **compartments**, which represent a defined space in nature. Compartments can be defined very broadly or very specifically. All the plants in an ecosystem could be defined as one compartment (as in Figure 16.2), but we could also recognize each species of plant as a separate compartment if we had more detailed data. A compartment contains a certain quantity, or **pool**, of nutrients in the standing crop. In a simple lake compartment model, such as shown in Figure 16.2, the phosphorus dissolved in the water is one pool and the phosphorus contained in the bodies of herbivores is another pool.

Q What would be the advantages and disadvantages of defining each species as a compartment in an ecosystem?

Compartments exchange nutrients, and thus we must measure the uptake and outflow of nutrients for each compartment. The rate of movement of nutrients between two compartments is called the **flux rate** and is measured as the quantity of nutrient passing from one pool to another per unit of time. The flux rates and pool sizes together define the nutrient cycle within any particular ecosystem, and the problem for the ecologist is to measure these component parts of the nutrient cycle. Ecosystems are not isolated from one another, and nutrients come into an ecosystem through meteorological, geological or biological transport mechanisms and leave an ecosystem via the same routes. Meteorological inputs include dissolved matter in rain and snow, atmospheric gases and dust blown by the wind; geological inputs include weathering and elements transported by surface and subsurface drainage; and biological inputs include movements of animals between ecosystems.

The nutrient cycle of phosphorus in lakes illustrates some of the consequences of nutrient cycles. Phosphorus and other nutrients tend to accumulate in the sediment of lakes so that continual nutrient inputs are required to maintain high productivity. Lake-fertilization experiments increase productivity only temporarily, and a continued input of phosphate is needed to sustain high production in the phytoplankton. Nutrient cycles are also critical for understanding how lakes and rivers can recover from the effects of nutrient additions from pollution.

Local Nutrient Cycles

Nutrient cycles may be subdivided into two broad types. The phosphorus cycle we have just described is an example of a sedimentary or **local cycle**, which operates within an ecosystem. Local cycles involve the less-mobile elements that have no mechanism for long-distance transfer (the non-volatile elements of Figure 16.1). By contrast, the gaseous cycles of nitrogen, carbon, oxygen and water are called **global cycles** because they involve exchanges between the atmosphere and the ecosystem (volatile elements). Global nutrient cycles link together all the world's living organisms in one giant ecosystem called the **biosphere**—the ecosystem of the whole Earth.

Local nutrient cycles—involving calcium, magnesium, potassium and phosphorus—derive all their elements from chemical rock weathering. As soils age, these nutrients should all be leached out of the topsoil by weathering, and so they should become limiting to primary production. If this loss of rock-derived nutrients continues, ecosystems should all face long-term decline. But this loss does not in fact seem to occur, and we need to see how nutrients cycle to understand why. A good place to start is with forested ecosystems subjected to logging.

■ 16.2 HARVESTING AFFECTS NUTRIENT CYCLES IN FORESTS

The harvesting of forest trees removes nutrients from a forest site, and this continued nutrient removal could result in a long-term decline in forest productivity unless nutrients are somehow returned to the system. Because of the economic importance of forest productivity, an increasing amount of research work is being directed toward the analysis of nutrient cycles in forests. Figure 16.3 shows the factors that must be quantified in order to describe the nutrient cycle in a forest ecosystem.

Nutrient budgets for forest ecosystems attempt to balance the inputs and outputs of nutrients to the system under study. The key to nutrient budgets is the mass balance approach described by the simple equation:

$$\text{Inputs} - \text{outputs} = \Delta \text{ storage}$$

If there is a change in storage (Δ storage) then the nutrient budget equation is not balanced, and nutrients must be accumulating somewhere in the ecosystem, such as in leaves or stems, or declining somewhere, such as in the soil. Tracing these inputs and outputs through the ecosystem is the essence of nutrient cycling.

During forest development, nutrients accumulate in leaves and wood. Figure 16.4 illustrates the rapid accumulation of five different nutrients in a stand of jack pine in eastern Canada. As trees increase in size during succession, the soil accumulates nutrients in the surface litter and in organic matter (humus) dispersed through the lower soil horizons. As forest stands age, there is a systematic change in the uptake of nutrients.

Figure 16.5 illustrates changes in the uptake rate of nitrogen in a spruce forest in Russia. In this forest the spruce canopy becomes more open after 70 years and understory vegetation increases in volume and importance in nutrient cycling. Not all forest successions will produce the same pattern of changes in nutrient cycling, but the general principle that nutrient cycling varies with forest age will be valid. In old growth forests there is little additional accumulation of nutrients and ecosystem inputs and outputs should be balanced (*Essay 16.1*).

Nutrient Pools in Forests

Where are the nutrients in forest ecosystems? Are they tied up in the soil or in the above-ground living vegetation? The location of the pools of different nutrients varies with the forest ecosystem. Nutrients that are in short supply are recycled more efficiently than those present in excess of requirements. In most forest sites, nitrogen is a major limiting factor to tree growth. Table 16.1 gives the organic matter and nitrogen content in the above-ground component of 32 forests from different climatic zones. In the boreal forests of Alaska only about 20% of the organic matter is present in the trees above ground. Low decomposition rates in these cold Alaskan forests cause most of the nitrogen and organic matter to be tied up in the soil. Coniferous forests have the largest forest floor accumulation of organic matter of all forests—on average about four times the biomass of tropical forests.

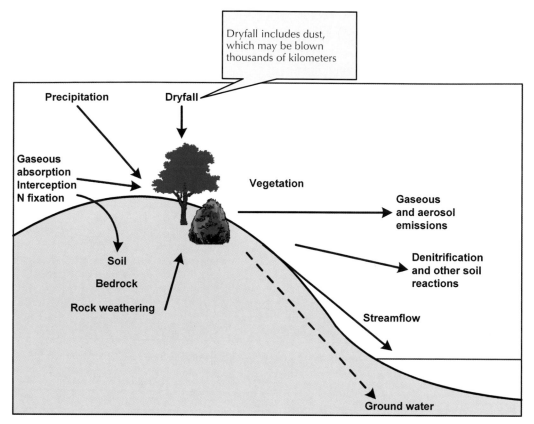

Figure 16.3 A schematic illustration of the pathways of nutrient movement through undisturbed forest ecosystems. To quantify nutrient cycling, all these pathways must be measured. (Modified from Waring and Schlesinger 1985.)

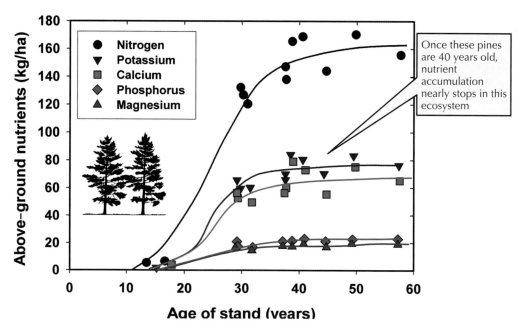

Figure 16.4 Accumulation of nitrogen, potassium, calcium, phosphorus and magnesium during the post-fire succession of jack pine (*Pinus banksiana*) stands in New Brunswick, Canada. Trees accumulate more nitrogen than other nutrients, and nitrogen is often the limiting factor for forest tree growth. (Data from MacLean and Wein 1977.)

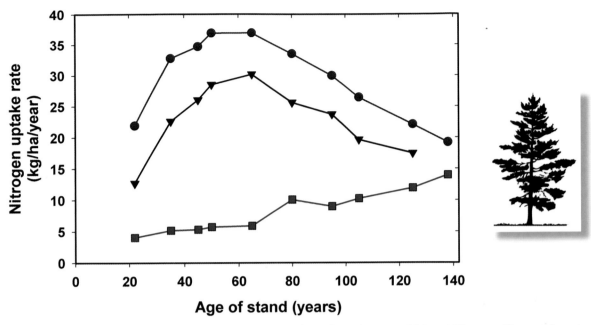

Figure 16.5 Uptake and cycling of nitrogen in spruce (*Picea abies*) from the age of 22 to 138 years. The uptake rate of nitrogen is highest at around 60 years of tree age when growth is maximal and drops off rapidly as trees age. (Data from Kazimirov and Morozova 1973.)

Nutrient cycles operate more quickly in warmer forests than in colder ones. If we assume as an approximation that forest soil nutrients are in equilibrium for the short time they can be studied (3–10 years), we can calculate average **turnover time** for each nutrient. Turnover time represents the time an average atom will remain in the soil before it is recycled into the trees or shrubs. Table 16.2 gives the mean turnover times for five elements. All northern forests have very slow turnover of nutrients. On average, a boreal conifer forest retains nitrogen 100 times longer than a Mediterranean evergreen oak forest. Deciduous forests turn over nutrients more rapidly than coniferous forests. Coniferous forests use nutrients more efficiently

Table 16.1 Accumulation of organic matter and nitrogen in the above-ground stems of trees for various forest regions. Amounts not above ground in tree stems are underground in roots and in the soil. Warmer forests have more organic matter and nutrients above ground in the tree stems.

Forest region	No. of sites	Organic matter (kg/ha)			Nitrogen (kg/ha)		
		In trees	Total	% above ground	In trees	Total	% above ground
Boreal coniferous	3	51,000	226,000	19	116	3,250	4
Boreal deciduous	1	97,000	491,000	20	221	3,780	6
Temperate coniferous	13	307,000	618,000	54	479	7,300	7
Temperate deciduous	14	152,000	389,000	40	442	5,619	8
Mediterranean	1	269,000	326,000	83	745	1,025	73
Average	32	208,000	468,000	45	429	5,893	7

Source: From Cole and Rapp (1981).

ESSAY 16.1 WHY DOES PRIMARY PRODUCTION DECLINE WITH AGE IN TREES?

Forest managers have known for many years that tree growth and wood production decline with tree age. Net primary production in forests reaches a peak early in succession and then gradually declines by as much as 76% from the peak (Gower *et al.* 1996). For example, primary production of 140-year-old Norway spruce in Russia declines 58% from the peak reached around 70 years of age:

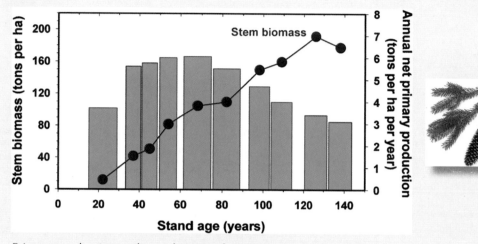

Primary production and stem biomass for Norway spruce in Russia

Why should this occur? It is not surprising that growth declines with age, because this happens in virtually all plants and animals. The main items of interest are the physiological mechanisms that produce this decline. There are three hypotheses to explain age-related decline in tree growth

1. The first, classical explanation is that there is a change in the balance of photosynthesis and respiration. As trees grow larger with age, they have more tissues that respire and lose energy, and proportionally less leaf area to photosynthesize. But this explanation is not supported by recent measurements showing that respiratory losses do not increase very much with tree age, because most of the sapwood uses little energy.

2. The second hypothesis is nutrient limitation by nitrogen as the forest ages. Nitrogen is commonly found to be the limiting factor to tree growth and, as forests age, more woody litter accumulates on the soil surface. Woody litter decomposes very slowly compared with fine litter from leaves, so that nitrogen becomes locked up in woody debris on the forest floor.

3. The third explanation, and the newest idea, is that as trees grow larger, water transport to the leaves becomes limiting. Increased hydraulic resistance is associated with the greater distance of travel of water from the roots to the stomata of the leaves. Trees close the stomata in their leaves to conserve water in their tissues and, because

photosynthesis is tightly coupled with the flow of CO_2 through the stomata, production declines. This hypothesis is consistent with the observation that leaf stomata close earlier in the day in older trees compared with young trees.

Current forest growth models suggest that nutrient limitation (hypothesis 2) and water flow limitations (hypothesis 3) are of nearly equal importance in reducing net primary production as trees age. An increase in respiration seems to contribute little to decreasing production (Gower *et al.* 1996). Knowing what limits primary production in forests is critical for understanding the impacts of climate change on ecosystems.

because they retain their needles and do not need to replace all their foliage each year.

Nutrients are lost from forest ecosystems in several ways

- Streams transport both dissolved and particulate matter—measurements of stream water chemistry can provide a good way to monitor overall forest function.
- Anaerobic soil bacteria produce methane and hydrogen sulfide gases.
- Plants release hydrocarbons, such as terpenes, from their leaves, and these compounds may add to atmospheric haze in summer.
- Both ammonia and hydrogen sulfide can be released from plant leaves.
- During forest fires nutrients are released in both gases and in particles.

- Finally, forest harvesting removes nutrients in wood from the ecosystem.

Q Are there any basic differences in the nutrient cycles operating in agricultural fields and those in forest plantations?

Hubbard Brook Experimental Forest

One of the most extensive studies of nutrient cycling in forests was begun by Gene Likens, F. Herbert Bormann, and Noye M. Johnson in 1963 at the Hubbard Brook Experimental Forest in New Hampshire. The Hubbard Brook forest is a nearly mature, second-growth hardwood ecosystem. The area is underlain by rocks that are relatively impermeable to water, so all run-off occurs in small streams. The area is subdivided into several small watersheds that are distinct yet

Table 16.2 Mean turnover time in years for the forest floor and its mineral elements by forest regions. Turnover time is very slow in subarctic forests and much more rapid in warmer forest stands. A steady state condition is assumed.

Forest region	No. of sites	Organic matter	Mean turnover time (years)				
			N	K	Ca	Mg	P
Boreal coniferous	3	353	230	94	149	455	324
Boreal deciduous	1	26	27	10	14	14	15
Temperate coniferous	13	17	18	2	6	13	15
Temperate deciduous	14	4	6	1	3	3	6
Mediterranean	1	3	4	0.2	4	2	1
Average—All stands		12	34	13	22	61	46

Source: From Cole and Rapp (1981).

Gene E. Likens (1935–) Director, Institute of Ecosystem Studies, Milbrook, New York

support similar forest communities, and these watersheds are good experimental units for study and manipulation (http://www.hubbardbrook.org/).

Nutrients enter the Hubbard Brook forest ecosystem in precipitation, and the precipitation input was measured in rain gauges scattered over the study area. Nutrients leave the ecosystem primarily in stream run-off, and this loss was estimated by measuring stream flows (Figure 16.6b). For most dissolved nutrients, the stream water leaving the system contains more nutrients than the rainwater entering the system. About 60% of the water that enters as precipitation turns up as stream flow—most of the remaining 40% is transpired by plants or evaporated. The chemical composition of the precipitation and the stream discharges changed very little from year to year.

Annual nutrient budgets for watersheds in the Hubbard Brook system can be estimated from the difference between precipitation input and stream outflow.

Eight ions show net losses from the ecosystem: calcium, magnesium, potassium, sodium, aluminum, sulfate, silica and bicarbonate. Three ions showed an average net gain: nitrate, ammonium and chloride. If we assume that these nutrient budgets should be in equilibrium in this ecosystem, the net losses must be made up by chemical decomposition of the bedrock and soil.

With this background, the ecologists studied the effect of logging on the nutrient budget of a small watershed at Hubbard Brook. One 15.6-hectare watershed was logged in 1966, and the logs and branches were left on the ground so that nothing was removed from the area (Figure 16.6e). Great care was taken to prevent disturbance of the soil surface to minimize erosion. For the first three years after logging the area was treated with herbicide to prevent any regrowth of vegetation. This deforested watershed was then compared with an adjacent intact watershed.

Run-off in the small streams increased immediately after logging, and annual run-off in the deforested watershed averaged 32% more than the control watershed in the 3 years after treatment. Detritus and debris in the stream outflow increased greatly after deforestation, particularly 2 to 3 years after logging. Correlated with this increase was a large increase in stream water concentrations of all major ions in the deforested watershed. Nitrate concentrations in particular increased 40- to 60-fold over the control values (Figure 16.7). For 2 years the nitrate concentration in the stream water of the deforested site exceeded the health levels recommended for drinking water. Average stream water concentrations increased 417% for calcium, 408% for magnesium, 1558% for potassium, and 177% for sodium in the 2 years after deforestation. A massive loss of nutrients resulted from this deforestation—a result no one had expected.

The net result of deforestation in the Hubbard Brook forest is that the ecosystem is simultaneously irrigated and fertilized, so that for a short time after normal logging occurs, primary production could be stimulated. An array of species has evolved to exploit these transient nutrient-rich situations following a disturbance by fire or logging. These transients help to prevent further nutrient losses and to restore some of the nutrient capital lost by logging or fire.

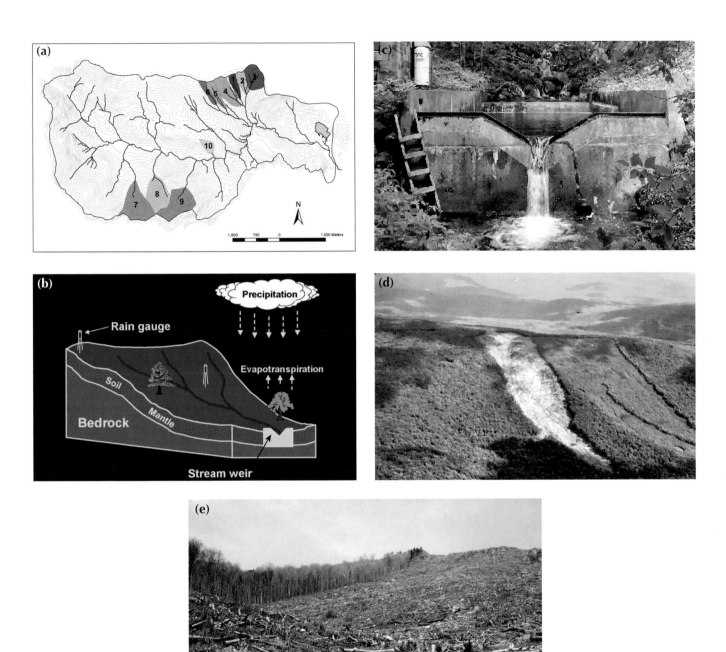

Figure 16.6 The Hubbard Brook Ecosystem Study. (a) Map of the watersheds in the Hubbard Brook Forest in New Hampshire. Small watersheds are numbered. Contour lines are at 15 m intervals. (b) The small watershed concept was the key to setting up experiments in the Hubbard Brook Forest. Because the bedrock is impervious, nutrients going out could be measured in streams leaving each watershed, providing the data shown in Figure 16.7. (c) One of the weirs measuring water flow and water chemistry at Hubbard Brook. (d) Aerial view of watershed 5 at Hubbard Brook, cleared of vegetation in an experiment. (e) Ground view of watershed 2 in 1970, 4 years after experimental logging. Nutrient cycling was studied in watersheds that were logged and adjacent watersheds that were left undisturbed in this northern hardwood forest. (Photos courtesy of Hubbard Brook Ecosystem Study.)

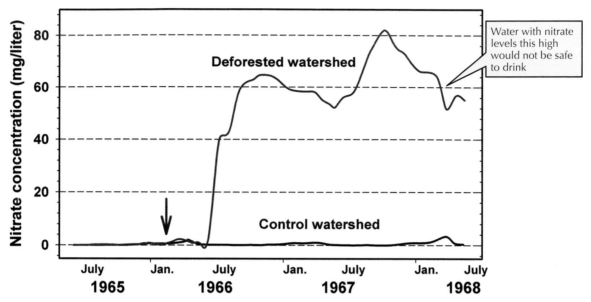

Figure 16.7 Stream water concentrations of nitrate in two watersheds at the Hubbard Brook Experimental Forest, New Hampshire. The blue arrow marks the completion of the cutting of the trees on the deforested watershed 2 (see Figure 16.6e). The control watershed was not disturbed. Nitrate loss was greatly accelerated after logging. Once vegetation grew back, nitrate loss returned to control levels. (Modified after Likens *et al.* 1970.)

The Hubbard Brook experiment has stimulated research on the sustainability of forestry practices throughout the world. Given the nutrient content of a forested ecosystem, two key processes are driving changes in nutrient cycles. First, removal of wood involves a loss of nutrients to the local ecosystem. This loss of nutrients can be minimized if tree branches and bark are left on site, rather than being discarded or burnt at a distant timber mill. Second, nutrients are being added to all ecosystems from the atmosphere in precipitation and dryfall and by the weathering of rocks (Figure 16.3). If forestry is to be sustainable, the losses due to harvesting must be made up by the gains from the atmosphere and rock weathering. At present it is not clear if forestry is sustainable in the long term. Given the current levels of nitrogen deposition, the forest at Hubbard Brook will not recover from its nitrogen losses in less than 200 years. In tropical rainforests in Malaysia, harvesting of trees removes up to 20% of the calcium from the whole ecosystem. This amount of calcium cannot be replaced within 200 years by weathering in soils that are on rocks that contain little calcium, and fertilization is needed to sustain productivity in these tropical ecosystems. If harvesting occurs more frequently than nutrient replacement time, the site would deteriorate unless fertilizers are used.

Q Is any long-term harm possible by using fertilizers to replace nutrients lost to ecosystems by harvesting?

There is considerable controversy over whether or not harvesting of trees induces a long-term decline in forest productivity. Unless nutrient input equals output in any ecosystem, productivity will decline. There is little evidence so far that this is happening, but most experiments are short term, and the key is the long-term response of the ecosystem over a time scale of 500 years. Many management practices in forestry, such as slash removal or burning, can help or hinder nutrient cycling, and at present we can tentatively conclude that there is as yet no evidence of long-term productivity declines in temperate forests utilized for tree harvest. Commercial forestry in the tropics is only a recent form of land use, and there are few data available to judge sustainability for tropical forest harvesting.

The work on nutrient cycling in forests has shown the need for guidelines to specify sound management procedure in forestry. For example, bark is relatively rich in nutrients, and so lumbering operations ought to be designed to strip the bark from the trees at the field site, not at some distant processing plant. The conservation of nutrients in forest ecosystems can be done intelligently only when we understand how nutrient cycles operate in these systems.

■ 16.3 NUTRIENT CYCLING DIFFERS IN TROPICAL AND TEMPERATE FORESTS

Large areas of the Northern Hemisphere have been previously glaciated, and the soils—derived from till in which the bedrock has been pulverized—are very fertile, with a high availability of nutrients. Areas of volcanic activity can also have rich soils. But, in much of the world, soils are very old, highly weathered and basically infertile. For example, the continents derived from Gondwana—Australia, South America and India—have large areas covered with very old, poor soils. The vegetation supported on these soils has adapted remarkably well by using nutrients efficiently—by recycling within the plant and by leaf

fall and reabsorption. One consequence is that many tropical forests store nutrients in vegetation and not in the soil. Most nutrients in temperate forests are generally in the soil.

We might expect individual plants growing in nutrient-poor soils to contain fewer nutrients than plants growing in fertile soils. In fact, the opposite is true. Plants from infertile habitats consistently have higher nutrient concentrations than plants from fertile habitats when grown under the same controlled greenhouse conditions. Plants from nutrient-poor habitats may achieve this nutrient-rich status by being more efficient than plants from nutrient-rich habitats.

Nutrient Use Efficiency

Nutrient use efficiency is a measure of how much biomass a plant can produce from a given amount of a specific nutrient. Productivity in tropical forests illustrates the law of diminishing returns—as nitrogen becomes abundantly available, productivity reaches a plateau at which some other resource becomes limiting (Figure 16.8). Similar relationships occur for phosphorus in tropical forests. Forest productivity is limited either by nitrogen or phosphorus levels in a variety of

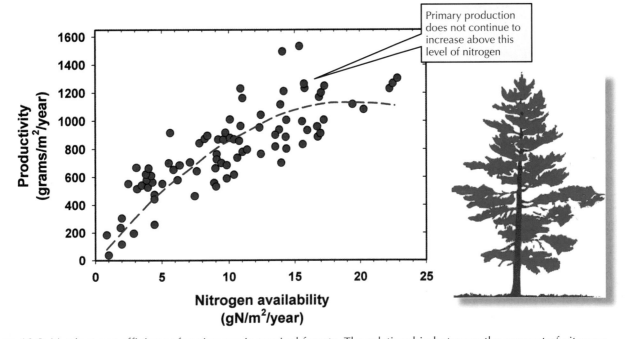

Figure 16.8 Nutrient use efficiency for nitrogen in tropical forests. The relationship between the amount of nitrogen available and the productivity of the site follows a curve of diminishing returns. Adding more and more nitrogen becomes less and less effective in increasing forest productivity. Nitrogen is typically a limiting factor to tree growth in these forests. Tropical forests growing on low nutrient sites thus use nitrogen more efficiently. (Data from Vitousek 1984.)

tropical and temperate forests, and the pattern of nutrient use efficiency is similar in all forests studied.

One consequence of this is that forest productivity may be high on soils with low nutrient levels. A classic example is the tropical rainforest of the Amazon Basin, which represents one type of nutrient cycling pattern.

Oligotrophic and Eutrophic Ecosystems

We can recognize two extreme types of nutrient cycling in forest ecosystems—a similar pattern to freshwater lakes. The **oligotrophic** pattern of nutrient cycling occurs on nutrient-poor soils, such as the Amazon Basin, and the **eutrophic** pattern occurs on nutrient-rich soils. In the temperate zone, where most forest research has been done, forests are usually of the eutrophic type that grows on rich soils. The tropics have a much higher proportion of oligotrophic ecosystems on poor soils (*Essay 16.2*). But exceptions occur—not all tropical forests are oligotrophic, nor are all temperate forests eutrophic. There are some striking differences between these forest types. Oligotrophic systems have a large biomass in the humus layer of the soil, and this layer of fine roots and humus is critical for nutrient cycling and nutrient conservation in these systems.

Productivity and nutrient cycling do not differ greatly in oligotrophic and eutrophic forests, as long as these ecosystems are not disturbed. But when the forest is cleared for agriculture, as is happening now in the Amazon, nutrient-poor systems quickly lose their productive potential, while the nutrient-rich ones do not. Once the humus and root layer on top of the mineral soil is disturbed in oligotrophic systems, the mechanism of efficient nutrient recycling is lost, and nutrients are leached out of the system in rainwater. Oligotrophic ecosystems cannot be used for crop production unless critical nutrients are supplied in fertilizers.

■ 16.4 THE SULFUR CYCLE IS DRIVEN BY HUMAN ACTIVITY AND PRODUCES ACID RAIN

Through the combustion of fossil fuels, human activity has altered the sulfur cycle more than any other nutrient cycle. While human-produced emissions of carbon dioxide and nitrogen are only about 5 to 10% of the level of natural emissions, we produce about 160% of the level of natural emissions of sulfur. One clear manifestation of this alteration of the sulfur cycle is the widespread problem of **acid rain** in Europe and North America (Figure 16.9). Acid precipitation is defined as rain or snow that has a pH of less than 5.6. Low pH values are caused by strong acids (sulfuric acid and nitric acid) that originate as combustion products from fossil fuels.

Acid rain emerged as a major environmental problem in the 1960s when damage to forests and lakes in Europe and eastern North America began to be noticed on a wide scale. It was one of the first wide-scale environmental problems because oxides of sulfur and nitrogen could be carried hundreds of kilometers and then deposited in rain and snow. Lakes in eastern Canada were dying because of air pollution from the mid-western states. Lakes in southern Norway were losing fish because of acid rain from England. By 1980 over large areas of Western Europe and eastern North America, annual pH values of precipitation averaged between 4.0 and 4.5, with individual storms producing acid rain of pH 2 to 3.

Sources of Sulfur Emissions

Sulfur released into the atmosphere is quickly oxidized to sulfate (SO_4) and redeposited rapidly on land or in the oceans. Figure 16.10 illustrates the sources and sinks of the global sulfur cycle. Short-term events, such as volcanic eruptions, contribute to the global sulfur cycle and make it difficult to estimate the equilibrium state of the atmosphere. Human emissions are the largest component of additional sulfur to the atmosphere. Ore smelters and electricity-generating plants have increased emissions during the past 100 years. To offset local pollution problems, smelters and generating plants have built taller stacks, which reduce pollution at ground level. Tall stacks (over 300 m) now are the standard, and they have exported the pollution problem downwind. Ice cores from Greenland show large increases in SO_4 deposition from the atmosphere in the last 50 years.

The destructive impact of sulfate pollution on vegetation has been known for over a century. Queenstown,

ESSAY 16.2   HOW DOES PHOSPHORUS GET TO HAWAII?

Soils are formed from rock weathering and, because phosphorus does not occur as a gas, once a rock is laid down, it contains all the phosphorus that the subsequent soil will ever have. As rock weathers, some phosphorus is lost to insoluble forms, and so soils should continually lost this critical element needed for plant growth, unless there is some outside source of input. On isolated oceanic islands, such as the Hawaiian Islands, we would expect older soils to have less and less phosphorus. Since this chain of islands was formed over a time span of 5 million years, it presents ecosystem ecologists with a near-perfect laboratory to analyze nutrient cycling and nutrient limitation (Vitousek 2004).

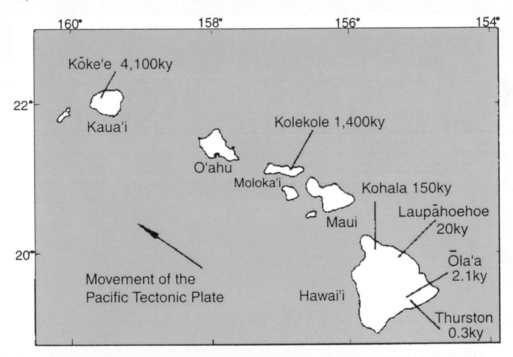

Map of the Hawaiian Islands. Because these are volcanic islands at the edge of the Pacific Plate, the age of the soils on the different islands varies from 300 years on new volcanic soils in the south-east to over 4,000,000 years in the north-west.

The guiding model for plant growth is that nitrogen should be the limiting major nutrient on newly-formed soils (which have relatively large amounts of phosphorus), but old soils should show phosphorus limitation (because as time passes nitrogen is fixed by organisms).

Vitousek (2004) used a transect across the Hawaiian Islands to test this model. By carrying out fertilizer trials with a tree that was common on all the islands, Vitousek (2004) showed that plant growth on very young soils was limited primarily by nitrogen and secondarily by phosphorus (because rock weathering was slow), but on older soils, nitrogen limitation disappeared and phosphorus was limiting, just as the model had predicted.

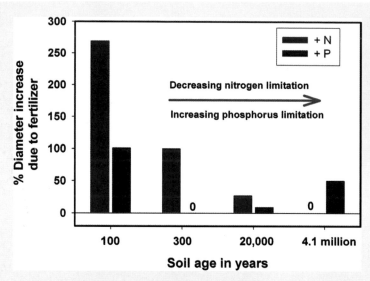

Nitrogen and phosphorus limitation across the Hawaiian Islands

But rock weathering turned out not to be the only source of inputs of phosphorus to the Hawaiian system. Dust from Asia has accumulated in ocean sediments and could be measured in ocean cores to show the long-term pattern of dust deposition over the last million years in the Pacific Ocean:

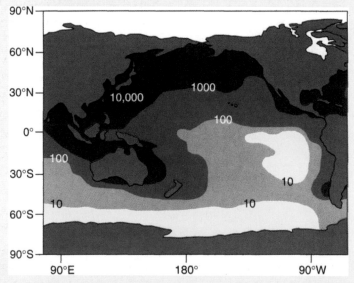

Inputs of dust from Asia to the Pacific Ocean over the last million years. The units of dust deposition are mg per square meter per year. A relatively low amount of dust reaches the Hawaiian Islands (Nakai et al. 1993).

In Hawaiian sites that are more than 150,000 years old, dust from Asia contributes 80% or more of the phosphorus that is available in these old soils (Chadwick *et al.* 1999). Nevertheless, the rate of phosphorus input in dust from Asia is relatively small, so that plants on these older soils are still phosphorus limited.

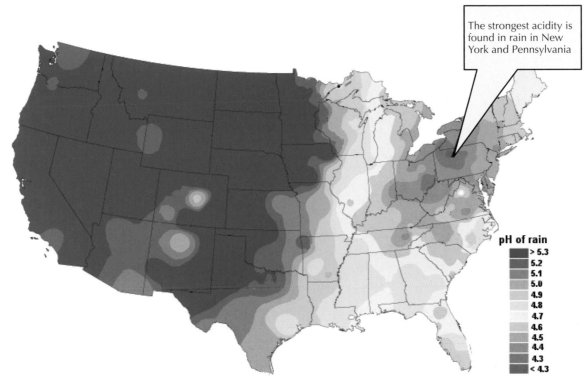

Figure 16.9 Distribution of acid precipitation in the United States in 2003—the lower the pH, the more acid the precipitation. Acid rain is a serious pollution problem in the eastern states. (Data from the National Atmospheric Deposition Program at the University of Illinois, 2005.)

Tasmania offers a striking example (Figure 16.11). From 1896 to 1922 sulfur-rich smoke poured from the copper smelter of the Mount Lyell Mining and Railway Company that created Queenstown. A typical ton of ore from the mine consisted of 48% sulfur, 40% iron, and less than 3% copper. More than half of the sulfur was lost in the emissions from the stacks, and the fumes from the smelter were toxic enough to kill plants on bad days. The prevailing winds carried the pollution to the east and south-east. Once the vegetation was killed, soil erosion from heavy rains, along with fire, removed all the topsoil, and the landscape now resembles the moon, 80 years after the smelter closed (Figure 16.11). Acid rain is not a new phenomenon.

There is a net transport of SO_4 from the land to the oceans. The ocean is also a large source of aerosols that contain SO_4. Dimethylsulfide [$(CH_3)_2S$] is the major gas emitted by phytoplankton in the sea, and it is oxidized quickly to SO_4 and then redeposited in the ocean. Sulfate is abundant in ocean waters (12×10^{20} g of elemental S), and the mean residence time for a sulfur molecule in the sea is over 3 million years.

The United States and most developing countries have reduced sulfur dioxide emissions during the past 30 years. In the United States sulfur dioxide emissions were reduced 50% between 1980 and 1999. Reduced emissions should reduce surface deposition of acid rain. However, the effects of acid rain on the environment do not go away immediately once sulfur dioxide emissions fall, and the key question remains: 'Will forest and aquatic ecosystems recover from the effects of acid rain, and at what rate?' At Hubbard Brook, the impact of acid rain has been to leach calcium from the soil to the extent that available calcium appears to limit forest growth, rather than nitrogen. Stream water chemistry at Hubbard Brook is slowly recovering from acid rain, and at least another 10–20 years will be needed for full recovery, even if sulfur dioxide emissions continue to decrease.

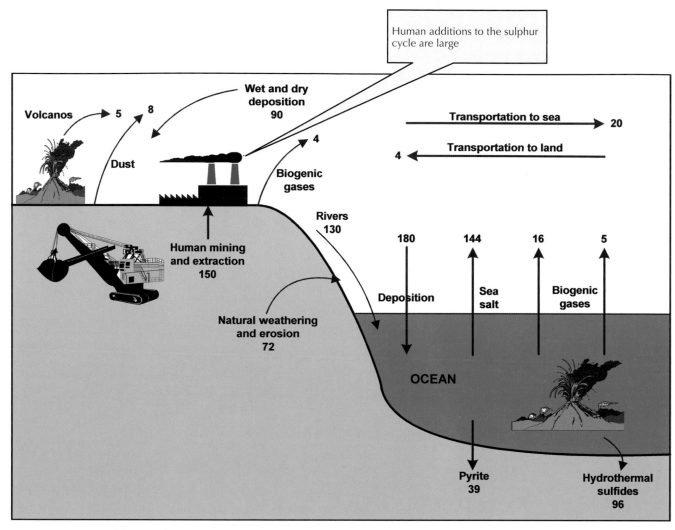

Figure 16.10 The global sulfur cycle. Burning of fossil fuels is the major component of atmospheric input of sulfur. Humans have affected the sulfur cycle more than any other nutrient cycle. All values are 10^{12} g S/yr. (Modified from Schlesinger 1997.)

Impact on Freshwater Ecosystems

Freshwater ecosystems are particularly sensitive to acid rain. In areas underlain by granite and granitoid rocks, which are highly resistant to weathering, the acid rain is not neutralized in the soil, so lakes and streams quickly become acidified. Lakes in these bedrock areas typically contain soft water. Thus knowing the bedrock geology can be an initial guide to locating sensitive areas (*Essay 16.3*). Areas such as the Precambrian Fennoscandian Shield in Scandinavia, the Canadian Shield, New England and the Rocky Mountains are thus potential trouble spots for acid rain.

The clearest effects of acid precipitation have been on fish populations in Scandinavia and eastern Canada. Fish populations were reduced or eliminated in many thousands of lakes in southern Norway and Sweden once the pH in these waters fell below pH 5. In Canada, lakes containing lake trout have been the principal focus of research on the impacts of acid rain. Lake trout disappear in lakes once the pH falls below 5.4. The cause is reproductive failure—newly-hatched trout die. Lake trout are a keystone predator in many Canadian lakes, and they disappear slowly in lakes of low pH. Adult trout do not seem to be affected by low

Figure 16.11 Effects of acid rain from a copper smelter on native vegetation at Queenstown, Tasmania. The copper smelter operated from 1896 to 1922. Acid rain from the sulfate emissions of the smelter killed all the vegetation and left bare soil to erode away. Photos taken October 1992. After 60 years there is little sign of any vegetation recovery. (Photos courtesy of A.J. Kenney.)

ESSAY 16.3   ACID RAIN AND THE SUDBURY EXPERIENCE

The restoration of ecosystems around the nickel and copper smelter at Sudbury, Ontario are an encouraging sign that ecosystems can restore themselves after damage by air pollution. From 1900 to 1970 the Sudbury smelters spewed over 100 million tons of sulfur dioxide into the atmosphere, as well as thousands of tons of toxic trace metals such as lead and cadmium. At its peak, Sudbury alone accounted for 4% of the global total emissions—an amount equal to the current emissions of the entire United Kingdom (Schindler 1997). Within 30 km of the smelter, vegetation was destroyed and thousands of lakes were acidified.

A giant smokestack in Sudbury, Ontario

One solution to the problem of emissions has been to build taller smokestacks, and in Sudbury the mining company Inco constructed a 380 m stack in 1972—the tallest smokestack in the western hemisphere. This spread the air pollution over a much larger area of Ontario downwind of the Sudbury smelter.

One way to combat acidification of lakes is to add lime (calcium carbonate) to them to raise the pH. In Europe, the Swedes have been particularly active in liming lakes affected by acid rain. But no one would pay for the enormous cost to lime the lakes around Sudbury, which form a natural experiment in acidification. Since 1980 Inco has reduced emissions so that the Sudbury smelters now emit less than 10% of their original air pollution. Sources of sulfur dioxide in eastern Canada have been reduced by more than half since that time. During the last 25 years, the ecosystem around Sudbury has been recovering via natural processes. Trees and shrubs have recolonized the surrounding areas, and the pH in most lakes has risen. Lake trout have colonized many lakes, but not all of them. The original concept that acidified lakes could never recover, except in geological time, has been shown to be false, and there is room for some optimism. But if acid rain continues to fall, even in reduced amounts, lakes will not be able to fully recover. The longer we wait to implement strict air pollution controls, the more damage will accumulate. Acid rain leaches cations, such as calcium, out of the soil and thus reduces the soil's natural ability to absorb acidity.

The important messages that Sudbury gives are: firstly, that ecosystems can recover from disturbances, although it may take longer than a few years; and, secondly, that we should not assume we can achieve a technological fix for ecological damage. Adding lime to acidified lakes solves some problems, but creates another whole set in its wake (Steinberg and Wright 1994).

pH, nor is there any food shortage at low pH. However, the impact on the small juveniles causes a slow decline in the trout population over 10–20 years. Once lake trout are gone, acid-tolerant fish, such yellow perch and cisco, become more abundant, and the food web shifts dramatically (Figure 16.12).

Q Could these changes in the fish community of lakes affected by acid rain be an example of multiple stable states?

The changes that humans have made to the sulfur cycle have the potential to change nutrient cycling in natural ecosystems in a great variety of ways we cannot yet understand, much less predict. We cannot continue this aerial bombardment of ecosystems in the naive belief that nutrient cycles have infinite resilience to human inputs. Recent efforts to curtail sulfate emissions from fossil fuels have reduced the emissions of SO_4, and we must continue to press for further reductions. Once acidic precipitation is reduced, both forest and lake ecosystems can begin to recover from the damage inflicted.

■ 16.5 THE NITROGEN CYCLE IS AFFECTED BY FERTILIZERS USED IN AGRICULTURE

The availability of nitrogen is often limiting for both plants and animals, and net primary production is often limited both on land and in the oceans by the amount of nitrogen available. Nitrogen is abundant in air (78% nitrogen), but few organisms can use N_2 directly. A small number of bacteria and algae can take nitrogen from the air and fix it as nitrate or ammonia. Many of these organisms work symbiotically in the root nodules of legumes to fix nitrogen, and this is a major source of natural nitrogen fixation. Human additions to the global nitrogen cycle have become substantial, particularly with the use of nitrogen fertilizers for agriculture.

Nitrogen Emissions

Figure 16.13 shows the global nitrogen cycle. Human activities add about the same amount of nitrogen to the biosphere each year as do natural processes, but this human addition is not spread evenly over the globe. The impact of human additions of nitrogen has shown up particularly as changes in the composition of the atmosphere. Nitrogen-based trace gases—nitrous oxide, nitric oxide and ammonia—have major ecosystem impacts. Nitrous oxide is relatively inert chemically and long-lived in the atmosphere. It traps heat and thus acts as a greenhouse gas to change climate. Nitrous oxide is increasing in the atmosphere at 0.25% per year. Nitric oxide by contrast is highly reactive and contributes significantly to acid rain as well as smog. Nitric oxide can be converted to nitric acid in the atmosphere, and in western United States acid rain is caused more by nitric acid than by sulfuric acid. In the presence of sunlight, nitric oxide and oxygen react with hydrocarbons from auto exhaust to form ozone—the most dangerous component of smog in cities and industrial areas. Nitric oxide is produced by burning fossil fuels and wood. The third nitrogen-based trace gas in the atmosphere is ammonia. Ammonia neutralizes acids and thus acts to reduce acid rain. Most ammonia is released from organic fertilizers and domestic animal wastes. Domestic feedlots for the fattening of cattle are a major source of ammonia.

Human Additions of Nitrogen to Ecosystems

The result of human activities on the nitrogen cycle has been an increased deposition of nitrogen on land and in the oceans. Because nitrogen additions are typically coupled with phosphorus additions, the result is eutrophication of freshwater lakes and rivers and coastal marine areas. Phosphorus additions to freshwater typically increase primary production, while nitrogen addition to estuaries increases primary production in marine environments. The adverse effects of eutrophication on aquatic systems are listed in Table 16.3. The North Atlantic Ocean Basin receives nitrogen from many rivers that transport excess nitrogen into the ocean. Nitrogen input to the North Atlantic has increased 2 to 20-fold since 1750, and inputs from northern Europe are the highest. Levels of nitrates in rivers are rising throughout the Northern Hemisphere in proportion to the growing human population along the rivers. In the Mississippi River, the levels of nitrates have more than doubled since

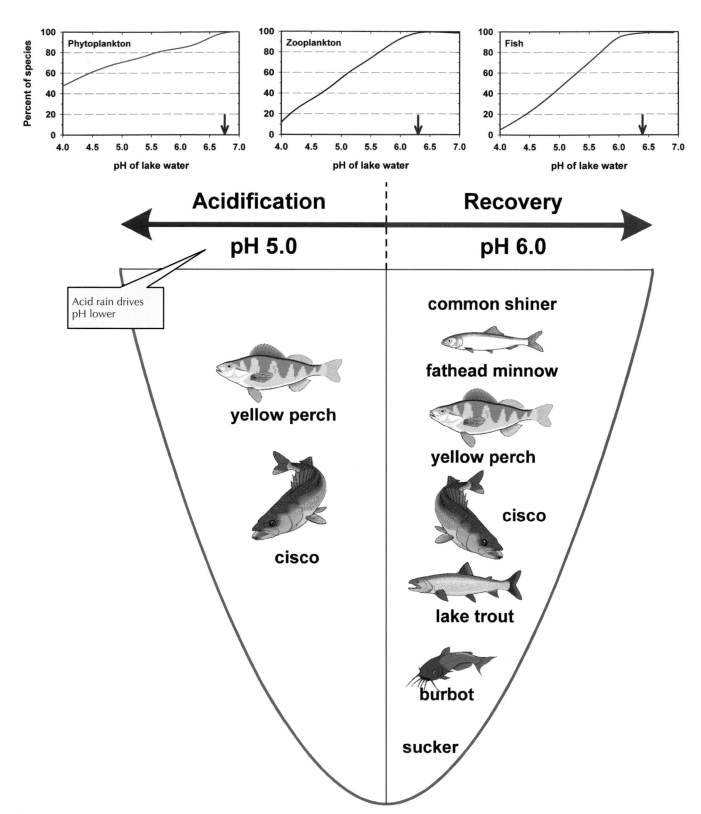

Figure 16.12 Schematic illustration of the impacts of acidification on eastern Canadian lakes dominated by lake trout. Once the input of acid rain is curtailed, these lakes recover slowly, but the full reversibility of the degradation has not yet been seen. Red arrows on graphs indicate the pH at which species loss begins. (Modified after Gunn and Mills 1998.)

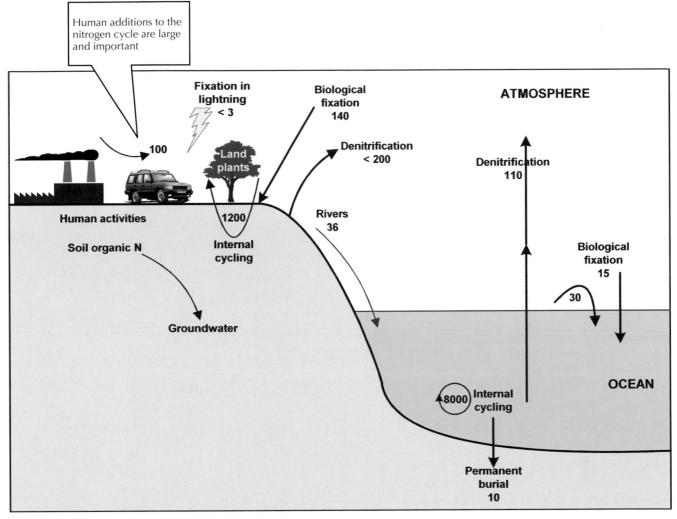

Figure 16.13 The global nitrogen cycle. Humans are having an impact on the nitrogen cycle by burning fossil fuels and wood. Nitrogen is abundant in the air, but cannot be used directly by plants in its gaseous form. Because nitrogen is often a limiting factor on primary production both on land and in the oceans, internal cycling mechanisms retain much of the nitrogen in the biosphere. All fluxes are in units of 10^{15} g N/year. (Modified from Schlesinger 1997.)

1965. Groundwater concentrations of nitrates are also increasing in agricultural areas—in some areas approaching the maximum safe level of nitrate in drinking water (10 mg per liter). The results of nitrogen additions to aquatic systems are nearly all negative—reducing water quality.

By contrast, the addition of nitrogen to terrestrial ecosystems can have positive effects. Nitrogen deposition on land can relieve the nitrogen limitation of primary production that is common in many terrestrial ecosystems. Swedish forests are all nitrogen limited and have averaged 30% greater growth rates in the 1990s compared with the 1950s. The important concept here is that of the **critical load**—the amount of nitrogen that can be added to the ecosystem and absorbed by the plants without damaging ecosystem integrity. When the vegetation can no longer respond to further additions of nitrogen (see Figure 16.8), the ecosystem reaches a state of nitrogen saturation, and all new nitrogen moves into groundwater or stream flow or back into the atmosphere. Nitrate is highly water soluble in soils, and excess nitrate carries away with it positively charged ions of calcium, magnesium and potassium. Excess nitrate can thus result in calcium,

Table 16.3 Adverse effects of nutrient additions of nitrogen and phosphorus on freshwater and coastal marine ecosystems.

Increased biomass of phytoplankton
Shifts in phytoplankton communities to bloom-forming species that may be toxic
Increase in blooms of gelatinous zooplankton in marine ecosystems
Increased biomass of benthic algae
Changes in macrophyte species composition and biomass
Death of coral reefs
Decreases in water transparency
Taste, odor and water treatment problems for domestic water supplies
Oxygen depletion
Increased frequency of fish kills
Loss of desirable fish species
Reductions in harvestable fish and shellfish
Decrease in aesthetic value of water body

Source: From Carpenter *et al.* (1998).

magnesium or potassium deficiency, which limits plant growth—this is why most commercial garden fertilizers contain more than just nitrogen.

Increasing nitrogen in terrestrial ecosystems can have undesirable impacts on biodiversity. In most cases adding nitrogen to a plant community reduces the biodiversity of the community. Figure 16.14 illustrates the impact of experimentally adding nitrogen for 12 years to grasslands in Minnesota. Species that are nitrogen responsive, often grasses, can take over plant communities enriched in nitrogen. The Netherlands has the highest rates of nitrogen deposition in the world—largely due to intensive livestock operations—and a consequence of this is a conversion of species-rich heathland to species-poor grasslands and forest. The mix of plant and animal species adapted to sandy, infertile soils is being lost because of nitrogen enrichment.

The nitrogen cycle, like the sulfur cycle, has been heavily impacted by human activities during the last 50 years. It is urgent that national and international efforts be directed to reversing these changes and moderating the adverse impacts on ecosystems. The most obvious direct impact on humans that flows from these changes in nutrient cycling is global climate change, which we will discuss in Chapter 21.

SUMMARY

Nutrients cycle and recycle in ecosystems, and tracking nutrient cycles is an important way of studying fundamental ecosystem processes. Human activities are changing cycles on a global scale, with consequences for biodiversity, ecosystem function and climate change. Nutrients reside in compartments and are transferred between compartments by physical or biological processes. Compartments can be defined in any operational way to include one or more species or physical spaces in the ecosystem. Nutrient cycles may be local or global. Global cycles, such as the nitrogen cycle, include a gaseous phase involving transport in the atmosphere. Less mobile elements, such as phosphorus, tend to have more local cycles.

Nutrient cycles in forests have been studied to determine nutrient losses associated with logging. The input

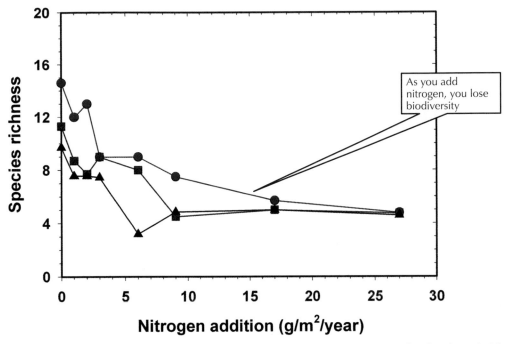

Figure 16.14 Vegetation responses to 12 years of nitrogen fertilization in Minnesota grasslands. Three fields were used, and six replicates were used for each level of nitrogen addition. Biodiversity declined dramatically as more nitrogen was added to these grasslands. A few competitive dominants took over the grassland when nitrogen was not limiting. (Data from Wedin and Tilman 1996.)

of nutrients must equal the outflow for any ecosystem, or it will deteriorate over the long term. Logging can result in high nutrient losses even if soil erosion is absent. An undisturbed forest site recycles nutrients efficiently. Nutrient use efficiency is important to ecosystem functioning—plants in poorer soils are more efficient in their nutrient use, so that more grams of plant tissue are produced per unit of nutrient.

The sulfur cycle is an example of a global nutrient cycle that is strongly affected by human activities. Burning of fossil fuels adds a large amount of SO_2 to the atmosphere, resulting in acid rain. Acid rain, in combination with other airborne pollutants, has caused forest declines in Europe and has eliminated fish populations from many lakes in Eastern Canada and Scandinavia. Controls on SO_2 release in North America and Europe have caused emissions to decline during the last 30 years, and ecosystems can recover once the inputs are stopped.

The nitrogen cycle is critical because primary production in many terrestrial ecosystems and in coastal waters of the oceans is limited by nitrogen. Nitrogen emissions by human activity have doubled the input of nitrogen to the air and waters of the globe. Smog in cities and acid rain are two impacts of this added nitrogen. Nitrogen and phosphorus leach from agricultural fertilizer and are a major cause of algal blooms in lakes and rivers. Unless we can curb these emissions of critical nutrients, global ecosystems will continue to be degraded.

SUGGESTED FUTHER READINGS

Aber JD, Goodale CL, Ollinger SV, Smith M-L, Magill AH, Martin ME, Hallett RA and Stoddard JL (2003). Is nitrogen deposition altering the nitrogen status of north-eastern forests? *BioScience* **53**, 375–389. (How nitrogen limitation of terrestrial primary production is being affected by air pollution.)

Helfield JM and Naiman RJ (2001). Effects of salmon-derived nitrogen on riparian forest growth and implications for stream productivity. *Ecology* **82**, 2403–2409. (An extraordinary tale of how salmon bring nitrogen from the ocean into freshwater lakes and streams when they spawn and die.)

Krupa SV (2003). Effects of atmospheric ammonia (NH_3) on terrestrial vegetation: a review. *Environmental Pollution* **124**, 179–221. (How ammonia damages leaves and interacts with other nitrogen pollutants to damage vegetation.)

Newman EI (1997). Phosphorus balance of contrasting farming systems, past and present. Can food production be sustainable? *Journal of Applied Ecology* **34**, 1334–1347. (The most important question of the 21st century, evaluated from farming systems in Britain, Egypt, China, and the USA, none of which are sustainable.)

Rabalais NN, Turner RE and Scavia D (2002). Beyond science into policy: Gulf of Mexico hypoxia and the Mississippi River. *BioScience* **52**, 129–142. (How can nitrogen pollution in this large river system be brought under control?)

Schindler DW (1997). Liming to restore acidified lakes and streams: a typical approach to restoring damaged ecosystems? *Restoration Ecology* **5**, 1–6. (Why one technological fix to an air pollution problem is not an adequate approach.)

Schlesinger WH (1997). *Biogeochemistry: An Analysis of Global Change.* 2nd edn. Academic Press, Inc., San Diego. (The best book on the details of nutrient cycling in ecosystems.)

QUESTIONS

1. Slash-and-burn agriculture is common in many tropical countries. Forests are cut and burned, and crops are planted in the cleared areas. Yields are usually good in the first year, but decrease quickly thereafter. Why should this be? Compare your ideas with those of Tiessen *et al.* (1994), and evaluate the sustainability of slash-and-burn agriculture.

2. Discuss the relative merits of making a compartment model (Figure 16.2) of a nutrient cycle very coarse (with only a few compartments) versus making it very fine (with many compartments).

3. A key question in restoration ecology is how long it will take for an ecosystem to recover from disturbance caused by humans. Discuss how we might find out what the timeframe is for ecological recovery from a disturbance such as acid rain.

4. Soils in Australia contain very low amounts of phosphorus—from a half to one-tenth the amount in North American soils (Keith 1997). Would you predict that eucalypts growing in Australia soils would be phosphorus limited instead of nitrogen limited? What adaptations might plants evolve to achieve high nutrient use efficiency when growing on soils of low nutrient content?

5. Coral reefs in the Great Barrier Reef off north-eastern Australia have been degrading during the last 10 years, with declining coral cover and increasing algal cover. This has been blamed on increasing sediment loads from rivers draining coastal sugar cane farms, with accompanying increases in nitrogen and phosphorus inputs to the reef. Discuss how you could evaluate this idea that river sediment pollution is the cause of coral reef decline. Szmant (2002) and Koop *et al.* (2001) discuss this issue.

6. Soils are formed by the breakdown and weathering of rocks. Compare the source of nutrients needed by plants in relation to the elemental composition of rocks. Discuss why all soils should not reach an irreversible state of nutrient depletion as they age, which would have dire consequences for plant growth. What external inputs of nutrients might occur in old soils? Chadwick *et al.* (1999) discuss this issue.

Chapter 17

LANDSCAPE ECOLOGY—INTERMINGLED ECOSYSTEMS

IN THE NEWS

Dryland salinity is a major problem in Colorado, North Dakota, Montana, Alberta, and southern Australia, as well as in the Middle East. Salt originates from the breakdown of minerals in rocks and from salt deposits in areas where the ocean once covered dry land. Near coastlines, salt is blown in from the sea. Areas with old, highly weathered soils, such as much of Australia, contain large amounts of salt. Most plants will not grow in soils that are highly saline—and no agricultural crops will grow in saline soils—so, as more areas become saline, agricultural production falls, deserts advance and biodiversity is reduced. About 400,000 ha of Colorado soils are saline-affected, as are 650,000 ha of Alberta soils. In Australia about 2 million hectares are already affected by salinity, and another 12 million hectares are at risk of succumbing during this century. Why has this problem of saline soils arisen, and what can be done about it?

The salt that threatens agricultural production dissolves in rainwater as it percolates through the soil, and accumulates in the water table. In areas with saline soils, the water table is contaminated with high levels of salt, and the key to salinity management is to prevent the saline water table from rising to the surface of the soil profile. Before agriculture began, native trees and shrubs in the landscape transpired large quantities of soil water, and kept the water table deep down in the soil profile. But native vegetation has been replaced over the last 200 years by crops such as wheat that have a shallow root system and use much less water than the original native vegetation. The result is that the water table rises, bringing the salt up toward the surface, as shown below.

As the water table rises closer to the surface, salt is washed into streams and rivers, causing salinity problems in freshwater ecosystems. Similar soil contamination with salt can occur in crops that are irrigated. The excess water used in irrigation can raise the water table, bringing salt to the surface (see http://www.nlwra.gov.au/).

There are three solutions to saline soil problems. The technological fix is to grow crops that are salt tolerant or, in the longer term, to select crop strains that are more and more salt tolerant. This is a temporary fix that can help, but it does not solve the underlying problem. The second solution is to

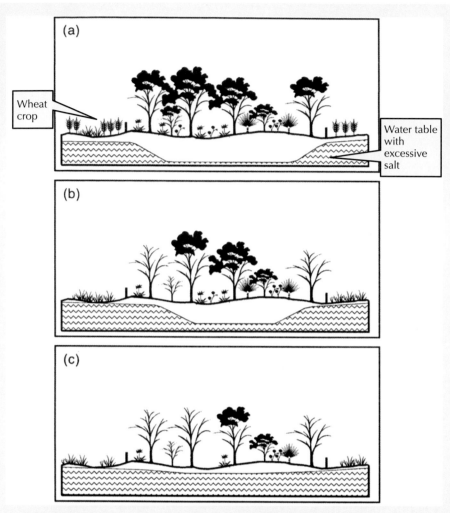

A shallow saline water table can kill native vegetation. (a) The water rises under crops that do not use as much water as native vegetation. (b) The rising water table kills native vegetation at the edge of the remnants, and the water table is now so high that crops cannot be grown. (c) As more native vegetation is killed, the water table rises almost to the ground surface so the whole area is degraded and few plants can survive (from Cramer and Hobbs 2002).

irrigate with more and more water so that the salts are dissolved out of the soil profile and move off into rivers and streams. This merely exports the salinity problem downstream. The third, ecological solution, to the problem of dryland salinity is to replant native vegetation to lower the water table. The ecological questions then become how much native vegetation needs to be replanted, how much of the landscape can be planted with crops and how much in native vegetation, and where on the landscape should the planting occur. In some areas, as much as 70% of the landscape may have to be replanted with native vegetation to lower the water table. Other ecological suggestions are to use farm forestry as an alternative land use, or to alternate agricultural crops with rows of trees or shrubs (Davidson 2002). These ecological solutions have the disadvantage that they will all produce land use that is less economically profitable to the farmers. In addition, the changes to salinity levels

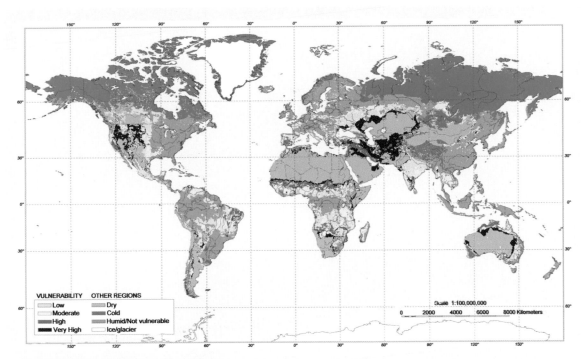

Areas of the world affected by salinity

are slow to operate and it may take decades before positive results become evident. But, without action at the landscape level, dryland salinity will continue to turn usable agricultural land of the world into deserts with long-term negative consequences.

Effects of salinity at the edge of remnant eucalypt woodland in western New South Wales. A rising water table has brought salt to the surface, killing trees and ground vegetation at the edge of the woodland. Few plant species can survive in the salty soil, so the water table remains close to the surface of the soil. (Photo courtesy of Sue Briggs.)

■ 17.1 LANDSCAPES INCLUDE SEVERAL ECOSYSTEMS

Landscape ecology is the ecology of regions—of large blocks of country that contain many different ecosystems. We have reached landscape ecology by moving from individuals to populations to communities to ecosystems and finally to landscapes, which are one step below the biosphere, or whole-Earth, ecosystem. Landscape ecology is thus an integrated discipline that brings to bear all the knowledge of ecological systems to address large-scale issues of land use and land planning. Landscapes are mosaics of habitat patches of varying size and shape. Patches may or may not be connected with similar patches, and how they are connected has important ecological effects.

Landscape ecology has two divergent orientations. One is based on scientific analysis that asks purely analytical questions about how the spatial arrangements of habitats affect the distribution and abundance of species, and how landscape patterns affect ecosystem processes. This first facet of landscape ecology is about asking how the ecological world works on a large scale. A second facet of landscape ecology is more practical—oriented to landscape planning and land use issues in regional development and conservation. In this chapter we will look at both these aspects of landscape ecology, but will emphasize the scientific questions that underlie the practical solutions to environmental problems (Figure 17.1). There are no ecological questions that are specific to landscape ecology, which attempts to view all the questions about distribution and abundance that we have considered in the previous 16 chapters within the spatial framework of a landscape.

There are three principal elements in landscape ecology. **Patches** of habitat are the first element that we need to analyze. A patch is a small area that is relatively homogeneous. All patches have **edges**, and the outer edge of a patch has a physical and biotic environment different from the center of a patch. Our second focus will be on the ecological impacts of edges. The third element is the **connectivity** of habitats, which brings in the concept of corridors and barriers to species movements. The combination of these three elements produces the **mosaic**, or how the patches and

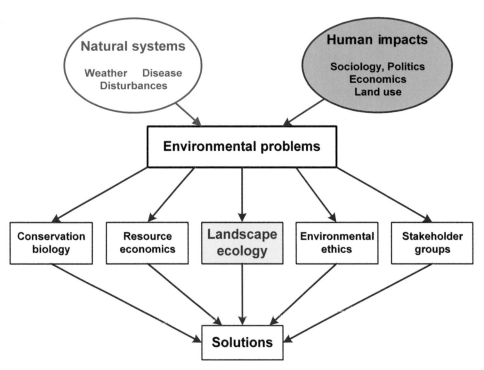

Figure 17.1 The role of landscape ecology in contributing to solutions of environmental problems. Environmental problems arise from the conflict between natural systems and human land use, and are affected by social, as well as ecological, constraints. Landscape ecology integrates ecological knowledge to help find solutions to these environmental problems, in concert with perspectives from many other approaches. (Modified from Wiens 1999.)

corridors of the landscape are arranged in space. Mosaics focus on the largest scale of landscapes and are the key to integrating all the information about patches, edges and corridors.

One of the difficulties in dealing with issues in landscape ecology is that we are unfamiliar with how attributes of landscapes can be measured. We begin our discussion with an illustration of a habitat mosaic and some simple measures of landscapes.

Measurements of Landscape Attributes

The earliest considerations of how to measure landscape attributes arose in the 1940s when large-scale aerial photography became available. Since the 1980s, satellite images have become easy to obtain and commonplace in landscape analysis. The principles of measurement are identical no matter what the scale of the landscape being analyzed. Figure 17.2 is a map of the forest landscape of Yellowstone National Park in western USA and illustrates a landscape that we can use as an example of what we might wish to measure.

Given a particular problem, the first decision a landscape ecologist must make is to define the scale of the analysis. For example, if you wish to determine the ecological and environmental factors that determine butterfly biodiversity, you must decide if the landscape should be studied at the scale of meters, kilometers or tens of kilometers. Snails use the landscape at the scale of meters, whereas eagles use the landscape at the scale of kilometers, so it is important to decide if you wish to study butterflies or eagles. There is no one 'correct' scale of landscape ecology, and the appropriate scale will depend on exactly the question or problem to which you wish to apply landscape analysis. In defining the scale of the analysis, a landscape ecologist must set boundaries to the landscape being analyzed.

The second decision a landscape ecologist must make is how to define a patch, and this will depend on the scale and the focus of the study. The scale of habitat maps must relate to the species or problem under study. **Fine-grained landscapes** are landscapes with many small patches, whereas **coarse-grained landscapes** have

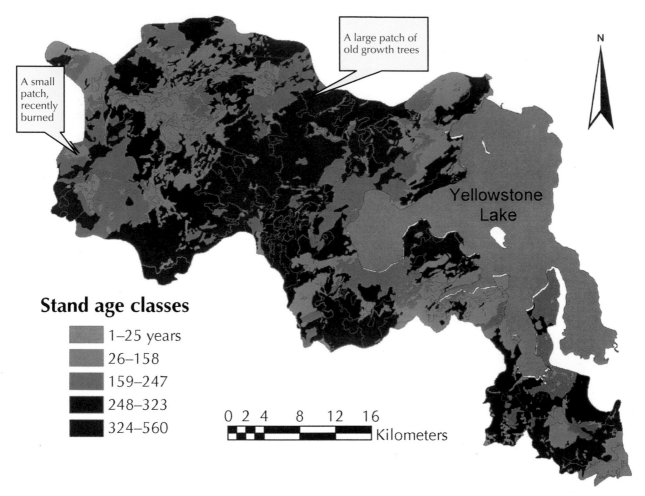

A large patch of old growth trees

A small patch, recently burned

N

Yellowstone Lake

Stand age classes

- 1–25 years
- 26–158
- 159–247
- 248–323
- 324–560

0 2 4   8   12   16
Kilometers

Figure 17.2 Lodgepole pine stands in the central part of Yellowstone National Park in 1985, before the extensive forest fires of 1988. Patches of different age vary greatly in size, edge and shape—a result of forest fires in the past four centuries. The whole landscape is a mosaic of these patches. This map covers 1290 km^2. (Modified from Tinker *et al.* 2003.)

only a few large patches. Different species view landscapes at different scales, and consequently there is no universal 'correct' grain size at which to study landscapes, so fine-grained and coarse-grained landscapes must be defined with respect to the species you are studying. If you are interested in lions and their prey in the Serengeti of East Africa, you would map habitat patches on a scale of tens or hundreds of meters, and these would be your patches. In a similar manner, if we were studying nutrient cycling in a region, we could map the water drainages in a region in relation to geology, and these large drainages would be our patches.

A landscape is typically composed of several types of patches, and the most common type of patch is usually defined as the **matrix**. The definition of the matrix depends on the purpose of the study. For example, mature coniferous forest could be the matrix in which patches disturbed by fire are embedded, or agricultural land could be the matrix in which patches of remnant forest are embedded.

What types of measurements can be taken on landscapes? Consider a hypothetical landscape composed of four kinds of habitats (Figure 17.3). Each patch of habitat can be described by a measure of the *area* of the patch, the *distance* around the edge, the longest *length* of the patch, and some measures of the *shape* of the patch. In addition the distance between patches could be measured, as well as the fraction of the landscape that is occupied by each particular kind of patch. There are many possible landscape measurements and many

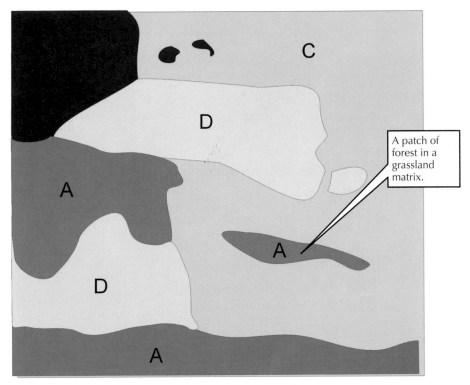

A patch of forest in a grassland matrix.

Figure 17.3 A simple landscape with four types of habitat—deciduous forest (A), conifer forest (B), grassland (C) and lakes (D). In this simple illustration we could measure patch area, distance along the patch edge, minimum distance between patches and the fraction of the total region (the square) occupied by each type of patch. Many other measures could be taken, but all should be relevant to the problem being analyzed.

computer programs to analyze images of landscapes (McGarigal 2002; see also http://www.umass.edu/landeco/pubs/Fragstats.pdf). The important point is to use measurements that are relevant to the problem at hand.

How do landscape patterns influence populations, communities and ecosystems? Let us consider some examples of ecological studies at the landscape level of analysis.

Patches and Landscape Fragmentation

Many species of both plants and animals live in habitat patches that are broken up among less suitable habitat. This fragmentation of patches occurs even in ecosystems that are undisturbed by humans (Figure 17.2), and we begin with the observation that fragmentation is a natural situation produced by disturbances such as fire, floods, windstorms and insect outbreaks. The problem is that human activities have added a great deal more fragmentation to native habitats, so that agricultural fields, clear-felled areas, suburbs and

highways now break up habitats that were once more continuous. Because many species of large mammals and birds cannot maintain viable populations in small habitat patches, fragmentation can lead to extinctions.

Fragmentation has two components when it refers to human alterations of the landscape (Haila 2002). Fragmentation results in habitat loss and a change in the spatial configuration of the remaining patches. The process of habitat loss needs to be kept separate from the changes in landscape configuration when we analyze the consequences of human alterations to the landscape. We will deal with habitat loss in Chapter 20 when we discuss conservation biology. In this chapter we will deal with fragmentation in the strict sense of changes in the spatial configuration of patches in the landscape.

Two general hypotheses have been proposed to describe the impact of fragmentation on ecological communities. The first hypothesis is that species richness increases with area, so patches that are larger should have more species in them (*Essay 17.1*). This

ESSAY 17.1   WHAT ARE AREA-SENSITIVE SPECIES?

When landscapes are broken up by agriculture, or other human activities, an array of habitat fragments is typically left. Species vary dramatically in their sensitivity to this fragmentation, and some species can be classified as area-sensitive species. The key question for each species of concern is 'how small a fragment will it occupy success-fully?' The best data available on area-sensitive species are from birds. In south-eastern Australia the eastern yellow robin is highly sensitive to the area of remnant woodland left after clearing for agriculture. In central New South Wales, Briggs, Seddon and Doyle (1999) obtained these data for 36 woodlots:

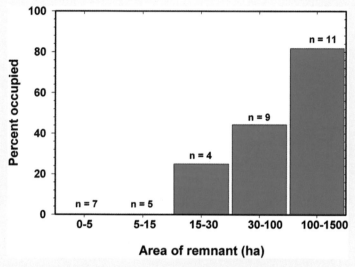

The level of occupation of remnants by eastern yellow robins is directly related to the remnant area

Eastern yellow robin. (Photo courtesy of Peter Fullagar.)

Eastern yellow robins do not occur in remnant woodlands less than 15 ha in area. Consequently, because of this threshold, protecting many small woodlots will not help preserve this declining species. The cause of this area-dependent occupancy is not known, but excessive predation or food shortages are commonly suspected causes (Zanette *et al.* 2000).

In the corn-belt of central Illinois only a few grasslands remain, and there are only nine grassland remnants larger than 40 ha (Walk and Warner 1999). Greater prairie-chickens, savannah sparrows and upland sandpipers nest only in grasslands larger than 40 ha. These birds do not recognize smaller grassland remnants as suitable nesting habitat, even though all the plants present are the same as those in larger fragments—this habitat selection has implications for reserve design to conserve grassland birds in agricultural areas.

The general rule is that the larger the body size of a species, the larger area it requires to survive and breed. Within this general framework it is important in analyzing landscapes to know how each species responds to the area of the patches available to it.

idea follows from the species-area curve (Chapter 14, page 329) and ecological work on islands, and is strongly supported by ecological data.

The second hypothesis is that population abundance will increase with area, so that animals and plants in small patches will suffer more mortality, or have lower reproductive rates, than individuals in larger patches. This idea arises from the general notion that small patches suffer from increased disturbances, predator concentration and chance effects on small populations. The second hypothesis is probable, but needs more detailed field study for confirmation.

In some studies, species richness increases with patch area, but in other studies the opposite effect occurs—the responses observed depend on the taxonomic group studied (Debinski and Holt 2000). In the tropical rainforests of central Brazil, bird diversity follows the first rule that larger areas have more species, but isolated patches of forest lose species. This means that fragmented rainforest stands have fewer birds than similar sized patches that are part of a continuous forest (Figure 17.4). However, fragmentation did not affect small mammals in this tropical rainforest, and the diversity of frogs and butterflies actually increased in the fragments because they were invaded by species that could live in the new grassy habitats of the matrix created for cattle grazing (Laurance *et al.* 2002). The

main point is that the effects of fragmentation depend on the species traits. Fragmentation may favor some species, but have detrimental effects on others.

Birds have been particularly well studied to determine the impacts of forest fragmentation on nesting success (Stephens *et al.* 2003). Fragmentation of large blocks of habitat into smaller blocks can reduce reproductive and survival rates of birds, and condemn some species to extinction. Wood thrushes are a common bird in eastern North American forests, and they have been declining in numbers during the last 20 years. Nesting success in wood thrushes is reduced in small forest blocks, compared with large blocks (Figure 17.5)—reduced reproductive output has led to fewer wood thrushes in fragmented forests. Nest predators, such as raccoons, cats and chipmunks, attack nests, and more than half the nests in small forest fragments (<100 ha) were attacked by predators.

The capercaillie is a large grouse that lives in northern Europe and, because of its size, has been a favorite bird for hunting. Capercaillie populations have declined across northern Europe from Russia to Scotland for the last 30 years, and considerable research has gone into determining the causes of this population decline. Reduced breeding success is the immediate cause of the population decline. The population decline coincides with large-scale forest harvesting in this large

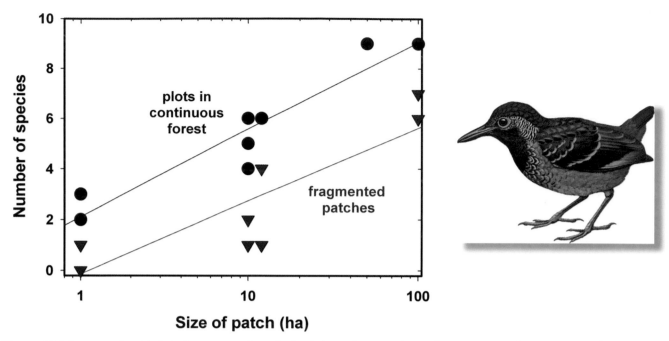

Figure 17.4 Insect-eating birds in fragments of tropical rainforest in central Brazil in 1994–95. Plots were sampled in continuous forest (blue) and in patches of remnant rainforest of equal size (red) that were fragmented by clearing for cattle ranches during the 1980s. For these birds larger patches contain more species, and fragmented patches contained fewer species than similar sized plots in continuous rainforest. The white-banded antbird (*Myrmornis torquata*) (illustrated) was present in all the continuous forest plots, but absent from all the fragmented patches. (Data from Stratford and Stouffer 1999.)

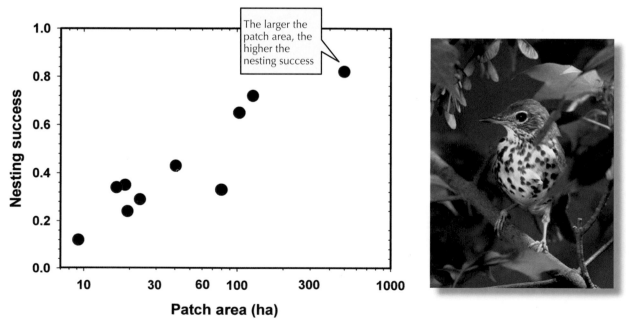

Figure 17.5 Nesting success of wood thrushes in Pennsylvania in deciduous forest patches of different sizes. Forest patches were isolated from other forest areas by agricultural land. Predation by bird and mammal predators reduced nest success in small fragments. A total of 171 nests were studied over two years in 10 different patches. (Data from Hoover *et al.* 1995.)

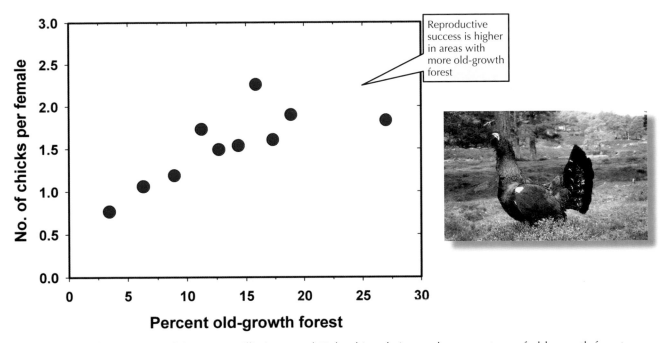

Figure 17.6 Breeding success of the capercaillie in central Finland in relation to the percentage of old-growth forest remaining in the landscape. Breeding success is measured by the number of chicks seen per female grouse in August. A total of 796 females were observed. The male capercaillie is the largest of the grouse family, weighing up to 5 kg. (Data from Kurki *et al.* 2000, photo courtesy of Robert Moss.)

area, so that landscapes with extensive old-growth forests have been replaced with younger, fast-growing, but smaller, trees. An extensive survey of reproductive success of capercaillie in central Finland identified the percentage of old growth forest in the landscape as a key to successful reproduction (Figure 17.6). However, this shortage of old-growth forests is only one cause of the capercaillie decline (Moss *et al.* 2001). Climate change has also played a role, and in this part of the world has resulted in a delay in the onset of spring weather. Cold springs result in capercaillie hens having poorer nutrition in early spring (fewer insects to eat), and chicks having to survive bouts of cold weather in late May and early June. Some of the capercaillie decline may also be due to increased predation.

Fragmentation impacts on landscapes vary in different ecosystems, but there are some general rules which apply to most situations and all species. When large blocks of habitat are broken up into smaller pieces, some species are lost, but other generalist or weedy species may colonize. Biodiversity is typically greater in larger patches. What happens to the species left in fragmented patches depends on the ecological interactions that occur at the edges of the patches.

■ 17.2 HABITAT EDGES ARE AREAS OF CONCENTRATED ECOLOGICAL INTERACTIONS

'Edge effects' is a general term covering all the ecological effects that stem from the creation of habitat edges, either naturally or by human activities. Edges are the outer boundaries of patches, and are differentiated with the assumption that the physical and biotic environments of edges differ from that experienced in the interior of the patch. Edge dimensions can be defined only with respect to a particular species or group of species, and edge effects can spread far into the center of a patch. In natural ecosystems, edges are often highly convoluted and may have a suite of species that are edge-specialists, which are more common in the edges than in the interior of patches. Humans typically replace convoluted natural edges with

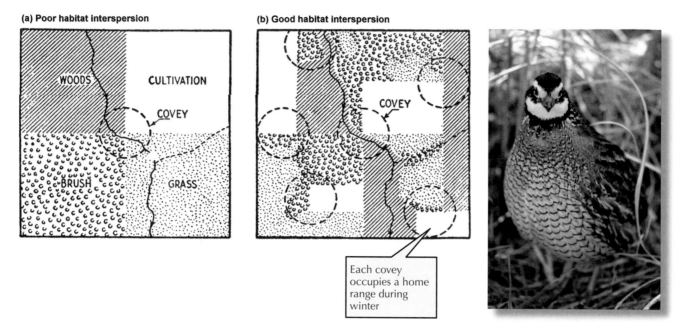

Figure 17.7 Interspersion of four habitat types or patches in relation to bobwhite quail landscape requirements. Aldo Leopold published this illustrative map in 1933 to show the importance of edges for the management of farm wildlife species, such as quail, which live in coveys during the winter. The two maps have equal amounts of each habitat type but differ greatly in the amount of edge and habitat interspersion. Wildlife managers have used this general principle of habitat interspersion to increase the abundance of some game species. But not all species increase in abundance with more edge habitat. (After Leopold 1933, page 130.)

straight-line 'administrative' edges that change the boundary dynamics in patches. Habitat edges are areas of concentrated ecological interactions. Edges may favor some species and hinder others, changing community dynamics.

Edge Effects in Wildlife Management

Wildlife ecologists have long recognized the importance of edges to many game species. Aldo Leopold in a classic 1933 book *Game Management* discussed the 'law of interspersion' and illustrated it with bobwhite quail (Figure 17.7). Leopold argued that many game species prefer edges, so that, for example, bobwhite quail coveys prefer to occupy three or four patches—grasslands, corn fields, woodland and shrubland—because they feed in one habitat, shelter from severe weather in another and nest in a third. Bobwhite quail feed on the seeds of grasses, weeds, and crops such as corn, soybeans, and wheat. Leopold recognized that some species, such as bobwhite quail, were farmland game species that preferred edges, whereas other species, such as antelope and buffalo, preferred only one

habitat and avoided edges. He recognized that it was important to know the natural history of each species to determine how it would respond to edges created by agricultural development. Leopold was using the ideas of landscape ecology long before the subject became a separate scientific discipline (Silbernagel 2003).

Edge effects can affect insect populations. Forest insect outbreaks occur periodically in many forests of the temperate regions, and can cause severe defoliation of trees. The forest tent caterpillar (*Malacosoma disstria*) is one of the most dramatic of North American forest pests, attacking many tree species, but especially trembling aspen. There is much variation in how long the high-density phase continues in these insect outbreaks, ranging from 2 years to 9 years in different forest patches. One key variable in predicting the length of the outbreak is the amount of forest edge habitat (Figure 17.8). As land has been cleared for agriculture, more and more edge habitat is produced in the remaining forest. Roland (1993) showed that the longer outbreaks occurred in regions with more forest edge. There are several mechanisms that could explain this

Aldo Leopold (1887–1948) Father of Wildlife Management in the USA

edge effect. Predators, parasites and disease organisms could be less effective in fragmented landscapes if their movement patterns are disrupted. In addition, edges of forest facing the sun create warmer microclimates that speed caterpillar development. Faster development means a shorter exposure time to the risks of predation and disease. In this case, forest fragmentation favors the pest species and reduces the effectiveness of the pest's natural enemies.

Rural Dieback in Australia

Fragmentation, and the edges that result, in agricultural landscapes can have unexpected detrimental consequences. For the past 40 years eucalypts in many areas of south-eastern Australia have been dying prematurely—a problem known as 'rural dieback'. Many species of trees are involved in dieback, and the worst affected areas are those where livestock production is the dominant land use (Landsberg 1990, Landsberg and Cork 1997). The chain of events that leads to dieback in agricultural areas begins with soil improvements (Figure 17.9). Fertilizer use and cattle droppings increase nitrogen in the soil, which in turn leads to the trees being stimulated to improve their leaf chemistry by increasing their nitrogen content. Herbivorous insects zero in on the high-nitrogen leaves, and the enriched

trees are attacked by a suite of native, leaf-eating insects. At the same time the insect-eating birds that would normally feed on these insects have been reduced in number because of habitat loss, so fewer insects are eaten by predators. As the trees become defoliated and stressed, they are susceptible to a variety of fungal diseases. The end result is typically tree death.

Rural dieback can be reversed by fencing trees from cattle in pastures and planting additional trees and shrubs to encourage more birds to use the site. One problem with small (<20 ha) fenced sites, where you would expect tree health to improve, is that these small sites can be taken over by bell miners, which are colonial, co-operatively-breeding honeyeaters that aggressively exclude all other insect-eating birds from their nesting areas (Clarke and Schedvin 1999). By excluding other insect-eating birds, bell miners release psyllids (small, sap-sucking insects) living on eucalyptus leaves from bird predation, which leads to a large build-up of psyllids, providing more food for the bell miners when they breed later in the season. Psyllids can damage and kill eucalyptus trees when they become abundant. Experimental removal of bell miners from one section of a eucalypt forest was followed by an immediate colonization of the site by other insect-eating birds, and a dramatic drop in insect infestations on the eucalypt leaves. At least in some cases, small patches of eucalypt forest may be impossible to save, and land managers are recommending to farmers that they retain, and if necessary rebuild, patches of trees larger than 20 ha to avoid the bell miner problem.

Bell miner (*Manorina melanophrys*). (Photo courtesy of Geoffrey Dabb.)

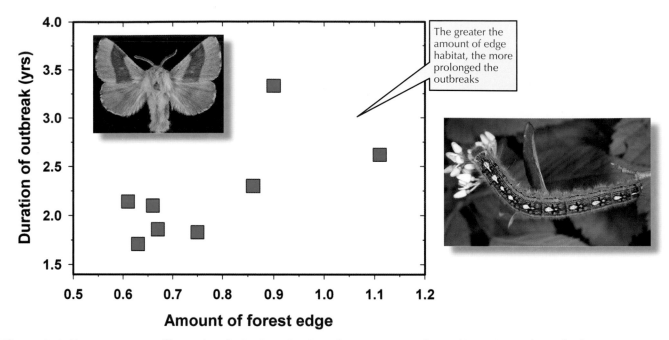

Figure 17.8 Forest tent caterpillar outbreaks in Ontario, Canada, are more prolonged in regions where the forest is more fragmented by agriculture, creating a larger amount of edge habitat (edge measured as km of edge per km^2). Insect parasites do not move about as efficiently in fragmented landscapes, and tent caterpillars lay more eggs on edge trees facing the sun. The warmer microclimate of these edge trees leads to faster larval development, and escape from natural enemies. (Data from Roland 1993, photo of larva courtesy of Dave Wagner.)

An increase in edges often causes a reduction in native biodiversity and an increase in introduced species, which may upset plant–herbivore or predator–prey interactions. Increasing edges improves habitats for bobwhite quail and forest tent caterpillars, but adversely affects aspen trees and eucalypt trees. Some species win and some species lose in the fragmentation process. The effects of human alterations to landscapes must be investigated using a species-by-species approach to find out what changes in community dynamics can result from fragmentation.

■ 17.3 CORRIDORS CONNECTING PATCHES OF SUITABLE HABITAT FACILITATE THE MOVEMENTS OF ORGANISMS

Patches of habitat in human-affected landscapes are often widely separated from one another, and the provision of corridors between patches has been one of the key recommendations of landscape ecologists to land planners (Wiens 2002). If a landscape is a mosaic of patches of different types, like Figure 17.3, the move-ments of individuals or nutrients through the landscape will depend on exactly how these patches are arranged in space (*Essay 17.2*). This can be summarized in an ecological maxim that the shortest distance between two points may not be a straight line. For example, if a mouse is moving from one woodlot to the next, it may not pay to cross a bare agricultural field and be exposed to bird predators, but rather to move along weedy fence lines under cover and travel a longer distance more safely.

Connectivity in Landscapes

Connectivity is a concept that must be focused on the particular problem of concern. In many cases connectivity is defined relative to particular species or groups of species, such as passerine birds or beetles. Movement of the species between patches is thus a critical variable defining the mosaic structure. How do individuals move through mosaics of more and less favorable habitats? If landscape managers preserve habitat corridors between patches of habitat, will these corridors be used for inter-patch movements?

Much discussion in conservation agencies has focused on providing corridors between refuges so that

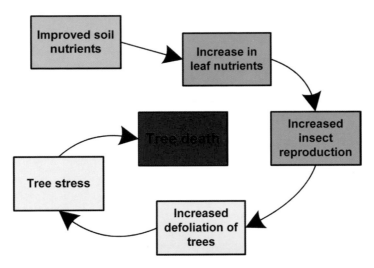

Figure 17.9 Rural dieback in eucalyptus trees in the New England Tablelands of New South Wales. Isolated trees in pastures have been dying at a high rate, while the same species of trees in nearby forest fragments were not affected. Improved soil nutrients come from agricultural fertilization and cattle manure, and this leads to higher levels of nitrogen in the leaves, making them better food for insect defoliators. Tree stress can be accelerated by drought. (Modified from Landsberg and Wylie 1991; photo by C. Krebs.)

species may disperse from one patch to the next. Many benefits have been suggested for providing corridors for species movements. But there are also potential costs to corridors, because they may transmit diseases, create fire corridors and expose individuals to increased predation risk (Table 17.1). The Florida panther (*Felis concolor*) has been reduced from approximately 1400 individuals to about 30 animals, which have been isolated in various undeveloped areas of South Florida. By providing a corridor system between wildlife refuges, managers hope to maintain panther populations in Florida (Simberloff *et al.* 1997). However, there are no data to determine how wide a corridor must be before large mammals such as the panther will use them.

Movements along Corridors

There are few detailed studies of the movements of individuals between patches and along corridors. A

ESSAY 17.2   AGRICULTURAL LANDSCAPE ECOLOGY

Modern agriculture has increasingly made landscapes more homogeneous, with mono-cultures of crops grown in large fields. One consequence of this has been to reduce the abundance of generalist insect predators that can help to reduce pest outbreaks. If the source areas for generalist predators are known, it may be possible to arrange crops across a landscape in ways that will maximize the potential for pest control.

The principle of strategic crop placement can be illustrated in a simple hypothetical landscape of three kinds of patches—a high value crop, a crop that serves as a source of generalist insect predators, and a third crop of neutral value for pest management:

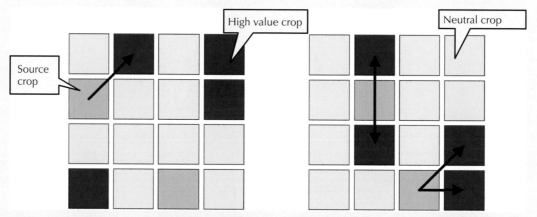

The arbitrary placement of crops on the farm to the left restricts the movements of insect predators. The strategic placement of crops on the farm to the right facilitates insect movement into the high value crop, reducing pest damage.

Cotton fields in southern Texas are shown in green as numbered fields, sorghum fields are shown in black and uncultivated land is shown in white. Other crops such as wheat or fallow fields are shown in grey. Dashed lines indicate radii of 1.6 km and 3.2 km from the central cotton field # 32. (From Prasifka et al. 2005.)

410

The source crop exports insect predators into the high value crop, which reduces the need for chemical pest control. But these insect predators cannot move long distances, and so by strategically organizing the farm landscape, the farmer can use the landscape more effectively.

A good example of the use of these principles of landscape management comes from information gathered on cotton farms in southern Texas (Prasifka *et al.* 2005). In this farming system, sorghum fields were the source for generalist predators and cotton fields were the high value crop. Other crops were wheat fields and fallow land. An illustration of the geometry of these fields is shown opposite:

Given this agricultural landscape, the ecological question is whether or not insect predator numbers were increased in cotton fields that were adjacent to sorghum fields or uncultivated land. Prasifka *et al.* (2005) showed that there was a significant effect of the areas of uncultivated land and sorghum within 1.6 km of a cotton field on the insect predator abundance in the cotton crop. The important conclusion is that the strategic arrangement of agricultural landscapes can benefit agricultural production and help reduce pesticide use.

Table 17.1 Advantages and disadvantages of corridors connecting patches in a mosaic landscape. (After Noss 1987.)

Potential Advantages of Corridors	Potential Disadvantages of Corridors

1. Increase immigration rate to a reserve, which could
  (a) increase or maintain species richness and diversity
  (b) increase population sizes of particular species and decrease probability of extinction or permit re-establishment of extinct local populations
	1. Increase immigration rate to a reserve, which could
	  (a) facilitate the spread of epidemic diseases, insect pests, exotic species, weeds and other undesirable species into reserves
	  (b) disrupt local social organization within the reserve

2. Provide increased foraging area for wide-ranging species	2. Facilitate spread of fire and other abiotic disturbances ('contagious catastrophes')
3. Provide predator-escape cover for movements between patches	3. Increase exposure of wildlife to hunters, poachers, and other predators
4. Provide a mix of habitats and successional stages accessible to species that require a variety of habitats for different activities or stages of their life cycles	4. Riparian strips, often recommended as corridor sites, might not enhance dispersal or survival of upland species
5. Provide alternative refuges from large disturbances (a 'fire escape')	5. Cost, and conflicts with conventional land preservation strategy to preserve endangered species habitat (when inherent quality of corridor habitat is low)
6. Provide 'greenbelts' to limit urban sprawl, abate pollution, provide recreational opportunities, and enhance scenery and land values	

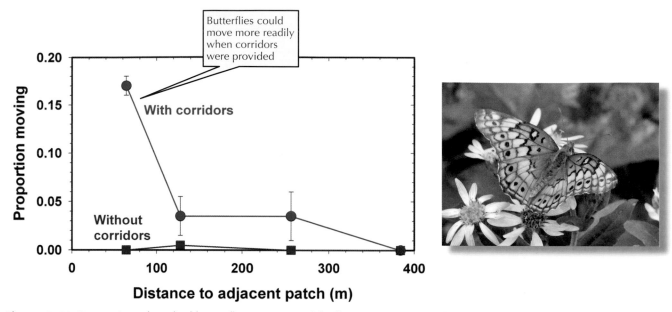

Figure 17.10 Proportion of marked butterflies (variegated fritillary *Euptoieta claudia*) that moved between patches connected by corridors or not connected and separated by varying distances of unsuitable forest habitat. Corridors generally increased butterfly movements except when patches were too far (>300 m) apart. (Modified from Haddad 1999, photo courtesy of Mike Reese.)

critical assumption of conservation biology is that corridors increase animal and plant movement between fragments of habitat. Haddad (1999) tested this idea with two butterfly species in pine plantations in South Carolina. Butterflies live in the open habitats between closed pine forests, and by harvesting the forest in patches he could construct butterfly habitat in two kinds of 1.64 ha square blocks—isolated blocks and blocks connected by a corridor of habitat. Corridors facilitated movements of butterflies (Figure 17.10), validating the presumed value of corridors for conservation of these species. Corridors also increased butterfly density. The variegated fritillary (*Euptoieta claudia*) was more than twice as dense in connected patches (0.45 individuals/ha) than it was in isolated patches (0.22 individuals/ha) (Haddad and Baum 1999). For these butterfly populations corridors were a positive benefit.

Corridors may benefit a whole suite of species. Tewksbury *et al.* (2002) created experimental patches of early successional habitat within a landscape of mature pine forest in South Carolina. They used a one-hectare central patch and four patches equidistant from the central patch (Figure 17.11). A 25 m wide corridor connected the central patch to one of the peripheral patches. The unconnected patches were either rectangular or

with 'wings' of habitat. This whole design was replicated eight times. They measured the movements of butterflies, pollen and seeds among these patches. The corridor connection not only increased butterfly movements between the connected patches, but also facilitated pollen transfer and seed dispersal among patches. Corridors again provided a positive benefit for the plants and animals in these ecological communities.

The value of corridors that connect patches of similar habitat within landscapes is now widely appreciated, so that it has become one of the cardinal rules of land planning for the conservation of biodiversity. No one has yet found any of the potential costs that might flow from corridors (such as fire or disease spread), although little work has been done on large mammals such as the Florida panther (Bowne and Bowers 2004).

■ 17.4 THRESHOLD EFFECTS COMPLICATE LANDSCAPE PLANNING AND MANAGEMENT

The continued alteration of natural landscapes by humans has produced a focus within landscape ecology on the planning and management of human-affected landscapes. The important concept introduced

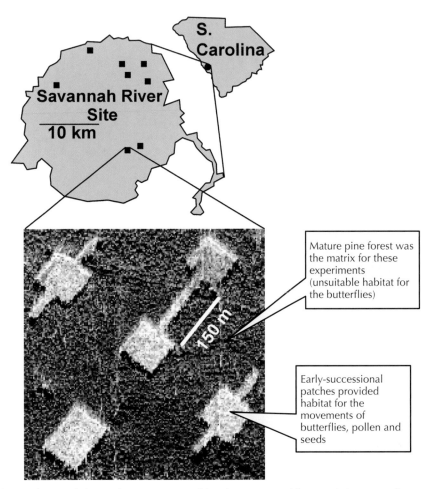

Figure 17.11 Aerial photo of one of the eight experimental landscapes used by Tewksbury *et al.* (2002), showing the configuration of the patches of early successional habitat (green squares) created within the matrix of mature pine forest (red color in this satellite image) in South Carolina. Each of the peripheral patches was 150 m distant from the central patch, and one of the peripheral patches was connected with a corridor to the central patch to see if this corridor facilitated movements between patches. The winged patches were thought to be able to intercept more dispersers, but, in this system, all the unconnected patches performed equally no matter what their shape. (Modified from Tewksbury *et al.* 2002.)

here is that of **threshold effects**. The impact of landscape change is not necessarily linear (Figure 17.12). Species responses do not necessarily rise or fall smoothly as landscape configuration is altered. If there are thresholds operating in an ecosystem, a species may do well up to some point and then suddenly disappear from the landscape.

Impacts of Oil and Gas Exploration

The impact of oil and gas exploration on woodland caribou in Canada is a good example of how small changes to forested landscapes may have large effects on animal populations. Woodland caribou have been declining in numbers over the last 100 years and have

become a threatened species in Alberta. Oil and gas exploration in northern Alberta has produced a series or roads, seismic lines and oil and gas wells in a mosaic habitat of otherwise undisturbed coniferous forest (Figure 17.13). The important point to note is that only 1% of the forest is disturbed by these gravel roads, seismic lines (5–8 m wide cut lines for exploration), and well sites (1 ha gravel pads). No one expected that a 1% land use change would affect these caribou.

By putting satellite radio collars on 36 woodland caribou, Dyer *et al.* (2001) could follow the movements of each animal with several locations per day through an annual cycle. The question they asked was whether caribou used the areas adjacent to the roads, seismic

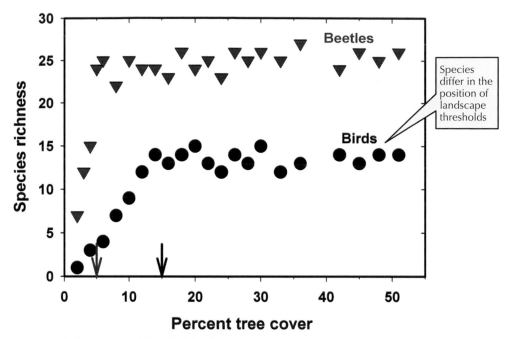

Figure 17.12 Hypothetical illustration of threshold effects in landscape mosaics. The loss of species when landscapes are changed by humans is not necessarily a straight line. In this hypothetical example, bird diversity reaches a plateau at 15% tree cover, and if tree cover falls below this threshold (arrow), species are lost. In contrast, beetle diversity has a threshold around 5% tree cover. Not all species in an ecosystem will show a threshold effect.

lines and well sites as often as they used areas away from disturbed sites. The researchers found that caribou avoided human developments, staying up to 1000 m away from well sites and 250 m away from roads and seismic lines. Avoidance of disturbed sites was maximal in late winter—a time of food stress. Caribou reduced their use of 22–48% of the entire study area due to these human disturbances on 1% of the landscape. Roads and seismic lines have also facilitated the travel of predators like wolves in the region—hunters have increased access with the road network. The net result of what would appear to be a minor loss of habitat translates into a major impact on woodland caribou and a declining population from these combined stresses.

Greenway Planning in Urban Environments

Landscape management is most evident to us around cities and towns, and the interactions between landscape ecologists and landscape planners have produced important habitat corridors that are starting to benefit biodiversity in cities and towns. The movement has been called 'greenway planning'. Greenways have the important function of providing corridors for wildlife, as well as recreational trails and waterways for people to use. The principle of greenway planning is to design a community-wide conservation network such that every new development contributes a segment of land to the greenway. These greenways should comprise 40–70% of the land in each new neighborhood in a coordinated way (Arendt 2004). Figure 17.14 gives an illustration of the ideas of greenway planning principles for a 46 ha site in North Carolina. The first step in greenway planning is to identify the conservation areas in the site—both those areas that form wildlife corridors and areas such as flood plains where no buildings should be sited. Once this has been mapped, the locations of roads and houses can be laid out and development can proceed with adequate protection of the conservation network in the region. Approaches such as greenway planning can help to minimize the impact of human building on wildlife.

What rules of thumb have arisen from landscape ecology to assist landscape planners and land

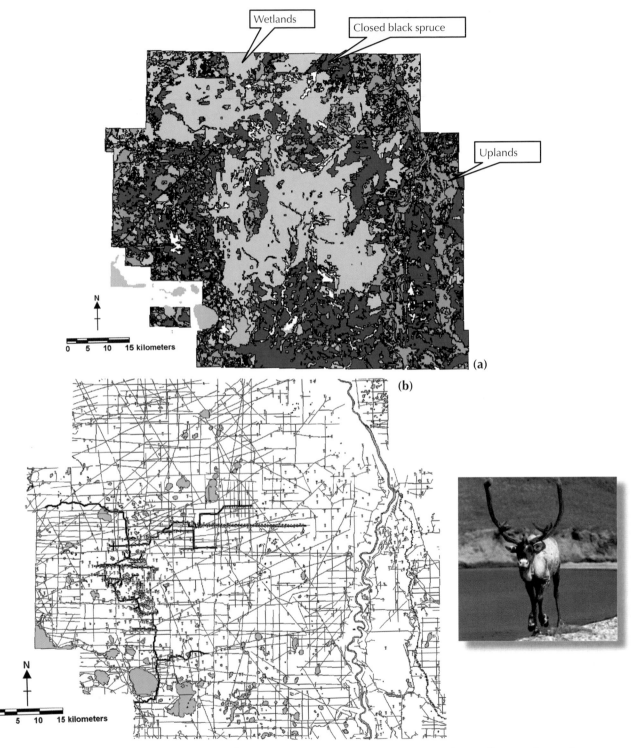

Figure 17.13 (a) Habitat mosaic of the 6000 km² study site for woodland caribou in northern Alberta. The study site is dominated by coniferous black spruce forest in wetlands (light green), closed black spruce wetlands (dark green) and uplands (orange) with aspen and white spruce. (b) Human development in this study area, showing well sites, roads, and seismic lines. Only 1% of the land area is occupied by these human developments, yet, because woodland caribou avoid these areas of human disturbance, up to half of the potential habitat is not used by the caribou. (Modified from Dyer *et al.* 2001, photo by A.J. Kenney.)

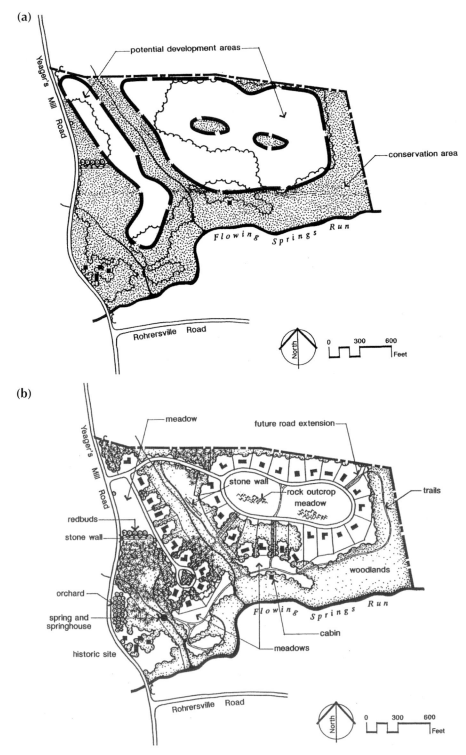

Figure 17.14 Greenway landscape planning illustration from a 46 ha site in North Carolina (Arendt 2004). Greenway planning links landscape ecology to housing development in a creative way to preserve corridors for wildlife. (a) The first step is to identify the part of the 46 ha property that has important conservation value. The greenway identified can serve both for wildlife conservation corridors within a region and human recreation. (b) Once this is achieved, the streets and house sites can be located on land that has low value for conservation. In this particular property the area along the creek is also a flood plain where no construction should occur.

management authorities? Haila (2002) lists some rules for general guidance:

1 Avoid clearing native habitats, particularly in regions where little native habitat remains. Protect large areas of native habitat.
2. Restore native vegetation by active measures, such as replanting.
3. Preserve corridors and other connecting habitat routes in landscapes that are heavily cleared of native vegetation.
4 Maintain environmental heterogeneity at the landscape level.
5 Identify particularly important microhabitats and preserve these.

These could be considered the basic rules of landscape management, and to apply them to any given situation will require specific data on the populations, communities and ecosystems that occupy the particular landscape.

SUMMARY

Landscape ecology is the ecology of large regions containing several ecosystems. Large regions are defined on the basis of the species or process one is studying. A large region to a butterfly is a tiny one to an eagle. The first principle of landscape ecology is to define the spatial scale of interest. Landscapes contain four elements: patches, edges, corridors and mosaics. The focus of landscape ecology is often on the effects that human uses of the landscape are having on natural ecosystems, communities and populations.

Patches vary in quality as well as in size and shape. The ability of species to move between patches is the key to understanding how a particular landscape structure will affect a species. Species with low dispersal powers are unable to colonize distant patches if corridors in the landscape are absent. Fragmentation is the breaking apart of formerly continuous blocks of habitat into pieces or patches. Fragmentation affects species interactions and thus community structure. Some species specialize in edge habitats, and fragmentation, which increases edges, will favor these species.

Connectivity is a basic principle of landscape ecology—the more patches are connected by corridors, the better will be the preservation of biodiversity and ecosystem function. Isolated patches are vulnerable to many destructive processes, and often lose species rapidly. A landscape ecologist interested in conservation typically prefers large patches connected by corridors.

Where possible the principles of landscape ecology can be applied to nature reserve design and suburban land planning. Landscape ecology forces us to look at the broad picture of how species populations, communities and ecosystems fit together in the modern Earth's landscapes, and how humans are changing these landscapes—often to the detriment of these natural ecosystems.

SUGGESTED FURTHER READINGS

Arendt R (2004). Linked landscapes. *Landscape and Urban Planning* **68**, 241–269. (How landscape ecology can be used to design greenway corridors for conservation in urban planning.)

Haila Y (2002). A conceptual genealogy of fragmentation research: from island biogeography to landscape ecology. *Ecological Applications* **12**, 321–334. (A critical review of the history of ecological research on habitat fragmentation.)

Silbernagel J (2003). Spatial theory in early conservation design: examples from Aldo Leopold's work. *Landscape Ecology* **18**, 635–646. (An example of the roots of landscape ecology from the work of the father of modern wildlife management.)

Stephens SE, Koons DN, Rotella JJ and Willey DW (2003). Effects of habitat fragmentation on avian nesting success: a review of the evidence at multiple spatial scales. *Biological Conservation* **115**, 101–110. (A comprehensive review of the question of whether birds reproduce less well in fragmented landscapes.)

Stratford JA and Stouffer PC (1999). Local extinctions of terrestrial insectivorous birds in a fragmented landscape near Manaus, Brazil. *Conservation Biology* **13**, 1416–1423. (The devastating impact of land clearing on tropical bird species in the Amazon.)

Wiens JA (2002). Riverine landscapes: taking landscape ecology into the water. *Freshwater Biology* **47**, 501–515. (An overview of how the principles of landscape ecology can be applied to rivers and streams.)

QUESTIONS

1. Rivers, lakes and the ocean are not usually thought of when there are discussions about landscape ecology. Discuss why this might be, and how the concepts of patches, edges and corridors might be applied to these aquatic systems. Do rivers differ from lakes and the oceans in their 'landscape' structure? Wiens (2002) discusses this issue.

2. Many studies in eastern North America have found that forest fragmentation reduces bird breeding success, but a similar study of fragmented populations in western North America found the opposite result, with birds in fragments having higher breeding success (Tewksbury *et al.* 1998). Discuss some reasons why fragmentation might benefit some bird species and disadvantage other species.

3. In Scotland the capercaillie population has declined during the last 30 years. Suggested explanations for this decline are forest fragmentation, landscape change from overgrazing by deer of nesting and chick-rearing habitat, increased predation, over-harvesting and climate change. Discuss what data you would like to have to test each of these hypotheses. Moss *et al.* (2001) discuss this issue.

4. List some of the possible difficulties of being able to recognize a threshold effect, such as that shown in Figure 17.12 (page 414) with field data.

5. In early studies of landscape ecology, fragments and patches were compared to oceanic islands (c.f. Figure 14.12, page 331). Contrast the fragments illustrated in Figures 17.2 and 17.3 with oceanic islands from the viewpoint of a forest tree and from the viewpoint of a bird. Haila (2002) discusses this issue.

6. With many ecological questions, the recommendations for solution typically involve the use of manipulative experiments. Discuss the applicability of this advice to landscape ecological problems. How would you determine what a 'control' landscape is for a study in landscape ecology? McGarigal and Cushman (2002) discuss these issues.

7. Sockeye salmon in western North America spawn in freshwater lakes and migrate to the ocean as small fingerlings and complete 99% of their growth in the North Pacific Ocean. They return to spawn, die and decompose in the freshwater lakes where they originated. Coastal grizzly bears ('brown bears') feed on salmon migrating to the spawning grounds. Discuss how this simple food chain can link together the ocean, freshwater and terrestrial ecosystems. Can all these ecosystems be included as one landscape?

Chapter 18

HARVESTING POPULATIONS—HOW TO FISH SUSTAINABLY

IN THE NEWS

The North Pacific Ocean and the Bering Sea have suffered a major collapse of marine mammal populations during the last 30 years, and there has been considerable controversy over the cause of this problem. The northern fur seal and the harbor seal began the collapse around the mid 1970s and the collapse has continued ever since. This was followed closely by a collapse of Stellar sea lions and sea otters during the 1980s and early 1990s. What could be the cause of this major, sequential collapse of marine mammals in the North Pacific?

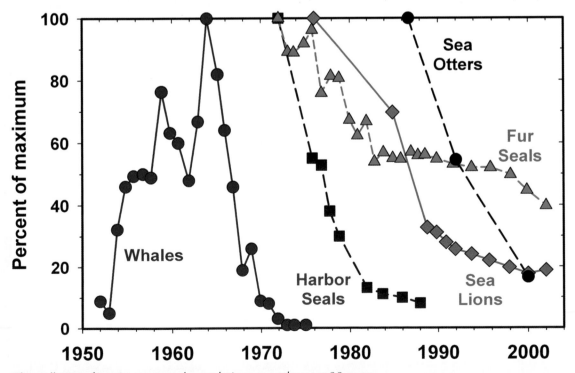

The collapse of marine mammal populations over the past 30 years

Two ideas have been suggested. The first suggested cause is food limitation that is being driven by physical changes in ocean conditions, which has reduced the prey base for these predatory mammals. A second suggestion is that a change in predation pressure by killer whales is the key behind all these declines. What does the evidence suggest?

The first hypothesis is not simple. Food limitation could have been exacerbated by the extensive commercial fisheries that have operated in Alaskan waters. But many of the species of fish eaten by seals and sea lions are not commercially fished, so it seems unlikely that the commercial fishery directly affected the survival of these marine mammals by taking away their food. Instead, the composition of the fish community changed and the diet of marine mammals, such as sea lions, switched to the more common species like pollock, which are lower quality food. Whether this kind of nutritional explanation can explain the decline of the other seals and sea otters is not yet known because no data are available.

The alternative hypothesis of changing predation has been stimulated by a recent analysis of the impact of commercial whaling in the North Pacific from 1946 to 1980. At least 500,000 great whales (sperm whales, fin whales and sei whales) were harvested in the North Pacific Ocean and in the Bering Sea from 1946 to 1975, and by the mid 1970s all great whale stocks were severely reduced. This rapid decline in great whale numbers must have caused major changes to both the prey of the whales and to their predators. Before commercial whaling, the great whales were probably a major food source for killer whales in the North Pacific. As the numbers of great whales collapsed from over-fishing, killer whales in the North Pacific shifted their predation to harbor seals, fur seals and sea lions, and finally to the smallest species, the sea otter—producing the cascade of losses shown in the graph. One problem with this hypothesis is that killer whales have always been thought to be too rare to cause such widespread losses to large populations of seals and sea lions. There is good information

Killer whales circle outside a sea lion colony in Alaska (Photo courtesy of Andrew Trites, UBC)

linking killer whale predation to the decline of sea otters during the 1990s, but killer whale predation on seals has been less well studied. To explain this collapse quantitatively, we need to know killer whale numbers and their diet—the problem is that killer whale population estimates for the North Pacific are difficult to obtain.

Consequently there is ecological uncertainty about the impacts of harvesting of whales on the seals and sea lions of the North Pacific. There is no question about the collapse of these populations, and great whale numbers were certainly driven down to low numbers by over-harvesting. But the role of killer whales is controversial and will be resolved only by more studies on these top predators (Williams *et al.* 2004).

■ 18.1 HARVESTING A POPULATION REDUCES ITS ABUNDANCE

Applied ecology meets economic reality in the harvesting of fish and forests. Fisheries have existed for thousands of years and wood has been used for fuel and building material for the same time period. But the growth of the human population has meant that harvesting has now reached levels that are not sustainable. It is critical for us to understand how harvested populations should be managed to prevent the loss of sustainable production. It is easiest to work these ideas out for fish populations, but the ideas are equally applicable to forestry, agriculture or mushroom harvesting.

To manage any population effectively, we must have some understanding of its dynamics. Most of human history might be said to illustrate this idea in graphic detail—a list of populations of fish destroyed by inadequate management should be both a warning and a stimulus for us to achieve some understanding of harvesting principles. The central problem of economically oriented practices, such as forestry, agriculture, fisheries and wildlife management, is how to produce the greatest crop without endangering the resource being harvested. This is the simple idea of **sustainability**. The problem we face in harvesting may be illustrated with a simple example from a farm pond. If you were raising fish in a small farm pond, you would obviously not remove the fish when they are small fingerlings because this would result in little fish production and no profit. At the other extreme, you would not let the fish grow too old because they would stop growing and some would start to die, so you would have few fish to sell. Somewhere between these two extremes will be some optimum point at which to harvest the fish—the problem is how to locate this optimum.

Next to forestry and agriculture, the greatest amount of work on the problem of optimum harvesting has been carried out in fishery biology. This is because of the tremendous economic importance of marine fisheries in particular. Many marine fisheries have dwindled in size since the 1920s because of over-fishing, and this has stimulated a great deal of research on 'the over-fishing problem.' All populations can sustain a harvest if it is not too severe, but over-harvesting can cause a population to collapse and never recover.

For any harvested population, the important unit of measure is the crop or **yield**. The yield may be expressed in **numbers** or **weight** of organisms and always involves some unit of time (often a year). We are interested in obtaining the optimum yield from any harvested population. The concept of **maximum sustainable yield** has been the basis of scientific resource management since the 1930s. Let us consider first the simple situation in which maximum yield in biomass is defined as the optimum yield. Implicit in this concept is the idea of a sustained yield over a long time period. Harvesting is not like mining—harvested resources must be renewable resources.

A Simple Harvesting Model for Fisheries

Concerns about over-fishing began to be raised in the early part of the 20th century. In 1931 the English fishery scientist E.S. Russell became one of the first to deal in detail with the conceptual basis of the harvesting problem in fisheries. In any exploited fish population, there will be a portion of the population that cannot be caught by the type of gear used (because they are too small) or that is purposely not harvested because it lives in an area that is not easily fished or is protected by law. The harvestable sector of the population is called the **stock.** For a fishery, interest normally centers on yield in weight, so instead of counting the individuals making up the stock, we will deal in biomass units. Russell pointed out that two factors decrease the weight of the stock during a year: **natural mortality** and **fishing mortality.** Similarly, two factors increase the weight of the stock: **growth** and **recruitment.** Consequently, one can write a simple equation to describe this relationship:

$$S_2 = S_1 + R + G - M - F$$

where S_2 = weight of the stock at the end of the year
S_1 = weight of the stock at the start of the year
R = weight of new recruits
G = growth in weight of fish remaining alive
M = weight of fish removed by natural deaths
F = yield to fishery

If we wish to balance the fish population, $S1 = S2$, and hence:

$$R + G = M + F$$

This means that in an unexploited stage, in which the stock biomass remains approximately constant from one year to the next, all growth and recruitment is on average balanced by natural mortality. When exploitation begins, the size of the exploited population is reduced, and the loss to the fishery is made up by compensatory changes such as (1) greater recruitment rate, (2) greater growth rate or (3) reduced natural mortality. In some populations, none of these three occurs, and the population is exploited to extinction because the right side of this equation always exceeds the left side.

Note that stability at *any* level of population density or stock size is described by this equation:

Recruitment + growth = natural losses + fishing yield

Thus the crucial question arises: 'What level of population stabilization provides the greatest weight of catch to the fishery?' One of the early attempts to solve this problem was made by another British fishery scientist, Michael Graham, who in 1935 proposed the **sigmoid curve theory.**

Sigmoid Curve Theory

Graham started by considering a very small stock of fish in an empty area of the sea. At what rate will such a stock increase in size? Graham suggested that the growth of this population would follow an S-shaped, or sigmoid, curve (Figure 18.1). Initially, the population grows slowly in absolute size, reaches a maximum rate of increase near the middle of the curve and grows slowly again as it approaches the asymptote of maximal density. The rate of increase reflects the biomass of stock added per year.

Graham pointed out that, if you wish to obtain the maximal yield from such a population described in Figure 18.1, you should keep the stock around point $S4$ of the curve. The important point here is that the highest production from such a population is not near the top of the curve, where the fish population is relatively dense, but at a lower density. This result is not immediately

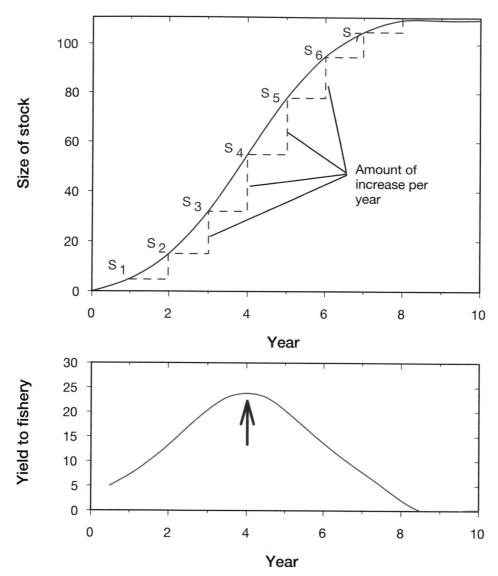

Figure 18.1 The sigmoid curve theory describes the growth of a population that could be harvested. The amount of increase per year is the yield that could be taken by the fishery, as shown in the bottom graph. Maximum yield (red arrow) is obtained by keeping the population at the point $S4$ on the graph, which is half the maximum population size or carrying capacity. (Modified after Graham 1939.)

intuitive, and Graham's result is now expressed as the first rule of exploitation: **maximum yield is obtained from populations at less than maximum density.**

Q Would this principle of exploitation also be applicable to forestry operations?

Fishery scientists have developed a variety of techniques for finding out the size of stock that will produce maximum yield. Their results can be summarized as four principles of exploitation:

1. Exploitation of a population reduces its abundance, and the greater the exploitation, the smaller the population becomes.
2. Below a certain level of exploitation, populations are resilient and compensate for removals by surviving or growing at increased rates.
3. Exploitation rates may be raised to a point where they cause extinction of the resource.
4. Somewhere between no exploitation and excessive exploitation is a level of maximum sustainable yield.

426

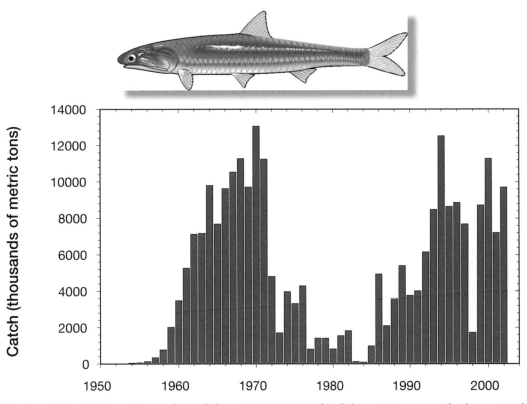

Figure 18.2 Total catch for the Peruvian anchovy fishery, 1950–2004. This fishery in Peru was the largest in the world until it collapsed in 1972 during an El Niño. In spite of reduced fishing, it took 25 years for the fishery to recover. (Data from FAO Yearbooks of Fishery Statistics.)

Let us look at one example of a fishery to see how this ecological understanding has been applied.

Over-fishing the Peruvian Anchovy

The Peruvian anchovy (*Engraulis ringens*) is restricted to the area of upwelling of cool, nutrient-rich water along the coasts of Peru and northern Chile. The upwelling causes very high productivity in the coastal zone. The Peruvian anchovy (also called the Peruvian anchoveta) is a short-lived fish, spawning first at about 1 year of age and rarely living beyond 3 years. It is a small fish—about 12 cm in length at age 1 and seldom reaching 20 cm in length. Young anchovies enter the fishery at only 5 months of age (8 to 10 cm). Anchovies occur in schools and are caught near the surface.

The Peruvian anchovy fishery started from a small level in 1950 and grew rapidly during the 1960s to become the largest fishery in the world by 1970. In 1972 it collapsed. From 1955, when the major fishery first began, the anchovy catch doubled every year until 1961. In 1970, 12.3 million metric tons were harvested,

and this single-species fishery comprised 18% of the total world harvest of fish. Figure 18.2 shows the total catch. The leading fishery scientists of the day fitted a sigmoid curve model to the fishery, and predicted a maximum sustainable yield around 10 million tons for the fishery. Anchovy are taken both by fishermen and by large colonies of seabirds, and these two had to be combined to measure the total 'catch.' From 1964 to 1971 the catch was close to the supposed maximum sustainable yield. But there was one problem: the sigmoid curve theory assumes that the environment is relatively constant and the fishery is in equilibrium.

In 1972 average conditions disappeared, and the Peruvian anchovy fishery collapsed. Early in 1972 the upwelling system off the coast of Peru weakened, and warm tropical water moved into the area. This phenomenon—known as 'El Niño' (The Child) because it often happens around Christmas—occurs about every 5 years and greatly changes the ecosystem of the area. The productivity of the sea drops, seabirds starve and anchovies move south to cooler waters and may

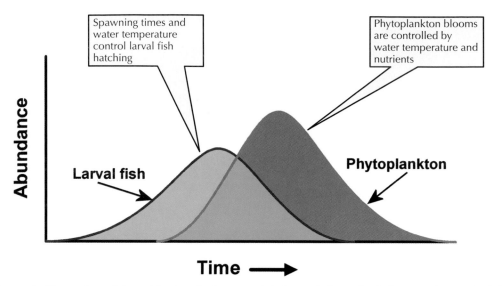

Figure 18.3 Schematic illustration of the critical period or match/mismatch hypothesis for recruitment in fish populations during one year. If fish spawning occurs too early (blue) there is inadequate food and the larval fish starve. If later spawning (green) overlaps the spring phytoplankton bloom (red), there is adequate food, and larval fish survive well. The timing of fish spawning and phytoplankton increase determines year-class recruitment of juvenile fish. In a good year these two curves would overlap completely, and in a bad year they would hardly overlap at all.

concentrate. In early 1972 very few young fish were found—the spawning of 1971 had been very poor, being only one-seventh of normal. Adult fish were highly concentrated in cooler waters in early 1972, and these concentrations produced large catches for the fishermen. By June 1972 the anchovy stocks had fallen to a low level, catches had declined drastically and no young fish were entering the population. The fishery was suspended to allow the stocks to recover, but from 1972 to 1985 there was little sign of a return of the anchovy toward its former abundance. Catches fell to low levels and began to recover only during the late 1990s after 25 years of low catches (Figure 18.2). The economic consequences of the fishery collapse of 1972 were very great, and might have been avoided if the fishery had been closed a few months earlier or if the fishing intensity had been slightly less than the maximum of 10 million tons. The Peruvian anchovy has become a model case of over-fishing and has raised the important question about how to manage fisheries in a sustainable manner when the environment is fluctuating. It illustrates all too well the dangers of assuming the sigmoid curve theory of fishing is correct all the time. The false assumption that fish populations are in a state of equilibrium, and that average conditions never change, are shown very clearly by the collapse of this fishery.

One of the key features in the failure of the Peruvian anchovy fishery was the collapse of recruitment in 1971. Reproduction, and subsequent recruitment, in many fish populations varies greatly from one year to the next, and cannot be assumed to be a constant. A very strong year-class will have an important impact on a fishery. This poses a general question for fishery scientists: Why should year-classes vary so much in their success? Most of the variation in fish recruitment is believed to occur in the early life-cycle stages—in the first few weeks or months of life. What happens in those early life-cycle stages and why do year-classes succeed or fail? This is a difficult problem being addressed by fisheries ecologists. The critical period or **match/mismatch hypothesis** postulates that early in the life of most fishes there is a short time period of maximum sensitivity to environmental factors (Figure 18.3). The match or mismatch between the timing of spawning and the food supply available at that time and place limits recruitment of many fish populations. It is commonly assumed that food availability is the key limiting factor, and food availability is determined by oceanographic effects—current patterns, winds

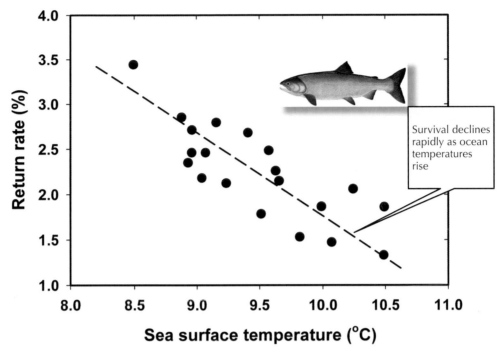

Figure 18.4 Relationship of spring sea surface temperature in the central North Pacific Ocean and the return rates of Japanese chum salmon, 1966 to 1987. Chum salmon juveniles are released from freshwater hatcheries in Japan, so the exact number released is known, as well as the number of adults that return from the ocean 4 years later. The return rate is the number of adults returning divided by the number of juveniles released. Possible explanations for this relationship are that oceanic predators of juvenile salmon are more effective at higher ocean temperatures, or that the juveniles' food supply is reduced in warmer waters. (Data from Ishida *et al.* 1995.)

and water temperature. Because the fate of newly hatched fry is critical in determining recruitment in marine fishes, studies have concentrated on these early life-history stages.

One way to search for explanations of year-class failures in fish is to look for correlations between environmental factors, such as water temperature, and the relative success of recruitment of young fish. Fisheries ecologists have found a great number of correlations between environmental factors and recruitment success. Let us look at two examples. Figure 18.4 illustrates a relationship between recruitment and ocean temperature for chum salmon from Japan. Salmon spawn in fresh water, but grow up out in the ocean. Sea surface temperature changes in the North Pacific seem to have dramatic effects on the abundance and growth rates of Pacific salmon in North America and in Asia. The two most likely mechanisms causing these correlations are food abundance and predation. If juvenile salmon do not reach a critical size in their first year at sea, they do not survive the following

winter. Food abundance during their first year at sea is the main factor limiting growth rates. The critical period determining year-class strength for these salmon seems to be the first summer and autumn of ocean life.

Satellite images of primary production in the ocean have now made it possible to measure phytoplankton blooms and to associate them with juvenile fish survival. Haddock are an important target species for North Atlantic fisheries, with a total catch over 200,000 tons in 2001. Since 1970, regular surveys have estimated the number of juvenile haddock off the coast of Nova Scotia in eastern Canada. Haddock mature at 4–5 years of age, and females lay 50,000 to 2 million eggs each spring. Some year-classes are much more successful than others—Figure 18.5 shows the relationship between the intensity of the spring algal bloom and the survival of juvenile haddock. As predicted by the match/mismatch hypothesis, when the spring plankton bloom is early, haddock juveniles survive well, and a good year-class is the result.

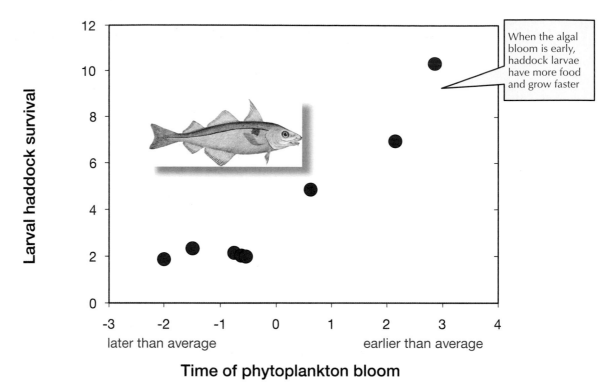

Figure 18.5 Recruitment of haddock off eastern Canada depends on the timing of the spring phytoplankton bloom, as predicted by the match/mismatch hypothesis shown in Figure 18.3. The timing of the spring phytoplankton bloom was determined from satellite images of ocean surface color and is measured as a deviation in weeks from the average. Recruitment was measured from the numbers of juvenile haddock captured in plankton nets in relation to the number of eggs laid. Each point represents one year of data. (After Platt *et al.* 2003.)

Q Would you expect recruitment to vary with the number of eggs a female fish lays?

The key point is that if the recruitment of juvenile fish is highly variable, yields to the fishery will also be highly variable, and we are back to the question of how the concept of maximum sustainable yield can apply to highly variable fish populations. We need to consider in more detail the concept of maximum sustainable yield.

■ 18.2 MAXIMUM SUSTAINABLE YIELD IS THE HARVESTING GOAL, BUT MAY NOT BE ATTAINABLE

Every industry that operates by harvesting renewable resources, such as fish or trees, is interested in obtaining the maximum yield without endangering the resource. In this way, we should be able to be efficient both economically and environmentally. But this goal has been elusive in both fisheries and forestry (*Essay 18.1*), and it is important for future sustainability to find out why it should be so difficult to achieve maximum yields.

The concept of maximum sustainable yield has dominated fisheries management since the 1930s. In many harvesting situations, maximum yield in biomass is not the desired goal. In sport fisheries, for example, the object is to maximize recreation, and the most desirable fish are often the largest ones. Hunters of large mammals may place more emphasis on the trophy status of the animals they harvest, and the harvesting of wildlife populations is often carried out without the goal of maximum biomass yield. So, not all harvesting has maximum yield in biomass as its goal and it is important as a first step to define the goals of harvesting, what exactly do we wish to maximize? For most commercial fisheries the object is to maximize yield in biomass, and we will assume for the

ESSAY 18.1   PRINCIPLES OF EFFECTIVE RESOURCE MANAGEMENT

Renewable resource management has historically failed, as many examples from fish harvesting show us. Yet we are now committed to the general principle of sustainable use of resources. How can we do a better job in the future? Five principles should underlie good resource management:

1. **Include humans as part of the system.** Human motivation, short-sightedness and greed can underlie many of the problems of resource management. Instead of thinking of humans managing resources, we should think of resources managing human behavior, often with a short-sighted timeframe.

2. **Act before scientific consensus is achieved.** For many management problems we do not need additional research to decide on management policies. Examples would include pollution impacts in the Great Lakes, tree harvesting on slopes subject to erosion, and harvesting of undersized fish. Calls for additional research on many topics are often just delaying tactics.

3. **Rely on scientists to recognize problems, but not to remedy them.** Good science is important for resource management, but it is not enough. The management of human activities is what is essential and this is a sociological, psychological and political problem.

4. **Distrust claims of sustainable resource use.** Because we have failed in the past to harvest sustainably, any new plan that claims to be sustainable should be suspect, and subject to detailed scrutiny. The linkage between basic research on fish populations and sustainable fisheries policies is a loose one, and good basic research does not automatically lead to better management.

5. **Confront uncertainty.** We often operate under the illusion that if we do enough research with enough funding we will be able to determine a solution to harvesting problems. But the large levels of natural variation found in most populations preclude any exact predictions about future dynamics. We need to favor management actions that allow for uncertainty, and to favor actions that are reversible if found to be damaging.

Sustainable development is the buzzword of the moment, and we must not pretend that scientific or technological advances will be sufficient to solve resource management problems. These are human problems that we have created many times in the past and under many types of political systems, and they will not necessarily be solved by more scientific data.

moment that this is the goal of harvesting. The same principles apply no matter what the harvesting goal.

The first problem is that in any fishery that harvests several species at the same time—it is impossible to harvest at maximum yield for all species. One species may be over-harvested, while another caught in the same nets is under-harvested. Even within a single species, there are often sub-populations, or **stocks**, that have different resilience to harvesting. Harvesting of Pacific salmon operates on mixtures of stocks from different river systems and different spawning areas within one system. The result is that less productive salmon stocks are over-fished, and even driven to extinction, while other more productive stocks are not fully utilized.

Economics of Harvesting

A second problem is that any specification of maximum yield must include economic factors. The real yield from fisheries is not fish but dollars, and economists have long recognized that it is poor business to operate a fishery at maximum yield as suggested by the sigmoid curve theory shown in Figure 18.1. Economists have pointed out that there is a level of harvesting associated with maximum sustainable economic revenue, and that this is usually at a lower fishing intensity than the harvesting rate that will give the maximum sustainable yield in biomass. What is optimal to an economist is not necessarily optimal to a biologist. In a simple fishery that follows the sigmoid curve theory, the maximum economic profit will always occur at a lower fishing intensity than maximum yield. If this simple model prevailed in the real world, the economic management of fisheries would always be a safe biological management strategy. Alas, it is not always such. Economists suggested many years ago that in an unmanaged fishery the only social equilibrium that will be reached is at the point where total costs equal total revenue, which is beyond the point of maximum sustainable yield. Under some situations, it will pay fishermen to deplete the fishery to extinction, as might have happened to whales had the International Whaling Commission not intervened. The key economic idea in these cases is that of **discounting future returns**. If a fisherman can make $1000 today by over-fishing, or $1500 in 10 years time by delaying the harvest, most fishermen will take the money now and not wait—the future is discounted as economists say (because of inflation, a dollar in 10 years time will be worth much less than a dollar today). This type of exploitation makes perfect economic sense under our current economic theories, but leads to ecological disaster and over-exploited populations. The principle is critical: **sustained yields cannot be achieved without strong social or political controls on the allowable harvest.**

Most of the world's fisheries are over-exploited and operate beyond the limits of sustainability (*Essay 18.2*) and a historical perspective on fisheries management leads to pessimistic conclusions about the future. The problem is that the concept of maximum sustainable yield is an equilibrium concept, and works well when a harvestable population is stable over time. But if the harvestable resource fluctuates, an effect called **Ludwig's ratchet** begins to operate, which continually pushes exploitation rates above their sustainable level (Figure 18.6). Estimates of sustainable harvest rates are nearly always too high and, if profit margins are good, additional investment in fishing equipment is made, and the harvesting industry becomes increasingly susceptible to a sequence of poor years, which inevitably arrive. Because of job losses in the poor years, governments will typically step in to subsidize harvesting during the poor years, which then encourages even more over-harvesting. The long-term result is a heavily subsidized industry with too much capacity that over-harvests the resource until it completely collapses. This description fits almost every fishery in the world today.

The Tragedy of the Commons

The **tragedy of the commons** was the term coined by Garrett Hardin in 1968 to describe the over-exploitation of resources that are open to anyone to use. Whenever a resource such as a marine fishery is held in common by all the people, the best policy for every individual is to over-harvest the resource or 'beggar your neighbor'. There can be no reason to stop harvesting at some optimum point because you can always make more money by over-harvesting and if you do not over-harvest, your neighbor will. This tragedy of the over-exploitation of common property resources can be averted only by some form of management that restricts fishing pressure or by converting a common property resource to a private resource through private ownership. Social control of harvesting is required for all large-scale fisheries, and for this reason good resource management is a creative mix of ecology, economics and sociology. Alas, this mix does not always produce good fisheries management, as the next two examples show.

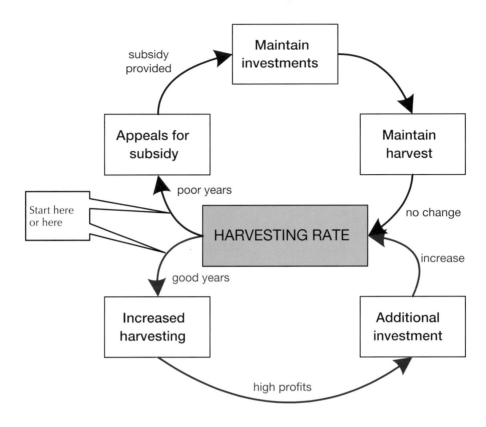

Figure 18.6 Ludwig's ratchet. For a fluctuating resource like a fishery, continuing economic investment and ecological optimism fuel a positive feedback that ratchets up the harvest rate to unsustainable levels and the eventual collapse of the fishery. Government subsidies in poor years are the key reason for this problem because they keep the harvesting rate high when it should be reduced. (Modified from Ludwig *et al.* 1993.)

■ 18.3 EXPLOITATION RATES MAY BE RAISED TO A POINT WHERE THEY CAUSE EXTINCTION OF THE RESOURCE

In most harvested systems, the species being harvested is a common species, so that over-harvesting causes commercial or economic extinction rather than true extinction. Because so many of the world's fisheries have been over-harvested, and have therefore collapsed, there are a great many examples that we can use to illustrate the folly of over-exploitation— we will discuss only two cases.

The Collapse of the Northern Cod Fishery

The Atlantic cod (*Gadus morhua*) is a marine fish that occurs in cool northern waters off eastern Canada, living in near-shore areas and out on the continental

shelf to a depth of 600 m. Cod have played a major role in the early colonization of North America by Europeans. When John Cabot came from England to Newfoundland in 1497 he found the sea 'swarming with fish—which can be taken not only with nets, but in baskets let down with a stone'. Basque fishermen from northern Spain had preceded Cabot to the Grand Banks off Newfoundland to catch cod, salt them and transport them back to Europe. Salted cod was a delicacy in Europe during the 15th and 16th centuries, and the need to dry cod and salt it on land drove the settlement of the north-eastern part of North America. Since Cabot's time, the Atlantic cod has been the dominant commercial species of the North-west Atlantic. Now it borders on extinction—a victim of over-fishing—and the collapse of the cod fishery has been a social, economic and ecological disaster for the people of Newfoundland.

ESSAY 18.2   HOW LARGE IS THE GLOBAL FISHERIES CATCH?

Since the 1950s, fisheries scientists have been trying to estimate the maximum sustainable yield of all the marine resources in all of the globe's oceans, but it has only been possible to gather reliable data for the past 30 years via the Food and Agriculture Organization of the United Nations. In 1969 in a prescient paper, John Ryther estimated that the maximum sustainable yield from the oceans would be about 100 million tones per year. Recent analysis of the catches reported by the United Nations from 1970 to 1999 by Reg Watson and Daniel Pauly have produced a slightly less optimistic forecast, shown in the following graph:

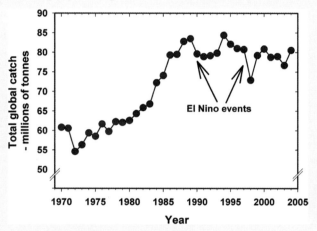

Total global fisheries catch from 1970 to 2005

The total global catch peaked in 1988 around 83 million tones and has been steady or declining for the past 15 years. In particular, El Niño events during the 1990s have reduced the catch below this maximum.

Marine resources provide a significant fraction of the protein needs of the world's population. Because the population has continued to grow, and the global catch appears to be at a maximum level, the amount of seafood available per person has fallen since 1988, as shown on the following graph:

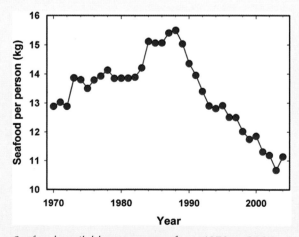

Seafood available per person from 1970 to 2004

434

Seafood per person has fallen 30% in the last 19 years and, if this trend continues, will be only half of its peak level by 2018.

The world's marine fisheries are in a deep crisis due to over-fishing and, if we are to maintain these renewable resources for the future, we must learn how best to harvest sustainably lest we repeat on a global scale the tragedy of the Peruvian anchovy fishery.

Cod populations of the North-west Atlantic can be subdivided into stocks, and fishery managers have tried to place boundaries on the stocks, or subpopulations, so that they could be managed separately (Figure 18.7). Our analysis here will concentrate on the northern cod stock occupying areas 2J, 3K and 3L on this map. This cod stock undergoes extensive spawning migrations of up to 800 km, moving north to the coast of Labrador to spawn in winter, and then south to the inshore areas of Newfoundland in summer. Cod are generalist predators and feed on herring, capelin, sand lance and a great variety of smaller fishes, as well as shrimp and crabs. Young cod feed on small crustaceans in the plankton and, as they grow, feed on shrimp and amphipods, as well as other juvenile fishes. Cod will eat almost anything. Female northern cod reach maturity at 6–7 years of age, and an 11-year-old female will lay about 2 million eggs each year. Fecundity increases geometrically with size—a 16-year-old female cod lays about 11 million eggs each year. Fertilized eggs rise to

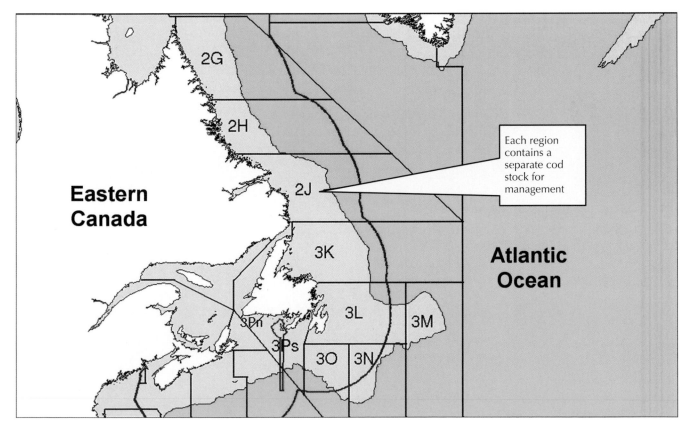

Each region contains a separate cod stock for management

Figure 18.7 Boundaries of the North Atlantic Fisheries Organization regions for the cod fishery off eastern Canada. The northern cod stock largely occupies the Grand Banks of Newfoundland—regions 2J, 3K and 3L. The edge between the pink and blue areas marks the edge of the continental shelf. The red line marks the 200-mile limit over which Canada has fisheries jurisdiction. (Map from Fisheries and Oceans Canada 1999.)

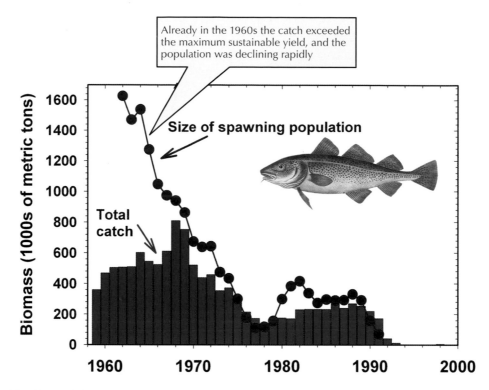

Figure 18.8 Total landings of the fishery for the northern cod (red histogram) from the 2J, 3K and 3L regions off Newfoundland from 1959 to 1998, and the estimated biomass of the spawning population (blue line). The fishery completely collapsed from 1989 to 1992, when it was closed and it remains closed in 2007. The size of the remaining population after 1992 is too small to show on the graph. (Data from Fisheries and Oceans Canada 1999.)

the ocean surface, where they are fed upon by a variety of predators. Only about 1 egg in a million will survive to complete the life cycle—mortality of eggs and larval cod is extremely high.

The cod fishery operated sustainably for nearly 500 years. During the 1600s the annual catch of cod was about 100,000 metric tons per year, and this rose as high as 200,000 tons in the 1700s due to increased demand in Europe for salted cod. During the 1800s the catch ranged from 150,000 tons to 400,000 tons annually. Until 1900 all the cod was salted and dried. After 1900 fishing boats became larger and had more efficient nets, and the efficiency of the fishing fleet continued to increase owing to technological enhancements. Frozen fish now replaced the older methods of preservation and cod fishing intensity continued to grow. During the 1950s about 900,000 tons of cod were harvested in the North-west Atlantic, and this increased to 2 million tons in the 1960s. Of this total harvest, the northern cod stock contributed about 40%. Figure 18.8

shows the harvest of northern cod from 1959 to 1998. The very high catches of the 1960s followed a series of good years and, in hindsight, were clearly unsustainable. In 1977 Canada extended control over coastal fishing out from 12 to 200 nautical miles, and most of the cod fishery off Newfoundland came under Canadian management. Canadian management has been a disaster. Spawning biomass of cod in 1991 was 4% of what it was in 1962. The cod fishery collapsed and the dominant commercial fishery of the North-west Atlantic was closed in 1992, throwing 35,000 Newfoundlanders out of work. It was still closed in 2006 and no recovery of the cod population is evident. What went wrong with the management of this fishery?

Two major scientific errors occurred in the management of northern cod stocks during the 1980s and early 1990s. Firstly, the estimates of the size of the stock were far too high. Cod stocks are estimated by measuring the catch in a series of fishing surveys off the coast. There appeared to be a change in the behavior of cod as their numbers collapsed. Cod were not randomly spread over the fishing area, but became concentrated in high density aggregations as their numbers fell, and fishing in these aggregations, once located, produced large catches. Fishermen do not fish at random, and a change in the behavior of the cod could be mimicked by a concentration of fishing boats, so that the catch in these areas of aggregation was not a good index of the total size of the fish population. Secondly, the mortality rate of cod from fishing was grossly underestimated because there are sources of mortality that were not measured by fishery scientists. The incidental catch—young cod that are too small are illegal to sell—is usually discarded at sea, causing mortality that is due to fishing, but does not contribute to the yield to the fishery. As the abundance of older fish was reduced, more and more undersized fish would have been caught. The size of this discarded catch can be enormous, with reports from some fishing trawlers of having to catch 500,000 cod, and discard 300,000 undersize fish, to get 200,000 legal sized cod. This increase in mortality rates of young fish would impact directly on how many cod reached the adult age of 6–7 years and begin to reproduce.

There has been enormous controversy over the causes of the northern cod collapse because it has devastated the economy of Newfoundland and cost the Canadian taxpayers at least $4 billion. Why did the northern cod stock collapse? There are two major hypotheses: changed oceanic conditions and over-fishing.

There is for almost every fishery crisis a potential explanation that the crisis is due to a change in environmental conditions. In particular, for marine fisheries, the favorite culprits are postulated to be changes in ocean water temperature and salinity. This particular hypothesis attributes the collapse of the northern cod stock to changes in oceanographic conditions during the late 1980s and early 1990s. If this idea were correct, it would remove all the blame for the collapse from the fishery managers and the fishermen. But while this hypothesis may be true for some fish species, it does not appear to be correct for the northern cod. Long-term data on ocean temperatures and salinity off Newfoundland, ice data from Labrador and air temperature data for the last century do not suggest that these recent years experienced extraordinary temperatures or salinity.

A variant of this hypothesis attributes the cod decline to predation by seals. Harp seals (*Phoca groenlandica*), which eat primarily 1- and 2-year-old cod, increased in abundance in the north Atlantic during the 1980s. But surveys of these small cod (too small for the commercial fishery to catch) have shown no change in their survival rates during the 1980s and early 1990s. There is no evidence that predation by harp seals had anything to do with the cod collapse. Many politicians have called for culling the harp seal population to 'save' the cod, but there is no scientific support for such a policy (*Essay 18.3*).

All the scientific evidence points to the second hypothesis of over-fishing as the cause for the collapse of the northern cod. The decline occurred throughout the 1980s at the same time that fishing effort was increasing and fishing gear was continually improving through technology. The mortality rates exerted on cod by fishing were excessive, so that few fish reached reproductive size (age 6–7) and there were few large cod left to lay large numbers of eggs. The commercial fishery for cod was highly efficient and able to find the last concentrations of cod as the population declined. Finally, many undersized cod were discarded as the fishery went into decline, and this unreported mortality added to the collapse of the stock.

The recovery of the northern cod will take decades, yet there is continuing political pressure to re-open the fishery. There has been only limited evidence of stock recovery during the last 13 years, and there is concern that the cod stocks may never recover. If we wish to define recovery as a spawner biomass of 1.5 million tons and a recruitment of 1 billion 3-year old cod each year, it will take approximately 35 years (until around 2030) to reach that population size. The next 30 years look bleak for the fishing industry of Newfoundland—a fishery that was sustained for 500 years has been destroyed in our lifetime. The collapse of the northern cod is well described by Ludwig's ratchet.

Antarctic Whaling

The exploitation of whale populations was the subject of vigorous and heated debate during the 1970s and

ESSAY 18.3   SHOULD WE CULL PREDATORS TO IMPROVE FISHERIES?

Fishermen want to harvest cod off Newfoundland and grey seals eat cod. Should we kill grey seals to relieve the pressure on cod so that we can catch more in the fishery? Resource management conflicts often occur in situations like this when humans exploit a resource, such as cod or salmon, that are also eaten by natural predators. Fishermen in particular passionately believe that seals are taking away from their livelihood and so shoot seals whenever possible, whether legally or illegally. It would seem to be common sense that if we cull a predator the prey species will become more abundant and we will be able to harvest more of them. But, like most simple ecological wisdom, this conclusion can often be wrong for several reasons.

Consider two simple food webs that involve a fishery:

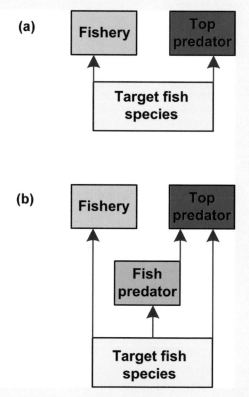

Possible food webs in a fishery

In the first case (a) we have the simple food web model, and, if this is the prevailing food web, culling the top predator (seals) will indeed allow more of the target species to be caught by the fishery. But consider case (b) in which there are other fish predators in the system. Now the energy from the target fish species can flow in three directions, and there is no simple prediction of what will happen if we cull the top predator. For example, if we cull the top predator, this will release the intermediate fish predator population, which can then eat more of the target species. The result of this potential change is to reduce the abundance of the target fish species rather than increase its abundance.

Predator culling management can thus produce the opposite effect that is desired and the fish catch may go down instead of up. The key here is to know the strengths and the magnitude of the arrows in these food webs. What is the diet of the top predator and what fraction of the target species is it taking each year? What is the diet of the fish predator and what fraction of their population is eaten by the top predator? These are difficult questions in quantitative fish population dynamics, but until they are answered there is no reason to implement a seal cull or any cull of top predators in the hope that the ecological effects will be simple. Yet the political pressures in these situations are strong, and many predator culls are implemented in the hope that the ecological interactions are simple.

1980s. At the present time all commercial whaling has been stopped and most whales are protected, and there is controversy over whether or not some whale stocks have recovered enough to sustain a limited fishery. The large whales comprise 10 species divided into two unequal groups. The sperm whale was the only toothed whale hunted commercially. The other nine species were all baleen whales, which have bony plates (baleen) in the roof of the mouth. Baleen whales are filter feeders whose principal food in the Antarctic is krill (shrimp-like crustaceans) and other plankton.

The history of whaling is characterized by a progression from more valuable species to less attractive species as stocks of the original targets were reduced. Modern whaling dates from 1868, when a Norwegian, Svend Foyn, invented the harpoon gun and the explosive harpoon. In about 1905 whalers pushed south into the Antarctic and discovered large populations of blue whales and fin whales. Blue whales dominated the catches through the 1930s, but by 1955 few were being taken (Figure 18.9). Attention was turned to the fin whale—originally the most abundant whale in the southern oceans. Fin whale numbers collapsed in the early 1960s. Sei whales were ignored as long as the bigger species were available and were not harvested until 1958. Sei whale catches were restricted after 1972 by the International Whaling Commission to prevent the collapse of these populations. All whale harvesting has historically followed the tragedy of the commons—sustained over-harvesting of the resource—until an international treaty was adopted to control exploitation.

Q **What data would you need to design a harvesting plan for whale species that have recovered in abundance?**

Harvesting models for whales have been developed extensively since 1961. The sigmoid curve-type models have proved inadequate (Figure 18.10). Maximum sustainable yield seems to occur at a density about 80% of equilibrium density, rather higher than the 50% predicted in the simple sigmoid curve model. Complications with these simple models are not difficult to find. Figure 18.10 assumes that all fin whales in the southern ocean belong to one population, but it is now known that several subpopulations occur. Whales may interact with other whale species and with seals, but most whale harvesting models are single-species models that do not recognize that many different species of whales and seals feed on krill in the Antarctic.

The present management of whales is directed to measuring the recovery rate of the depleted whale populations. Paradoxically, most of the data we now have on whales came from whaling operations and, now that commercial whaling has stopped, additional research has to be mounted to monitor how whale populations respond. Whale populations change slowly—even 10 years is a short time to estimate accurately a population's response to protection from exploitation.

The principal food of the baleen whales, krill—a group of 85 species of shrimp-like crustaceans—is now being commercially harvested in the Antarctic. Krill

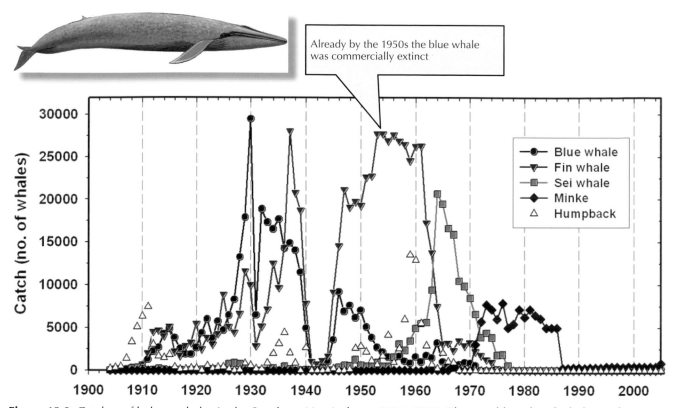

Figure 18.9 Catches of baleen whales in the Southern Hemisphere, 1904–2005. The usual lengths of whales in the commercial catches were: blue, 21–30 m; fin, 17–26 m; sei, 14–16 m; humpback, 11–15 m; and minke, 7–10 m. The blue whale is illustrated (courtesy of Frank Knight). As each species was over-harvested, whalers moved to catching the next largest species. (Data from FAO Fishery Statistics and Allen 1980.)

are on average about 6 cm long and weigh about 1–2 grams. Krill are so abundant in Antarctic waters that they have been considered for potential harvest for many years. Estimates of the sustainable harvest for krill are extremely large—almost equal to the total production of all other fisheries on the planet. Commercial harvesting began in the 1970s, but has been hampered by the remote location of the Antarctic and by processing problems with krill once they are captured. Krill have powerful digestive enzymes that tend

to spoil the catch by breaking down the edible tissues immediately after death. Krill also contain high amounts of fluoride, which must be removed before they can be used for human food. One of the emerging conservation problems of the southern oceans is to estimate the impact of krill harvesting on the recovery of whale populations in the Antarctic.

The history of whaling, and the decline of the large whales from over-harvesting, is another unfortunate example of the pursuit of economic gains to the detriment of ecological integrity in the oceans.

■ 18.4 BELOW A CERTAIN LEVEL OF EXPLOITATION, POPULATIONS ARE RESILIENT

While many marine fisheries have been over-exploited, others have been managed successfully and have avoided the pitfalls we have just described for the

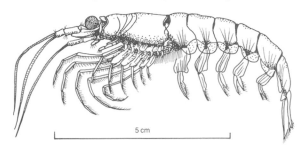

Krill (*Euphausia superba*)

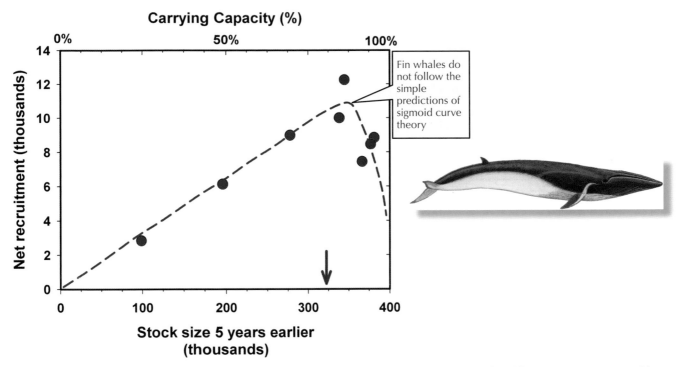

Figure 18.10 Sigmoid curve-type model for Antarctic fin whales. Estimated stock size and yield at maximum sustainable yield are indicated by arrows. The sigmoid curve model would predict maximum yield at 50% of carrying capacity (about 200,000 whales), but these data suggest maximum yield around 80–85% of carrying capacity, indicated by the red arrow on the X-axis (approximately 330,000 whales). (Drawing courtesy of Frank Knight; data from Chapman 1981.)

northern cod and whales. These success stories can serve as a basis for designing harvesting protocols that can be sustained.

Western Rock Lobster Fishery

Some fisheries are successful and sustainable—the western rock lobster fishery off Western Australia is one of the success stories. It was awarded a Marine Stewardship Council certification in March 2000—the first fishery in the world to receive this award.

Female western rock lobsters (*Panulirus cygnus*) reach sexual maturity at 6–7 years of age, and spawn in deep water in summer. The first larval stages are carried by oceanic currents for 9–11 months before the currents bring them onshore at almost one year of age in the last larval stage, called the puerulus stage, when they are about 25 mm long. Juveniles feed and grow on shallow reefs for 4–5 years and as they mature, move to deeper water to spawn. Rock lobsters are caught close to shore when they are 3–4 years old (about 500 g), before they have reached maturity (Figure 18.11).

The annual commercial catch of rock lobsters rose from very low numbers in 1944 to a peak around 11,000 tonnes by 1980, and has a commercial value of $200 to $300 million (Figure 18.12). For the past 20 years the catch has fluctuated around this average. Lobsters are captured in pot traps, and part of the successful management of this fishery has centered on strong restrictions on fishing effort. The number of pot traps that can be used and the number of days on which fishing is allowed are strongly controlled. The critical biological information that is needed for management is the size of the breeding stock, and this is estimated from the numbers of egg-carrying females captured per pot trap (Figure 18.13).

When the breeding stock declined in the early 1990s, stringent restrictions were applied to reverse this decline. These restrictions included a reduction of 18% in the number of pots to be used and an increase in the minimum legal size of lobster that could be taken in the fishery. The western rock lobster fishery has the advantage of being able to monitor the abundance of the late

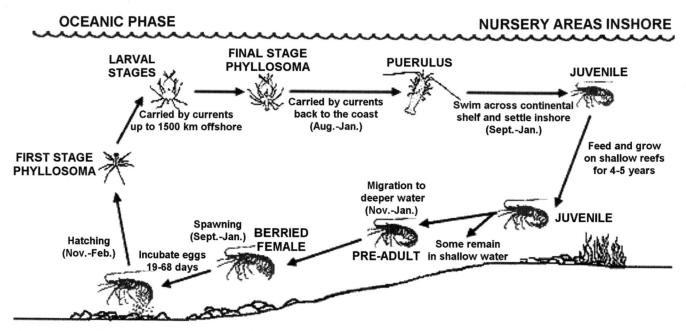

Figure 18.11. Life cycle of the western rock lobster off Western Australia (courtesy of Bruce Phillips).

larval stages each year. These small larvae settle inshore and must grow for 3-4 years before they reach catchable size for the fishery. By monitoring the abundance of these juveniles, managers can predict the allowable catch 4 years in the future, and set regulations accord-

ingly. The allowable catch fluctuates from year to year because recruitment in the rock lobster is set by oceanographic conditions during larval life, and high water temperatures (>22°C), along with strong westerly winds, increase larval growth rates and increase successful

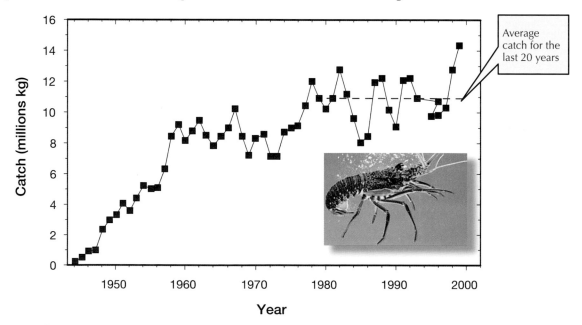

Figure 18.12 Trend in the catch of western rock lobsters since 1944. Over the last 20 years the catch has been sustainable and has averaged about 11 million kg, with fluctuations in the catch being caused by variation in survival and recruitment of juvenile lobsters. (Data from Caputi *et al.* 2003.)

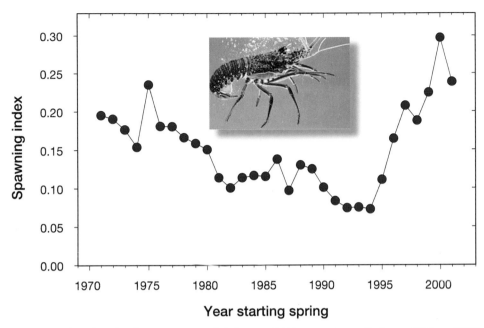

Figure 18.13. Spawning stock index for the western rock lobster off Western Australia from 1971 to 2001. This index is closely related to the number of reproductive females in the population. The drop in the spawning index in the early 1990s triggered management actions to reduce the catch and thereby reverse the population decline. The spawning stock index is a count of the number of eggs on females captured by commercial fishermen in one pot trap, relative to potential egg production per female. (Data from Caputi *et al.* 2003.)

settlement on shallow reefs close to the shore. The fishing industry co-operates with management because, with advance warning of good and poor years, they can adjust their effort and expectations of profit. Management decision rules have now been developed in consultation with the stakeholders in this fishery.

The success of this fishery is based on several factors. It is a limited-entry fishery in which each license holder is allowed to use a fixed number of pot traps, and this fishing effort can be adjusted annually to the recruitment success 3–4 years earlier. There is close governmental control of the industry, and, because all the management is located in one state body, there are no competing jurisdictions that have to be consulted and satisfied. All in all, it is a remarkable success story of a sustainable fishery.

Risk-aversive Management Strategies

Because of the many failures of fisheries management in the past, resource managers have begun to search for strategies of management that are designed to minimize risk and safeguard resources. Daniel Pauly from

the University of British Columbia has been at the forefront of the push towards the proper management of the world's fishery resources. He, along with many other fishery ecologists, has brought the ecological concepts of sustainable management into the discussion of practical fishery management.

Two different approaches have been suggested to minimize risk in harvesting. The first is to redirect management toward a harvest strategy that does not

Daniel Pauly (1946–) Director, Fisheries Centre, University of British Columbia

simply try to achieve the maximum sustainable yield. We try to find a harvesting strategy that maximizes yield averaged over a long time period during which the population fluctuates naturally. At the same time as we try to maximize long-term yields, we need to minimize the risk of resource collapse or extinction. Two popular strategies for risk-aversive harvesting are:

1. to impose a constant percentage harvest on a population
2. to harvest all individuals above a threshold population size, with no harvest allowed below the threshold.

In both cases there is a threshold or 'escapement' level below which harvesting stops. The problem with many fisheries is that this threshold is set too low to protect the resource, and often the management authorities do not know very accurately where the population is with respect to the threshold. The problem with threshold harvesting is that in some years when the population is below the threshold there will be no harvest, with the attendant economic losses and unemployment. For this reason some fishery ecologists prefer a constant harvest rate strategy.

A second general strategy is to impose protected areas, or 'no-take' zones, on the resource. This is a bet-hedging strategy in which we reduce the catch (a cost) for the benefit of a reduced risk of catastrophic collapse of the fishery. This strategy has been discussed particularly for marine fishes. The idea of a protected area in the aquatic realm is equivalent to the idea of national parks on land. They are most useful for bottom-dwelling fish that inhabit large areas of the ocean floor, and are non-migratory. The idea is simple: set aside a large enough area of a 'no-take' zone to ensure that the stock will remain at greater than 60% of carrying capacity over a given time horizon (for example, 20 years). Fishermen could harvest at a specified rate outside the 'no-take' zone, but would not be permitted to fish inside this protected area. The details of how to achieve these simple goals need to be worked out for each resource, and the detailed trade-off of costs and benefits needs to be measured if the protected area strategy is to obtain practical support among fishermen. By defining and protecting marine parks that are no-fishing zones in the ocean, it is possible to prevent over-exploitation of many marine aquatic resources.

Marine reserves are a recent concept in fisheries management and there is already enough accumulated data to suggest that they work very effectively to increase the biomass of aquatic organisms in adjacent areas. By measuring the density of marine organisms inside and outside marine reserves of varying size, marine ecologists have shown that marine reserves do indeed work to increase local density of marine organisms (Figure 18.14). Density inside the reserves was up to three times that outside the reserve. Surprisingly, this percentage increase in fish and invertebrate density was only slightly affected by the area of the reserve. But, clearly, larger reserves are much better in a quantitative sense because increasing a fish population from 10 to 20 in a small reserve has less overall impact than increasing the population from 1000 to 2000 in a large reserve.

Q **What types of marine fish species might not profit from a marine protected area strategy?**

A protected area strategy for marine fishes is clearly an important means of preventing over-fishing. Two things are essential for such a protected strategy—firstly, the area must be protected from poaching and, secondly, long-term studies need to be implemented to study the population changes that may not be apparent in a short-term analysis of the success of marine reserves. The idea of marine reserves is an important new strategy for trying to prevent the kinds of disasters we have seen in the northern cod fishery and in the history of whaling during the last century.

■ 18.5 HARVESTING CAN BE GENETICALLY SELECTIVE AND RESULT IN UNDESIRABLE EVOLUTIONARY CHANGES

Harvesting for sport has always concentrated on catching the largest fish or hunting for the largest grizzly bear. Wildlife and fishery ecologists have been concerned that such harvesting could be genetically

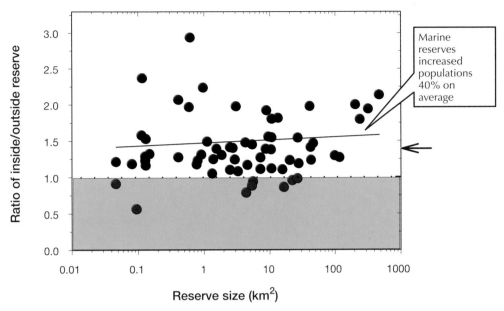

Figure 18.14 Impact of marine protected areas on the density of marine invertebrates and fishes. The ratio of density inside the reserve to that outside the reserve should be 1.0 if there is no effect. The zone below a ratio of 1.0 is shaded pink. Most reserves of all sizes showed large increases in density in the protected areas. Only 8 reserves (red dots) failed to gain from protection. (Data from Halpern 2003.)

selective, and fishery scientists in particular have been apprehensive that sustained fisheries may cause undesirable evolutionary trends. Both sport and commercial fisheries typically target larger individuals, and one possibility is that this selection by the fishery removes the fittest individuals and leaves behind individuals with slow growth rates and a reduced size at sexual maturity. Fishery scientists have been reluctant to consider evolutionary changes in harvested stocks because there has been no hard proof of genetic changes in population productivity. However, recent laboratory studies demonstrate that such genetic changes can occur rapidly.

The Atlantic silverside, *Menidia menidia*, is a common marine fish along the east coast of North America. Silverside have one generation per year, so they are suitable for laboratory selection experiments. To test for genetic changes associated with harvesting, fishery ecologists raised six experimental populations of silverside in large tanks and subjected them to three treatments: large fish harvest, small fish harvest and random size harvest. In each case they removed 90% of the fish in an artificial fishery and studied the offspring of each group to measure the possible impact of

Darwinian selection for size. Figure 18.15 shows the results after five generations of laboratory selection. There was a rapid divergence in size, with the large and small stocks differing almost two-fold both in weight and yield to the laboratory fishery.

Q **What management regulations might you introduce for a fishery that is size-selective to prevent selection against genetically large fish?**

The importance of these considerations is that management plans for fisheries that aim at maximum sustainable yield in the short term may produce the opposite effect of reducing yields in the long run because of Darwinian selection for size. How could we reduce these unwanted side effects of fishing? One possible way is to establish marine reserves where there is no fishing and thus no selection for size. A second way would be to set a maximum limit on size, instead of the usual minimum size limit, so that large fish would be protected. Harvesting of fish, wildlife and timber needs to adopt a Darwinian perspective on the possible consequences of natural selection operating through the harvesting process.

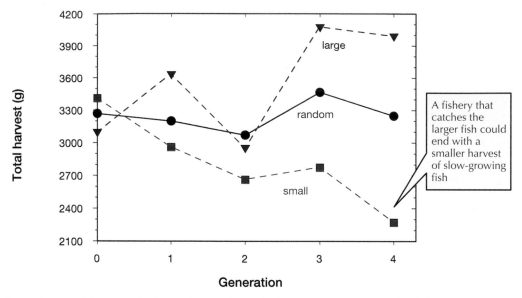

Figure 18.15 Size-selective fishing applied to Atlantic silverside for five generations in the laboratory. Small populations had 90% of the largest fish removed by the fishery each generation (= select for small fish and select against large fish), and large populations had 90% of the smallest fish removed. Random populations had random removal of 90% of the fish. The biomass harvested by generation 5 was 75% higher in the large-selected fish because they were growing faster. Growth rates are partly an inherited trait in many fish populations, and most marine fisheries operate to remove the largest fish, in effect selecting genetically for small or slow-growing individuals. (After Conover and Munch 2002.)

SUMMARY

To harvest a population in an optimal way, we must understand the factors that limit the abundance of that population. That humans so frequently mismanage exploited populations, such as the northern cod, is partly a measure of our ignorance of population dynamics. When humans harvest any population, its abundance must decline and the losses caused by harvesting must be compensated for by increased growth, increased reproduction or a reduced natural mortality (or the population will become extinct). Harvested populations typically lose the older and larger individuals, and often respond by a reduction in the age at sexual maturity. Species vary greatly in the amount of harvesting they can sustain.

Maximum sustainable yield is often the goal of resource managers. Simple and complex models have been developed to estimate the maximum sustainable yield for both fisheries and forestry. Most models contain the hidden assumption that the environment remains constant and, for this reason, they often fail in practice to prevent over-exploitation and collapse.

Economics and politics add further difficulties to achieving maximum sustainable yield for valuable populations. Harvesting is subject to a ratchet effect in which the exploitation rate is pushed by ecological optimism, economic and social pressures toward over-exploitation and collapse.

Harvesting often takes out the largest and fastest growing individuals and may shift the genetic composition of the harvested population toward smaller, slow-growing genotypes. The Darwinian consequences of harvesting have not been considered in management strategies, and managers need to adopt a broader ecological framework for their management decisions.

Management of forestry, fishery and wildlife resources is at present based more on political and economic pressures than on scientific knowledge and forecasting. Because of the inherent ecological uncertainty in anticipating future changes in populations, resource management must adopt more risk-aversive strategies. The imposition of a protected-area strategy or 'no-take' zones is one approach that can work for some fisheries. One of the great challenges of modern ecology is to help place resource management on a sustainable basis.

SUGGESTED FURTHER READINGS

Conover DO and Munch SB (2002). Sustaining fisheries yields over evolutionary time scales. *Science* **297**, 94–96. (Can natural selection affect fishery yields?)

Finney BP, Gregory-Eaves I, Douglas MSV and Smol JP (2002). Fisheries productivity in the north-eastern Pacific Ocean over the past 2200 years. *Nature* **416**, 729–733. (How oceanographic changes linked to climate have produced large changes in many fish species in the last 22 centuries.)

Halpern BS (2003). The impact of marine reserves: do reserves work and does reserve size matter? *Ecological Applications* **13**, S117–S137. (An analysis of the utility of reserves for sustaining marine fisheries.)

Hilborn R, Orensanz JM and Parma AM (2005). Institutions, incentives and the future of fisheries. *Philosophical Transactions of the Royal Society of London, Series* B **360**, 47–57. (A capsular summary of the most successful fisheries and the worst failures.)

Ludwig D, Hilborn R and Walters C (1993). Uncertainty, resource exploitation, and conservation: lessons from history. *Science* **260**, 17, 36. (A now classic analysis of why resources are rarely exploited in a sustainable manner and the origin of Ludwig's ratchet.)

Myers RA, Hutchings JA and Barrowman NJ. (1996). Hypotheses for the decline of cod in the North Atlantic. *Marine Ecology Progress Series* **138**, 293–308. (The collapse of the northern cod fishery and why it was permitted to occur.)

Pauly D, Christensen V, Guenette S, Pitcher TJ, Sumaila R, Walters CJ, Watson R and Zeller D. (2002). Towards sustainability in world fisheries. *Nature* **418**, 689–695. (Why fisheries have not been sustainable and how this might be reversed.)

Springer AM, Estes JA, van Vliet GB, Williams TM, Doak DF, Danner EM, Forney KA and Pfister B (2003). Sequential megafaunal collapse in the North Pacific Ocean: An ongoing legacy of industrial whaling? *Proceedings of the National Academy of Sciences of the USA* **100**, 12223–12228. (One hypothesis why seals, sea lions and otters have been declining in the North Pacific.)

QUESTIONS

1. Forest resources are another major natural resource subject to harvesting regimes. Are forest resources in your area being harvested in a sustainable manner? How could you determine if this was true or not? Which of the admonitions given in this chapter for fisheries would apply equally well to forest harvesting?

2. Ricker (1982) showed that, since 1950, sockeye salmon caught in British Columbia have decreased in size by 140 to 180 grams on average, which is about 5% of their weight. Discuss mechanisms by which a fishery might select in favor of smaller salmon.

3. What policy initiatives might you use to disrupt Ludwig's ratchet, shown in Figure 18.6, for a specific fishery?

4. Tropical coral reef subsistence fisheries operate on many different species of fish and are often over-harvested. What rules-of-thumb might you use to manage a tropical coral reef subsistence fishery in which you have limited funding and no detailed data on the individual fish species populations?

5. Compare marine fisheries, freshwater fisheries, agriculture and forestry with respect to ecological measures that would indicate they are operating in a sustainable manner. Are there any differences in the difficulties these four areas face when they try to become more sustainable?

6. The Peruvian anchovy fishery (Figure 18.2) is still among the largest fishery in the world. What happens to this large biomass of fish once it is caught? A start in answering this question may be found on the web site of the Royal Society for the Protection of Birds (RSPB) at http://www.rspb. org.uk/ourwork/policy/marine/fisheries/sustainable/sustainable.asp.

Chapter 19

PEST CONTROL: WHY WE CANNOT ELIMINATE PESTS

IN THE NEWS

Gorse is a perennial evergreen shrub with prickly stems and leaves and sharp spines. It is native to Britain and Western Europe, but has been introduced into Australia, New Zealand, North America, Hawaii and Chile, where it has become a serious weed. Gorse forms thickets that are impenetrable to cattle and crowd out native plants. Its leaves cannot be eaten by domestic animals other than goats.

Gorse (Ulex europaeus) *covering a hillside near Hamner Springs on the South Island of New Zealand.*
(Photo courtesy of Alice Kenney.)

Gorse was originally introduced into Australia and New Zealand as an ornamental hedge plant that was stock-proof and served as a wind shelter for cattle and sheep. It has spread widely by seed, and the result is that gorse has become one of the top ten weeds in temperate regions of the world. Gorse colonizes disturbed areas and, because it is a legume and fixes nitrogen, it can grow in almost any soil. Every year it produces many seeds that can lie dormant in the soil for 20 years and individual plants can live for 30–40 years. The result is a seed bank under gorse of up to 100 million seeds per hectare (10,000 seeds per m^2).

How can a successful weed such as gorse be controlled? Chemical, mechanical and biological methods can be used, but no single method can be used exclusively. Gorse burns well—and fire will destroy adult plants—but on the burnt area many seedlings emerge from the seed bank, potentially increasing the infestation. Sheep and goats can graze these seedlings or the small seedlings can be killed by herbicides. Bees use gorse flowers for honey—one of the positive values of this weed—so care must be taken not to use herbicides during the flowering period. Mowing seedlings is not effective. Gorse seedlings compete poorly with pasture plants such as clover and ryegrass, so pastures can be burned and then fertilized and sowed with pasture species, which eliminate gorse seedlings by competition.

Biological control has been attempted using an array of introduced insects. A seed weevil, *Exapion ulicis,* was introduced from England to New Zealand, Australia, the United States and Chile as early as 1931. Larvae of this weevil destroy 90% of the gorse seeds formed in spring, but they do not attack the seeds produced in autumn, so that annual seed production by gorse is reduced by only about 50%. Other seed predators, and five species that attack gorse foliage, have been introduced to New Zealand and Australia in recent years. The spider mite *Tetranychus lintearius,* introduced from Europe, has caused much damage to gorse foliage but does not appear to kill adult plants. Other insects are being screened for possible release in New Zealand and Australia, but, so far, biological control agents have weakened, but not eliminated, gorse infestations.

A successful strategy for gorse control must include all the techniques that reduce gorse abundance—from fire to grazing to herbicides to biological control. The main problem has been the high cost of this effort, which for farmers has averaged from $700 to $1500 per ha—a cost exceeding the value of the farmland. Weeds are opportunistic plants that are difficult and expensive to control, and part of any overall weed strategy must be to prevent the spread of weeds both within countries and between continents.

Gorse leaves, flowers, seeds, and spines. The pea-like, fragrant flowers are 2–2.5 cm long.

CHAPTER 19 OUTLINE

■ 19.1 PEST CONTROL IS APPLIED ECOLOGY THAT ASKS WHAT FACTORS LIMIT THE AVERAGE DENSITY OF A PEST

Some species interfere with human activities, in which case they are assigned the label '**pests**.' The most damaging pests we have are introduced species. The first response to pests is to **control** them. Control used in this context means to **control damage** and is not used in the engineering sense of regulating the pest population around some equilibrium density. One of the obvious ways of controlling damage is to reduce the average abundance of the pest species, but there are other ways of reducing damage by pests without affecting abundance (such as using insect repellents).

Economic Pests and Ecological Pests

In the field of pest control, a pest population is defined as being controlled when it is not causing excessive economic damage, and as uncontrolled when it is. The boundary between these two states will depend on the particular pest. An insect that destroys 4 to 5% of an apple crop may be insignificant biologically, but may destroy the grower's margin of profit. Conversely, forest insect pests may defoliate whole areas of forest without bankrupting the lumbering industry. The concept of an **economic threshold** must be applied to all questions of pest control. This includes the cost of the damage caused by the pest, cost of control measures, profit to be gained from the crop, and interactions with other pests and their associated cost. Pest control is an ecological problem as well as a social and economic problem. In this chapter we will discuss mainly the ecological aspects of pest control, but it is important to realize that the economic and social dimensions are equally important when it comes to developing pest-control strategies.

Pest control in most agricultural systems is achieved by the use of toxic chemicals, or **pesticides**. An estimated 2.5 billion kilograms (nearly 5 billion pounds) of toxic chemicals are being used annually throughout the world to control plant and animal pests. Despite the use of these pesticides, about 48% of the world's crops are lost to pests before and after harvesting. In spite of increasing pesticide use in the last 60 years,

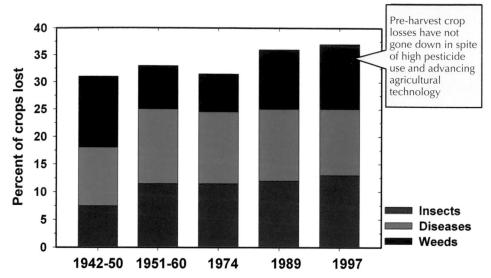

Pre-harvest crop losses have not gone down in spite of high pesticide use and advancing agricultural technology

Figure 19.1 Percentage of crops estimated to have been lost before harvest in the United States because of insect pests, plant diseases and weeds, 1942–1997. Pesticide use increased 33-fold over this period. About 20% more of the crop is lost to other pests after harvest. (Data from Pimentel *et al.* 1992, and Pimentel 1997.)

crop losses have gone up or remained constant rather than gone down as one would expect (Figure 19.1). Pesticides are only a short-term solution to the problem of pest control for several reasons. Firstly, toxic chemicals have strong effects on many species other than pests. Rachel Carson was the first naturalist to point out to the public at large the ecological consequences of toxic chemicals. The effects of DDT on bird populations (*Essay 19.1*), which Rachel Carson highlighted in *Silent Spring*, is a good example of how pesticides can degrade environmental quality. Secondly, many pest species are becoming genetically resistant to toxic chemicals that formerly killed them. Insects that attack cotton have evolved resistance to so many pesticides that it is no longer possible to grow cotton in parts of Central America, Mexico and southern Texas. Thirdly, the use of toxic chemicals in some situations can actually produce a pest problem where none previously existed. This is perhaps the most surprising effect of toxic chemicals. Rice paddies sprayed with insecticides show more pest individuals after spraying than paddies left unsprayed (Figure 19.2).

When sprayed with DDT, lemon trees in southern California became infested by massive outbreaks of a scale insect. Toxic chemicals, such as DDT, destroy many insect parasites and predators that cause mortality in the pest species and, after treatment, the few pest individuals that survive can multiply without limitation.

Strategies for Pest Control

How can we achieve pest control without these problems? There are five primary strategies for dealing with pests:

1. **Natural control**: pest populations are exposed to naturally occurring predators, parasites, diseases and competitors.

Rachel Carson (1907–1964) Author of *Silent Spring* (1962)

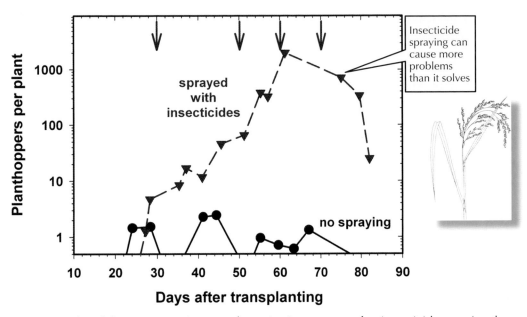

Figure 19.2 Rice crops in the Philippines contain more damaging insect pests after insecticide spraying than control areas in which no spraying was done. Broad-scale spraying of toxic chemicals (red arrows) kills many of the insect predators that limit the abundance of the critical rice pests, so that insect pest abundance and damage to the crop is worse after spraying. These data are counts of brown plant hoppers (*Nilaparvata lugens*) on individual rice plants in unsprayed paddies (black) and sprayed paddies (red). (After Way and Heong 1994.)

2. **Pesticide suppression**: pest populations are treated with herbicides, fungicides, insecticides or other chemical poisons to reduce their abundance.
3. **Cultural control:** pests are reduced by agricultural manipulations involving crop rotation, strip cropping, burning of crop residues, staggered plantings or other agricultural practices.
4. **Biological control:** pests are reduced by biological introductions of predators, parasites or diseases, by genetic manipulations of crops or pests, by sterilizing pests, or by mating disruption by the use of pheromones (sex attractants).
5. **Integrated control**, or integrated pest management (IPM), means the use of all four of these strategies to reduce pest damage with the aim of minimizing pesticide use and maximizing natural control.

In this chapter we shall discuss the principles used in biological control and cultural control and relate these to ecological theory. We will not discuss the details of chemical pesticide control.

Q Would you expect introduced species to become pests more often than native species?

Biological and cultural controls aim to reduce the average density of a pest population, and may be viewed as a practical application of the problem of what determines average abundance (Figure 19.3). The aim of pest management is not to eradicate the pest, which is usually impossible, but to reduce its impact to an acceptable level. **Eradication** is commonly thought to be the object of pest control, but, in all but a few cases, it cannot be achieved even with much effort (*Essay 19.2*). Pests are pests particularly because they are so resilient and that is what makes them so difficult to control.

The general procedure for biological control is as follows:

- a pest, often an introduced species, is causing heavy damage
- efforts are then made to find highly specialised predators and parasites in the pest's home country that can be introduced to its new country
- if the efforts are successful, the pest population is reduced to a level at which no economic damage occurs

ESSAY 19.1 DDT AND BIRD POPULATIONS

The modern environmental movement can be said to have begun with the publication in 1962 of Rachel Carson's classic *Silent Spring*, which described the results of the misuse of DDT and other pesticides. In the fable that began that volume, she wrote:

> 'It was a spring without voices. On the mornings that had once throbbed with the dawn chorus of robins, catbirds, doves, jays, wrens and scores of other bird voices there was now no sound; only silence lay over the fields and woods and marsh.'

Silent Spring was heavily attacked by the pesticide industry, but its scientific foundation has stood the test of time. Misuse of pesticides is now widely recognized to threaten not only bird communities, but the entire ecosystem as well.

The potentially lethal impact of DDT on birds was first noted in the late 1950s when spraying to control the beetles that carry Dutch elm disease led to extensive poisoning of robins in Michigan and elsewhere. Researchers discovered that earthworms were accumulating the persistent pesticide and that the robins eating them were being poisoned. Other bird species were also poisoned and gradually, thanks in no small part to Rachel Carson's book, gigantic 'broadcast spray' programs were greatly reduced.

But DDT, its breakdown products and other chlorinated hydrocarbon pesticides (and PCBs) posed a more insidious threat to birds. Because these poisons are persistent they tend to concentrate as they move through the food chains. Chlorinated hydrocarbons accumulate in fatty tissues and, for example, when seabirds feed on contaminated fish, most of the DDT from the fish ends up in a relatively few birds. With several steps in the food chain concentrating the chemical, slight environmental contamination in the water can be turned into a heavy pesticide load in birds at the top of the food chain. In one

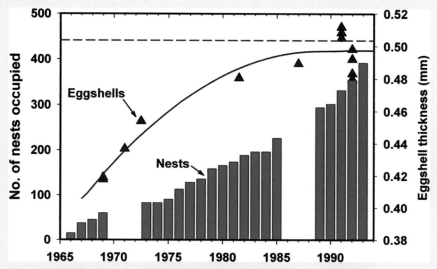

Recovery of osprey populations in Wisconsin and improving eggshell thickness since 1965. DDT and other pesticides reduced eggshell thickness until DDT was banned in 1972. Dotted red line indicates normal eggshell thickness.

Long Island estuary, concentrations of less than one-tenth of a part per million (PPM) of DDT in aquatic plants and plankton resulted in concentrations of 3–25 PPM in gulls, terns, cormorants, mergansers, herons and ospreys.

Osprey

The accumulation of large concentrations of chlorinated hydrocarbons in birds does not usually kill them outright. Rather, DDT and similar compounds, alter the bird's calcium metabolism in a way that results in thin eggshells, which crack during incubation. With little successful reproduction, bird populations plummeted.

Eggshell-thinning resulted in a dramatic decline of the osprey and brown pelican population in much of North America and the extermination of peregrine falcons in eastern United States and south-eastern Canada. Eggshell-thinning caused smaller declines in golden eagles, bald eagles and white pelicans. Fortunately, the cause of the breeding failures was identified in time, and the use of DDT was banned almost totally in the United States and Canada in 1972.

The reduced bird populations started to recover quickly after DDT was banned, with species such as ospreys and robins returning to pre-DDT levels of breeding success within 10–20 years. The peregrine falcon has been re-established in the eastern United States and brown pelican populations have now recovered. The concerns about DDT have raised awareness of the problems of persistent pesticides in the food chain, and have increased the pressure to develop pesticides that cannot accumulate in ecosystems.

Let us look at two examples of successful biological control.

■ 19.2 SUCCESSFUL PROGRAMS OF BIOLOGICAL CONTROL POINT TO PRINCIPLES FOR USE IN FUTURE ATTEMPTS

Not all biological control programs are successful, and we need to look at the details of programs that have worked well to see if we can decipher the keys to success.

Prickly Pear Cactus

Prickly pear is native to North and South America. There are several hundred species of prickly pear, about 26 of which have been introduced to Australia for garden plants. One species, *Opuntia stricta*, has become a serious weed in Australia. In 1839, *O. stricta* was brought to Australia as a plant in a pot from the southern United States and was planted as a hedge plant in eastern Australia. It gradually got out of control and was recognized as a pest by 1880. By 1900 it occupied some 40,000 km^2 (15,600 mi^2) and spread rapidly in Queensland and New South Wales (Figure 19.4):

	Area Infested with *Opuntia*	
	km^2	**mi^2**
1900	40,000	15,600
1920	235,000	90,600
1925	243,000	93,700

About half this area was dense growth completely covering the ground, rising 1 to 2 meters in height, and too dense for anyone to walk through (Figure 19.5).

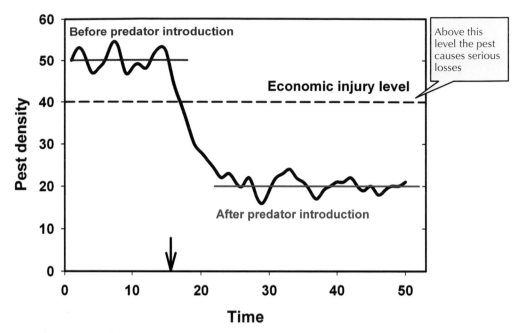

Figure 19.3 Classical pest control by biological or cultural methods. This schematic figure illustrates biological control in which the average abundance of an insect pest (blue line) is reduced after the introduction of a predator (marked by the arrow). The economic threshold or economic injury level (black dashed line) is determined by the cropping system. The position of the economic threshold is not changed by biological or cultural control programs, whose object is to reduce the average density of the pest below the economic injury level.

Prickly pear is propagated by seeds and by segments. The cactus pads, when detached from the parent plant by wind or people, can root and begin a new plant. Seeds are viable for at least 15 years. The problem of eradicating this weed was largely one of cost. The grazing land it occupied in eastern Australia was worth only a few dollars an acre, but poisoning the cactus cost about $25 to $100 an acre. Consequently, homesteads had to be abandoned to this invasion.

In 1912 two entomologists were sent from Australia to visit the native habitats of *Opuntia* and to suggest possible biological control agents that could be introduced. They sent back a mealybug from Sri Lanka, *Dactylopius indicus*, which was released and, in a few years, it had destroyed a minor pest, *Opuntia vulgaris*. But the major pest, *O. stricta*, continued to spread and, after World War I, it was subjected to a more intensive effort of biological control. Beginning in 1920, investigations in the United States, Mexico and Argentina resulted in 50 species being sent back to Australia for possible control. Of these, only 12 species were released—three were of some help in controlling

O. stricta, but only one, the moth *Cactoblastis cactorum*, was capable of eradicating it.

Cactoblastis cactorum is a moth native to northern Argentina. Two generations occur each year. On average, the females lay about 100 eggs and the adults live about 2 weeks. The larvae damage the cacti by burrowing and feeding inside the pads and by introducing bacterial and fungal infections via their burrows. Two introductions of *Cactoblastis* were made. The first introduction in 1914 failed. For the second introduction, approximately 2750 eggs were shipped from Argentina in 1925, and two generations were raised in cages until March 1926, when 2 million eggs were set out at 19 localities in eastern Australia. The moth was immediately successful, and further efforts were expended from 1927 to 1930 in spreading eggs and pupae from one field area to another.

By 1928 it was obvious that *Cactoblastis* would control *O. stricta*, so further parasite introductions were curtailed. *Cactoblastis* multiplied rapidly up to 1930 and between 1930 and 1931 the *Opuntia* stands were ravaged by an enormous *Cactoblastis* population (Figure 19.4).

ESSAY 19.2   WHEN CAN WE ERADICATE PESTS?

When introduced pest species are discovered in a country, typically a cry goes out to eradicate them. We can eradicate some pest species, but we ought to be careful about declaring eradication as a goal of any control program. Eradication implies the removal of all individuals of a species from an area to which reintroduction will not occur. If a species has spread over a wide area, it is unlikely that eradication is going to be possible, no matter how much money is available. Six factors are needed for a successful eradication program:

1. Sufficient resources to complete the project.

2. Clear lines of authority for decision making during the eradication work.

3. A target species that can be eradicated because it is easy to find and kill.

4. Effective means to prevent reintroduction.

5. Easy detection of the species when it is scarce.

6. Plans for restoration management if the species has become dominant in the community (lest one pest be replaced by another bad pest).

At present, few countries besides New Zealand and Australia have operational plans for dealing with new pests for which eradication is a possibility.

Rats introduced to islands have become serious pests of native wildlife species, especially seabirds, and some of the most successful eradication programs have been applied to rats on islands. Ninety New Zealand islands, ranging in size from 1 to 11,300 ha, have been cleared of Pacific rats, Norway rats and black rats, which were originally introduced by shipwrecks. Poisoning with anticoagulant rodenticide baits has been the major technique used in these eradication programs. Poison baits were distributed by hand, or by helicopter on larger islands. The cost of these eradication programs for rats has declined to about $100–200 per ha in 2004 and, for 10 islands, an added conservation benefit has been the eradication of feral cats along with the rats.

Larger animals, such as goats, can be readily detected and eliminated on small islands, but on larger islands with complex vegetation eradication becomes more expensive and difficult. Feral goats have been eliminated from 22 islands off New Zealand by shooting, but eradicating large animals on mainland areas is far more difficult because of immigration.

No one had expected that significant pests like rats could be eradicated on small or large islands just 10 years ago—these success stories have encouraged scientists to work toward eradication of some pests, particularly in island situations. But while eradication is the 'holy grail' of pest control, it will be possible only in a small number of situations, and we should not expect that most pest control problems will be solved so easily.

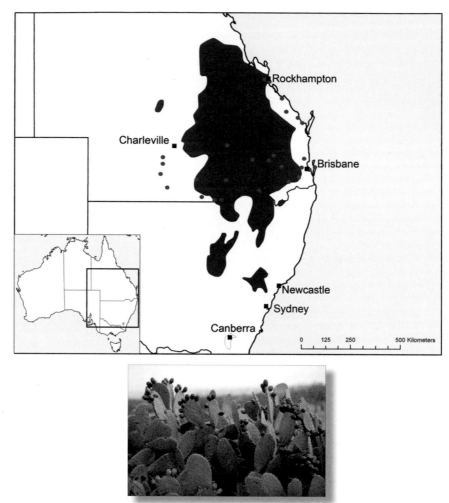

Figure 19.4 Distribution of prickly pear cactus (*Opuntia stricta*) in eastern Australia in 1925—at the peak of infestation—and in 1965–1975 (areas of small, local infestation indicated by blue dots). Black squares are cities. (After White 1981.)

Figure 19.5 (a) Dense stand of prickly pear prior to the release of *Cactoblastis*, October 1926, Chinchilla, Queensland, Australia. (b) The same stand is shown three years later after attack by the biological control agent, *Cactoblastis*, October 1929. Biological control of this cactus was achieved extremely rapidly. (After Dodd 1940; photographs courtesy of A.P. Dodd and Commonwealth Prickly Pear Board.)

This collapse of the prickly pear caused the moth population to fall steeply in 1932–1933, and the cactus then began to recover in some areas. Between 1935 and 1940 *Cactoblastis* recovered and completely controlled the cactus. Prickly pear survived after 1940 only as a scattered plant in the community. The present picture is that *Opuntia* exists in a stable metapopulation at low density maintained by *Cactoblastis* grazing. The eggs of *Cactoblastis* are not laid at random but are clumped on some plants, while other plants escape infestation entirely. Plants heavily loaded with larvae are subsequently completely destroyed, and many *Cactoblastis* larvae thus starve and die. Larvae cannot move from one plant to another if cacti are 2 meters or more apart. The clumping of the eggs of *Cactoblastis* thus both destroys *Opuntia* plants and ensures that not all plants are killed so that the entire population does not become extinct, although local populations do disappear.

Q Why should female moths lay eggs in clumps?

Most of the areas where prickly pear is now considered a periodic pest are outside the original area of dense cactus infestation (Figure 19.4) and plants in these areas seem to be partly resistant to *Cactoblastis* attack. Without *Cactoblastis*, prickly pear would make a rapid recovery.

Why was *Opuntia* such a successful plant in eastern Australia? Three important physiological properties of *Opuntia* determine its success. Firstly, the tissues of this cactus are almost entirely photosynthetic. There is minimal investment in structural tissues, and the root system is shallow and small. Secondly, *Opuntia* is capable of crassulacean acid metabolism (CAM)—a process in which CO_2 fixation is largely done at night, when minimal water vapor is lost to the atmosphere. Thus photosynthesis can thus be carried out with minimal water loss. Third, CAM plants retain photosynthetically competent tissues throughout periods of stress. When the rains arrive, CAM plants can immediately begin to photosynthesize and grow. Because of this combination of characteristics, *Opuntia* proved a near perfect opportunist with superior competitive ability over the native plants that lacked CAM metabolism. Its Achil-

les heel was found among its herbivores in the moth *Cactoblastis*.

There is one important general lesson that the *Opuntia*–*Cactoblastis* story illustrates: many different species of herbivores had to be introduced before the best candidate for control emerged. There was no way to know beforehand that *Cactoblastis* would be a better control agent than the mealybug *Dactylopius*. Researchers could only try it and see.

In an ironic flip side to the successful biological control of *Opuntia* in Australia, *Cactoblastis* is not welcome everywhere. In Florida and the Caribbean islands, where *Cactoblastis* had never occurred, there are 99 species of native *Opuntia* cacti. They are now threatened by an accidental introduction of *Cactoblastis* into Florida, and there are now widespread efforts to wipe out this moth in these areas where it is considered a pest.

Floating Fern

The floating fern *Salvinia* is a plant native to South America. It was introduced to Sri Lanka in 1939 through the Botany Department at the University of Colombo, and then spread over the next 50 years to Africa, India, South-east Asia and Australia. It is a serious aquatic weed, forming mats up to 1 m thick and covering lakes and canals, rivers and irrigation channels. Because *Salvinia* clogged waterways (Figure 19.6), all water transport and fishing was disrupted causing major problems.

Salvinia was incorrectly identified until 1972, when it was recognized to be a new species. Because of this taxonomic uncertainty, ecologists were not able to look for specialized herbivores of this plant in its native habitat until the plant was found in south-eastern Brazil in 1978. *Salvinia molesta* is unusual in being sterile, and all ramets (individual plants) appear to be genetically identical no matter where they occur in its geographical range. Plants of *Salvinia* are colonies held together by horizontal branches under the water surface. The rate of growth of *Salvinia* is limited by temperature and nitrogen concentration of the water.

Three insect species (a weevil, *Cyrtobagous singularis*, a moth and a grasshopper) found attacking *Salvinia auriculata* in Trinidad in the 1960s were introduced in

Figure 19.6 Biological control of the floating fern *Salvinia molesta*. (a) Wappa Dam, Nambour, Queensland, completely covered by *Salvinia*, October 1982. (b) The same scene in September 1983 after the population explosion of the weevil *Cyrtobagous salviniae* and subsequent crash of both *Salvinia* and the weevil. (c) The biological control agent for *Salvinia*: the adult weevil *Cyrtobagous salviniae* (1.5 mm long). (Photos courtesy of Dr. P.A. Room, CSIRO Division of Entomology and Scott Bauer, USDA.)

Sri Lanka, India, Africa, and Fiji in the 1970s. None of these introductions had any impact on the weed. Once the correct species, *S. molesta,* was located in Brazil, what were thought to be the same insect species were collected and the weevil was released in north Queensland, Australia in 1981. The weevil increased dramatically and destroyed the *Salvinia* within one year (Figure 19.6). It was then discovered to be a new species of weevil *Cyrtobagous salviniae,* and proved to be highly successful at controlling *Salvinia* in Australia, India, Sri Lanka, Botswana and Namibia. The success of the weevil is partly explained by its tolerance of high population densities before it shows interference competition and emigration occurs. The weevil reaches densities of 1000 adults per square meter and, by feeding on the buds as adults and on roots and rhizomes as larvae, the weevil either kills the plants or reduces greatly their size.

Q **How long should you wait to judge a biological control operation as a success or a failure?**

Salvinia molesta has become a significant problem in the 12 states of the southern USA from Florida to Texas and has colonized the Lower Colorado River. The introduced weevil *Cyrtobagous salviniae* has been successful in controlling these populations—the biological control of floating fern appears to be a worldwide success story.

The biological control of floating fern highlights the need for proper taxonomy of both the pest species and their potential biological control agents. Closely related species are not ecologically interchangeable, and the adaptations that determine success may be determined by small differences.

Successful Control Agents

Most biological control has operated empirically with few rules-of-thumb—an approach that has achieved some spectacular successes. But, if we are to avoid a case-by-case approach, we need to develop some general theory of biological control that could guide empirical work. General models of predator–prey interactions suggest a set of properties needed by successful biocontrol agents:

1. host specific so that they do not attack desirable species
2. life-cycle events synchronous with the pest species
3. high intrinsic rate of increase
4. ability to survive with few prey available
5. a high searching ability.

These properties are typical of many insect predators, but are not typical of predators in general. Most predators are considered by these properties to be poor candidates for biocontrol because they are generalists and not host-specific, they are rarely synchronized with the pest, they have relatively low rates-of-increase values, and they may feed on other, beneficial prey species.

The difficulty of using these general properties to develop a predictive model of selecting biological control agents is that they are still too vague, and many insect species have been introduced for biological control without success. A considerable effort has been made to develop mathematical models for biological control, particularly for predator–prey kinds of interactions. These models have been most useful for developing a general understanding of key species interactions, but they have not been able to predict what agents to use in a biocontrol operation.

■ 19.3 SELECTING FOR CROP PLANTS THAT ARE RESISTANT TO PESTS IS EFFECTIVE FOR BIOLOGICAL CONTROL

Biological control can involve introducing herbivores, predators and parasites to attack pest species, as the previous two examples have illustrated, but it can also operate by manipulating the genetic pool of the pest species being attacked. **Genetic control** is a variant of biological control that uses two strategies to reduce pest problems. Agricultural crop plants can be manipulated to increase their resistance to pests by genetic selection of strains that are better able to withstand pest attack. A second strategy is to change the pest species genetically so that they become sterile or less vigorous and thus decline in numbers, reducing damage.

The use of crop varieties resistant to attack by pests is one of the oldest and most useful techniques of pest control. In 1861 the grape phylloxera—an aphid that feeds on the roots of grape plants—was accidentally introduced into Europe from North America. The European grape (*Vitis vinifera*) was extremely susceptible to the phylloxera, and the winemaking industry of France was on the brink of collapse by 1880. The American grape (*Vitis labrusca*) is resistant to phylloxera attack, so European grape vines were grafted onto American rootstocks to produce artificial hybrid grape plants that were resistant to phylloxera attack. All European grape plants are now grafted on American grape rootstocks to maintain this resistance to phylloxera—an industry was saved by using plants with natural resistance to a pest species.

The mechanisms of resistance of American grape rootstocks seems to involve both chemical and mechanical means. Resistant rootstocks contain more phenols, which inhibit protein digestion in insects, and more tannins and plant hormones that affect the phylloxera aphids. Roots of resistant grape plants also develop a corky layer around phylloxera feeding sites, which reduces phylloxera survival by mechanical means. Presumably there is an evolutionary war going on between resistant grape plants and phylloxera, and increased virulence of phylloxera might be expected. But resistant rootstocks have controlled phylloxera for more than 120 years—a remarkable case of plant resistance to a pest insect.

Breeding Resistant Plant Varieties

Resistant varieties of many crop plants have been developed by selective breeding. In principle, the method used is very simple. Individual plants that are not being damaged are sought in an area where the pest species is common, and these plants are removed to the greenhouse for selective breeding. If resistance is inherited in the greenhouse lines, the new selected variety may be used for commercial production. The development of resistant varieties of crop plants is one of the most important continuing developments in modern agriculture.

Q **If resistant genes are so useful to a plant species, why are they not common in the population already?**

Resistant plants do not necessarily have chemical defenses, such as those against grape phylloxera. Morphological defenses can be highly effective. Soybeans are still a major crop in the midwestern United States despite the presence of a serious potential pest—the potato leafhopper (*Empoasca fabae*). The potato leafhopper will not attack soybean varieties that have leaves covered with short hairs, whereas they attack, and nearly destroy, soybeans that have smooth leaves. The hairs are a mechanical defense against insect movement and are a highly effective defense mechanism.

Breeding resistant varieties of plants has been an important factor in limiting pest damage in many crops, but the rapid adaptability of plant pathogens has compromised much effort. For example, potato blight is caused by a fungus—*Phytophthora infestans*. This disease first appeared in the 1840s in Europe, where it spread rapidly, causing the potato famine in Ireland (*Essay 19.3*). An attempt has been made to introduce single genes, derived from wild species closely related to the cultivated potato, to give a high level of resistance. Four genes have been used in this way, but, after the commercial introduction of each new gene for resistance, new races of the fungus appeared that could attack the 'resistant' potatoes. Sexual recombination or asexual mutation of fungal pathogens results in rapid evolutionary changes in field populations, so that crop resistance breaks down over time.

Genetic Engineering for Resistance

One promising area of intense development at the present time is the production of resistant crop plants by means of genetic engineering. Genes that produce resistance in one species can be moved into a crop plant to make it genetically resistant to specific pests. Alternatively, bacteria may be used as vehicles to carry biopesticide genes. Currently, *Bacillus thuringiensis* (Bt) is the main focus for developing insect-resistant crops. This bacteria normally lives in the soil and carries a gene for a toxic protein that kills the larvae of butterflies and moths. By splicing this gene into bacteria that normally live on crop plants, genetic engineers can produce insect-resistant crops. Insect pests ingest the bacteria while feeding on the plant and thereby are

ESSAY 19.3   THE POTATO FAMINE IN IRELAND, 1846-48

The Potato Famine in Ireland was a defining moment in Irish history. Before the famine began, Ireland had a population of 7 million people; when it was over there were only 4 million left. About 1 million people starved to death, and 2 million emigrated to North American and Australia. The famine was caused by a fungal disease—potato blight—that wiped out the Irish potato crop of 1846 and 1847. The *Times* of London reflected on the deaths of one million Irish with this editorial in 1847:

'All this was natural, and might have been expected from the original character and antecedent conditions of the Irish people. It was the same roote and innate disposition which thwarts and baffles and depresses them whithersoever they turn their steps. On the banks of the Liffey or the Liver, the Thames or the St. Lawrence, the Murray or the Mississippi, it's the same thing. It is this that prevents them from working when they can idle; from growing rich when they work; from saving when they receive money. It seems a law of their being—a hard, a pitiable, a saddening law; but one hitherto unaltered, and—we hope only to external appearance—unalterable. But why is it that in Manchester or Leeds or Stockport when he works and is well paid, the Irishman never thrives? The Englishman and the Scotchman from small beginnings struggled into comfort, respectability, competence; nay, sometimes, even into wealth and station. The Scotch or English spinner in no few cases has become a manufacturer and a capitalist; the Irish hardly in any. Thrown among mechanics of the two nations—receiving the same wages they do—he rarely attains the same position, or improves his condition in any degree.

All these things are facts beyond doubt and denial. We repeat them not for reproach or contumely, but to show that there are ingredients in the Irish character which must be modified and corrected before either individuals or Government can hope to raise the general condition of the people. It is absurd to prescribe political innovations for the remedy of their sufferings or the alleviations of their wants. Extended suffrage and municipal reform for a peasantry who have for six centuries consented to alternate between starvation on a potato and the doles of national charity! You might as well give them bonbons and ratafas.

We have great faith in the virtues of good food. Without attributing the splendid qualities of the British Lion wholly to the agency of beef steaks, we may pronounce that a people that has been reared on sold edibles will struggle long and hard against the degradation of a poorer sustenance … Le ventre gouverne le monde

For our own parts, we regard the potato blight as a blessing. When the Celts once cease to be potatophagi, they must become carnivorous. With the taste of meats will grow the appetite for them. With this will come steadiness, regularity, and perseverance; unless indeed the growth of these qualities be impeded by the blindness of Irish patriotism, the shortsighted indifference of petty landlords, or the random recklessness of Government benevolence. The first two may retard the improvement of Ireland; the last, continued in a spirit of thoughtless concession, must impoverish both England and Ireland. But nothing will strike so deadly a blow, not only at the dignity of Irish character, but also the elements of Irish prosperity, as a confederacy of rich proprietors to dun the national Treasury, and to eke out from our resources that employment for the poor which they are themselves bound to provide, by every sense of duty, to a land from which they derive their incomes. It is too bad that the Irish landlord should come to ask charity of the English and Scotch mechanic, in a year in which the export of produce to England has been beyond all precedent extensive and productive. But it seems that those who forget all duties forget all shame. The Irish rent must be paid twice over.'

In such a way ecological disasters can take on social and political significance, compounded with a threat to vegetarian lifestyles. This historical diversion also provides a cautionary hint not to believe everything we read in the newspapers.

Potato blight was fortunately rare during the remainder of the 19th century, and it has been only in this century that resistant genes to this fungus have been crossed into potatoes to make resistant varieties that helped to eliminate this serious disease.

poisoned. Alternatively, the Bt genes that produce the toxins can be transferred directly into the plant's genome, so the plant would protect itself.

Transgenic plants, such as cotton with Bt genes, are now put forward as the future model of crop protection—by providing plants that are genetically resistant to insect pests. One anticipated problem with this technology is that pest insects will become resistant to the biopesticide, just as they became resistant to chemical pesticides. More than a dozen species of insects are already resistant to the toxins produced by Bt. For example, Diamondback moth larvae have evolved resistance in field crops. To avoid resistance developing farmers have been advised to plant 20% of cotton or corn crops as non-transgenic plants, so that the majority of the pests develop in unmanipulated crops. But this strategy results in farmers losing money because of pest losses in the unprotected non-transgenic plants. One approach to get around this problem is to combine two toxins into transgenic plants, so that if one toxin does not kill the pest, the second one will. This strategy will work if there is no cross-resistance between the two toxins so they act independently in the pest insects. The development of resistance to both natural and artificial pesticides will continue to be a major problem in pest management in agriculture.

■ 19.4 THE FERTILITY OF PESTS CAN BE REDUCED THROUGH STERILIZATION AND IMMUNOCONTRACEPTION

Sterilization

In addition to changing the genetic make-up of the plants, we can attempt to alter the genome of the pest species. The simplest genetic manipulation that can be carried out on a pest species is sterilization. Sterility can be produced in several ways, but the usual procedure is to sterilize large numbers of pest individuals by radiation or by chemicals and then to release them into the wild, where they can mate with normal individuals. Because of sterile matings, the number of progeny produced in the next generation is greatly reduced, and control can be achieved. The sterile-insect technique cannot be used on all pest populations because it requires the rearing and sterilizing of large numbers of individuals and a situation in which immigration of fertile individuals is greatly reduced.

Mosquitoes that transmit malaria and other serious diseases are important candidates for sterile insect control methods. In a typical mosquito control program, millions of male mosquitoes are produced in a rearing factory, sterilized and released into the field. At least 28 attempts since 1960 have been carried out to

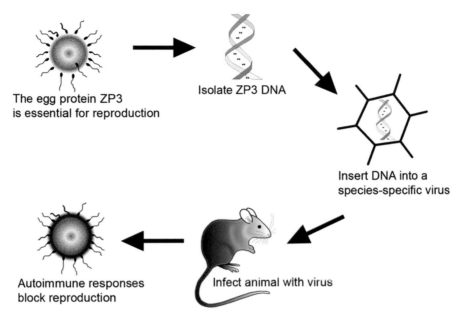

The egg protein ZP3
is essential for reproduction

Isolate ZP3 DNA

Insert DNA into a
species-specific virus

Infect animal with virus

Autoimmune responses
block reproduction

Figure 19.7 General procedure for immunocontraception in mammals, illustrated for the house mouse. The aim is to immunize the pest species against its own sperm or egg proteins, so that fertility is blocked and the reproductive rate declines to zero. The same general procedure could potentially be applied to any mammal or bird pest. (After Pech *et al.* 1997.)

use this method to reduce mosquito numbers, but many problems have plagued these trials and few have shown any success. One problem has been that sterilization has been achieved by radiation, and irradiated male mosquitoes are often not competitive with wild males. New transgenic methods of sterilizing mosquitoes are now available and may be able to overcome the problems of sterilizing individuals without reducing their vigor. The biotechnology revolution can be useful in developing improved methods of pest control, as illustrated in the next section.

Immunocontraception

A new method of biological control for vertebrates has emerged with recent developments in biotechnology—**immunocontraception**. While much of biological control has aimed at ways of increasing the mortality rate on a pest population, an alternative approach is to reduce the fertility of the pest. Immunocontraception is a biological birth-control method achieved by genetic engineering that makes a species immune to its own gametes, so that fertilization cannot occur. Fertility can be reduced by the use of immunocontraceptive vaccines delivered in a bait, or by a virus or other

contagious agent that spreads naturally through the target pest population (Figure 19.7). There are many different sperm and egg surface proteins that could be used for immunocontraception. One of the most commonly used set of proteins are the zona pellucida glycoproteins (ZPG) which facilitate sperm penetration of the egg. Antigens against zona pellucida proteins prevent sperm from attaching to the surface of the egg, and thus prevent fertilization.

Immunocontraception works. Figure 19.8 illustrates the decline in fertility of white-tailed deer in a suburban Maryland area in the United States. After the start of the contraception treatments, the population of deer declined by 8% per year. White-tailed deer are superabundant in many areas of eastern North America where all their predators have been removed and agricultural areas provide superabundant food. Land management agencies have been searching for methods of population reduction for deer that do not involve direct killing of animals in highly populated areas. The only disadvantage of this approach to deer management is that it is relatively expensive, because the immunocontraceptive must be delivered by darts. Improved methods of delivery are needed before this

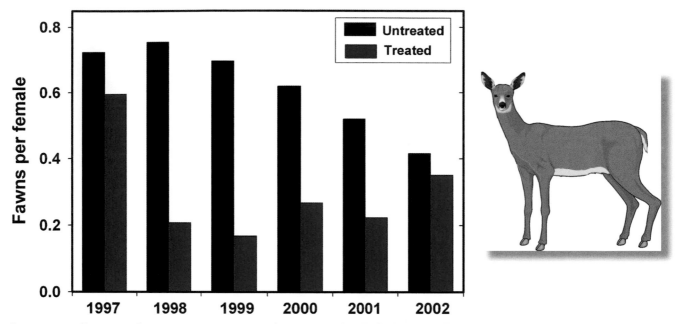

Figure 19.8 Efficiency of contraception against a free-ranging herd of white-tailed deer (*Odocoileus virginianus*) in suburban Maryland, 1997–2002. Deer were inoculated with darts containing 65 µg of porcine zona pellucida protein. As many adults as possible were treated, starting in 1997, and the bars show the fertility in subsequent years of the study. The fertility rate was cut dramatically—this could be one approach to control overabundant deer populations. (Data from Rutberg *et al.* 2004.)

technique can be applied to larger deer herds in agricultural areas.

Given that immunocontraception works, we next need to ask what impact this will have on the productivity of the pest species and what level of sterilization needs to be achieved to reduce numbers. We would expect that the higher the level of sterilization, the fewer the litters would be produced, but this is an oversimplification for mammals, which have a social system and can compensate for sterilization. Consider the case in which there is a pecking order among females. If only the top dominant one breeds, and sterilization removes her from the pecking order (allowing subordinate females to breed), the effect of sterilization on productivity is much less pronounced. So sterilizing even 50–75% of the females may have little impact on reproductive output in these social species. The main point is that imposing sterility on part of a population may not have a simple 1:1 impact on the population.

Q Would you expect to see the evolution of genetic resistance to immunocontraception in a species?

A good illustration of the impact of sterilization of pests on population trends comes from studies on the European rabbit in Australia. Female rabbits in wild populations were trapped and surgically sterilized at two sites: one in Western Australia and one in southeastern Australia. Populations with 0%, 40%, 60% and 80% sterility were then followed for 4 years. The reproductive output of rabbits decreased by the amount of sterility imposed (Figure 19.9 a). The juveniles produced, however, survived much better in the sterility treatments (Figure 19.9 b), compensating partly for the imposed sterility. The adult rabbits that were sterilized also survived much better than fertile females (Figure 19.9 c), again compensating for the sterility imposed. The net result was that population density hardly changed, in spite of sterilization, until a level of 80% imposed sterility was reached. The practical message from these experiments is that immunocontraception will not be effective in reducing rabbit numbers in Australia unless it can reach about 80% of the rabbit population annually.

Immunocontraception is an innovative idea that can assist in a coordinated plan of pest control for a

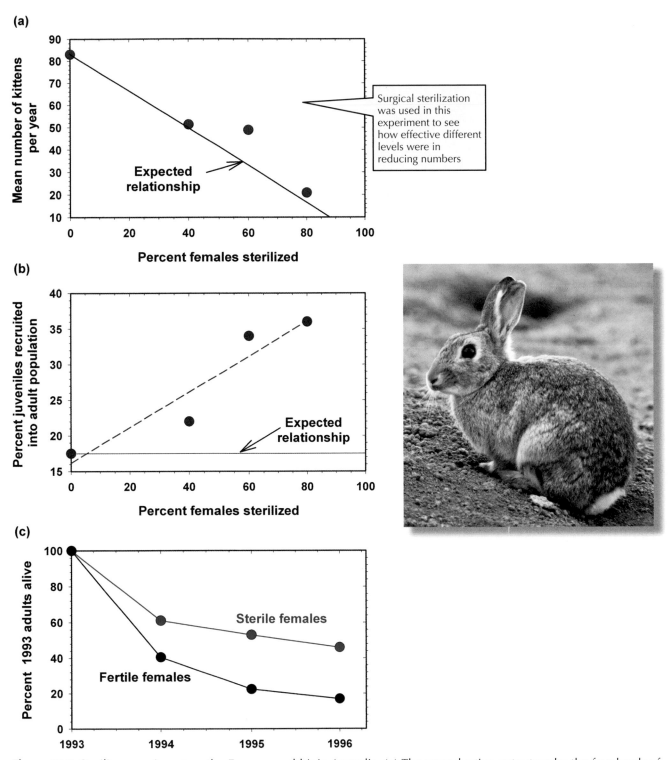

Figure 19.9 Sterility experiment on the European rabbit in Australia. (a) The reproductive output under the four levels of imposed female sterility. The number of rabbit kittens emerging from burrows is close to that expected by the percentage of sterility (blue line). (b) Percentage of juvenile rabbits surviving to become adults in the four treatments. Juvenile survival compensated for the sterility by increasing above the expected (= control) line shown in black. (c) Adult female survival during the experiment. Adult rabbits sterilized surgically in the first year of the study survived much better than females that bred, so there is a survival cost of reproduction in females that compensates for the imposed sterility. (Data from Twigg and Williams 1999.)

467

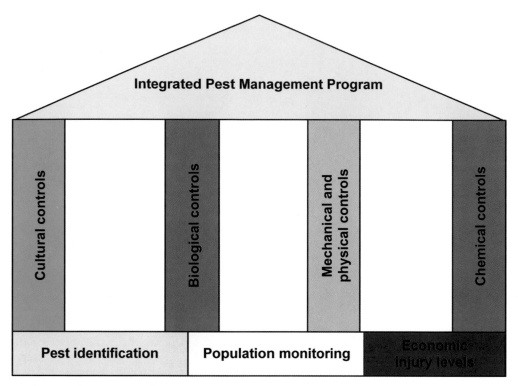

Figure 19.10 Foundations of an integrated pest management (IPM) system. No single type of pest control will be sufficient for many agricultural and forestry pests—the various alternative methods of control need to be integrated ecologically. Detailed taxonomic identification of the pest is an important foundation, as is careful monitoring of the pest population and damage. (Diagram courtesy of U.S. Department of Agriculture.)

variety of wildlife species. Much work is now under way in the United States, Australia and New Zealand to determine its potential.

■ 19.5 INTEGRATED PEST MANAGEMENT ADOPTS A SYSTEMS APPROACH AND USES ALL AVAILABLE CONTROL METHODS

Many important pests cannot be controlled by any one technique, so biologists concerned with pest management have been forced to take a wider view of pest problems. A unified approach, called **integrated control** or **integrated pest management** (IPM), uses biological, chemical and cultural methods of control in an orderly sequence (Figure 19.10). The objective of integrated pest management is to minimize economic, environmental and health risks. Integrated control can be achieved only if the population ecology of the pest and its associated species and the dynamics of the crop

system are known. Integrated control systems are ecologically sound because they rely on natural biological control as much as possible and resort to chemical treatments only when absolutely necessary. A considerable amount of information is needed to permit the effective use of an integrated control program. Density levels of the potential pest populations, stage of plant development and weather data are often required to enable the pest manager to predict the future development of the crop and to judge the need for pesticide application. Two examples will illustrate the range of controls that can be used.

Cultural Control of Rice Blast Disease

One example of a large-scale integrated pest management program is found in the rice growing area of Yunnan Province of south-western China. Rice blast is a fungal disease that attacks many, but not all, varieties of rice. By inter-planting a mixture of two varieties of rice—a traditional one susceptible to rice blast and a new high-yielding variety selected to be resistant to rice blast (Figure 19.11)—Chinese farmers have been able to reduce the incidence of rice blast and increase rice yields by 10–15%.

This simple type of cultural control provides a way of making the agricultural landscape less of a monoculture in which pests thrive. Chinese farmers have

Figure 19.11 Integrated pest management of rice crops in Yunnan Province, south-western China. (a) Two rice varieties are inter-planted and, because they differ slightly in color, they give a striped appearance to the agricultural landscape. (b) A close-up photo shows the traditional taller rice variety separated by four rows of a high-yielding dwarf variety that is resistant to rice blast disease. The traditional variety of rice is preferred for its flavor, and provides higher income to the farmers, but if it is grown as a monoculture it is devastated by rice blast. This intercropping experiment was developed by the International Rice Research Institute in the Philippines, and is now being applied to over 4 million hectares of south-western China—one of the largest scale pest control experiments yet conducted. (Photos courtesy of International Rice Research Institute, Los Banos, Philippines.)

developed a variety of inter-planting methods to control pests and reduce pesticide use and have been important pioneers in the use of IPM.

Alfalfa Weevil Control

Another example of an integrated control program is the alfalfa pest management project developed in Indiana and now in use across much of North America. Alfalfa is an important crop because it produces high-quality feed for cattle and also improves the soil by fixing nitrogen. Alfalfa is a perennial crop and is relatively long-lasting. Several hundred species of insects can be found in alfalfa fields, yet only a few are serious pests. The alfalfa weevil (*Hypera postica*) is the most important single alfalfa pest in the world, and one for which an integrated control program has been designed.

The life cycle of the alfalfa weevil in the eastern United States is shown in Figure 19.12. Eggs are laid in the fall and winter, and they hatch in the spring. New adults that emerge in the spring feed for a short time and then move into wooded areas to aestivate during the summer. In the fall, adults return to the alfalfa fields and become sexually mature. When many eggs are laid over the winter, larvae hatch and start to feed just as the alfalfa plant begins to grow in the spring. Damage can be severe, and spring weather is critical: low temperatures retard larval growth more than plant growth so that little damage occurs; higher temperatures speed larval development and increase damage.

Weather conditions are critical for determining the timing of control procedures against the alfalfa weevil. Because temperature is the major variable for both the plant and insect populations, and temperature is so variable seasonally, the best way to assess control requirements is to count the larvae of the weevil on the alfalfa stems. For example, farmers can be given these rules for determining control action:

Collect 30 alfalfa stems at random and shake the alfalfa larvae off by vigorous shaking of the stems. Count the larvae and determine:

- if weevil abundance is less than 1 larvae per alfalfa stem, no action needed.

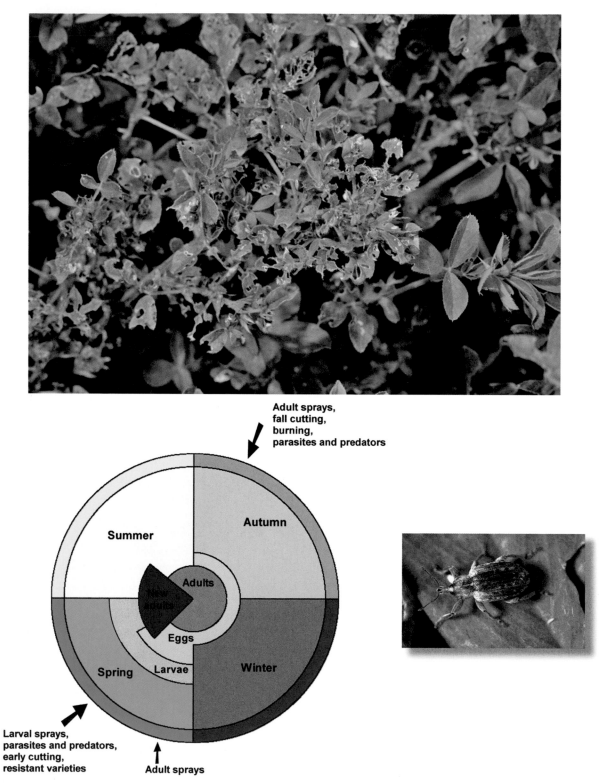

Figure 19.12 Life cycle of the alfalfa weevil in the eastern United States. Severe defoliation shown in the photo at top can occur in the spring from larval feeding. In the summer, adults move into woody habitats and then return to alfalfa fields in the fall. Possible seasonal control methods are listed. (After Armbrust and Gyrisco 1982, photo courtesy of Jack Kelly Clark, University of California Statewide IPM Project.)

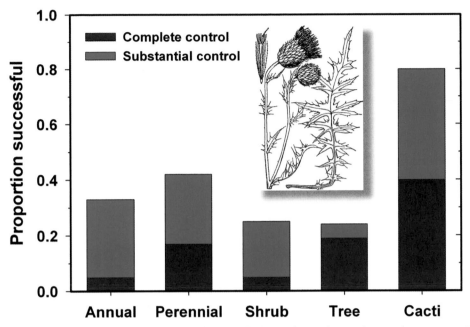

Figure 19.13 Proportion of biological control projects for weeds that achieved complete success or partial success. There were reliable data on 22 annual herb species, 36 perennial herbs, 21 shrubs, 8 trees and 20 cacti. (Data from Straw and Sheppard 1995.)

- if weevil abundance is greater than 3 larvae per stem, action is required immediately.
- if weevil abundance is between 1–3 larvae per stem, continued monitoring is needed, and the cost of treatment needs to be determined in relation to the value of the alfalfa crop.

Timing becomes a critical element in all integrated control programs—hence a detailed understanding of the life cycle of the pest and its host is necessary. The proper timing of an insecticide spraying can reduce the alfalfa weevil population, allowing the plants to grow, and delaying the onset of high weevil densities until later in the spring, when insect parasites can attack weevil larvae and maintain densities below the damage threshold. Emphasis in integrated control has shifted from trying to eradicate the pest to asking how much damage can be tolerated. This is a key point in pest control—to determine the damage. We cannot eliminate pests completely and must learn how best to live with them, taking advantage of all techniques available to keep them below the economic threshold of damage.

Integrated control programs derive their validity from field studies and are thus empirical ecology in action. They have not been developed as theoretical strategies, but as working programs, and they hold great promise for the future because they retain biological control as a core element of the integrated program.

■ 19.6 ECOLOGISTS HAVE A GENERAL THEORY OF BIOLOGICAL CONTROL BUT NOT A THEORY SPECIFIC ENOUGH TO PERMIT PREDICTION OF FUTURE SUCCESSES

Why can we not control all pests by biological control? Biological control is something akin to a gambling system—it works, sometimes. But how often? About one-third of the parasites and predators introduced get established more or less permanently after introduction. If we define success in biological control according to economic benefits, about 16% of classic biological control attempts qualify as complete successes. For weeds, about 17% of biological control attempts were completely successful (Figure 19.13). Why is this? What makes some biological control agents like *Cactoblastis* work so well, while others completely fail? A number of empirical generalizations have been suggested.

Generalizations About Successful Control

Most successful biological control programs have operated quickly. Three generations (or a maximum of three years) is suggested to be the upper limit—if definite control is not achieved in the vicinity of the colonization point within this time, the control agent will be a failure. This rule of thumb suggests that colonization projects should be discontinued after three years if no success has been achieved and that prolonged efforts at establishment are wasting money. Most of the successful biological control examples to date support this rule, which suggests that major evolutionary changes in the host–parasite system seldom occur in introduced pests. If a parasite is not already adapted to control the host, it will not evolve quickly into a successful control agent.

The unfortunate truth is that we can evaluate a biological control agent only in retrospect, and biological control programs are part gambling—we release a predator or parasite and hope for the best. A vital historical lesson is the frequency with which a critical species, such as the *Cactoblastis* moth, was released more on faith than on any evidence that it could control the pest. There is at the moment no evidence that biological control would not be just as successful if we were to release a random sample of the enemies of the pest species. But ecological theory can learn from the past successes and failures of control programs and develop new insights in how to select biological control agents.

Most successful biological control programs have resulted from a single species of parasite or predator (Figure 19.14), which raises the question 'If one parasite species is good, are two species better?' Some ecologists argue that only one species should be released at a time for pest control, because two parasites might interfere with each other when the pest is reduced to low numbers. This argument follows from the observation that native insect pests have a great number of predators and parasites. For example, the spruce budworm—a serious forest defoliator in eastern Canada—has over 35 species of parasites and many predators, yet it remains a serious pest. Is the spruce budworm a pest because it has many parasites? Or does it have many parasites because it is moderately abundant?

If competitive interactions occur between introduced parasitoids and predators, biological control should be more successful when fewer enemies are released—a prediction consistent with the data shown in Figure 19.14 for weeds. Because the importation and release of biological control agents is at least slightly risky, because non-target species may be affected, it is preferable that only a minimum number of control agents be released. Of course, this recommendation depends on the ability to identify the key agent that will be successful in biological control.

Biological control clearly works often enough that it is an important component of pest control management. Economic pressures run high in this field, because crops worth millions of dollars may be destroyed by a single pest. Consequently, states such as California have full-time bureaus devoted solely to searching the world for insects to control current agricultural pests. Candidates for control are carefully screened before they are released to make sure that they will not destroy the native fauna rather than the pest. However, once these control agents are introduced, little further work is usually done. Either the agents work, and the pest decreases, or they do not work and the entomologists look for other parasites or predators. Consequently, there is often a shortage of detailed ecological data on exactly why biological control has worked or why it has failed.

Q **Should we be able to predict which species will succeed as biological control agents?**

Resource Concentration Hypothesis

The contrast between the restricted fluctuations of natural ecosystems and the recurrent pest outbreaks in agricultural systems suggests another way of looking at pest control problems. Why do pest species thrive in our agricultural systems? Three reasons can be suggested. Firstly, agricultural systems are typically **monocultures**—often of genetically similar plant varieties—whereas natural ecosystems have a great deal of spatial complexity. The hazards of dispersal and habitat selection are greatly reduced when the habitat becomes a monoculture. Monocultures thus permit higher herbivore densities and more crop damage—an idea called the **resource concentration hypothesis.**

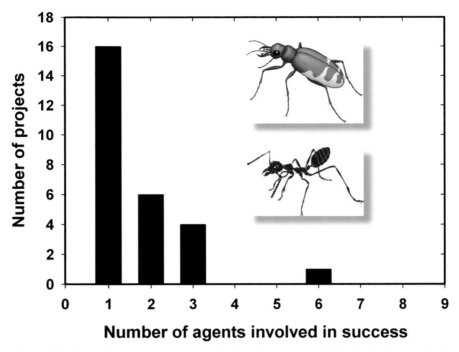

Figure 19.14 The number of biological control agents involved in the successful biological control of weeds for programs in which more than one species was released. In these 27 projects, 153 species were released, yet only one species was the main control agent in most of the successful programs. Biological control of weeds does not seem to be a cumulative result from many introduced control agents, but rather a result of one highly effective species. (Data from Denoth *et al.* 2002.)

Secondly, the plants, herbivores and predators of agricultural crops do not always form a co-evolved system, and hence the normal processes of evolutionary integration are never achieved in agricultural systems. Thirdly, the number of disturbances is much greater in agricultural systems than in natural systems. This leads to a reduced diversity of species in agricultural systems and makes natural communities and agricultural systems fundamentally different. If diversity produces stability, as we argue in Chapter 12, agricultural systems will be prone to instability and the resulting pest outbreaks. The practical message is that we should try to increase spatial complexity in crop systems.

Integrated pest management, through the use of biological control with other types of control tactics, is rapidly becoming one of the most important practical applications of ecological theory to modern problems of food production. We are gradually replacing an outmoded version of pest eradication, through toxic chemicals, with a new view of crop management with minimal environmental disturbance. To achieve this goal, we need to know the population biology of both the crops and their associated pests. The challenge is great because the pay-offs are so vital.

■ 19.7 INTRODUCED BIOLOGICAL CONTROL AGENTS MAY THEMSELVES BECOME PESTS

Introducing an alien species into an ecological system to control a pest is not without some ecological risks. The danger is that the introduced species will attack non-target organisms and cause more damage than it cures. The clearest examples are from generalized predators released for biological control. The small Indian mongoose (*Herpestes javanicus*) was introduced into Hawaii and many islands in the West Indies to control rats in sugar cane fields. It has become an important predator of native birds on all these islands, and is suspected of causing the extinction of some reptiles in the West Indies. The predatory snail (*Euglandina rosea*) has been introduced from Florida and Central America to many islands in the Pacific and Indian

Small Indian Mongoose (*Herpestes javanicus*). (Photo courtesy of Jack Jeffrey.)

Oceans to control another introduced snail, the Giant African snail *Achatina fulica*. The Giant African snail can reach 15 cm in length and was originally introduced for food. It eats hundreds of plants and is considered an agricultural pest on many islands. Biological control has consisted of introductions of predatory snails, but these snail predators are not restricted in their feeding, so they have driven native snail species to low numbers or extinction. Hawaii had 931 species of land snails, but 600 species have disappeared since European colonization (mostly because of forest clearing for agriculture). The snail *Euglandina rosea* introduced for biological control is now causing the extinction of many of the remaining native snails. The lesson learned from these mistakes is not to use generalist predators as biological control agents.

Three aspects of biological control programs have come under scrutiny because of these potential problems. Firstly, pest problems must be quantified before a biological control program is activated. How much damage is being caused, and is the problem more aesthetic than economic? Some species that are thought of as pests do not actually cause economic or environmental damage. Secondly, non-target species must be tested more widely before a potential biocontrol agent can be considered for release. A broad array of non-target species must be considered for potential harm, not just other agricultural crops, because the risk of introductions typically involves conservation problems with native species. The possible evolution of the biocontrol agent to attack new hosts also needs to be considered, for example, by determining the genetic basis of host preference. Thirdly, more research is required after a biocontrol agent has been released. To minimise costs, studies of the demographic processes by which biological control is achieved are typically severely limited, so we know only whether a release was a success or a failure and never know why. As more and more pest species are spread around the world, the necessity of strict guidelines for the release of alien species becomes stronger, so that a clear estimate of the costs, as well as the benefits, can be evaluated.

Giant African snail (*Achatina fulica*)

SUMMARY

Pests are species that interfere with human activities and hence need to be controlled. Most pest control in agricultural systems is achieved in a temporary manner with pesticides, but these toxic chemicals affect other important species and become ineffective because pests develop genetic resistance to the toxins. Biological control makes use of herbivores, predators and diseases to reduce the average abundance of the pest species. Cultural control uses crop rotation, inter-planting of crops and other agricultural practices to reduce pest damage. Pests can rarely be eradicated—the ecological objective is to limit pest damage to acceptable levels.

474

There are many cases of major reductions in numbers of introduced pests by herbivores, predators or insect parasitoids that are specially introduced for purposes of control. These successes have produced great economic benefits to farmers, and have encouraged further introductions for potential pest control. But more than 50% of biological control attempts have failed and have left the chemical control as the only means of controlling these pests. We can describe the general ecological attributes of most of the successful control agents, but we cannot predict in advance which agents might be best for control, nor can we explain why failure is so common.

Genetic techniques of biological control of pests can be accomplished by producing resistant crop plants or by interfering with the fertility or longevity of the pest. Many techniques for the genetic control of pests have been proposed, but only a few have been used successfully in the field. Immunocontraception is a new method of pest control being developed for mammalian species that are overabundant. All forms of pest control raise ecological questions of how the pest may compensate for increased mortality or reduced fertility. Genetic engineering holds great promise for producing both new methods of reducing pest numbers and crops that are more resistant to insect pests.

Integrated pest management (IPM) combines the best features of biological, cultural and chemical control methods to minimize the environmental degradation that has been typical of modern agriculture, which has relied heavily upon chemicals. To achieve integrated control, we need to understand the population dynamics of the pest species and this is currently one of the greatest challenges in applied ecology.

There are potential risks in biological control programs that the control agent will attack other native species in addition to the targeted pest. The cure must be better than the disease—extensive testing must be carried out before and after any biological control program is activated.

SUGGESTED FURTHER READINGS

Cerda H and Paoletti MG (2004). Genetic engineering with *Bacillus thuringiensis* and conventional approaches for insect resistance in crops. *Critical Reviews in Plant Sciences* **23**, 317–323. (An analysis of the best ways of making crops resistant to insect pests.)

Fenner F and Fantini B (1999). *Biological Control of Vertebrate Pests: The History of Myxomatosis, an Experiment in Evolution*. CABI Publishing, New York. (The classical case of biological control of the rabbit in Australia by a viral disease.)

Flint ML and van den Bosch R (Eds) (1981). A history of pest control. In *Introduction to Integrated Pest Management*, pp. 51–58. Plenum Press, New York. (The colorful history of the first days of biological control of pests.)

Hone J (2007). *Wildlife Damage Control*. CSIRO Publishing, Melbourne. (A good introduction to the principles of pest control for wildlife populations.)

Myers JH and Bazely D (2003). *Ecology and Control of Introduced Plants*. Cambridge University Press, Cambridge. (The definitive analysis of the successes and failures of controlling weeds.)

Secord D (2003). Biological control of marine invasive species: cautionary tales and land-based lessons. *Biological Invasions* **5**, 117–131. (The difficulties of using biological control in the oceans, and the risks that are involved.)

Thies C and Tscharntke T (1999). Landscape structure and biological control in agroecosystems. *Science* **285**, 893–895. (Cultural control of pests by arranging crops in a landscape that can help biological control in agriculture.)

QUESTIONS

1. Purple loosestrife (*Lythrum salicaria*) is a wetland plant introduced to North America from Europe in the early 1800s. It has been declared a severe environmental problem in Canada and the United States because it is believed to take over wetlands, displacing native vegetation and adversely impacting on wildlife species. Discuss how you would test the hypothesis that purple loosestrife has detrimental effects on other species in wetlands. Compare your action plan with the data presented by Hager and McCoy (1998), who questioned whether purple loosestrife does have adverse impacts.

2. Why does the bacterium *Bacillus thuringiensis* produce proteins that are toxic to insects? Review the biology of this bacterium and its geographical distribution and discuss the evolution of its toxic proteins. Lambert and Peferoen (1992) provide references.

3. Fire ants were introduced into the southern United States around 1918 and are now a serious pest. Efforts to control fire ants have been very controversial and of limited success. Review the biology of fire ants, and discuss the reasons for the poor success of control policies. Lewis *et al.* (1992) provide an overview of the problem in California.

4. Review the evidence for and against the idea that biological control is much more successful on islands such as Hawaii than on continental areas (see Myers and Bazely 2003, pp. 43–45.)

5. Why might it be difficult for seed predators to be effective biological control agents for weedy plants? What characteristics of a seed predator would you search for if you were in charge of find-

ing a good candidate species for introduction in order to control a weed? Myers and Risley (2000) discuss this problem.

6. While it has been possible to eradicate rats from small and large islands off New Zealand, it has been impossible to eradicate house mice (Towns and Broome 2003). Discuss several reasons why this might occur and how you might overcome these difficulties.

7. Explain in population dynamic terms the paradox of pest control: 'if you kill pests, they will not necessarily become less abundant.'

Chapter 20

CONSERVATION BIOLOGY: ENDANGERED SPECIES AND ECOSYSTEMS

IN THE NEWS

Seahorses are a group of 50 to 100 marine fishes that are highly adapted for living among seagrass beds, mangrove roots and coral reefs in shallow temperate and tropical waters. They occur between 50°N to 50°S latitude, with most species occurring in the Western Atlantic Ocean and Indo-Pacific region. Throughout most of this region they are under threat.

Many seahorse species are included on the IUCN Red List of Threatened Species. Because they live inshore in coastal regions, these intriguing fish are threatened by over-exploitation (for traditional medicines and aquarium display), accidental capture in fishing gear and habitat degradation. Their biology makes them particularly susceptible to over-fishing. They have a

Hippocampus kuda—*a tropical seahorse from northern Australia.* (Photo courtesy of David Harasti.)

small brood size and the male broods the young for up to 4 weeks. Most species seem to be monogamous, so mate loss reduces reproduction when population densities are low. Sea horses tend to remain in the same location, so that depopulated areas may not be recolonized quickly.

Seahorses are a classic conservation issue. Very little is known about their biology and populations in the field, yet they are exploited for trade, particularly in Asia. The majority of landed seahorses go to traditional Chinese medicine and its derivatives, such as Japanese and Korean traditional medicines. Traditional Chinese medicine is recognized by the World Health Organization as a viable health care option, and has a global following. Seahorses are used to treat a range of conditions, including respiratory disorders, such as asthma, sexual dysfunctions and general lethargy and pain. Traditional Jamu medicine in Indonesia and folk medicine in the Philippines also make use of seahorses.

China's economic growth since the mid-1980s is probably the principal cause of the great surge in current demand for seahorses. In response, subsistence fishermen in Asia have increasingly targeted seahorses, so that many now obtain the majority of their annual income from these fishes. The seahorse trade involves many fishermen and consumers, each catching or buying relatively few seahorses. Export routes are often through unofficial channels, such as personal luggage on commercial flights. The largest known net importers of seahorses are China, Hong Kong and Taiwan. The largest known exporters are Thailand, Vietnam, India and the Philippines.

The total global consumption of seahorses was at least 25 million seahorses in 2001 (more than 70 metric tonnes). Although the largest users are in Asia, many nations outside Asia also import dried seahorses for medicines and curios. In addition, the aquarium trade absorbs hundreds of thousands of live seahorses, with most destined for sale in North America, Europe, Japan or Taiwan.

The impact of removing millions of seahorses can be assessed only indirectly because global seahorse numbers are unknown, taxonomic identities are unclear, geographic ranges are undefined and fisheries undocumented. Nevertheless, local reports of declining catches are now common. Seahorse numbers that have been sampled in five countries suggest a 50% decline over the past five years. The best-studied populations in the Philippines declined 70% between 1985 and 1995. Large seahorses are considered increasingly rare, and even 'less-desirable' seahorses, such as juveniles, are now collected for traditional Chinese medicine and the aquarium trade.

Demand for seahorses exceeds supply, and Hong Kong traders have characterized potential seahorse sales as 'limitless.' Perhaps 30% of seahorses are now being used for patent medicines. Increasing demand, the specialised nature of the fishery, and the paucity of other options for many seahorse collectors, mean that the seahorse trade can be expected to persist even as seahorse numbers decline. Seahorses can be expensive. In 1995 one species of dried seahorse sold for $1200 per kilogram in Hong Kong. Live seahorses for aquaria sell for $12–$60 in the United States. As far as we know, no species has yet been driven to extinction by the seahorse trade, but this may indicate only that so little data are available for these fishes.

Seahorses present a classic conservation issue

In 2002 the 161 member countries of CITES (Convention on International Trade in Endangered Species of Wild Fauna and Flora) added seahorses to Appendix II of the convention, which means that the international trade in seahorses will have to be regulated to ensure it is not detrimental to the survival of wild populations. This took effect in May 2004 and is an important step forward in the conservation of seahorses.

Habitat degradation is a second threat to seahorse populations because they inhabit shallow, coastal areas, which are highly influenced by human activities. If trade in seahorses can be regulated, ecological information on species abundances and reproductive rates need to be gathered to determine how well populations are faring, and whether continued decline toward extinction can be prevented by good management of coastal waters and trade.

CHAPTER 20 OUTLINE

■ 20.1 CONSERVATION BIOLOGY IS THE APPLIED ECOLOGY OF ENDANGERED SPECIES

Conservation biology is concerned with population decline and scarcity, and is a central focus of much public concern. Much of population and community ecology has been focused on the abundant species in ecosystems and the factors affecting them. Many ecologists have pointed out that most species in a community are rare, and rarity itself ought to be a focus for research. Species that have become endangered or threatened are either rare or in sharp decline, and in this chapter we ask what are the causes of decline and rarity of species and what can we do to alleviate problems of threatened populations.

Conservation has become an important political issue during the last 15 years, and practical issues of conservation are continually in the newspapers and on television. The magnitude of the conservation issue can be illustrated most easily with data for well-known groups such as birds and mammals (Figure 20.1). Nearly one-quarter of all known mammals on Earth are classed as threatened species by the International Union for the Conservation of Nature. Many fewer species of reptiles, amphibians and fishes are classed as threatened, but this reflects merely the lack of study of these groups. For insects the problem is much worse because, of the one million described species of insects, less than 0.1% have been evaluated for their conservation status.

Given these serious problems of threatened and endangered species, what can ecologists do to help in their solution? Conservation biology divides rather cleanly into two separate approaches, called the **small-population approach** and the **declining-population approach**. When a species is declared to be at risk of extinction, the defining characteristics is that its population is small. Conservation biologists ask if there is a set of attributes of small populations that can help to define ways of reducing their risks of extinction—this is called the small population approach. Other conservation ecologists try to look for the endangered species of the future, and recognize these by declining population size. No matter how many individuals exist today, if a population

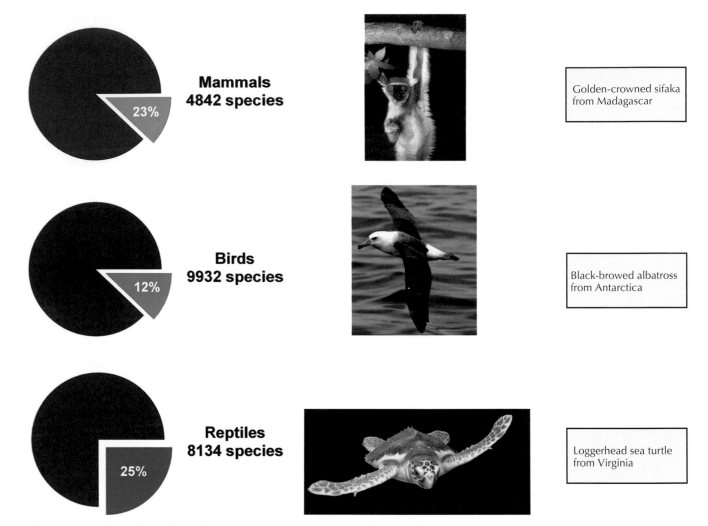

Figure 20.1 The total number of species, and the percentage of species threatened with extinction in 2003, among the mammals, birds, and reptiles. The species counts are the total number of described species in each group for the world. For most groups, only a small fraction of species have been evaluated for threatened status. Mammals and birds have been completely evaluated, but only about half of the reptiles have been evaluated, so the number threatened is an approximation. (Data from International Union for the Conservation of Nature Red List 2004; photo of sifaka courtesy of Lemur Center, Duke University; photo of albatross courtesy of Tony Palliser; photo of sea turtle courtesy of Turtles of Virginia.)

declines in size continually, it will eventually reach critical levels and become endangered. These ecologists look at ways of finding out why species decline in numbers, no matter the size of the current population. Like doctors, they hope that the diagnosis of the causes of population declines will suggest some management actions to alleviate the decline and restore the population to health. We will explore the characteristics of these two approaches to gain insight into how conservation problems might be solved.

■ 20.2 SMALL POPULATIONS CAN SUFFER FROM CHANCE EVENTS AS WELL AS INBREEDING DEPRESSION

This approach focuses on the population consequences of rarity and the abilities of small populations to deal with the genetic and demographic costs of having only a few individuals in the population. The ideal population to study is a small island population, or a small group of endangered species of animals or plants in a

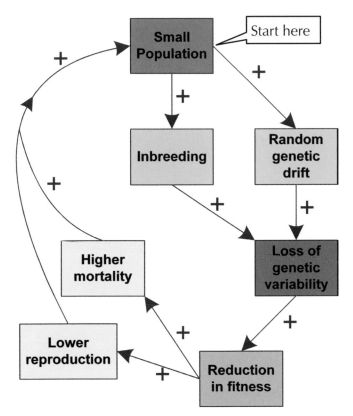

Figure 20.2 The extinction vortex for small-populations. Small populations, such as those on islands or in zoos, fall into a vortex of positive feedback loops in which small population size leads to inbreeding and genetic drift, and the loss of genetic variability. Because genetic variability is necessary for viability, fitness falls and the population size is reduced further because mortality goes up and reproduction goes down. The final result is extinction.

zoo or a botanical garden, and the questions arising from this paradigm deal largely with population genetics and demographic models of extinction in small populations.

The essence of the small-population paradigm is encapsulated in the **extinction vortex** (Figure 20.2). Small populations risk positive feedback loops of inbreeding depression, genetic drift, and demographic chance events that lead inexorably to extinction. An essential feature of the small population paradigm is a set of strong theoretical predictions that follow from population genetics theory. The key element is the maintenance of genetic variability, and the critical assumption is that species require genetic variability

for future evolution and thus long-term persistence. The flip side of this assumption is that species that have no genetic variability will never persist in evolutionary time. If this assumption is correct, conservation biologists must strive above all to maintain genetic variability, and the conservation of a species that has no genetic variability is a waste of resources because this species is doomed anyway. How can we define this goal of maintaining genetic variability? One way is to try to define a minimum viable population for species at risk.

Minimum Viable Populations

Above a certain population density, rare species will be able to sustain their numbers and not become extinct. This idea has been formalized as the concept of the **minimum viable population** (MVP)—defined as that population size that will insure at some acceptable level of risk that the population will persist for a specified time. For example, we might find that an endangered butterfly will have a 90% chance of surviving 100 years if the population size is 200 adults or more. The analysis of minimum viable populations is difficult because it involves the analysis of probabilities of extinction and these probabilities can never be exact. We need to explore the question of extinction and how it occurs.

What factors cause a species to become extinct? Once a population is small, it is subject to a variety of chance events. Some extinctions are caused by these chance events, three of which are:

1. **Demographic variability.** This source of variation reflects random variation in birth and death rates that can lead by chance to extinction. If only a few individuals make up the population, the fate of each individual can be critical to population survival. Consider the extreme case of an island population with only one male and one female. If the female produces only male offspring and then dies, this hypothetical population becomes extinct. In general demographic variability is critical to extinction only when populations are fewer than about 30–50 individuals.

2. **Genetic variability.** Because evolution cannot occur without genetic variability, any loss of genetic variation can be a cause of extinction. Many genetic studies have shown that individuals with more heterozygous loci are more fit than individuals with less genetic variation. Genetic variability is lost by **genetic drift**—the non-random assortment of genes during reproduction—and by **inbreeding**—the breeding of close relatives. Both drift and inbreeding are minimized when populations become large, but are classic examples of the problems faced by small populations.

3. **Environmental variability and natural catastrophes.** These include variation in population growth rates imposed by changes in weather and biotic factors, as well as by fire, floods, hurricanes and landslides, which can also be responsible for species declines. The key concept is how much variation the environment imposes on the birth and death rate of the population.

Small populations may become small because of habitat changes, but much of the small-population approach focuses on small populations of rare species. Rare species are particularly important in conservation biology, but not all rare species present conservation problems. Classic rare species are often those with small geographic ranges and narrow habitat specificity. Many plants of this type are restricted endemics, and are often endangered or threatened. Other rare species have very large geographical ranges and occur widely in different habitats, but always at low density. These species are ecologically interesting, but almost never appear on lists of endangered species.

- **What criteria should be specified to classify a particular species as rare?**

Detailed studies of small populations becoming extinct are rare and the few known cases illustrate how a mix of chance events can doom a species to extinction. The greater prairie chicken (*Tympanuchus cupido pinnatus*) was a common grouse from New England to Virginia when Europeans arrived in North America. The prairie chicken was originally distributed across the central plains of the United States, but has been fragmented by agriculture into scattered populations in central and western United States. In Illinois, prairie chickens numbered in the millions in the 19th century, but declined to 25,000 birds by 1933. By 1993 there were only 50 prairie chickens left in Illinois, but there were large populations remaining in Kansas, Minnesota, and Nebraska. Figure 20.3 shows the decline of one population in Jasper County of central Illinois from 1970 to 1997. The population decline was mirrored in a decline in the hatching rate of eggs, which was thought to be due to low levels of genetic diversity. In 1992 a translocation program was begun to move prairie chickens from Kansas and Nebraska and over the next 5 years a total of 271 prairie chickens were moved into Illinois. Egg viability immediately improved (Figure 20.3) and the population rebounded. Reduced genetic variability in the declining Illinois population was verified by analyzing microsatellite loci isolated from the roots of feathers from museum specimens and recent collections. The number of alleles found in the recent Illinois population is both less than the number found in other large populations and the number present in Illinois before the population decline of the last 50 years. These genetic data confirm that the collapse of the Illinois prairie chicken population followed the path of the extinction vortex until it was rescued from imminent extinction in 1992 by translocating new genetic stock.

Inbreeding and Fitness

A central problem of small populations is inbreeding depression. Inbreeding is the mating of close relatives, which has always been viewed as a serious problem in zoos as well as in agricultural animal breeding. Specific breeds of chickens, mice and dogs can suffer from inbreeding depression, evidenced by a reduction in their reproductive output and in their survival rates.

Do wild populations also suffer from inbreeding or are these effects confined to zoos and animal breeders? Extensive data now from field populations of plants, birds and mammals show that inbreeding acts to reduce the fitness of individuals and the performance of populations. In plants, seed production, germination and stress tolerance are all reduced in inbred populations. In mammals, birth weight, survival and

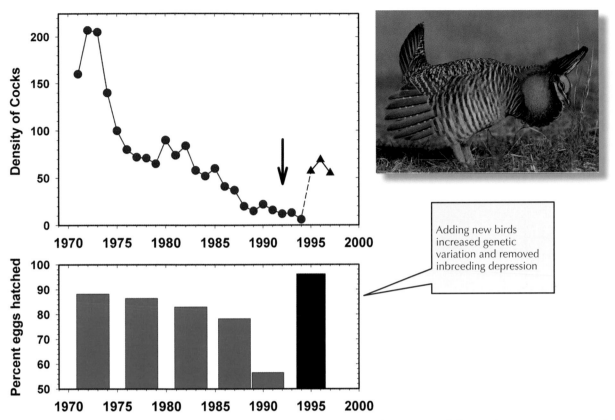

Figure 20.3 The decline of the Greater Prairie chicken (*Tympanuchus cupido*) in central Illinois from 1970 to 1997. The population collapse was mirrored in a reduction in fertility. In 1992 an experimental translocation of prairie chickens from Minnesota, Kansas and Nebraska was made to increase genetic variability (blue arrow). The population rebounded strongly after this introduction of new genes. (Modified after Westemeier *et al.* 1998, photo courtesy of Michael Zurawski.)

disease resistance are reduced by inbreeding. Let us look at two examples that illustrate the detrimental effects of inbreeding on populations in the field.

White-footed mice from the wild were taken into the laboratory, inbred and then released back into the field to measure the performance of inbred versus non-inbred mice. In the laboratory, survival of young from birth to weaning at 20 days of age was slightly reduced, but, once released into the field, adult survival was much poorer in inbred mice (Figure 20.4). In the field, inbred mice also lost body weight, while non-inbred mice gained weight. Inbreeding had more severe detrimental effects on survival in the field than in the laboratory.

Song sparrows on Mandarte Island in coastal British Columbia periodically suffer from severe winter losses during stormy weather. Survival during these storms was influenced by inbreeding—the stronger the inbreeding background of an individual, the lower the chances of survival over winter. High mortality is due to severe weather, but exactly which individuals survive storms is partly determined by inbreeding status. Small populations will always be subject to some inbreeding, but, as zoo managers have shown, inbreeding can be minimized by careful mate-selection.

The 50/500 Rule

One general principle is that the smaller the population, the greater the risk from inbreeding, as well as chance events that can lead to extinction. But how small is small, and can we use the genetic ideas associated with inbreeding to determine the minimum viable population for a particular species? Genetics and demography provide two very different approaches for estimating minimum viable populations. Population genetics provided a very simple rule for minimum

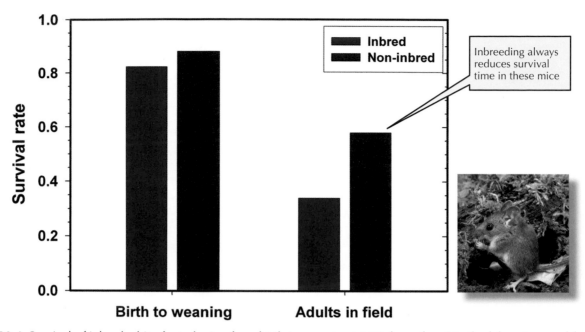

Figure 20.4 Survival of inbred white-footed mice from birth to weaning (at 20 days of age) in the laboratory and for 10 weeks after release in the field. Inbred mice survived much less well in field populations, and slightly less well in the laboratory. (Data from Jimenez *et al.* 1994, photo courtesy of Alice Kenney.)

viable populations—the 50/500 rule. Inbreeding can be kept to a low level with a minimum population of about 50 animals—this rule has worked well for animal breeders working in agriculture. This level of 50 breeders is high enough to prevent inbreeding depression, which is one factor in the extinction vortex (Figure 20.2). To prevent genetic drift, a larger population is needed, and a minimum of 500 organisms should be sufficient to allow evolution to proceed unimpeded. Real populations would often have to be 3–10 times larger than this, because not all individuals survive to breed. The 50/500 rule was put forward as a rule of thumb and should not be applied as a law for all species. Because they are based purely on genetic concepts, they cannot be applied to animals and plants that are subject to varying levels of environmental variation and different breeding systems. Simple rules for minimum viable populations need critical evaluation before we can use them in practical decisions about endangered species, and conservation biologists always strive to have much larger populations than this rule recommends.

The small population approach in conservation biology has a strong theoretical base in population genetics, and is useful in exploring the problems that small populations must deal with in order to survive in the short term and in the long term. While this approach is good at analyzing the problems faced by small populations, it does not always solve the problem of how to manage small populations, which is the focus of the next approach to conservation.

■ 20.3 DECLINING POPULATIONS NEED A DIAGNOSIS OF THE CAUSES OF DECLINE TO PREVENT EXTINCTIONS

The declining population approach takes a more activist route toward conservation problems. It focuses on ways of detecting, diagnosing and halting a population decline. In demographic terms, the problem is seen as a population in trouble and for this reason this paradigm is action oriented. Some external agent must be identified as the cause of the decline, and the research effort focuses on what can be done about it. Because it is action oriented, there is almost no theory in this paradigm. Research efforts are concentrated on each specific case study and, at least in the short term,

ESSAY 20.1   DIAGNOSING A DECLINING POPULATION

Much of practical conservation biology depends on the careful diagnosis of declining populations. Graeme Caughley in 1994 laid out a series of logical steps to determine what is driving a species toward extinction.

1. **Confirm that the species is presently in decline, or that it was formerly more widely distributed or more abundant.** This will require some qualitative or quantitative assessment of population trends and distribution. A species in decline may be common or rare. Both types can become conservation problems if the decline continues.

2. **Study the species' natural history and collect all information on its ecology and status.** For many species there is a considerable amount of background knowledge—formal and informal. Information on related species may be useful here.

3. **List all the possible causes of the decline, if you have enough background information.** This is the method of multiple working hypotheses, and you need to cast a wide net to consider all possible causes. Remember that direct human actions may be an agent of decline, but do not restrict your hypotheses to human causes.

4. **List the predictions of each hypothesis for the decline, and try to specify contrasting predictions from the different hypotheses.** Do not assume that the answer is already known by scientific or folk wisdom.

5. **Test the most likely hypothesis by experiment to confirm that this factor is indeed the cause of the decline.** Often factors are correlated with the decline, but not causing it. The best type of experiment is to remove the suspected agent of decline.

6 **Apply these findings to the management of the threatened species.** This will involve monitoring subsequent recovery until the problem of decline is resolved.

Applying this approach to an endangered species already low in abundance will be difficult, but there is no alternative. Several suspected agents of decline may have to be removed at once, and later studies undertaken to identify exactly which one was most responsible. It is better to save the species than to achieve scientific purity.

there will be no great theoretical advances in understanding the causes of extinction, but the declining population approach gets the job done when the research is carried out properly. This approach does not consider the current size of the population as important—it is the downward trend that is the main concern (*Essay 20.1*).

Paul and Anne Ehrlich of Stanford University have been instrumental in putting declining populations and extinction on the world agenda by books and popular writings on the biodiversity crisis. They have highlighted the many ways in which humans are contributing to population losses. Ehrlich's main point is that many extinctions are not due to 'chance' in the broad sense. Many population declines are completely determined by some inexorable process from which there is no escape without action. Deforestation is one such change; glaciation is another. If an area is deforested, all species that require trees are eliminated. Many extinctions occur when some essential resource is removed or when something lethal is introduced to the environment. Loss of habitat leads to deterministic extinctions, which is a major problem in almost every ecosystem on Earth.

Paul and Anne Ehrlich—conservation biologists, Stanford University

The great auk (*Pinguinis impennis*) became extinct by 1844. (Print courtesy of Natural History Museum, London.)

What processes lead to declines and ultimately extinctions? The four causes of extinction are:

1. Overkill
2. Habitat destruction and fragmentation
3. Introduced species
4. Chains of extinction

Let us consider each of these in turn.

Overkill as a Cause of Extinction

Overkill consists of fishing or hunting at a rate that exceeds a population's capacity to rebound. The species that are most susceptible to overkill are the large species with low rates of natural increase—elephants, whales, rhinoceros and other species that are considered valuable by humans. Species on small islands are also vulnerable to extinction. The great auk—a large flightless seabird—was hunted to extinction on islands in the Atlantic Ocean by the 1840s because of a demand for feathers, eggs and meat.

The decline of the African elephant is a classic example of the impact of hunting on the populations of a large mammal. The African elephant is the largest living terrestrial mammal, weighing up to 7500 kg. Sexual maturity is reached only after 10–11 years and a single calf is born every 3–9 years. The potential rate of increase is only 6% per year—a low population growth rate. Illegal hunting for ivory has been the major cause of the recent collapse of the elephant population. The demand for elephant ivory has been highly variable

and, once the price increased during the 1970s, the amount of poaching for ivory grew dramatically. There has been much controversy over the need to legalize the ivory trade and control it in those African countries, such as Botswana, that have protected their elephant population. Currently there is a ban on ivory trade, but this is having little impact in Central and East Africa where poaching is rampant.

Overkill—excessive human exploitation—will remain a problem for all animals and plants that are valuable or large.

Habitat Destruction and Fragmentation

The second factor in the evil quartet that promotes extinctions is habitat loss. Habitats may simply be destroyed to make way for housing developments or agricultural fields. Cases of habitat destruction would appear to provide the simplest examples of the declining population paradigm. One example will illustrate how subtle the impacts of habitat destruction can be.

The red-cockaded woodpecker is an endangered species endemic to south-eastern United States. It was once an abundant bird from New Jersey to Texas, inland to Missouri. It is now almost extinct in the northern and inland parts of its geographic range. The red-cockaded woodpecker is adapted to pine savannahs, but most of this woodland has been destroyed for agriculture and timber production. These birds feed on insects under pine bark and nest in cavities in old pine trees. Because old pines have been mostly cut down, the availability of nesting holes has become limiting.

Designing a recovery program for the red-cockaded woodpecker has been complicated by the social organization of this species. They live in groups of a breeding pair and up to four helpers, which are nearly all males. Helpers do not breed, but assist in incubation and feeding. Young birds have a choice of dispersing or staying to help in a breeding group. If they stay, they become breeders by inheriting breeding status following the death of older birds. Helpers may wait many years before they acquire breeding status.

From a conservation viewpoint, the problem is that red-cockaded woodpeckers compete for breeding vacancies in existing groups, rather than form new groups. New groups might occupy abandoned territories or start at a new site and excavate the cavities needed for nesting. The key problem is the excavation of new breeding cavities. Because of the time and energy needed to excavate new cavities—typically several years—birds are better off competing for existing territories than building new ones. Habitat loss appeared to be the main factor causing population decline.

To test this idea, conservation biologists artificially constructed cavities in pine trees at 20 sites in North Carolina. The results were dramatic—18 of 20 sites were colonized by red-cockaded woodpeckers and new breeding groups were formed only on areas where artificial cavities were drilled. This experiment showed clearly that much suitable habitat is not occupied by this woodpecker because of a shortage of cavities. Management of this endangered species should not be directed toward reducing mortality of these birds, but instead should focus on the provision of tree cavities suitable for nesting.

An additional complication of cavity nesting species is competition for cavities. The endangered red-cockaded woodpecker population at the Savannah River Site in South Carolina was rescued from near

Red-cockaded woodpecker. (Photo courtesy of J.L. Hanula.)

extinction by a combination of adding artificial cavities for nesting and translocating birds from larger populations nearby. To prevent competition for the artificial cavities, 2304 southern flying squirrels (*Glaucomys volans*) were removed over 10 years from 1986 to 1995. The woodpecker population responded dramatically by increasing from 4 to 99 individuals in response to these management actions.

The rescue of the red-cockaded woodpecker is a good example of how successful conservation biology must depend on a detailed understanding of population dynamics and social organization, so that limiting factors can be identified and alleviated. There are no general prescriptions for rescuing endangered species—we must operate on a case-by-case approach. Detailed information on resource requirements, social organization and dispersal powers are required before recovery plans can be specified for species suffering from habitat loss and fragmentation.

Humans have utilized a large fraction of the land surface of the Earth for agriculture, and many plants and animals cannot survive in agricultural landscapes. Of the remaining areas, many have been fragmented or broken up into small patches. We discussed the impacts of forest fragmentation in Chapter 17, and here we deal with the corollary of forest fragmentation—the loss of habitat. Habitat loss is occurring at a

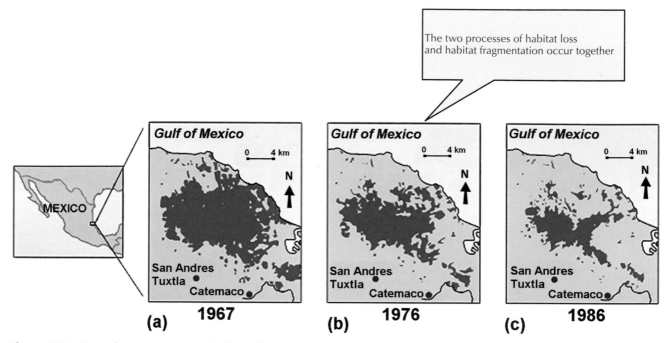

Figure 20.5 Forest loss associated with forest fragmentation in Veracruz, south-east Mexico, 1967–1986. Tropical rainforest (blue) has been removed at a rate of 4.2 percent per year. Not only is the continuous forest being fragmented, but extensive amounts of rainforest habitat have been lost to development. (Modified from Dirzo and Garcia 1992.)

rapid rate in tropical forests, which can be tracked using satellite imagery. Figure 20.5 shows tropical rainforest losses in the Sierra de Los Tuxtlas, Veracruz, Mexico, over 20 years from 1967 to 1986. Deforestation has advanced from the lowlands and, by 1986, about 84% of the original forest had been lost. By the year 2000 only 8% of the original forest remains in the form of an archipelago of small forest islands in this area. Deforestation in Veracruz is caused mostly by clearing for cattle ranches. The human population of this region has more than doubled in the last 25 years.

Habitat loss is an insidious agent of biodiversity loss because it is rare to have a good catalog of the species in a region before and after habitat clearing. Singapore is an exception. The British set up a good catalog of species present on the island of Singapore when they took command of the area in 1819. Since this time 95% of the original forest habitat has been cleared for city development, and 4 million people now live in Singapore. The original fauna and flora of Singapore remains on the 5% of habitats not yet destroyed. The result of habitat loss has been a catastrophic loss of species in all groups (Figure 20.6), averaging 70% loss of biodiversity. Habitat loss is without question a major reason for the biodiversity crisis in conservation.

Fragmentation creates patches of habitat that may be suitable for species but are never colonized. Isolated patches in tropical rainforest areas are particularly susceptible to species loss. A good illustration is the Bogor Botanical Garden, which was established in 1817 on 86 ha in west Java. Until 1936 the Botanical Garden was connected with other forest areas to the east, but for the last 60 years it has been isolated, with the nearest patch of forest 5 km away. Of the 62 bird species recorded as breeding in the Botanical Garden from 1932 to 1952, 20 species had disappeared by 1980–1985 and four more were close to extinction. The species that were lost were the less common species, and this low abundance, combined with the lack of recolonization from surrounding areas, has been the main cause of extinction. The result is that much of the conservation value of the Botanical Garden for birds has been lost because the patch of woodland is too small to support by itself a secure population of many tropical forest birds.

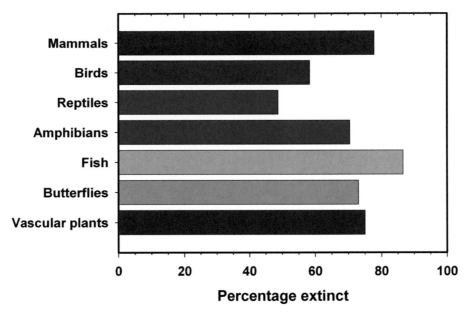

Figure 20.6 Extinction rates for some plant and animal groups from Singapore over the period 1819 to 2002. Singapore (540 km²) has lost 95% of its forests since the British first colonized the area in 1819. This habitat loss has resulted in a catastrophic loss of rainforest species after deforestation, and illustrates the kinds of losses that could be expected when habitats are extensively cleared. (Data from Brook *et al.* 2003.)

Q Is there such a thing as an 'extinction debt' associated with habitat loss?

In almost all cases, fragmentation leads to species loss. The prairies of North America are a good example. When Europeans first arrived, prairie covered about 800,000 ha of southern Wisconsin, but now occupies less than 0.1% of its original area. Plant surveys of 54 Wisconsin prairie remnants studied in 1948–54 were repeated in 1987–88. Between 8% and 60% of the plant species were lost during these four decades, at an average rate of 0.5% to 1.0% per year. At this rate of extinction approximately half the plant species would disappear in the next 50 to 100 years. This high rate of extinction can be traced to the elimination of fire from these prairie remnants. Losses were particularly high among the shorter plant species and the rare species. The control of fire in prairies seems to be the agent of decline for prairie plants, and controlled burns should be done to reverse these population declines.

Landscape ecology is an essential component in the analysis of the impacts of habitat fragmentation.

Studies of the dynamics of species in patches, and the value of corridors connecting patches are critically important for conservation, as we saw in Chapter 17.

Impacts of Introduced Species

Introduced animals are responsible for about 40% of historic extinctions. Most of these data come from mammals and birds—for which we have more detailed information—and these are no doubt biased. But no one doubts the adverse impacts of introduced species. The Nile perch was introduced into Lake Victoria in the early 1980s and caused the extinction of over 200 endemic species of cichlid fish between 1984 and 1997.

Nearly 50% of the mammal extinctions of the past 200 years are from Australia. Figure 20.7 shows the weight distribution of the threatened and extinct mammals of Western Australia. Neither the very small nor the very large mammals have been affected in these recent losses. There is a critical weight range from 35 to 4200 g which contains all the missing mammals. There are many causes that can be suggested to explain these extinctions—from habitat clearing associated with

The eastern hare wallaby (*Lagorchestes leporides*) became extinct in 1890

New Zealand eagle—the largest bird of prey known— became extinct by 1400 AD

agriculture, to changes in the fire regime, to introduced herbivores as competitors, and introduced predators. The main culprit seems to be the introduced predators, particularly the red fox. The detail of the loss of medium size marsupials in Australia mirrors closely the spread of the red fox. If the red fox can be controlled, some of the threatened species, which are now confined to fox-free offshore islands, could be reintroduced to their former range.

Q Could introduced species be a benefit to biodiversity rather than a risk?

Introduced species are currently one of the most serious conservation risks. As global trade increases, many inadvertent or deliberate introductions are being made with little regard for their conservation consequences.

Chains of Extinctions

The last of the evil quartet causing extinctions is a set of secondary extinctions that follow on from a primary extinction. If a species is lost and other species depend on it for survival, these species must also become extinct. Chains of extinctions require obligate specialist relationships that are more typical of the tropical areas than of temperate or polar zones. One obvious chain of extinctions would involve the loss of parasite species when their host becomes extinct. This has received little attention so far.

The clearest examples of chains of extinction involve large predators that disappeared when their prey became extinct. The extinct forest eagle of New Zealand (*Harpagornis moorei*), which weighed 10–13 kg, preyed on large ground birds and died out around 1400 AD when the moas became extinct in New Zealand. The decline of the black-footed ferret in North America was associated with the decline of its main food—prairie dogs—on the Great Plains. The black-footed ferret is currently being reintroduced into areas where the prairie dog colonies are safe, but its future is not secure because it is also highly susceptible to canine distemper, which is endemic in carnivores on the Great Plains.

Implicit in all these agents of extinction is the population decline that precedes the final extinction. If we can recognize declining populations, determine the processes causing the decline, and reverse these processes, we have a hope of preventing further losses of plants and animals.

■ 20.4 PARKS AND RESERVES CAN HELP PRESERVE SPECIES IF THEY ARE LOCATED PROPERLY, ARE LARGE AND WELL MANAGED

One way to conserve species in danger of extinction is to set up reserves or protected places. National parks in many countries have been viewed as protected areas

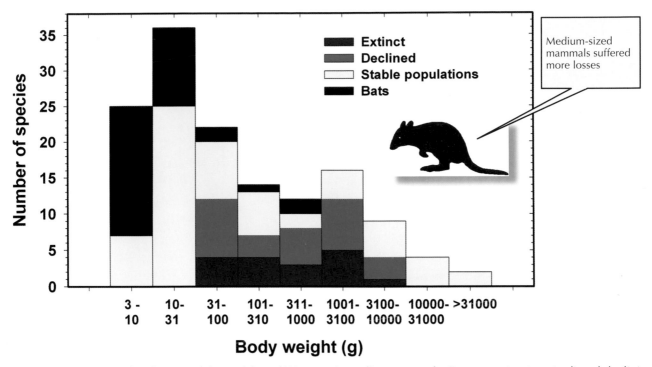

Figure 20.7 Frequency distribution of the weights of Western Australian mammals. Recent extinctions (red) and declining populations (green) have been concentrated in the medium-sized mammals between 35 and 4200 g, and the ecological question is why the smaller and the larger mammals were not affected. (After Burbidge and McKenzie 1989.)

for populations and communities. The selection and design of nature reserves is an important part of conservation biology, and much effort has gone into developing good methods of reserve selection and design. To begin, we need to specify exactly what the reserve is supposed to accomplish. Two quite divergent aims are often stated for reserves:

1. To conserve specific animal and plant communities subject to change because of fire, grazing, or predation. These reserves must be managed by intervention to set the permissible levels of fire, grazing and predation.
2. To allow the system to exist in its natural state and to change as governed by undisturbed ecological processes, so that no attempt will be made to influence the resulting changes in populations and communities.

Often reserves like national parks have both these aims, which often creates a recipe for conflict over what kinds of changes are acceptable and what kinds are unacceptable to the managers or the general public.

Few national parks and reserves are set up to exist in their natural state without any management, although some, such as Kluane National Park in the Yukon and Yellowstone National Park in the western USA, come close to this ideal. Two key questions about reserves are where to locate them, (if there is a choice) and what size of reserve is needed to achieve conservation goals.

Locating Reserves

Approximately 12.7% of the world's land is now set aside as some form of a reserve, but this is not evenly distributed, and the goal of many governments is to protect about 12% of terrestrial habitats in their jurisdiction. If we are given the job of selecting and locating reserves, how should we proceed? One way is to identify hotspots that are particularly rich in species, and to locate reserves in these areas. One problem with this approach is that areas that are hotspots for birds are typically not hotspots for, say, butterflies, so that one cannot choose reserves on the basis of only one taxonomic group and hope that it will protect other groups as well. Nevertheless, some small areas are much richer in species than others, and we should use this kind of information to help select reserves. How should land managers proceed in selecting reserves for conservation?

There are many different possible ways of selecting reserves, depending on conservation objectives. Systematic conservation planning should go through six stages (*Essay 20.2*). The conservation goal of some reserves may be to preserve rare species (such as the orang-utan in Borneo), or to preserve sites with many different species, or to preserve the largest number of higher taxonomic units, such as genera or families. Most reserve selection is carried out with presence/absence data, rather than species abundances, because it is easier to determine presence/absence than it is to estimate abundances for many species.

Q **How would conservation planning differ if you were concerned about protecting only plant species, rather than animal species?**

For a reserve system to be useful for conservation it is necessary to know the ecological requirements of the species of concern. A special problem exists for species that use temporary habitats. Many butterflies use areas temporarily for egg laying and larval development. If these areas are set aside in a reserve and the protected areas change, for example from a meadow to a forest, the butterfly would lose its host plants and disappear. Butterflies are often distributed as meta-populations, and movements between suitable patches of habitat are critical to survival.

Reserve Size

One of the most significant contributions of conservation biology has been to show that viable populations of some species are so large that it may be impossible to maintain the required number of animals in parks or sanctuaries. Figure 20.8 illustrates this problem for the grizzly bear in the Yellowstone–Grand Teton Park area. The Yellowstone Park region is among the largest park in the United States and everyone has always assumed that it was sufficiently large to preserve all of its biodiversity. Not so. If we draw a biotic boundary for the grizzly bear with a minimum viable population of 500 bears, the area needed to support this popula-

tion is 122,330 sq. km, which is about 12 times the actual park area of 10,328 sq. km. Our existing parks are far too small to maintain large mammals and birds on the scale we are now used to. Areas of private land outside of parks must also contribute to the preservation of diversity, and the integration of land use for agriculture and forestry with conservation is an important area of focus.

About 12.7% of the Earth's land area was protected in 2003, comprising nearly 102,000 areas occupying 18.8 million km^2. But most of the protected areas in the world are small (Figure 20.9), with 58% being less than 1000 ha in area and occupying only 0.2% of the total protected area of the globe. By contrast, the nine largest protected areas (including Greenland National Park at 972,000 km^2) comprise nearly 17% of the total area protected. Protected areas are not always protected from poaching and hunting, and setting aside land for conservation is an important first step, but not the end point.

The important message is that for the conservation of the Earth's biodiversity we cannot rely only on parks and reserves. We need to cultivate methods of preserving biodiversity in all areas of human use—agricultural regions, grazing areas and forest plantations.

■ 20.5 THE CONTINUED LOSS OF HABITAT AND THE HUMAN POPULATION INCREASE ARE THE ROOT CAUSES BEHIND THE CONSERVATION CRISIS

At present, conservation biology is ruled by case studies of species that are in peril, so the best way to learn about conservation biology is by looking in detail at examples of successful programs. The two examples of conservation problems that follow illustrate the practical realities of applying conservation principles to endangered plants and animals. The first example is a plant species endangered from natural events because it needs disturbed habitat. The second example is an owl endangered by excessive logging.

Furbish's Lousewort

Furbish's lousewort (*Pedicularis furbishiae*) is a small herb that was once thought to be extinct and, when

ESSAY 20.2 SYSTEMATIC CONSERVATION PLANNING

Systematic conservation planning can be separated into six stages. The process is not unidirectional—there will be many feedback loops. Begin by designating clearly the planning region, and then proceed as follows:

1. **Compile available data on the biodiversity of the planning region**

   Review existing data and decide on which data sets are sufficiently consistent to serve as surrogates for biodiversity across the region.

   If time allows, collect new data to augment, or replace, some existing data sets.

   Collect information on the localities of species considered rare or threatened in the region.

2. **Identify explicit conservation goals for the planning region**

   Set quantitative conservation targets for species, vegetation types or other features (for example, at least three occurrences of each species, 1500 ha of each vegetation type, or specific abundance targets tailored to the conservation needs of individual species).

   Set quantitative targets for minimum size, connectivity or other design criteria.

   Identify qualitative targets or preferences (for example, as far as possible, new conservation areas should have had minimal previous disturbance from grazing or logging).

3. **Review existing conservation areas**

   Measure the extent to which quantitative targets for representation and design have been achieved by existing conservation areas.

   Identify the imminence of any threat to under-represented features, such as species or vegetation types.

4. **Select additional conservation areas**

   Regard established conservation areas as focal points for the design of an expanded system.

   Identify sets of new conservation areas for consideration as additions to established areas. Options for doing this include reserve selection algorithms or decision-support software to allow stakeholders to design expanded systems that achieve regional conservation goals.

5. **Implement conservation actions**

   Decide on the most appropriate or feasible form of management to be applied to individual areas. The preferred form of management cannot always be implemented.

   If a selected area proves to be degraded or difficult to protect, return to stage 4 and look for alternatives.

   Decide on the relative timing of conservation management when resources are insufficient to implement the whole system in the short term.

6. Maintain the required values of conservation areas

Set conservation goals at the level of individual conservation areas (for example, maintain seral habitats for one or more species). Ideally, these goals will acknowledge the particular values of the area in the context of the whole system.

Implement management actions in and around each area to achieve the goals.

Monitor key indicators that will reflect the success of management actions in achieving goals. Modify management as needed.

(Source: from Margules and Pressey (2000).)

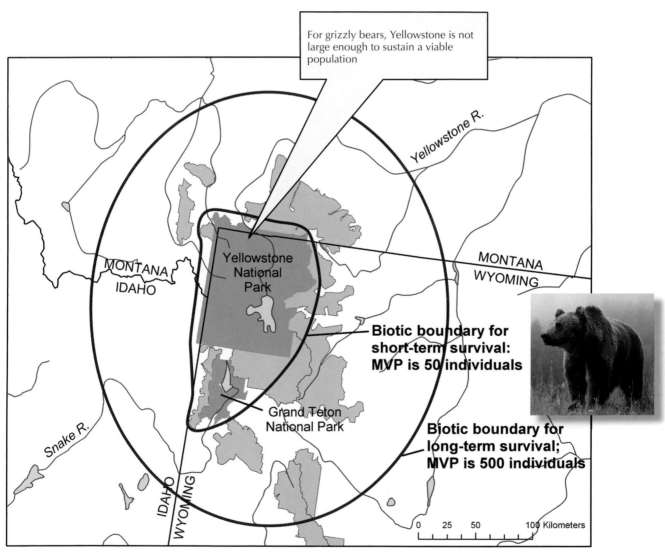

Figure 20.8 The legal and conservation-defined boundaries of the Yellowstone–Grand Teton National Park assemblage in the United States with respect to grizzly bears. The conservation-defined boundaries (red lines) are defined by the entire watershed for this park and the area necessary to support a minimum viable population (MVP, 50 individuals for short-term survival, 500 individuals for long-term survival) of the grizzly bear (*Ursus arctos*), which has a large home range (489 km^2). National parks are shown in green, federal wilderness areas in orange. (Modified from Newmark 1985.)

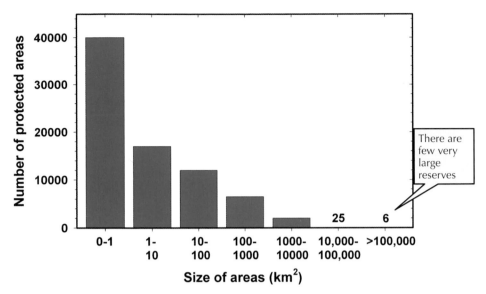

Figure 20.9 The number of protected areas in the world in 2003. There are few large areas over 10 000 km² (too small to show up on the scale, so the number of reserves is given). Most protected areas are small. The six largest protected areas include Greenland (972,000 km²), which has more ice and rock than biodiversity and Ar-Rub'al-Khali in south-western Saudi Arabia (640,000 km²), which is desert. The six largest areas comprise 13% of the total area of the globe that is protected in one form or another. (Data from IUCN 2004.)

rediscovered in northern Maine, it became one of the first plant species to be listed as endangered in the United States. Furbish's lousewort is a herbaceous perennial that reproduces only by seed. It lives along riverbanks and only in disturbed habitats, and is found along a single river in northern Maine (Figure 20.10). The key ecological process for this herb is the frequency of disturbances. Disturbance is frequent along the St. John River because of ice jams and ice scour in the spring. Ice scour is a benefit for Furbish's lousewort because it opens new habitat, but it is also a cost because it kills many plants (Figure 20.11).

To estimate population viability, conservation ecologists studied 6000 individually marked and mapped plants in 15 local populations along the St. John River. Furbish's lousewort is a poor competitor so the higher the overall vegetative cover, the less well the population does. Woody, shrub-dominated habitats also are poor for this lousewort, as are dry soils (Figure 20.11). The prediction from the study of these local populations is that open habitats with good soil moisture will support viable populations of Furbish's lousewort.

This prediction assumes that viable populations are not destroyed by catastrophic events such as ice scour.

Of 32 local populations that were regularly surveyed, 2–12% disappeared each year because of ice scour or riverbank collapse. New populations were established at a rate of 3% of empty sites per year. Population viability depends on the balance between extinction rates and establishment rates, and for the years of study extinction rates seemed to be higher than establishment rates.

• Are endangered animal species that depend on disturbances more or less difficult to conserve than endangered plant species that require disturbances?

Species such as Furbish's lousewort present a challenge for conservation biologists. We cannot simply protect the best local populations and ignore others because the disturbance regime of the river causes local extinctions by chance and new sites for colonization must be available. A reserve system that would protect only the existing populations of this plant may not provide enough recolonization sites. By contrast, too little disturbance would also doom this species because woody vegetation would take over the riverbank

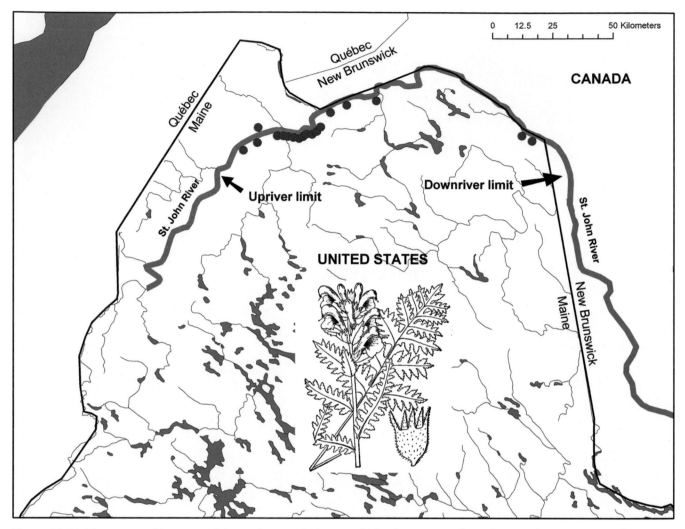

Figure 20.10 The geographic range of the endangered plant, Furbish's lousewort, along the St. John River in northern Maine. Red dots are cities and blue dots are populations of this rare plant. (Modified from Menges 1990.)

habitats. There is some optimum point of disturbance. Too much or too little could be detrimental to the long-term survival of this endangered plant.

Genetic models of population viability are not relevant for Furbish's lousewort because this species does not seem to have any detectable genetic variation. Our analysis dealt only with demographic and environmental variability in the form of natural catastrophes as potential causes of extinction for this plant species.

The Northern Spotted Owl

The northern spotted owl (*Strix occidentalis caurina*) has been the focus of intense debates and confrontations over how the remaining old-growth forests of western United States should be managed. The northern spotted owl is a territorial owl that lives in old growth conifer forests. Each pair of owls uses about 250–1000 ha (1–4 sq. miles) of valuable old-growth forest—nesting in hollow trees and feeding on small mammals, birds and insects. Heavy logging on private land in the last 40 years has destroyed most of the old-growth forest upon which these owls depend. Most of the remaining old growth is on land managed by the U.S. Forest Service and the National Park Service. The northern spotted owl is listed as an endangered species—the total population in the Pacific Northwest is approximately 1200 pairs.

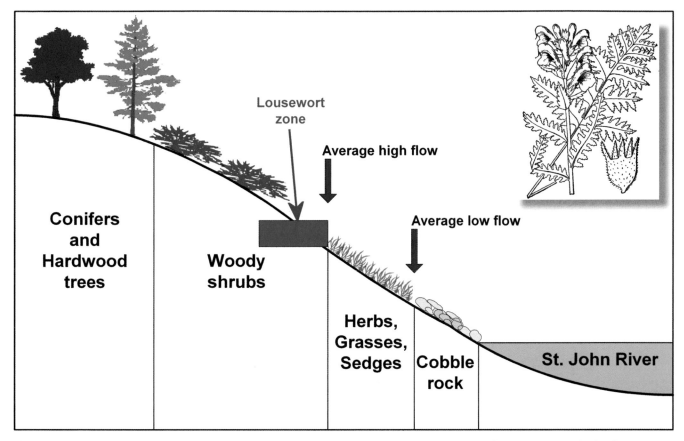

Figure 20.11 The profile of the St. John River bank in northern Maine, and the associated vegetation. Furbish's lousewort grows only in the green zone that is disturbed by ice scour in spring, just above the high water mark indicated by the red arrow. (Modified from Furbish's Lousewort Habitat Model, U.S. Fish and Wildlife Service, Gulf of Maine Program, 2001.)

Old-growth forests are being rapidly reduced in the Pacific North-west as they are elsewhere around the globe. A large part of the controversy over the northern spotted owl is over the question of what type of habitat this owl requires and how much its habitat can be fragmented by logging without causing a population decline. Northern spotted owls strongly prefer old-growth forests for feeding and for roosting. In fragmented forests owls move around more but still feed and roost only in old growth (Figure 20.12). The home range size of owls varies with the prey base. The most common prey in Washington and Oregon is the northern flying squirrel (*Glaucomys sabrinus*).

In Washington State, owls use about 1700 ha of old-growth forests, but in Oregon they use less than half that area. These differences in home ranges are directly related to the prey-base:

State	Home Range (ha)	Prey Available (g/ha)
Washington	~ 1700	61
Oregon—Douglas fir	813	244
Oregon—mixed conifer	454	338

Further diet studies in northern California showed that the wood rat (*Neotoma fuscipes*) was a major prey item, and that owls preferred larger prey (wood rat average weight = 230 g), when they were available, rather than flying squirrels (110 g), which were a second choice.

When ecologists surveyed 11,057 sq. km throughout the range of the northern spotted owl, they found no owls in forests 50–80 years old, and confirmed that owls occurred only where old-growth stands were

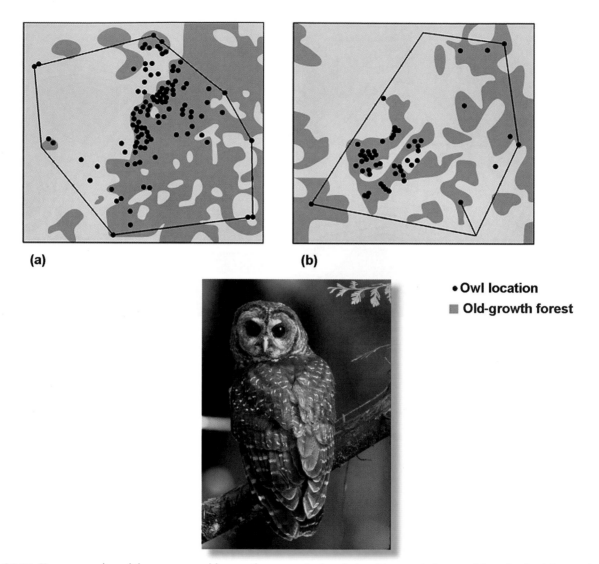

Figure 20.12 Two examples of the areas used by northern spotted owls in old-growth forests (blue shading) in south-western Oregon. (a) Lightly fragmented area (b) heavily fragmented old-growth forest. Very little use is made of the young forest. (Modified from Carey *et al*. 1992, photo courtesy of US Fish and Wildlife Service.)

present. Figure 20.13 shows that northern spotted owls were both more common and reproduced more successfully in old-growth forests. Landscapes with less than 20% old-growth forest rarely supported an owl population. Spotted owls nest in trees much larger and older than the average tree in old-growth stands. In northern California more than 80% of their nest trees were more than 300 years old and most were over 1.2 m diameter. If we wish to conserve spotted owls, there is a clear conflict with logging plans that use an 80–100 year rotation time for forest blocks.

One surprising result of studies on the northern spotted owl is that wilderness areas are not very suitable as habitat for the owls. Productivity inside protected wilderness areas was only 30–50% as high as productivity in old-growth forest outside these designated areas. A high proportion of the wilderness areas, as well as National parks in the Pacific Northwest, is at high elevations, which is less suitable for these owls. The surprising result is that currently protected stands of old-growth forest in parks and wilderness areas may be unable to sustain the northern spotted owl.

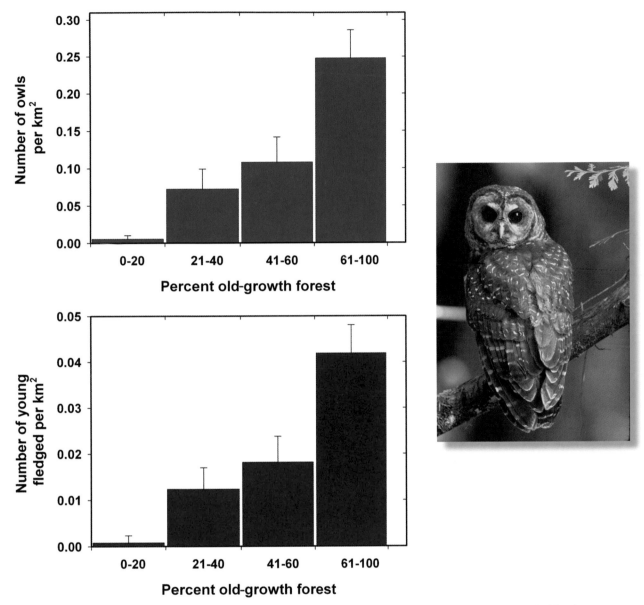

Figure 20.13 Density and reproductive success of northern spotted owls in relation to the amount of older forest on 145 forest areas in Washington, Oregon and northern California. These owls do well only in areas with a large fraction of old-growth forest remaining. (From Bart and Forsman 1992, photo courtesy of US Fish and Wildlife Service.)

How much old-growth forest must be kept to preserve the northern spotted owl? The key parameters needed to make this estimate are the dispersal and colonization success of young owls and the survival and reproductive rates of territorial owls living in landscapes with variable amounts of old forest. The projections of population growth rates of the northern spotted owl are most sensitive to the adult survival rate.

Q Why should the population growth rate of the spotted owl be more sensitive to adult survival instead of juvenile survival?

Are the current conservation plans for the northern spotted owl being successful in preventing population decline? All analyses of the northern spotted owl recognize that a large part of the remaining old-growth

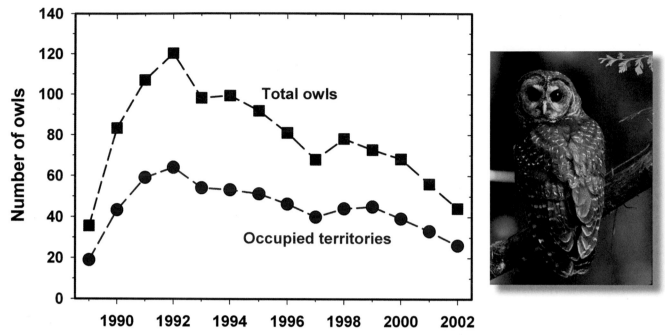

Figure 20.14 The number of northern spotted owls counted and the number of territories occupied in the Cle Elum Study Area of Okanogan–Wenatchee National Forest, Washington from 1989 to 2002. During the first 3 years the counting methods were being improved so the initial rise in numbers is not real. Since 1992 owl numbers have been steadily declining at 8% per year, implying that the current conservation plan is not working properly. (Data from E. Forsman, 2004.)

forests in the Pacific North-west must be preserved if we wish to save this species. The problem thus passes from the conservation biologist to the general public as a matter of policy. The competing land use for these forests is logging and the associated jobs in the timber industry. The conflict over the northern spotted owl is a conflict over short-term needs and long-term goals. At the current rate of harvesting, most of the old-growth forests in the Pacific North-west will be gone within 20 years—at that time the problems of the timber industry will still be with us, but the northern spotted owl may not (Figure 20.14). The present conflict over land use in old-growth forests is one example of a much broader issue of how human populations and the Earth's biota can coexist without serious disruptions. This is the central issue of the 21st century for conservation biology.

SUMMARY

Conservation biology is concerned with the ecology of rare and declining species. Two threads of conservation

biology are a focus on small populations and the consequences of being small (the small population approach) and a focus on declining populations (the declining population approach). Small populations are subject to an array of uncertainties from chance demographic events (such as having all male offspring) to chance environmental impacts (such as a flood) to chance genetic events (such as genetic drift). Not all small populations present conservation problems. Being small increases the chances of extinction for many populations, and can lead a species into an extinction vortex powered by positive feedbacks of chance processes. An elegant body of theory has given us a good description of the hazards of being small.

The declining population approach focuses on identifying the ecological causes of decline and designing alleviation measures to stop the decline. It contains almost no ecological theory, but is focused on individual action plans. Only by understanding the population biology of an endangered plant or animal can we provide a rescue plan for a declining population. In some cases, such as the African elephant, the causes of population decline are clear. In other cases we do not have the ecological understanding to recommend action, and we need to develop further insights so we can develop action plans.

Extinction is the ultimate conservation focus and four causes are prominent: excessive hunting or harvesting, habitat destruction and fragmentation, introduced species and chains of extinction. The major causes of recent extinctions have been habitat destruction and introduced species. Habitat destruction leads to population reductions that may trigger the extinction vortex, so protecting habitat is a major goal for all conservation efforts. At present about 12–13% of the world's land areas is protected, but most protected areas are small and enforcement is lax. Existing parks and reserves are seldom large enough to contain viable populations of larger vertebrates, so conservation efforts on private lands are essential to maintaining our flora and fauna.

The ecological challenge of conservation biology is to develop specific management plans for individual species, while the political challenge to the broader conservation movement is to protect large natural areas from destruction. Without parks and reserves there can be no conservation, but with them there is no guarantee of success unless conservation biology can solve the challenging ecological problems of endangered species.

SUGGESTED FURTHER READINGS

Caughley G and Gunn A (1996). *Conservation Biology in Theory and Practice*. Blackwell Science, Oxford, UK. (The clearest analysis of conservation problems yet written.)

Dayton PK (2003). The importance of the natural sciences to conservation. *American Naturalist* **162**, 1–13. (How the study of natural history is essential for success in conservation.)

Keller LF and Waller DM (2002). Inbreeding effects in wild populations. *Trends in Ecology and Evolution* **17**, 230–242. (A synopsis of how common inbreeding is in the wild and how much it can affect small populations.)

Macdonald DW and Service K (Eds) (2007). *Key Topics in Conservation Biology*. Blackwell Publishing, Oxford, UK. (An excellent series of papers on how conservation biology can move forward from its present achievements.)

Margules CR and Pressey RL (2000). Systematic conservation planning. *Nature* **405**, 243–253. (A careful and thorough analysis of what must be done to conserve biodiversity.)

Myers N, Mittermeier RA, Mittermeier CG, da Fonseca GAB and Kent J (2000). Biodiversity hotspots for conservation priorities. *Nature* **403**, 853–858. (Where on the Earth are the richest areas for biodiversity?)

Newmark WD (1995). Extinction of mammal populations in western North American national parks. *Conservation Biology* **9**, 512–526. (A critical assessment of why even our large national parks are too small.)

Wilson EO (2002). *The Future of Life*. Random House, New York. (A plea for conservation from the world's premier conservation biologist.)

QUESTIONS

1. Barro Colorado Island was formed 85 years ago in central Panama when Gatun Lake was created as part of the Panama Canal. Since that time 65 (of 394) species of birds have disappeared from the island, 21 of them in the last 25 years. Discuss what mechanisms might cause extinctions of birds that can fly in an undisturbed area of tropical forest. Robinson (1999) discusses these changes.

2. Much of conservation biology focuses on rare species, yet Tilman *et al.* (1994) suggested that the species that are the best competitors and the most abundant are at greatest risk from habitat loss. Discuss why more abundant species might be at more risk than rare ones when habitat area is reduced. McCarthy *et al.* (1997) discuss this problem.

3. Kirtland's warbler is an endangered species that breeds in northern Michigan jack-pine forests. Since 1951 the population of this species has been declining and it now numbers about 200 individuals. The most important factor in the population drop seemed to be increasing parasitism of nests by brown-headed cowbirds. Cowbirds were removed from the breeding area of Kirtland's warbler starting in 1971, but no change has occurred in warbler numbers (Ryel 1981). Read Walkinshaw (1983) and Ryel (1981) and discuss why this might be and what management plan you would now recommend for this endangered species.

4. Review the history of the successful rehabilitation of the endangered Lord Howe Island woodhen (*Tricholimnas sylvestris*) on Lord Howe Island in the Pacific (Caughley and Gunn 1996, pp. 75–81).

Discuss the reasons for the success of this project and the general principles it illustrates for conservation problems.

5. Captive breeding programs are one technique used to rescue endangered species. Under what conditions should captive breeding be used? Discuss the limitations of captive breeding as a conservation strategy. Western and Pearl (1989) provide references.

6. Amphibian populations have been declining in many parts of the world during the last 10 years (Storfer 2003). Discuss the hypotheses proposed to explain these declines and suggest a research plan to rescue these populations. Fisher and Shaffer (1996) discuss the amphibian declines in central California, and Collins and Storfer (2003) give an overview of the problem.

Chapter 21

ECOSYSTEM HEALTH AND HUMAN IMPACTS

IN THE NEWS

The news every day is filled with conflicts between 'greenies' and developers, and economists and environmentalists, over economic growth and the environment. One way to judge these controversies might be to obtain some index of the environment that we can call ecosystem health. The general notion of ecosystem health is a useful one to provide for society and decision-makers some indication of how the global environment is coping with stresses arising both from physical factors, such as hurricanes, fires and floods, and from human actions, such as burning fossil fuels. The problem is that these environmental issues have become polarized between two different world views (Costanza *et al.* 2000). For simplicity we will call these two the technological optimist view and the technological skeptic view. The technological optimist believes that through technological innovation humans will dominate nature and become independent of nature. All future challenges, whether water shortages or AIDS, will be overcome by further technological progress. This view is the dominant world view now, and is based partly on the fact that this approach to life has produced the great progress in human health and living standards over the last 250 years. By contrast, the technological skeptic is dubious that the trends of the past can be extrapolated into the future. The basis for this skeptical view is that the human population has increased so much in recent years that we risk destroying essential natural systems via pollution, climate change, over-harvesting, and habitat destruction (Rapport and Whitford 1999). Ecologists in general are technological skeptics, while many business economists are technological optimists. Because these are such different world views, it would be useful to move the debate to a new level to see if there could be some approach that could reconcile these two groups.

One suggestion is to view the environment as an investment portfolio, and to ask how we ought to manage this portfolio, particularly in situations of uncertainty. For example, there is much uncertainty about future climate change. How can we accommodate to ecological and environmental uncertainty? One way is to use game theory to construct a pay-off matrix of what might happen under these two views of the world (Costanza *et al.* 2000). The pay-off matrix is an over-simplified, right-or-wrong game that looks at our current beliefs and what will become future reality.

We ask in the pay-off matrix what will be the consequences of these two views of the world. If the optimists are correct and we adopt their policies—and they work—we reach the highest pay-off, and all is well. But, if the skeptics are in fact correct about the future state of the world, and we continue with the optimist policies, we reach the worst pay-off—disaster. By contrast, if we pursue the skeptics'

		Real state of the world	
		Optimists correct	**Skeptics correct**
Current policies to be used in environmental decisions	Technological optimists' policies	High	Disaster
	Technological skeptics' policies	Good	Very good

policies and the optimists are correct, we reach only a good outcome, because skeptical policies, such as reducing carbon dioxide emissions and other pollutants, are costly and will reduce economic growth. Finally, if the skeptics are correct about the future and we adopt skeptic policies, we have done the right thing. Which box do we wish to choose?

The pay-off matrix basically asks whether we humans wish to gamble with the biosphere. The prudent person will not wish to risk disaster and, even if you are a technological optimist, it would be sensible to adopt the skeptic's position to hedge your bets. The important point is that the skeptical viewpoint does not mean we should stifle new technology, but rather that we should adopt the precautionary principle. We need to acknowledge the great uncertainty we have about future events, and ask how we can manage our environmental affairs, while acknowledging the uncertainty of our scientific information. The simple principle of **not to implement changes that cannot be reversed** is a good summary of the skeptics' position on environmental policies.

Business portfolio managers have long ago worked out how to manage financial affairs in conditions of uncertainty, and it is useful to look at the four principles of financial management as they might be applied to the environment.

1. **Protect your capital.** The first rule of all financial managers is to live off interest and not erode capital. This principle focuses on the concept of natural capital and the services that humans obtain from the environment.

2. **Hedge your investments**. The simple rule of not to put all one's eggs in one basket applies to the environment as well. We should not assume that all the technological optimists' eggs are golden and that none will have adverse side effects on the environment (such as the use of DDT).

3. **Do not risk more than you can afford to lose**. Risk management is critical for all environmental policies, and the decisions that humans make about environmental actions must take into account the preservation of natural capital.

4. **Buy insurance**. All prudent persons protect their assets with insurance against unforeseen catastrophic events. Environmental insurance is obtained by setting aside national parks and marine reserves and protecting biodiversity. Who knows what species will be useful in 100 years?

Environmental problems are often large-scale, global problems and, while we might agree on policies in one country, others will disagree so that it then becomes difficult to achieve global action on issues such as climate change. The principle of sustainability needs to be an accepted principle of all governments, and we need to begin managing our global environmental problems with an acknowledgement of the uncertainty of our ecological knowledge and with the wisdom to be humble.

■ 21.1 PROBLEMS WITH HUMAN IMPACTS

'Humans, including ecologists, have a peculiar fascination with attempting to correct one ecological mistake with another, rather than removing the source of the problem'. (Schindler 1997, p. 4)

Human impacts are now degrading all the Earth's ecosystems. This cannot continue without creating environmental chaos for our children and grandchildren. During your lifetime these impacts will lead to the most significant problems facing the globe, and our present preoccupation with the stock market and political chicanery will be viewed as the latest example of Nero fiddling while Rome is burning. Throughout this book we have touched on many of these problems, from overfishing to pest control and the conservation of endangered species. This chapter brings these problems into central focus and asks both **what are the problems** and **what can we do about them?** The ecologist in this position is much like a medical doctor. As scientists, our job is to diagnose problems and suggest cures. But, just as a patient can ignore a doctor's advice to stop smoking, for example, the public through its political structures can ignore the advice of ecologists and all their recommendations to ameliorate problems. The current newspaper debate over the role of humans in causing the climate changes that are occurring is a clear example of how excuses are manufactured so that the public can comfortably ignore scientific findings. Translating science into policy is an art, and ecological scientists are slowly getting better at it. Our first objective must be to get the science right, and this book is a starting guide to achieving this goal. The central mandate of applied ecology is to assess these human impacts scientifically and to suggest ways of ameliorating them.

The goal of this analysis is to protect ecosystem health, both at the local level and at the global level. In a simple sense, the goal is to pass on to our children and grandchildren an Earth that has its biodiversity protected, sustainable systems of agriculture and forestry, and a healthy environment. The key question is 'What do we need to change to reach these goals?'

In this chapter I will first summarize the human population problem—the root cause of all the adverse human impacts on ecosystems. After estimating how much of the Earth's resources humans have appropriated, I will discuss three aspects of ecosystem health: the carbon cycle, climate change and species invasions.

■ 21.2 HUMAN POPULATION GROWTH

If growth is good, as most businesspeople seem to believe, we ought to be nearing perfection. The human population of the globe has been growing throughout most of recorded history, and probably for more than two million years before that. In 2004 there were an estimated 6.3 billion people on the Earth (*Essay 21.1*). The human population is growing at 1.2% per year, or 211,000 people per day. Figure 21.1 shows the growth of the human population over the last 500 years. The increase appears geometric, but is even more rapid than geometric (Cohen 1995). Because no population can go on increasing without limit, Figure 21.1 immediately raises two questions:

1. What is happening now?
2. Can we estimate the carrying capacity of the Earth for humans in the long term?

Current Patterns of Population Growth

The human population can exist in one of two stable configurations that will lead to population stability: zero population growth = high birth rates – high death rates, *or*

zero population growth =
low birth rates − low death rates   (21.1)

The movement between these two states has been called the **demographic transition** (see Figure 6.14 page 126). The demographic transition is a descriptive theory rather than a law of human population growth (Gelbard *et al.* 1999). The important point is that both birth and death rates have declined dramatically in most countries over the last 60 years. After 1950, mortality rates declined rapidly in all countries, but birth rates have declined in a more variable manner (Figure 21.2). Fertility decline has been most dramatic in China. In 1970, the average Chinese woman could

expect to have 5.9 children, but by 2006 the expected family size was 0.84 children. In India, fertility rates have fallen more slowly and irregularly. In the United States, the birth rate is nearly at replacement levels, while in Europe, Russia, Canada and Australia the birth rate is well below the replacement level of 2.1 children per female.

One consequence of variable fertility rates on the growth of the world's population is that the current rapid rise in human population is composed of two quite different elements (Figure 21.3). In the developed nations, populations are near to equilibrium with net reproductive rates near replacement (total fertility rate = 2.1 children per female). In many developed countries, such as Canada, Australia, and the United Kingdom, in which the total fertility rate is below the replacement level, populations will decline in the long term if there is no net immigration and the reproductive rate does not change. Most developed countries, however, are still increasing in population each year without immigration because the age structure is not in equilibrium, so births exceed deaths (Table 21.1). About 80% of the world's people now live in the less developed countries, and most of the population growth is occurring in these nations (Figure 21.3).

The projected human population of the globe depends on assumptions about future changes in fertility and mortality. In 2050, the United Nations projects a population that might range from 7.3 to 10.7 billion people (Gelbard *et al.* 1999). No matter what projection is used, without some catastrophe there will be at least 1.3 billion people added to the population in the next 25 years because of population momentum (Box 21.1 pages 507–8). The questions that arise from these projections are: 'What human population can the biosphere support? Is the world already over-populated? Will it be over-populated in 2050?'

Carrying Capacity of the Earth

What is the carrying capacity of the Earth for humans? This question has been asked for more than 300 years by scientists interested in demography (Cohen 1995). As a first step, we can ask what range of estimates has so far been produced for how many people the Earth can support. We have already

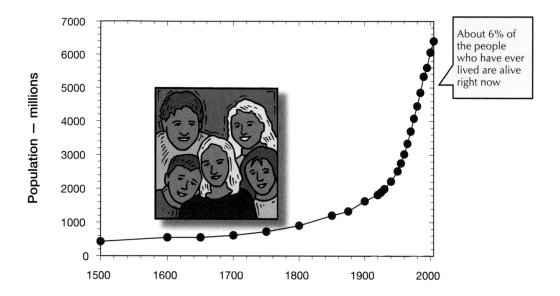

Figure 21.1 Human population growth in the last 500 years. The population increase appears to grow approximately geometrically. Population growth accelerated dramatically after about 1900. Note that the impact of the millions killed in World War I and World War II cannot even be seen on this graph. (Data from Cohen 1995, Appendix 2 and Population Reference Bureau 2006.)

ESSAY 21.1   WHAT IS A BILLION ANYWAY?

Humans are the supreme counters in the universe—we learn in school how to use decimals and exponents and how to manipulate them correctly. But, at some point our numerical system overwhelms use—such is the case with a billion. There are now over 6 billion people on Earth and in 2006 the U.S. Government borrowed $455 billion to balance its budget. How can we get a grasp on how big a billion is? Two simple games can be used to illustrate how big a billion is.

1. **Question**: how long would it take you to spend $1 billion if you were able to spend $10,000 a day, every day of the year on anything you wished?
   Answer: To spend $1 billion you would need 274 years with no days off.

2. **Question**: how far would you walk if you took a billion steps?
   Answer: If we assume each of your steps is 61 cm (about 2 feet), you would walk 610,000 kilometers (379,000 miles). This would mean you could walk to the moon and almost all the way back or if you are more conservative, you could walk around the equator about 15 times. Remember to take some water with you!

The main point is that while we can talk about and manipulate large numbers such as a billion, they exceed our ability to comprehend how large they really are. So when we note that the human population increased one billion from 1987 to 2000 and will increase another billion by 2012, the size of this increase is almost incomprehensible.

Table 21.1 Human population of the different continents and some selected countries as of July 2006. Population projections for 2025 and 2050 are based on United Nations estimates of projected future trends in reproduction and mortality. The rate of increase is the natural rate from births and deaths, and excludes immigration and emigration. Doubling times assume the rate of increase does not change in the future.

Country	Population in July 2006	Rate of increase (%/yr)	Doubling time (years)	Projected population	
	(millions)			2025	2050
World	6,555	1.2	58	7,940	9,243
More developed	1,216	0.1	693	1,255	1,261
Less developed	5,339	1.5	47	6,685	7,982
Africa	924	2.3	30	1,355	1,994
North America	332	0.6	116	387	462
Canada	32.6	0.3	231	38	42
United States	299.1	0.6	116	349	420
Central America	149	1.9	37	187	214
South America	378	1.4	50	465	528
Asia	3,968	1.2	58	4,739	5,277
Saudi Arabia	24	2.7	26	36	47
Yemen	22	3.2	22	39	68
Bangladesh	147	1.9	37	190	231
India	1,122	1.7	41	1,363	1,628
Pakistan	166	2.4	29	229	295
China	1,311	0.6	116	1,476	1,437
Japan	128	0.0	–	121	101
Europe	732	-0.1	–	717	665
Norway	5	0.3	231	5	6
Sweden	9	0.1	693	10	11
France	61	0.4	174	63	64
Germany	82	-0.2	–	82	75
Poland	38	0.0	–	37	32
Russia	142	-0.6	–	130	110
Italy	59	0.0	–	59	56
Spain	45	0.2	347	46	44

Source: Population Reference Bureau, 2006 World Population Data Sheet.

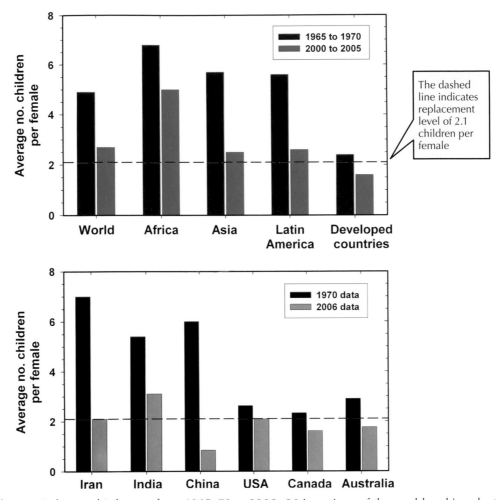

Figure 21.2 Changes in human birth rates from 1965–70 to 2000–06 in regions of the world and in selected countries. The transition from high birth rates to low birth rates has been particularly rapid in Asia and Latin America. (Data from Population Reference Bureau 2007.)

illustrated in Figure 6.15 (page 128) the range of estimates for the carrying capacity of the Earth presented by many authors over the last 300 years. These estimates vary so much that no sensible population policy could be based on them, and so ecologists have taken another approach to this question.

Carrying capacity is difficult to estimate. The equation that has been used for carrying capacity is:

$$\text{Carrying capacity} =$$
$$\frac{(\text{ha land})(\text{yield per ha})(\text{kJ per crop unit})}{\text{no. kJ needed per person per year}}$$

This equation is a definition and, if we could quantify the terms in it, we could establish the carrying capacity of the Earth. But, because there are many

crops and different soils, variable yields, losses to pests, and differences in assumed standards of living, it has been impossible to get anyone to agree on how these numbers should be constrained.

A more promising approach to estimating the carrying capacity of the Earth is to recognize that we have multiple constraints because we need food, fuel, wood and other amenities such as clothing and transportation. One promising approach is to express everything in the amount of land needed to support each activity, such as wood production, and then to sum these requirements. But this approach has limits also because it is difficult to express energy requirements directly as land areas, and water requirements may be more of a constraint than land areas.

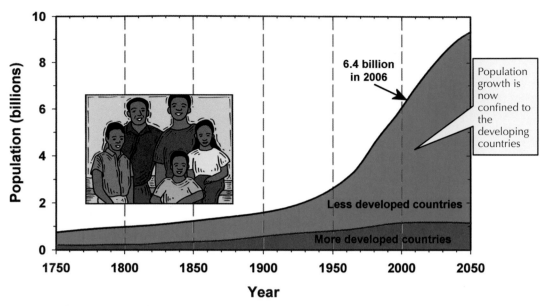

Figure 21.3 The contribution to world population growth of the more developed countries (red) and the less developed countries (green) since 1750. The major population growth we are now experiencing, and will experience for the next 50 years, will come from the developing countries of South-East Asia, Africa, and Latin America. (Data from Population Reference Bureau 2007.)

A recent advance in using multiple constraints to estimate carrying capacity is summarized in the concept of an **ecological footprint** (Wackernagel and Rees 1996). For each nation, we can calculate the aggregate land and water area in various ecosystem categories that is appropriated by that nation to produce all the resources it consumes and to absorb all the waste it generates. Six types of ecologically productive areas are distinguished in calculating the ecological footprint: arable land, pasture, forest, ocean, built-up land and fossil energy land. Fossil energy land is calculated on the basis of the land needed to absorb the CO_2 produced by burning fossil fuels. All measures are thus converted to land area per person. If we add up all the biologically productive land on the planet, we find there were about 2 ha of land per person alive in 1997. If we wish to reserve land for parks and conservation, we must reduce this to 1.7 ha per person of land available for human use. This is the benchmark for comparing the ecological footprints of nations (Figure 21.4). Because different countries differ in their agricultural capacity and resource base, the available ecological capacity for each country must be adjusted for its productivity in

each of the six types of productive ecosystems. We end up with a simple comparison of ecological footprints and available ecological capacity. Figure 21.4 illustrates these results for 14 countries and Figure 21.5 shows the history of how the world's ecological footprint has changed over the past 50 years.

Two things are evident from these two graphs.

Firstly, the world in general was already in ecological deficit by 1987. Secondly, countries vary greatly in their individual footprint size and in their available capacity. The usefulness of these estimates of ecological footprints is that they can direct us toward sustainability in our use of the world's resources by targeting the specific areas in which particular countries overutilize resources.

The analysis of human impacts via ecological footprints suggests that the world is already above its carrying capacity. We can check this conclusion by looking at two other calculations of human impacts—one for water and one for primary production. Fresh water is an important resource because it has no substitute for its uses and is difficult to transport more than a few hundred kilometers. The amount of water on Earth and the volumes moving around are so large that we

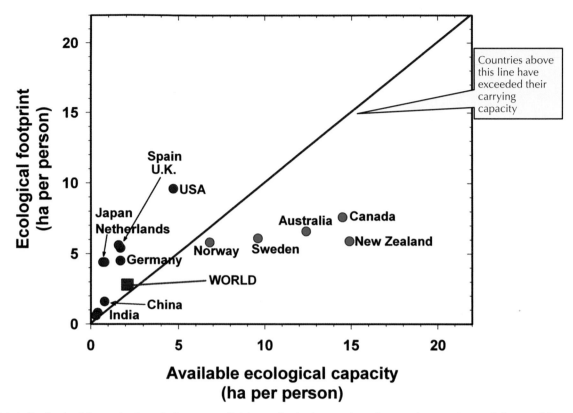

Figure 21.4 Ecological footprint in relation to available ecological capacity of several countries and the world in 2003. The ecological footprint expresses in hectares of land per person the current demand of global resources made by each country. The available ecological capacity measures in land area per person the resource base of each particular country. Countries in red above the diagonal are in an ecological deficit in 1997; countries in green below the diagonal still have resource surpluses. (Data from Halls 2006.)

must discuss the water cycle in units of km^3 (=10^{12} liters). If all the rainfall on land were evenly spread, each weather station would record about 70 cm of rain per year (Schlesinger 1997). Fresh water constitutes only about 2.5% of the total volume of water on Earth, and two-thirds of it is locked up in glaciers and ice.

Figure 21.6 shows the global water cycle. There is a net transport of water vapor from the oceans to the land that contributes about one-third of the rainfall to land areas. The mean residence time of a water molecule in the ocean is about 3100 years—a tribute to the large volumes of sea water in the ocean basins. The volume of groundwater is poorly known. It is known as fossil water, because it is deep in the ground and cannot be reached by plants, and has a mean residence time of over 1000 years (Schlesinger 1997). We depend on fresh water flowing through the hydrological cycle through precipitation for our needs.

Precipitation that falls on land then goes in two directions—evapotranspiration and run-off. Some precipitation is used for vegetative growth of forests, crops and pastures, and some evaporates back into the atmosphere. This component is called **evapotranspiration**. The remaining precipitation goes to **run-off**. Run-off is the source of water for cities, irrigated crops, industry and to sustain aquatic ecosystems. Postel *et al.* (1996) estimated that, over the whole Earth, humans now use 26% of all evapotranspiration on land and 54% of freshwater run-off.

Can we appropriate more of the terrestrial water supply for an increased population? Postel *et al.* (1996) argue that we cannot because most land suitable for rain-fed agriculture is already in production. We could build more dams on rivers to increase our use of run-off, but the maximum this would permit is estimated as 64% usage of run-off. Given an expected population

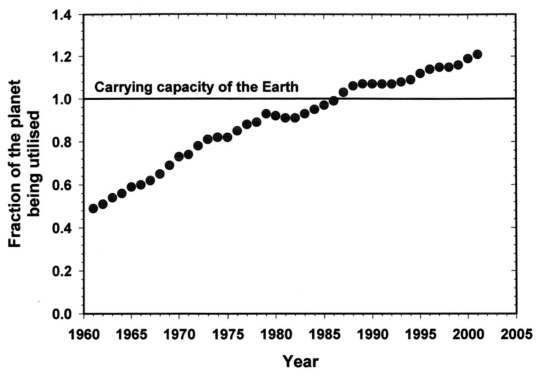

Figure 21.5 Changes in the ecological footprint of the entire world's population since 1960. Already by 1987 humans had exceeded the carrying capacity of the planet. (Data from Living Planet Report 2004, courtesy of World Wildlife Fund.)

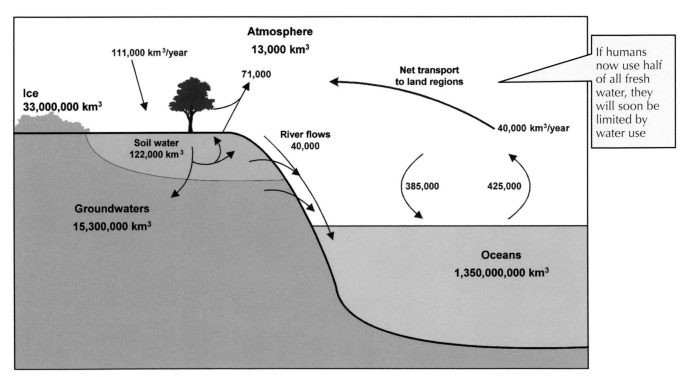

Figure 21.6 The global water cycle. Pools of water are in units of km^3 (black), and flows of water are in km^3 per year (red). About half of the pool of soil water is available to plants. Humans currently use about half of all the freshwater run-off. (Modified from Schlesinger 1997.)

growth of approximately 21% in the next 19 years (Table 21.1) something will have to change. If we do not wish to destroy completely all the freshwater aquatic ecosystems of the globe, we will have to increase our efficiency of water use and reduce the amount of water used for irrigation. Desalination of sea water is an expensive option because it is energy-intensive and is unlikely to provide a solution for the less developed countries. The message is similar to that which we received via ecological footprints—humans are close to or above the carrying capacity of the Earth.

21.3 THE CARBON CYCLE AND CLIMATE CHANGE

The most important global issue of our time is climate change and, to begin to understand climate change, we must uncover the impacts humans are having on the global carbon cycle. Plants and animals are primarily composed of carbon and the global carbon cycle is a reflection of primary and secondary production. The fixation of carbon by plants in photosynthesis over geological time accounts for the oxygen in the Earth's atmosphere. Humans have affected the global carbon cycle almost as much as they have the sulfur cycle, and intense public interest now focuses on the resulting greenhouse effect and climate change.

Figure 21.7 shows the global carbon cycle. The carbon cycle is mostly related to carbon dioxide. Dissolved inorganic carbon in the ocean is the largest pool of carbon.

The ocean contains about 56 times as much carbon as the atmosphere (Schlesinger 1997). The atmospheric pool of carbon is slightly larger than the total carbon bound up in vegetation. The largest fluxes of the global carbon cycle are between the atmosphere and land vegetation and the atmosphere and the oceans. These two fluxes are approximately equal, and the mean residence time of a molecule of carbon in the atmosphere is about 5 years.

The Global Carbon Budget

To understand the impacts of the global carbon cycle, we must quantify the inputs and outputs. This is complicated because the global carbon budget is not in equilibrium. The amount of CO_2 in the atmosphere has not been constant. Since 1750—the start of the Industrial Revolution—there has been a rapid and continuous rise in CO_2 levels in the atmosphere. The most accurate long-term measurements of CO_2 have been taken on Mauna Loa in Hawaii (Figure 21.8) and show a sustained increase of 0.4% per year (1.5 ppm). Superimposed on the long-term increase in CO_2, there is a seasonal trend in concentration. The seasonal oscillations of CO_2 are the result of the seasonal uptake of CO_2 by plants in photosynthesis. The majority of terrestrial vegetation occurs in environments that have a seasonal growth cycles, so that atmospheric CO_2 levels go down in summer when plants fix more CO_2 (Figure 21.8). Some of the long-term increase in CO_2 comes from the burning of fossil fuels. If all CO_2 from fossil fuels accumulated in the atmosphere, it would be increasing about 0.7% per year. Only about 56% of the CO_2 released from fossil fuel is accumulating in the atmosphere (Keeling *et al.* 1995). What happens to the remainder?

We cannot at the present time balance the global carbon budget to answer this question, but research has focused on the two most likely candidates for the disappearing CO_2—the ocean and the world's forests. Oceanographers believe that about one-third of the CO_2 from fossil fuels enters the ocean each year, particularly in higher latitude areas (Global Carbon Project 2003). CO_2 is exchanged only at the ocean's surface, but much of the carbon in the oceans is in the deeper waters. Exchange in the oceans between surface waters and deep waters occurs only very slowly. Turnover of carbon for the entire ocean occurs about every 350 years (Schlesinger 1997).

Terrestrial vegetation can both serve as a source and a sink for CO_2. One source of atmospheric carbon dioxide is the destruction of terrestrial vegetation for agriculture, often by clearing and burning especially in the tropics (Detwiler and Hall 1988, Schlesinger 1997). Shifting cultivation is a dominant form of land use in tropical countries, and about three-quarters of all land use changes fall under this heading. Shifting cultivation is less destructive of forest because after 1–3 years the farmers abandon the fields to secondary succession and move on. Such temporary clearing of land

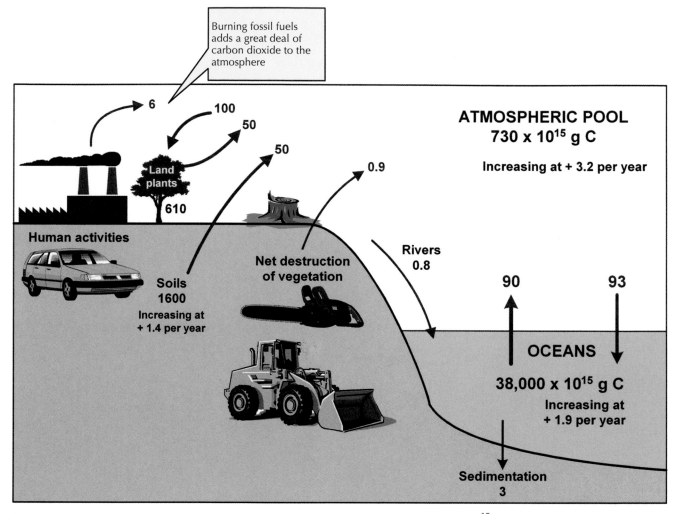

Figure 21.7 The present-day global carbon cycle. All pools are expressed in units of 10^{15} g C (black) and all annual fluxes in units of 10^{15} g C/yr (blue). (Modified after Schlesinger 1997 with data from the Australian Bureau of Meteorology, 2007.)

contributes less CO_2 to the atmosphere than does permanent conversion of forest land to pastures. In 1990, the best estimate was that tropical areas were a net source of 1.6×10^{15} g carbon because of deforestation (Dixon *et al.* 1994).

Terrestrial vegetation is a second important sink for the CO_2 that has increased from the burning of fossil fuels and land clearing. The re-growth of forests in the Temperate Zone has captured significant amounts of carbon during the last 20 years. The data now available provide the following global budget for carbon (Table 21.2, Schlesinger 1997): units = 10^{15} g carbon per year.)

There is considerable research at present that is aimed at increasing the precision of these estimates and in identifying sources and sinks.

In trying to balance the global carbon budget, it is important to remember that we are focusing on the annual movements of carbon, not on the amount stored in the various reservoirs. The oceans contain the largest pool of carbon, but most of this carbon turns over very slowly. Desert soils contain more carbon (in the form of carbonates) than all the terrestrial plants, but there is virtually no exchange of carbon between the atmosphere and desert soils.

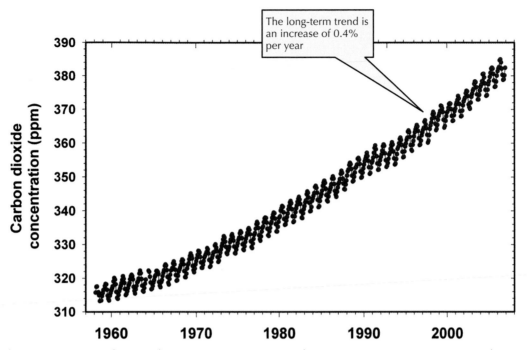

Figure 21.8 The concentration of atmospheric CO_2 at Mauna Loa Observatory in Hawaii since 1958. The annual oscillation reflects the seasonal cycles of photosynthesis and respiration by land biota in the Northern Hemisphere, while the overall increase is largely due to the burning of fossil fuels. (Data from Keeling and Whorf (2004) and Scripps Institute of Oceanography.)

Table 21.2 Global carbon budget for 2007. Accurate measurement of carbon sources and carbon sinks is now a high research priority.

Net emissions			=	Net changes in the carbon cycle				
fossil fuel	+	destruction of terrestrial plants	=	atmospheric increase	+	oceanic uptake	+	terrestrial plant growth
6.0	+	0.9	=	3.2	+	2.0	+	1.7

Terrestrial vegetation is taking up more CO_2 (Idso 1999, Norby *et al.* 1999, Gerber *et al.* 2004). If biomass is increasing in terrestrial vegetation because of a 'fertilization' of vegetation by CO_2, and if this increase is rapid enough, we have located an important sink for the global carbon cycle. The important question is how much do rising CO_2 levels stimulate plant growth?

Plant Community Responses to Rising CO_2

Plants need CO_2 for photosynthesis so we might predict that if CO_2 levels increase, plants will grow more, and this increased plant growth would sequester some of the rising CO_2 shown in Figure 21.8. A key question is how will plant communities respond to rising carbon dioxide levels? To answer this question, individual plants can be grown in greenhouses with enriched CO_2 levels, but these results are difficult to extrapolate to global ecosystems where CO_2 may not be the major factor limiting plant growth. Fortunately, we now have field data to analyze for the impact of CO_2 on primary production.

Long-term data on primary production are now available from satellites and long-term study plots, and are now providing the critical data to measure trends in carbon storage. Phillips *et al.* (1998) measured the gain in carbon in tropical forests over the past 25–40

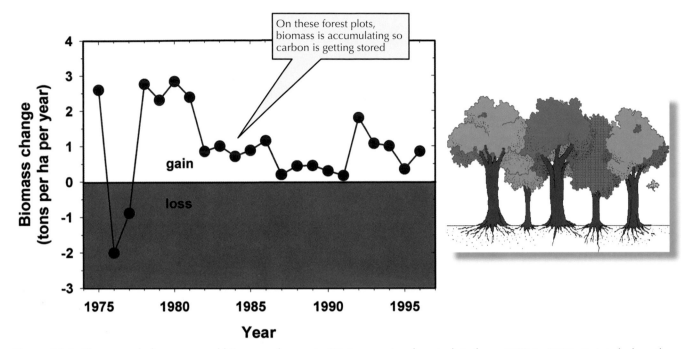

Figure 21.9 The annual above-ground biomass change in 97 Amazonian forest plots from 1975 to 1996. Points below the blue line indicate biomass loss and points above the line indicate biomass gain. All trees on these plots were measured each year. In almost all years biomass increased—thus carbon was being taken up from the atmosphere. Growing tropical forests could be a large part of the missing carbon sink. (Data from Phillips *et al.* 1998.)

years from basal area measurements on 600,000 individual trees scattered in 478 plots across the tropics. Figure 21.9 shows the biomass change in Amazonian forests from 1975 to 1996. On average in these 97 plots trees increased in biomass by 1 ton per ha per year. Biomass in trees is accumulating particularly in the neotropical forests of Central and South America. This biomass increase is equivalent to fixing 0.62 tons of carbon per ha per year, and this carbon sink may account for about 40% of the terrestrial storage of carbon dioxide in the global carbon cycle.

Satellite data with time frames of 15–20 years can now be used to measure the community responses to CO_2 enrichment (Nemani *et al.* 2003). Figure 21.10 shows the global picture of change in temperature, water availability and solar radiation from 1982 to 1999, derived from satellite data. The primary productivity of different parts of the globe can be broadly categorized as being limited by temperature, water, or solar radiation (Figure 21.10a) and, with this as a basis, we can examine the changes in these three variables on a global scale in Figure 21.10b, c, and d. Some areas, such

as western North America, have heated up since 1982, while other areas, such as Siberia, have cooled down. Rainfall has increased in Brazil and Europe, but decreased in Australia and South Africa. Solar radiation has increased in India and northern Canada, but decreased in West Africa and Australia. The pattern of climate change has not been simple for the globe, and the details of what is changing on a regional scale are important.

From changes in greenness indices, satellite data can derive estimates of net primary production on a global scale. Figure 21.11 shows that primary production has increased on average about 6% over 18 years— the largest increase in primary production occurred in the Amazon rainforest. Not all of this increase in primary production can be attributed to the rise in CO_2 levels over this time period. Decreased cloud cover in the Amazon region permitted an increase in solar radiation, which stimulated tree growth (Figure 21.9 and 21.10).

Much scientific effort is now under way to measure more precisely the changes in the global carbon cycle

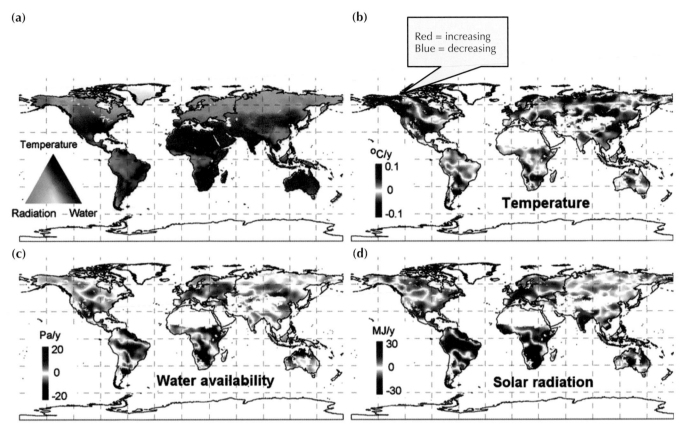

Figure 21.10 Global changes in the climatic factors that limit plant growth. (a) Geographic distribution of potential climatic constraints to plant growth. Recent changes from 1982 to 1999 in (b) growing season average temperature, (c) water availability and (d) solar radiation. All climatic data are expressed as changes relative to long-term averages, so positive changes indicate improved conditions for plant growth. (After Nemani *et al.* 2003.)

and to define more clearly and more locally the sources and sinks for CO_2. All of this is important because of the linkages between the carbon cycle and climate change, to which we now turn.

Climate Change

Global warming has become one of the major environmental issues of our time. Global warming is a function of the greenhouse effect—one of the most well-established theories of atmospheric chemistry (Schneider 1989). Figure 21.12 illustrates the greenhouse effect, which arises because the Earth's atmosphere traps heat near the surface. Water vapor, CO_2 and other trace gases absorb the longer, infrared wavelengths emitted by the Earth. An increase in the concentration of greenhouse gases thus tends to warm the Earth by re-radiation. None of this is controversial— the greenhouse principles apply equally well on Earth

as they do on Venus (dense CO_2 atmosphere, very hot) and Mars (thin CO_2 atmosphere, very cold). The controversy arises in how to predict from the greenhouse effect exactly how much the Earth's temperature will rise for a specified change in greenhouse gases, such as CO_2. We have already seen in Figure 21.10 the observed changes in climate during the last 20 years, but the more important issue is what will happen to the Earth's climate in future (*Essay 21.2*).

Paradoxically, one of the best ways to understand future climate change is to look backwards. Ice cores have provided one way of doing this. Air is trapped by snow as it is transformed into glacial ice, so by extracting ice cores we can sample the atmosphere back in time. The most spectacular example of this method is a 3623 m long ice core collected by the Soviet Antarctic Expedition at Vostok, Antarctica (Petit *et al.* 1999). This ice core spans 420,000 years. Temperature at the time

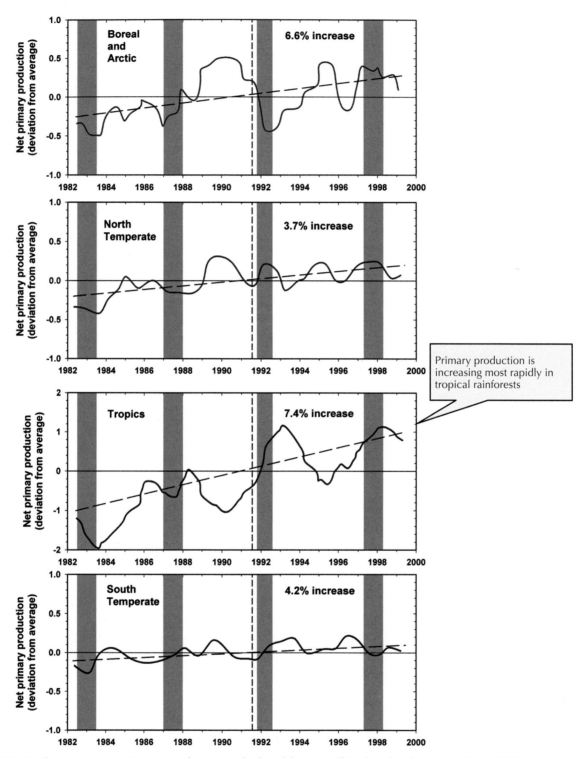

Figure 21.11 Changes in net primary production calculated from satellite data for the period from 1982 to 1999 (18 years) in relation to latitude. Primary production has increased most strongly in tropical areas, particularly in Amazonia, and also strongly in the higher latitudes of the northern hemisphere. Year-to-year variation in primary production follow El Niño events (grey bars) and volcanic eruptions (Mount Pinatubo eruption in mid 1991 = vertical dashed line), and the El Niño impact is particularly strong in tropical regions. Values of primary production are expressed as deviations from the long-term average value (0.0, horizontal lines) in units of Pg of carbon per year. (Modified after Nemani *et al.* 2003.)

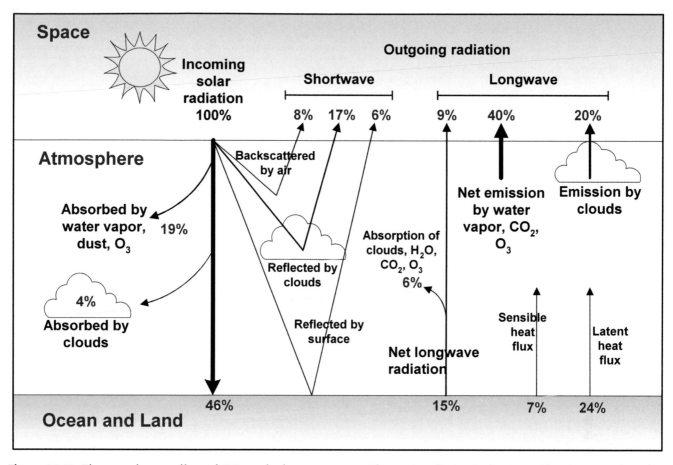

Figure 21.12 The greenhouse effect of CO_2 and other trace gases. The sun's radiation is dominated by short wavelengths, which are reflected or absorbed by the Earth's surface. The radiation absorbed is re-radiated at longer wavelengths, which can be absorbed by atmospheric gases including CO_2. Higher concentrations of these gases in the atmosphere reduce the net emission of long wave radiation to space, warming the Earth. (Modified from MacCracken 1985.)

of ice formation can be determined by the ratio of oxygen-18 to oxygen-16 in the ice (Lorius *et al.* 1985). The resulting time series of changes in the Vostok ice core are shown in Figure 21.13.

There is a close correlation between carbon dioxide in the air and global temperatures over the past 420,000 years. But there are three difficulties in extrapolating these correlations into the future. Firstly, our present CO_2 levels of 380 ppm exceed the levels found in nature during the last 400,000 years (Figure 21.14). We do not know if we can extrapolate the relationships found in the past. Secondly, we are changing CO_2 levels very rapidly from year to year, whereas the historical changes in CO_2 were very slow. We do not know the rates at which equilibrium can be established between CO_2 sources and sinks. Thirdly, Figure

21.13 shows a correlation between CO_2 and temperature, but we do not yet know which is cause and which effect (Stauffer 1999).

Other greenhouse gases besides CO_2 can contribute to global warming. The most important of these are methane, nitrous oxides, ozone and chlorofluorocarbons (CFCs)—these trace greenhouse gases may together be as important as, or more important than, CO_2 in the greenhouse effect of the 21st century. The atmospheric concentration of methane has increased about 150% since 1750 and is far above the average levels of the last 420,000 years. Fossil fuels, agriculture and cattle production are major components of methane increases. What is not under question now is that the world's average temperature has been increasing, particularly during the last 35 years (Figure 21.15).

ESSAY 21.2   EL NIÑO AND THE SOUTHERN OSCILLATION

Climate change on the time scale of months and years is what affects ecosystems and humans most directly. The most famous of these short-term changes is El Niño, which was first recognized by South American fishermen as an incursion of warm water off the coast of Peru. Once weather data began to be assembled, meteorologists discovered that El Niño events were correlated with the difference in atmospheric pressure between Tahiti and Darwin, Australia. The Southern Oscillation is the name given to this seesaw of change of atmospheric pressure

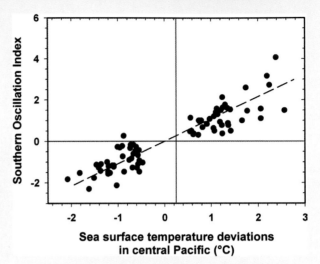

The Southern Oscillation Index correlates well with sea surface temperature deviations in the Central Pacific.

between these two stations. Once it was realized that these oceanographic and atmospheric processes were coupled, the joint name El Niño–Southern Oscillation (ENSO) was coined. The Southern Oscillation Index is correlated with sea surface temperatures in the central and eastern Pacific.

The Southern Oscillation Index (SOI) is a relative index, measured by the deviation of the atmospheric pressure at Darwin minus the atmospheric pressure at Tahiti, scaled to a long-term average of zero. When the Southern Oscillation Index is positive, ocean temperatures are warmer than usual in the eastern and central Pacific and colder than usual in the western Pacific and Indian Oceans.

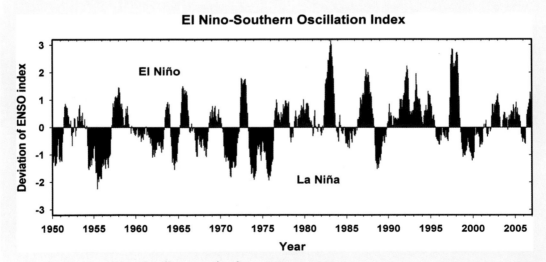

The El Niño–Southern Oscillation Index from 1950 to 2006

The impact of ENSO events are dramatic. When an El Niño event occurs, weather changes are triggered on a global scale, as the following maps show:

Maps showing the climatic effect of El Niño events

Mild winters in north-eastern USA and western Canada are typical of El Niño events, and severe droughts in Australia, India, Indonesia, Brazil and Central America are typical of El Niño years, but not all El Niño events produce the same results.

The ENSO cycle has an average period of 4 years, but varies from 2 to 7 years in length. There is much variation in the strength of the Southern Oscillation for reasons that are not yet clear, but are presumably driven by ocean temperatures.

The impacts of El Niño on global ecosystems are varied and significant. Warm surface temperatures lead to coral bleaching and the destruction of coral reefs. The pelagic upwelling ecosystem off Peru collapses because the warm water displaces the nutrient-rich cold water. Fisheries, such as the Peruvian anchovy, collapse and seabirds, which also depend on these fish, suffer high mortality during El Niño events. Pacific salmon production in the North Pacific is linked to similar changes in oceanographic events (Mantua *et al.* 1997). It is likely that many ecological changes are driven by large-scale weather changes caused by these oceanographic shifts, and the linkages between large-scale weather changes and ecosystem dynamics is a critical focus of current research.

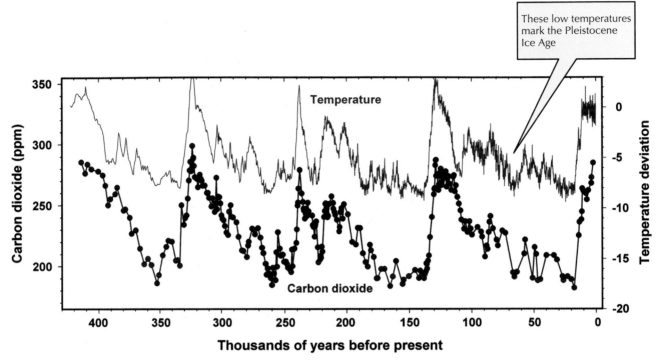

Figure 21.13 Long-term variations in global temperature (red) and atmospheric carbon dioxide concentration (blue) determined from the 3623 m long Vostok ice core, Antarctica, covering the last 420,000 years. Carbon dioxide can be measured from air trapped in the ice as it is formed. Temperature is measured as the deviation from present day temperatures (°C). The temperature changes show the last four ice ages—each about 100,000 years long. There is a high correlation between CO_2 levels and global temperature. (Modified after Petit *et al*. 1999 and Intergovernmental Panel on Climate Change Climate Change 2001 Report.)

How will changes in greenhouse gas levels and climatic warming affect the Earth's biota? There is an urgent need to answer this question, but it is not the only question ecologists face, and it is important to place this question within the perspective of all the human impacts on the Earth's ecosystems. We have discussed these impacts in various places throughout this book, and need to deal with three major human impacts in the remainder of this chapter—climate change, land transformation and biotic additions and losses. Of these three, climate change gets the most publicity, but the other two changes are more insidious and may even be more important (Vitousek *et al*. 1997).

The difficulty of providing specific answers to the question of the impacts of climate change for agricultural and natural ecosystems is very great (Walker *et al*. 1999). The sanguine and simple view that climate change will lead to more plant growth and more luxuriant natural ecosystems with no adverse consequences is certainly not true. We need to determine the impacts of climate change on plant communities and then on the animals that depend on them. What do we know about impacts of climate change on animals?

There is no evidence that animals will be affected *directly* by changing CO_2 levels in the next century. But there is concern that herbivorous animal populations and communities may be affected indirectly by changes in their food plants. Nitrogen content of plant leaves tends to fall with increased CO_2 levels, and this may result in a reduction of larval insect growth in nitrogen-limited insect species. Insect herbivores are often limited by plant nitrogen in natural communities (White 1974) and reduced nitrogen may reduce insect numbers on CO_2 enriched vegetation. To compensate for low nitrogen levels, insect herbivores feeding on

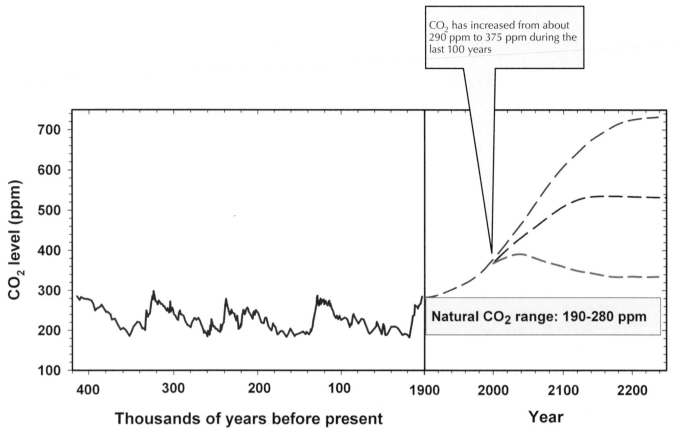

Figure 21.14 Long-term changes in atmospheric carbon dioxide concentration and the recent increases caused by burning fossil fuels. Note the change in scale of the year axis. Three scenarios for the next 250 years are outlined, depending on what action the world takes on energy use. The current level of 380 ppm CO_2 is already outside the observed range of CO_2 for the past 400,000 years, which makes the extrapolation of older climatic correlations difficult. (Data from the Intergovernmental Panel on Climate Change, Climate Change Report 2001.)

plants grown with higher CO_2 levels may increase their feeding rate by 20–80% (Fajer 1989). As CO_2 levels rise, plant damage could actually increase, even if insect numbers fall.

Climatic warming could also disrupt the timing of hatching in insects (Dewar and Watt 1992). Insects that feed on newly emerged foliage are very sensitive to the age of the foliage. Maximum overlap between bud burst and larval emergence results in good survival and growth of the insects. Consequently, in Scotland these larvae will emerge earlier if climatic warming occurs. By contrast, the vegetative buds of Sitka spruce, their food plant, are not greatly affected by spring temperatures and will open only slightly earlier when the climate warms. The net result will be a mismatch between the moth and its food plant that will reduce moth survival and growth (Dewar and Watt 1992). These short-term effects could be alleviated by natural selection in the longer term, but the disruption of life cycles in insect herbivores could be a major impact of global warming.

The concentrations of carbon-based secondary compounds (such as phenolics and terpenes) in plants also tend to increase under CO_2 enrichment (Peñuelas and Estiarte 1998). Both increased CO_2 and decreased nitrogen tend to increase the secondary compounds that act as deterrents to herbivore feeding. These compounds may also act to change the rate of plant decomposition through their effects on fungi and microbes (Ball 1997). There is no work yet available on the impact

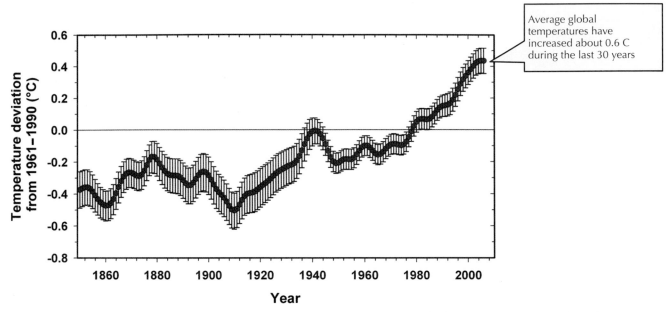

Figure 21.15 Changes in annual average global temperature during the period of meteorological recordings from 1850 to 2006. Both sea and land temperatures are included. In many countries the past 10 years have included five to seven of the warmest years recorded since systematic weather data were begun. (Data from Met Office Hadley Centre for Climate Change, 2007.)

of CO_2-enriched plants on herbivorous mammals or birds, but impacts through secondary chemicals might be expected.

■ 21.4 CHANGES IN LAND USE

Humans convert forests to pastures and agricultural fields into suburbs with relatively little thought given to the ecological consequences of these land use changes. But these changes, hectare by hectare, have among the most serious impacts on global ecosystems (Vitousek 1994). The problem is to quantify these impacts. The availability of satellite data ought to make global measures of land use changes relatively easy to establish. Unfortunately, this is not the case for two reasons. Firstly, interpreting satellite data is not a simple matter—for global measurements that demand high resolution, satellite images with associated ground verification are very expensive. Secondly, different definitions of land classes cause confusion in measuring land use changes. For example, two studies in India examined deforestation rates from 1981 to 1990

and produced widely different results. One study by FAO (Food and Agricultural Organization of the UN) estimated deforestation of 0.6% per year, while a simultaneous study by NRSA (National Remote Sensing Agency, India) estimated 0.04% deforestation per year (Menon and Bawa 1998). Having better satellites will not necessarily solve the second problem of defining ecosystem types that can be viewed using satellite imagery. Tropical deforestation is estimated at 8% per decade (Watson 1999), but this is an educated guess rather than a precise ecological measurement.

Changes in land use have a dual impact—in habitat lost for plant and animal communities and in habitat fragmentation with its associated problems (see page 401). The conversion of land from forest to agriculture leads to a large increase in CO_2 emissions. Houghton *et al.* (1999) estimated that until 1960 in the USA, carbon emissions from land use changes were greater than those from fossil fuel combustion. During the last 40 years, forests have been increasing in the United States, as a result of tree replanting, farms being abandoned and fire protection. During the 1980s Houghton *et al.* (1999) estimated that this increase in forest cover was a net sink for carbon amounting to about 10–30% of the emissions from fossil fuel burning. Changes in land management have long-term consequences for climate change as well as direct effects on biodiversity.

■ 21.5 BIOTIC INVASIONS AND SPECIES RANGES

Introduced species form another major impact on the biosphere. Human activity has moved many species across continents and oceans that they could not have otherwise crossed, with the resulting impacts we have discussed in Chapter 3. The introduction of the rabbit to Australia (page 164) and the zebra mussel to the United States are just two of many such examples. Table 21.3 gives some examples of the magnitude of these biotic invasions. The economic impact of invasions is one strong reason for attempting to reduce introductions through the imposition of strict quarantine laws (Ruesink *et al.* 1995, Vitousek *et al.* 1996).

One of the most difficult impacts of climate change to alleviate will be the resulting changes in species distributions. If species ranges are currently controlled by climate, a shift in climate implies a shift in geographic range. Shifts in range have occurred many times in the history of the Earth, and so at first glance one may think that this problem will take care of itself. Two factors argue against complacency. Firstly, the speed of change in climate is now many times greater than it has ever been in the past. The critical question then becomes how fast can species shift their geographic ranges? Secondly, human changes in land use have interrupted many possible corridors of movement for both plants and animals, so that historical means of dispersal may no longer be possible.

The geographical distribution of many trees is probably limited by climate, so trees are a good group to use for studying this question. Figure 21.16 illustrates the possible magnitude of range shifts that may occur for American beech in the next 100 years (Davis 1989). Two different climate change models were used to develop these predictions, both of which are coarse approximations. The range changes are very large and there is no precedent in the past for such a rapid shift.

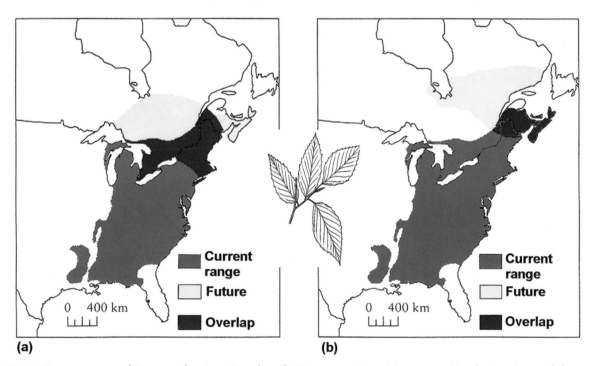

Figure 21.16 Current geographic range for American beech (*Fagus americana*) in eastern North America and the potential future range in 100 years according to two climatic warming models: (a) a milder scenario and (b) a more severe scenario. (Modified from Roberts 1989.)

Table 21.3 Biotic invasions of vascular plants, freshwater fish and birds. Island habitats are particularly vulnerable to invasions, but continental areas have also been strongly impacted.

Taxa	Country	Number of native species	Number of non-native species	Percentage of non-native species
Vascular plants	Germany	1718	429	20.0
	Finland	1006	221	18.0
	France	4200	438	9.4
	California	4844	1025	17.5
	Canada	9028	2840	23.9
	Greenland	427	86	16.8
	Australia	20,000	2000	10.0
	Tanzania	1940	19	1.0
	South Africa	20,263	824	3.9
	Bermuda	165	303	64.7
	Hawaii	956	861	47.4
	Fiji	1628	1000	38.1
	New Zealand	1790	1570	46.7
Freshwater fish	California	76	42	35.6
	Canada	177	9	4.8
	Australia	145	22	13.2
	South Africa	107	20	15.7
	Brazil	517	76	12.8
	Hawaii	6	19	76.0
	New Zealand	27	30	52.6
Birds	Europe	514	27	5.0
	South Africa	900	14	1.5
	Brazil	1635	2	0.1
	Hawaii	57	38	40.0
	New Zealand	155	36	18.8

Source: data from Vitousek et al. (1996).

At the end of the last Ice Age, North American trees dispersed north at rates ranging from 10 to 45 km per 100 years. Beech trees moved at an average rate of 20 km per 100 years (Woods and Davis 1989). But, if the predictions indicated in Figure 21.16 are correct, beech trees will have to move north about 40 times faster than they did at the end of the Ice Age. This could only occur with human help. One of the major conservation problems of the next century for North America and Europe could be the re-establishment of plant and animal communities at more northerly sites as climatic warming proceeds.

If plant communities are disrupted by climatic warming during the next century, the animal communities on which they depend will be equally disrupted. There are at present few data available on which to judge these potential effects—the estimation and monitoring of climatic effects on both plants and animals are part of the research agenda for all ecologists in the coming decade.

■ 21.6 ECOSYSTEM SERVICES

In our technological world, we have lost touch with the services that ecosystems provide to our human society. **Ecosystem services** is a collective phrase used to refer to all the processes through which natural ecosystems, and the species they contain, help sustain human life on this planet. If you ask someone watching TV what services natural ecosystems provide to him or her, you would probably get a blank look, so here is a list of a few things we take for granted as ecosystem services:

- purification of air and water
- mitigation of droughts and floods
- generation and preservation of soils and soil fertility
- detoxification and decomposition of wastes
- pollination of crops and natural vegetation
- dispersal of seeds
- nutrient cycling
- control of many agricultural pests by natural enemies
- maintenance of biodiversity
- protection of coastal shores from erosion
- protection from ultraviolet rays
- partial stabilization of climate
- moderation of weather extremes
- provision of aesthetic beauty

These ecosystem services are greatly underappreciated by human society, as they have no dollar value attached to them. Human life would cease to exist without these ecosystem services, and so they are of immense value to us. But can we quantify this value?

Robert Costanza and colleagues (1997) attempted to put a dollar figure on ecosystem services and came up with an estimate of US$33 trillion per year—nearly twice as much as the gross national product of all the countries of the globe ($18 trillion). Many of the valuation techniques that are used in economic analyses of ecosystem services are based on the 'willingness to pay' idea (*Essay 21.3*). If an individual owns a commercial forest and ecosystem services provide a $50 increment to timber productivity, that individual should be willing to pay up to $50 for these services. The problem is that willingness-to-pay may not be a good measure if individuals are ill-informed about the actual ecosystem services provided.

There are many uncertainties about this approach to quantify the value of ecosystem services. The largest service contribution is from nutrient cycling, which makes up about half the total value of ecosystem services. The key point is that the value of ecosystem services is large and, if these services were actually paid for in our economic system, the global market system would be completely different. Many projects, such as large dams or irrigation projects, would no longer be economical because their true cost would inevitably be seen to exceed the perceived social benefits.

Freshwater ecosystem services may be one of the simpler ones to quantify because humans need drinking water and therefore we value lakes and rivers that are kept free of serious pollution (Wilson and Carpenter 1999). Specific indicators of water quality, such as water clarity or the frequency of algal blooms, are simple and can be readily explained to the public as important indicators of ecosystem health. Most of the attempts to quantify freshwater ecosystem services have been site-specific and cannot readily be extrapolated to larger areas, such as a state or country. For example, lake-front property values can be shown to decline substantially as a lake becomes more polluted and algal blooms more frequent. The challenge is to extrapolate these very local economic evaluations to a larger scale, so that the value of ecosystem services can enter environmental policy debates.

One large-scale experiment illustrates all too well how little we understand and value ecosystem services. The Biosphere 2 in Oracle, Arizona was an attempt to experimentally construct a closed ecosystem covering 1.27 ha (Figure 21.17). From 1987 to 1989 a forest

ESSAY 21.3   ECONOMICS OF ECOSYSTEM SERVICES

Everyone agrees that the environment's services are valuable. The water we drink and the air we breathe are available to us only because of ecosystem services that we take for granted. How can we enroll market forces in the conservation of ecosystems? How can we get corporations and governments to invest in natural capital? Two examples show how we might proceed to do this.

New York City's water supply comes from a watershed in the Catskill Mountains and, until recently, water was purified by the natural processes of root systems, soil micro-organisms and filtration in the soils of this watershed and sedimentation in its streams. But, by the early 1990s, continued sewage additions to Catskill streams, as well as fertilizer and pesticide use in local agriculture, had degraded the Catskill water supplies to standards below those set for drinking water by the Environmental Protection Agency. In 1996 the City had two choices:

1. Build and operate a water filtration plant to purify the water at a cost of $6 billion to $8 billion and running costs of $300 million per year

2. Restore the integrity of the Catskill ecosystem.

The city chose to restore ecosystem integrity by buying land in and around the catchment area so that its use could be restricted, and subsidizing the construction of better sewage treatment plants. By investing $1 to $1.5 billion in natural capital, the City of New York has saved $6 to $8 billion investment in physical capital (Chichilnisky and Heal 1998).

Over 90% of plants are pollinated by animals, including about 70% of major crop plants, and agricultural pollination is another clear example of an ecosystem service that we take for granted. In this example the economic value of pollinators can be estimated very easily by looking at the crop yield in the absence or presence of pollinators (Daily 1997). Honeybees, which have provided much of the pollination of our crops, have been declining in the United States, partly because of introduced diseases. Between 1990 and 1994 there was a 20% decline in the number of honeybee colonies in the United States. The key question is whether or not wild species of pollinators can take up all the slack in pollination when honey bees decline. If they cannot do so, the losses to consumers would range from $1.6 to $5.7 billion per year. The important question becomes how to manage the landscape to maximize the abundance of non-honeybee pollinators.

Pollination is one of the least understood processes in ecosystem functioning and, because of its importance to human food supplies, will be in the spotlight as an ecosystem service with large yet unappreciated economic value.

with soil and a miniature ocean were constructed inside the airtight Biosphere with an investment of over $200 million. From October 1991 an experiment was begun with eight people living in isolation inside the Biosphere 2 ecosystem for two years. Unlimited energy and technology were available from the outside to support the project. But the system failed and the experiment had to be stopped after 15 months (Cohen and Tilman 1996). Atmospheric oxygen dropped to 14%, and CO_2 fluctuated irregularly. Most of the

Figure 21.17 View of part of the Biosphere 2 site near Tucson, Arizona. The original aim was to construct a completely enclosed, functioning ecosystem that humans could occupy indefinitely, but it was a complete failure. Oxygen levels collapsed and many species became extinct, so the experiment had to be stopped after just one year. (Photo courtesy of P.R. Schwob, The Classical Archives.)

vertebrate species in Biosphere 2 became extinct and all of the pollinators died out. Population explosions of pests such as cockroaches showed clearly that the ecosystem services were impaired even before the experiment stopped. The conclusion was clear: no-one yet knows how to engineer a system that will provide humans with all the life-support services that natural ecosystems produce for free. We have no alternative but to maintain the health of ecosystems on Earth—that is the ecological challenge for the 21st century.

SUMMARY

Human impacts on the Earth's ecosystems are rooted in the continuing increase in world population. Six billion people now inhabit the Earth and the population is growing by 214,000 people every day. The less developed countries of the globe contribute most of this increase, but much of the adverse impact of humans arises in the developed world with high demands for energy and materials.

The carrying capacity of the Earth is difficult to estimate. One approach is to estimate the ecological footprint of a nation in terms of the amount of land it uses to produce the commodities it consumes. By this measure, the Earth is already at its full carrying capacity, and many countries have exceeded their available ecological resources and are thus in deficit. By no known measure can the Earth sustain its current population if we all live the high consumption lifestyle of the developed world.

Human impacts strongly influence the carbon cycle. Rising levels of CO_2 in the atmosphere have occurred for 100 years, owing to fossil fuel burning and the destruction of native vegetation. The global carbon budget is out of balance, with the rising CO_2 levels only partially taken up by the phytoplankton in the oceans and by increased plant growth, particularly in tropical areas.

Climate change is one practical side effect of the carbon cycle. Greenhouse gases, such as CO_2, methane and nitrous oxide, trap heat at the Earth's surface and increase global temperatures. How increasing greenhouse gases will affect the Earth's biota is a focus of intensive current research. Increased plant growth and increased primary production have accompanied increasing CO_2 during the last 20 years. The long-term ecosystem consequences for plants are still far from clear. Animal communities will be affected indirectly through their food plants.

Because many geographical distributions are limited by climate, climatic warming in the next century will have dramatic effects on the distribution of native animals and plants, including disease-causing organisms. Ecosystem restoration may be the major global conservation problem of the next century.

Ecosystems provide services in air and water purification, pollination and nutrient cycling that are greatly undervalued by society because they are not traded in the marketplace. We do not yet know how to construct intact ecosystems that can support human life, so should take care of the Earth and vigorously guard the ecosystem services it provides us for free.

SUGGESTED FURTHER READINGS

Barbraud C and Weimerskirch H (2001). Emperor penguins and climate change. *Nature* **411**, 183–186. (The impacts of climate change on the iconic bird of the Antarctic.)

Berteaux D, Humphries MM, Krebs CJ, Lima M, McAdam AG, Pettorelli N, *et al.* (2006) Constraints to projecting the effects of climate change on mammals. *Climate Research* **32**, 151–158. (An analysis of how difficult it will be to predict the effects of climate change for some groups of animals.)

Harvell CD, Mitchell CE, Ward JR, Altizer S, Dobson AP, Ostfeld RS and Samuel MD (2002). Climate warming and disease risks for terrestrial and marine biota. *Science* **296**, 2158–2162. (A sober discussion of how certain diseases will spread to new areas as the climate warms.)

Moss R, Oswald J and Baines D (2001). Climate change and breeding success: decline of the capercaille in Scotland. *Journal of Animal Ecology* **70**, 47–61. (Effects of climate change on an iconic grouse species.)

Parmesan, C and Yohe G (2003). A globally coherent fingerprint of climate change impacts across natural systems. *Nature* **421**, 37–42. (A comprehensive synthesis of the effects of climate change that have already been measured.)

Thomas CD, Franco AMA and Hill JK (2006). Range retractions and extinction in the face of climate warming. *Trends in Ecology & Evolution* **21**, 415–416. (The increasing concern about species loss in the future.)

Wackernagel M, Onisto L, Bello P, Callejas Linares A, Lopez Falfan IS, Mendez Garcia J, *et al.* (1999) National natural capital accounting with the ecological footprint concept. *Ecological Economics* **29**, 375–390. (A discussion of how we can measure the impact of humans in different countries.)

Wilson MA and Carpenter SR (1999). Economic valuation of freshwater ecosystem services in the United States: 1971–1997. *Ecological Applications* **9**, 772–783. (An important attempt to put a dollar figure on aquatic ecosystem services.)

Worm B, Barbier EB, Beaumont N, Duffy JE, Folke C, Halpern BS, *et al.* (2006) Impacts of biodiversity loss on ocean ecosystem services. *Science* **314**, 787–790. (How the extinction crisis associated with global change will affect the oceans.)

QUESTIONS

1. In agricultural landscapes, farmers have the choice of managing roadside verges by mowing, burning, grazing or doing nothing to them. Discuss the implications of these four treatments for the global carbon cycle.

2. Discuss the economic meaning of the word 'value' and the ecological meaning of this word. Is there a difference? What activities might an economist include in a valuation of the use of fresh water? Does economics value the non-use of a resource such as lake water? Wilson and Carpenter (1999) discuss these issues.

3. The concept of *ecosystem health* has been criticized as a concept that applies well to individuals but poorly to a whole ecosystem. Discuss the application of the idea of *health* to populations, communities and ecosystems. Rapport *et al.* (1998) discuss this question.

4. Peatlands in boreal and subarctic regions store a large pool of organic matter not yet decomposed. Review the distribution of peatlands and their potential role in climatic change scenarios. Gorham (1991) provides an overview.

5. Phosphorus is a major nutrient needed by crops, but it does not occur in the atmosphere and is a sparse element in most rocks. Discuss how you could determine if modern agriculture is sustainable with respect to phosphorus. How could you determine if pre-industrial agriculture was sustainable? Newman (1997) evaluates these questions.

REFERENCES

Aber JD, Goodale CL, Ollinger SV, Smith M-L, Magill AH, Martin ME, Hallett RA and Stoddard JL (2003). Is nitrogen deposition altering the nitrogen status of Northeastern forests? *BioScience* **53**, 375–389.

Ahlgren I, Frisk T and Kamp-Nielsen L (1988). Empirical and theoretical models of phosphorus loading, retention and concentration vs. lake trophic status. *Hydrobiologia* **170**, 285–303.

Allen KR (1980). *Conservation and Management of Whales.* University of Washington Press, Seattle.

Alroy J (1999). The fossil record of North American mammals: evidence for a Paleocene evolutionary radiation. *Systematic Biology* **48**, 107–118.

Anderson DM (1994). Red tides. *Scientific American* **271**, 52–58.

Anderson KG, Kaplan H and Lancaster J (1999). Paternal care by genetic fathers and stepfathers I: Reports from Albuquerque men. *Evolution and Human Behavior* **20**, 405–432.

Anderson RM and May RM (1979). Population biology of infectious diseases: Part I. *Nature* **280**, 361–367.

Anderson S (1985). The theory of range-size (RS) distributions. *American Museum Novitates* **2833**, 1–20.

Arendt R (2004). Linked landscapes. *Landscape and Urban Planning* **68**, 241–269.

Armbrust EJ and Gyrisco GG (1982). Forage crops insect pest management. In *Introduction to Insect Pest Management.* (Eds RL Metcalf and W Luckman) pp. 443–463. John Wiley & Sons, New York.

Arsenault R and Owen-Smith N (2002). Facilitation versus competition in grazing herbivore assemblages. *Oikos* **97**, 313–318.

Austin MP (1999). A silent clash of paradigms: some inconsistencies in community ecology. *Oikos* **86**, 170–178.

Austin MP, Nicholls AO, Doherty MD and Meyers JA (1994). Determining species response functions to an environmental gradient by means of a ß-function. *Journal of Vegetation Science* **5**, 215–228.

Ayres DR, Smith DL, Zaremba K, Klohr S and Strong DR (2004). Spread of exotic cordgrasses and hybrids (*Spartina* sp.) in the tidal marshes of San Francisco Bay, California, USA. *Biological Invasions* **6**, 221–231.

Ball AS (1997). Microbial decomposition at elevated CO_2 levels: effect of litter quality. *Global Change Biology* **3**, 379–386.

Barbraud C and Weimerskirch H (2001). Emperor penguins and climate change. *Nature* **411**, 183–186.

Bart J and Forsman ED (1992). Dependence of northern spotted owls *Strix occidentalis caurina* on old-growth forests in the western USA. *Biological Conservation* **62**, 95–100.

Baskin Y (2002). *A Plague of Rats and Rubbervines: the Growing Threat of Species Invasions.* Island Press, Washington, D.C.

Begon M, Mortimer M and Thompson DJ (1996). *Population Ecology: A Unified Study of Animals and Plants.* Blackwell Science, Oxford, UK.

Beisner BE, Haydon DT and Cuddington K (2003). Alternative stable states in ecology. *Frontiers in Ecology and the Environment* **1**, 376–382.

Belsky AJ (1986). Does herbivory benefit plants? A review of the evidence. *American Naturalist* **127**, 870–892.

Bennett WA, Currie RJ, Wagner PF and Beitinger TL (1997). Cold tolerance and potential overwintering of the red-bellied piranha *Pygocentrus nattereri* in the United States. *Transactions of the American Fisheries Society* **126**, 841–849.

Bergelson J (1996). Competition between two weeds. *American Scientist* **84**, 579–584.

Bergerud AT (1980). A review of the population dynamics of caribou and wild reindeer in North America. *Proceedings of the 2nd International Reindeer/Caribou Symposium*, 556–581.

Berryman AA (1981). *Population Systems: A General Introduction.* Plenum Press, New York.

Berteaux D, Humphries MM, Krebs CJ, Lima M, McAdam AG, Pettorelli N *et al.* (2006). Constraints to projecting the effects of climate change on mammals. *Climate Research* **32**, 151–158.

Bertness MD and Callaway R (1994). Positive interactions in communities: a post cold war perspective. *Trends in Ecology and Evolution* **9**, 191–193.

Bertness MD and Leonard GH (1997). The role of positive interactions in communities: lessons from intertidal habitats. *Ecology* **78**, 1976–1989.

Billings WD (1938). The structure and development of oldfield shortleaf pine stands and certain associated physical properties of the soil. *Ecological Monographs* **8**, 437–499.

Blumstein DT and Daniel JC (2002). Isolation from mammalian predators differentially affects two congeners. *Behavioral Ecology* **13**, 657–663.

Bock CE and Fleck DC (1995). Avian response to nest box addition in two forests of the Colorado Front Range. *Journal of Field Ornithology* **66**, 352–362.

Bond WJ, Woodward FI and Midgley GF (2005). The global distribution of ecosystems in a world without fire. *New Phytologist* **165**, 525–538.

Boutin S, Krebs CJ, Boonstra R, Dale MRT, Hannon SJ, Martin K *et al.* (1995). Population changes of the vertebrate community during a snowshoe hare cycle in Canada's boreal forest. *Oikos* **74**, 69–80.

Bowne DR and Bowers MA (2004). Interpatch movements in spatially structured populations: a literature review. *Landscape Ecology* **19**, 1–20.

Boyer JS (1982). Plant productivity and environment. *Science* **218**, 443–448.

Brady MJ, Risch TS and Dobson FS (2000). Availability of nest sites does not limit population size of southern flying squirrels. *Canadian Journal of Zoology* **78**, 1144–1149.

Brett JR (1970). Temperature. In *Marine Ecology. Vol. I, Environmental Factors, Part I*. (Ed. O Kinne) pp. 515–560. Wiley-Interscience, New York.

Briggs SV, Seddon J and Doyle S (1999). Predicting biodiversity of woodland remnants for on-ground conservation. *National Heritage Trust (Australia), Special Report AA 1373.97*, Canberra, September 1999.

Brook BW, Sodhi NS and Ng PKL (2003). Catastrophic extinctions follow deforestation in Singapore. *Nature* **424**, 420–426.

Brooks JL and Dodson SI (1965). Predation, body size, and composition of plankton. *Science* **150**, 28–35.

Brown CR and Brown MB (1996). *Coloniality in the Cliff Swallow: The Effect of Group Size on Social Behavior*. University of Chicago Press, Chicago.

Brown JH (1984). On the relationship between abundance and distribution of species. *American Naturalist* **124**, 255–279.

Brown JH and Lomolino MV (1998). *Biogeography*. Sinauer Associates, Sunderland, Massachusetts.

Buesseler KO, Andrews JE, Pike SM and Charette MA (2004). The effects of iron fertilization on carbon sequestration in the Southern Ocean. *Science* **304**, 414–417.

Burbidge AA and McKenzie NL (1989). Patterns in the modern decline of Western Australia's vertebrate fauna: causes and conservation implications. *Biological Conservation* **50**, 143–198.

Buss LW and Jackson JBC (1979). Competitive networks: nontransitive competitive relationships in cryptic coral reef environments. *American Naturalist* **113**, 223–234.

Calderini DF and Slafer GA (1998). Changes in yield and yield stability in wheat during the 20th century. *Field Crops Research* **57**, 335–347.

Caley MJ and Schluter D (1997). The relationship between local and regional diversity. *Ecology* **78**, 70–80.

Caputi N, Chubb C, Melville-Smith R, Pearce A and Griffin D (2003). Review of relationships between life history stages of the western rock lobster, *Panulirus cygnus*, in Western Australia. *Fisheries Research* **65**, 47–61.

Carey AB, Horton SP and Biswell BL (1992). Northern spotted owls: influence of prey-base and landscape character. *Ecological Monographs* **62**, 223–250.

Cargill SM and Jefferies RL (1984). Nutrient limitation of primary production in a sub-arctic salt marsh. *Journal of Applied Ecology* **21**, 657–668.

Carpenter SR, Caraco NF, Correll DL, Howarth RW, Sharpley AN, and Smith VH (1998). Nonpoint pollution of surface waters with phosphorus and nitrogen. *Ecological Applications* **8**, 559–568.

Caughley G, Grigg GC, Caughley J and Hill GJE (1980). Does dingo predation control the densities of kangaroos and emus? *Australian Wildlife Research* **7**, 1–12.

Caughley G and Gunn A (1996). *Conservation Biology in Theory and Practice.* Blackwell Science, Oxford, UK.

Caughley G, Shepherd N and Short J (1987a). *Kangaroos: Their Ecology and Management in the Sheep Rangelands of Australia.* Cambridge University Press, Cambridge, UK.

Caughley G, Short J, Grigg GC and Nix H (1987b). Kangaroos and climate: an analysis of distribution. *Journal of Animal Ecology* **56**, 751–762.

Cerda H and Paoletti MG (2004). Genetic engineering with *Bacillus thuringiensis* and conventional approaches for insect resistance in crops. *Critical Reviews in Plant Sciences* **23**, 317–323.

Chadwick OA, Derry LA, Vitousek PM, Huebert BJ and Hedin LO (1999). Changing sources of nutrients during four million years of ecosystem development. *Nature* **397**, 491–497.

Chapman DG (1981). Evaluation of marine mammal population models. In *Dynamics of Large Mammal Populations.* (Eds CW Fowler and TD Smith) pp. 277–296. John Wiley & Sons, New York.

Chase JM (2000). Are there real diffcrences among aquatic and terrestrial food webs? *Trends in Ecology and Evolution* **15**, 408–412.

Chichilnisky G and Heal G (1998). Economic returns from the biosphere. *Nature* **391**, 629–630.

Clark JS, Fastie C, Hurtt G, Jackson ST, Johnson C, King GA, *et al.* (1998). Reid's paradox of rapid plant migration. *BioScience* **48**, 13–24.

Clarke A (1990). Temperature and evolution: Southern Ocean cooling and the antarctic marine fauna. In *Antarctic Ecosystems.* (Eds KR Kerry and G Hempel) pp. 9–22. Springer-Verlag, Berlin.

Clarke MF and Schedvin N (1999). Removal of bell miners *Manorina melanophrys* from *Eucalyptus radiata* forest and its effect on avian diversity, psyllids and tree health. *Biological Conservation* **88**, 111–120.

Clausen J, Keck DD and Hiesey WM (1948). *Experimental Studies on the Nature of Species. III. Environmental Responses of Climatic Races of* Achillea. Carnegie Institute, 581, Washington, D.C.

Coale KH, Johnson KS, Chavez FP, Buesseler KO, Barber RT, Brzezinski MA, *et al.* (2004). Southern Ocean iron enrichment experiment: carbon cycling in high- and low-Si waters. *Science* **304**, 408–414.

Cohen JE (1995). *How Many People Can the Earth Support?* W.W. Norton, New York.

Cohen JE and Tilman D (1996). Biosphere 2 and biodiversity: the lessons so far. *Science* **274**, 1150–1151.

Cole DW and Rapp M (1981). Elemental cycling in forest ecosystems. In *Dynamic Properties of Forest Ecosystems.* (Ed. DE Reichle) pp. 341–409. Cambridge University Press, Cambridge, UK.

Collins JP and Storfer A (2003). Global amphibian declines: sorting the hypotheses. *Diversity & Distributions* **9**, 89–98.

Connell JH (1961a). Effects of competition, predation by *Thais lapillus*, and other factors on natural populations of the barnacle *Balanus balanoides*. *Ecological Monographs* **31**, 61–104.

Connell JH (1961b). The influence of interspecific competition and other factors on the distribution of the barnacle *Chthamalus stellatus*. *Ecology* **42**, 710–732.

Connell JH, Hughes TP and Wallace CC (1997). A 30-year study of coral abundance, recruitment, and disturbance at several scales in space and time. *Ecological Monographs* **67**, 461–488.

Connell JH and Lowman MD (1989). Low-diversity tropical rain forests: some possible mechanisms for their existence. *American Naturalist* **134**, 88–119.

Conover DO and Munch SB (2002). Sustaining fisheries yields over evolutionary time scales. *Science* **297**, 94–96.

Cook RE (1969). Variation in species density of North American birds. *Systematic Zoology* **18**, 63–84.

Costanza R, Daly H, Folke C, Hawken P, Holling CS, McMichael AJ, Pimentel D and Rapport DJ (2000). Managing our environmental portfolio. *BioScience* **50**, 149–155.

Costanza R, Darge R, de Groot R, Farber S, Grasso M, Hannon B, *et al.* (1997). The value of the world's ecosystem services and natural capital. *Nature* **387**, 253–260.

Côté IM (2000). Evolution and ecology of cleaning symbiosis in the sea. *Oceanography and Marine Biology: Annual Review* **38**, 311–355.

Côté IM and Sutherland WJ (1997). The effectivness of removing predators to protect bird population. *Conservation Biology* **11**, 395–405.

Cramer VA and Hobbs RJ (2002). Ecological consequences of altered hydrological regimes in fragmented ecosystems in southern Australia: Impacts and possible management responses. *Austral Ecology* **27**, 546–564.

Crocker RL and Major J (1955). Soil development in relation to vegetation and surface age at Glacier Bay, Alaska. *Journal of Ecology* **43**, 427–448.

Croxall JP, Trathan PN and Murphy EJ (2002). Environmental change and Antarctic seabird populations. *Science* **297**, 1510–1514.

Currie DJ (1991). Energy and large-scale patterns of animal- and plant-species richness. *American Naturalist* **137**, 27–49.

Currie DJ and Paquin V (1987). Large-scale biogeographical patterns of species richness of trees. *Nature* **329**, 326–327.

Cyr H and Pace ML (1993). Magnitude and patterns of herbivory in aquatic and terrestrial ecosystems. *Nature* **361**, 148–150.

Daily GC (Ed.) (1997). *Nature's Services: Societal Dependence on Natural Ecosystems.* Island Press, Washington, D.C.

Darwin C (1859). *The Origin of Species, by Means of Natural Selection or the Preservation of Favoured Races in the Struggle for Life,* John Murray, London, UK.

Dauphin G, Zientara S, Zeller H and Murgue B (2004). West Nile: worldwide current situation in animals and humans. *Comparative Immunology, Microbiology and Infectious Diseases* **27**, 343–355.

Davidson S (2002). Reinventing agriculture. *Ecos* **111**, 18–25.

Davis MB (1989). Lags in vegetation response to greenhouse warming. *Climatic Change* **15**, 75–82.

Dawkins R (2006). *The Selfish Gene.* Oxford University Press, New York.

Dayton PK (2003). The importance of the natural sciences to conservation. *American Naturalist* **162**, 1–13.

de Blij HJ, Muller PO and Williams RS Jr. (2004). *Physical Geography: The Global Environment.* Oxford University Press, New York.

DeAngelis DL (1992). *Dynamics of Nutrient Cycling and Food Webs.* Chapman & Hall, New York.

DeAngelis DL and Waterhouse JC (1987). Equilibrium and nonequilibrium concepts in ecological models. *Ecological Monographs* **57**, 1–21.

Debinski DM and Holt RD (2000). A survey and overview of habitat fragmentation experiments. *Conservation Biology* **14**, 342–355.

del Moral R and Jones C (2002). Vegetation development on pumice at Mount St. Helens, USA. *Plant Ecology* **162**, 9–22.

DeLucia EH, Hamilton JG, Naidu SL, Thomas RB, Andrews JA, Finzi A, *et al.* (1999). Net primary production of a forest ecosystem with experimental CO_2 enrichment. *Science* **284**, 1177–1179.

Denoth M, Frid L and Myers JH (2002). Multiple agents in biological control: improving the odds? *Biological Control* **24**, 20–30.

Detwiler RP and Hall CAS (1988). Tropical forests and the global carbon cycle. *Science* **239**, 42–47.

Dewar RC and Watt AD (1992). Predicted changes in synchrony of larval emergence and budburst under climatic warming. *Oecologia* **89**, 557–559.

Diamond JM (1999). *Guns, Germs, and Steel: The Fates of Human Societies.* W.W. Norton & Company, New York.

Dirzo R and Garcia MC (1992). Rates of deforestation in Los Tuxtlas, a Neotropical area in southeast Mexico. *Conservation Biology* **6**, 84–90.

Dixon RK, Brown S, Houghton RA, Solomon AM, Trexler MC and Wisniewski J (1994). Carbon pools and flux in global forest ecosystems. *Science* **263**, 185–190.

Dobson A and Meagher M (1996). The population dynamics of brucellosis in Yellowstone National Park. *Ecology* **77**, 1026–1036.

Dobson AP, Bradshaw AD and Baker AJM (1997a). Hopes for the future: restoration ecology and conservation biology. *Science* **277**, 515–522.

Dobson AP, Rodriguez JP, Roberts WM and Wilcove DS (1997b). Geographic distribution of endangered species in the United States. *Science* **275**, 550–553.

Dodd AP (1940). *The Biological Campaign Against Prickly-pear.* Commonwealth Prickly Pear Board, Brisbane.

Downing JA, Osenberg CW and Sarnelle O (1999). Meta-analysis of marine nutrient-enrichment experiments: variation in the magnitude of nutrient limitation. *Ecology* **80**, 1157–1167.

Dublin HT, Sinclair ARE and McGlade J (1990). Elephants and fire as causes of multiple stable states in the Serengeti-Mara woodlands. *Journal of Animal Ecology* **59**, 1147–1164.

Dyer SJ, O'Neill JP, Wasel SM and Boutin S (2001). Avoidance of industrial development by woodland caribou. *Journal of Wildlife Management* **65**, 531–542.

Ebert D (1998). Experimental evolution of parasites. *Science* **282**, 1432–1435.

Ebert D and Bull JJ (2003). Challenging the trade-off model for the evolution of virulence: is virulence management feasible? *Trends in Microbiology* **11**, 15–20.

Egerton FN, III (1973). Changing concepts of the balance of nature. *Quarterly Review of Biology* **48**, 322–350.

Elton CS (1958). *The Ecology of Invasions by Animals and Plants*. Methuen, London, UK.

Enright JT (1976). Climate and population regulation: the biogeographer's dilemma. *Oecologia* **24**, 295–310.

Estes JA, Tinker MT, Williams TM and Doak DF (1998). Killer whale predation on sea otters linking oceanic and nearshore ecosystems. *Science* **282**, 473–476.

Ewald PW (1995). The evolution of virulence: a unifying link between parasitology and ecology. *Journal of Parasitology* **81**, 659–669.

Fajer ED (1989). The effects of enriched CO_2 atmospheres on plant–insect herbivore interactions: growth responses of larvae of the specialist butterfly, *Junonia coenia* (Lepidoptera: Nymphalidae). *Oecologia* **81**, 514–520.

Fenner F and Fantini B (1999). *Biological Control of Vertebrate Pests: The History of Myxomatosis, an Experiment in Evolution*. CABI Publishing, New York.

Fenner F and Myers K (1978). Myxoma virus and myxomatosis in retrospect: the first quarter century of a new disease. In *Viruses and Environment*. (Eds E Kurstak and K Maramorosch) pp. 539–570. Academic Press, New York.

Finegan B (1996). Pattern and process in neotropical secondary rain forests: the first 100 years of succession. *Trends in Ecology and Evolution* **11**, 119–124.

Finney BP, Gregory-Eaves I, Douglas MSV and Smol JP (2002). Fisheries productivity in the northeastern Pacific Ocean over the past 2200 years. *Nature* **416**, 729–733.

Fisher RN and Shaffer HB (1996). The decline of amphibians in California's Great Central Valley. *Conservation Biology* **10**, 1387–1397.

Flint ML and van den Bosch R (1981). A history of pest control. In *Introduction to Integrated Pest Management*. pp. 51–81. Plenum Press, New York.

Flux JEC (1993). Relative effect of cats, myxomatosis, traditional control, or competitors in removing rabbits from islands. *New Zealand Journal of Zoology* **20**, 13–18.

Forero MG, Tella JL, Hobson KA, Bertellotti M and Blanco G (2002). Conspecific food competition explains variability in colony size: a test in Magellanic penguins. *Ecology* **83**, 3466–3475.

Forman RTT (1964). Growth under controlled conditions to explain the hierarchical distributions of a moss, *Tetraphis pellucida*. *Ecological Monographs* **34**, 1–25.

Fraser RH and Currie DJ (1996). The species richness-energy hypothesis in a system where historical factors are thought to prevail: coral reefs. *American Naturalist* **148**, 138–159.

Freed LA, Conant S and Fleischer RC (1987). Evolutionary ecology and radiation of Hawaiian passerine birds. *Trends in Ecology and Evolution* **2**, 196–203.

Garshelis DL and Johnson CB (2001). Sea otter population dynamics and the *Exxon Valdez* oil spill: disentangling the confounding effects. *Journal of Applied Ecology* **38**, 19–35.

Gaston KJ (1988). Patterns in the local and regional dynamics of moth populations. *Oikos* **53**, 49–59.

Gaston KJ (1991). How large is a species' geographic range? *Oikos* **61**, 434–438.

Gaston KJ, Quinn RM, Blackburn TM and Eversham BC (1998). Species–range size distributions in Europe. *Ecography* **21**, 361–370.

Gelbard A, Haub C and Kent MM (1999). World population beyond six billion. *Population Bulletin* **54**, 1–44.

Gerber S, Joos F and, Prentice IC (2004). Sensitivity of a dynamic global vegetation model to climate and atmospheric CO_2. *Global Change Biology* **10**, 1223–1239.

Gibbons DW, Reid JB and Chapman RA (1993). *The New Atlas of Breeding Birds in Britain and Ireland: 1988–1991*. T. & A.D. Poyser, London, UK.

Gintis H, Bowles S, Boyd R and Fehr E (2003). Explaining altruistic behavior in humans. *Evolution and Human Behavior* **24**, 153–172.

Gitay H and Noble IR (1997). What are functional types and how should we seek them? In *Plant Functional*

Types: Their Relevance to Ecosystem Properties and Global Change. (Eds TM Smith, HH Shugart and FI Woodward) pp. 3–19. Cambridge University Press, Cambridge, UK.

Gleason HA and Cronquist A (1964). *The Natural Geography of Plants.* Columbia University Press, New York.

Gliwicz ZM (1990). Food thresholds and body size in cladocerans. *Nature* **343**, 638–640.

Global Carbon Project (2003). *Science Framework and Implementation: Earth System Science Partnership (IGBP, IHDP, WCRP, DIVERSITAS).* Global Carbon Project Report No. 1, Canberra.

Gomez JM and Zamora R (2002). Thorns as induced mechanical defense in a long-lived shrub (*Hormathophylla spinosa*, Cruciferae). *Ecology* **83**, 885–890.

Gorham E (1991). Northern peatlands: role in the carbon cycle and probable responses to climatic warming. *Ecological Applications* **1**, 182–195.

Gower ST, McMurtrie RE and Murty D (1996). Aboveground net primary production decline with stand age: potential causes. *Trends in Ecology & Evolution* **11**, 378–382.

Graham M (1939). The sigmoid curve and the overfishing problem. *Rapport du Conseil International pour l'Exploration de la Mer* **110**, 15–20.

Great Lakes Fishery Commission (2005). Commercial fish production in the Great Lakes 1867–2000. http://www.glfc.org/databases/commercial/commerc.php.

Greene E (1987). Individuals in an osprey colony discriminate between high and low quality information. *Nature* **329**, 239–241.

Greenwood RJ (1986). Influence of striped skunk removal on upland duck nest success in North Dakota. *Wildlife Society Bulletin* **14**, 6–11.

Grenfell BT and Dobson AP (Eds) (1995). *Ecology of Infectious Diseases in Natural Populations.* Cambridge University Press: Cambridge, UK.

Grime JP (1979). *Plant Strategies and Vegetation Processes.* John Wiley and Sons, New York.

Grubb PJ (1987). Global trends in species-richness in terrestrial vegetation: a view from the Northern Hemisphere. In *Organization of Communities Past and Present.* (Eds JHR Gee and PS Giller) pp. 99–118. Blackwell, Oxford, UK.

Guay JC, Boisclair D, Rioux D, Leclerc M, Lapointe M and Legendre P (2000). Development and validation of numerical habitat models for juveniles of Atlantic salmon (*Salmo salar*). *Canadian Journal of Fisheries and Aquatic Sciences* **57**, 2065–2075.

Gunn JM and Mills KH (1998). The potential for restoration of acid-damaged lake trout lakes. *Restoration Ecology* **6**, 390–397.

Haddad NM (1999). Corridor and distance effects on interpatch movements: a landscape experiment with butterflies. *Ecological Applications* **9**, 612–622.

Haddad NM and Baum KA (1999). An experimental test of corridor effects on butterfly densities. *Ecological Applications* **9**, 623–633.

Hager HA and McCoy KD (1998). The implications of accepting untested hypotheses: a review of the effects of purple loosestrife (*Lythrum, salicaria*) in North America. *Biodiversity and Conservation* **7**, 1069–1079.

Haila Y (2002). A conceptual genealogy of fragmentation research: from island biogeography to landscape ecology. *Ecological Applications* **12**, 321–334.

Hairston NGJ and Hairston NGS (1993). Cause–effect relationships in energy flow, trophic structure, and interspecific interactions. *American Naturalist* **142**, 379–411.

Halls C (2006). *Living Planet Report 2006.* World Wildlife Fund and Global Footprint Network, Gland, Switzerland.

Halpern BS (2003). The impact of marine reserves: do reserves work and does reserve size matter? *Ecological Applications* **13**, S117–S137.

Hamilton WD (1971). Geometry of the selfish herd. *Journal of Theoretical Biology* **31**, 295–311.

Hanski I, Kuussaari M and Nieminen M (1994). Metapopulation structure and migration in the butterfly *Melitaea cinxia*. *Ecology* **75**, 747–762.

Hanski I, Pakkala T, Kuussaari M and Lei G (1995). Metapopulation persistence of an endangered butterfly in a fragmented landscape. *Oikos* **72**, 21–28.

Harrison S (1991). Local extinction in a metapopulation context: an empirical evaluation. *Biological Journal of the Linnean Society* **42**, 73–88.

Harvell CD, Mitchell CE, Ward JR, Altizer S, Dobson AP, Ostfeld RS and Samuel MD (2002). Climate warming and disease risks for terrestrial and marine biota. *Science* **296**, 2158–2162.

Hawkins BA, Field R, Cornell HV, Currie DJ, Guégan J-F, Kaufman DM, *et al.* (2003). Energy, water and broad-scale geographic patterns of species richness. *Ecology* **84**, 3105–3117.

Hawkins BA, Porter EE and Diniz-Filho JAF (2003). Productivity and history as predictors of the latitudinal diversity gradient of terrestrial birds *Ecology* **84**, 1608–1623.

Hawksworth DL and Kalin-Arroyo MT (1995). Magnitude and distribution of biodiversity. In *Global Biodiversity Assessment*. (Ed. VH Heywood) pp. 107–191. Cambridge University Press, Cambridge, UK.

Hayward TL (1991). Primary production in the North Pacific Central Gyre: a controversy with important implications. *Trends in Ecology and Evolution* **6**, 281–284.

Heesterbeek JAP and Roberts MG (1995). Mathematical models for microparasites of wildlife. In *Ecology of Infectious Diseases in Natural Populations*. (Eds BT Grenfell and AP Dobson) pp. 90–122. Cambridge University Press, Cambridge, UK.

Heide-Jørgensen M-P and Härkönen T (1992). Epizootiology of the seal disease in the eastern North Sea. *Journal of Applied Ecology* **29**, 99–107.

Heinsohn R and Legge S (1999). The cost of helping. *Trends in Ecology & Evolution* **14**, 53–57.

Helfield JM and Naiman RJ (2001). Effects of salmon-derived nitrogen on riparian forest growth and implications for stream productivity. *Ecology* **82**, 2403–2409.

Henke SE and Bryant FC (1999). Effects of coyote removal on the faunal community in western Texas. *Journal of Wildlife Management* **63**, 1066–1081.

Henry HAL and Aarssen LW (1997). On the relationship between shade tolerance and shade avoidance strategies in woodland plants. *Oikos* **80**, 575–582.

Herrera CM (1982). Defense of ripe fruit from pests: its significance in relation to plant-disperser interactions. *American Naturalist* **120**, 218–241.

Hilborn R, Orensanz JM and Parma AM (2005). Institutions, incentives and the future of fisheries. *Philosophical Transactions of the Royal Society of London, Series B* **360**, 47–57.

Hixon MA (1998). Population dynamics of coral-reef fishes: controversial concepts and hypotheses. *Australian Journal of Ecology* **23**, 192–201.

Holland HD (1995). Atmospheric oxygen and the biosphere. In *Linking Species and Ecosystems*. (Eds CG Jones and JH Lawton) pp. 127–136. Chapman and Hall, New York.

Holway DA (1999). Competitive mechanisms underlying the displacement of native ants by the invasive Argentine ant. *Ecology* **80**, 238–251.

Holzapfel C and Mahall BE (1999). Bidirectional facilitation and interference between shrubs and annuals in the Mojave Desert. *Ecology* **80**, 1747–1761.

Hone J (2007). *Wildlife Damage Control*. CSIRO Publishing, Melbourne.

Hoover JP, Brittingham MC and Goodrich LJ (1995). Effects of forest patch size on nesting success of wood thrushes. *Auk* **112**, 146–155.

Horn HS (1975). Markovian properties of forest succession. In *Ecology and Evolution of Communities*. (Eds ML Cody and JM Diamond) pp. 196–211. Harvard University Press: Cambridge, Mass.

Houghton RA, Hackler JL and Lawrence KT (1999). The U.S. carbon budget: contributions from land-use change. *Science* **285**, 574–578.

Houston DB (1982). *The Northern Yellowstone Elk: Ecology and Management*. Macmillan, New York.

Huppert A and Stone L (1998). Chaos in the Pacific's coral reef bleaching cycle. *American Naturalist* **152**, 447–459.

Huston M (1979). A general hypothesis of species diversity. *American Naturalist* **113**, 81–101.

Huston M and Smith T (1987). Plant succession: life history and competition. *American Naturalist* **130**, 168–198.

Idso SB (1999). The long-term response of trees to atmospheric CO_2 enrichment. *Global Change Biology* **5**, 493–495.

Irwin RE (2000). Hummingbird avoidance of nectar-robbed plants: spatial location or visual cues. *Oikos* **91**, 499–506.

Ishida Y, Welch DW and Ogura M (1995). Potential influence of North Pacific sea-surface temperatures on increased production of chum salmon (*Oncorhynchus keta*) from Japan. In *Climate Change and Northern Fish Populations*. (Ed. RJ Beamish) pp. 271–275. Canadian Special Publication of Fisheries and Aquatic Sciences, Ottawa, Ontario.

Iverson LR, Prasad AM, Hale BJ and Sutherland EK (1999a). *An Atlas of Current and Potential Future Distributions of Common Trees of the Eastern United States*. pp. 1–245. General Technical Report NE-265. Northeastern Research Station, USDA Forest Service.

Iverson LR, Prasad AM and Schwartz MW (1999b). Modeling potential future individual tree-species distributions in the Eastern United States under a climate change scenario: a case study with *Pinus virginiana*. *Ecological Modelling* **115**, 77–93.

Jackson SM and Claridge A (1999). Climatic modelling of the distribution of the mahogany glider (*Petaurus gracilis*), and the squirrel glider (*P. norfolcensis*). *Australian Journal of Zoology* **47**, 47–57.

James CD and Shine R (2000). Why are there so many coexisting species of lizards in Australian deserts? *Oecologia* **125**, 127–141.

Janzen DH (1970). Herbivores and the number of tree species in tropical forests. *American Naturalist* **104**, 501–528.

Jenkins B, Kitching RL and Pimm SL (1992). Productivity, disturbance and food web structure at a local spatial scale in experimental container habitats. *Oikos* **65**, 249–255.

Jimenez JA, Hughes KA, Alaks G, Graham L and Lacy RC (1994). An experimental study of inbreeding depression in a natural habitat. *Science* **266**, 271–273.

Jolliffe PA (2000). The replacement series. *Journal of Ecology* **88**, 371–385.

Jones GP (1990). The importance of recruitment to the dynamics of a coral reef fish population. *Ecology* **71**, 1691–1698.

Kalleberg H (1958). Observations in a stream tank of territoriality and competition in juvenile salmon and trout (*Salmo salar* L. and *S. trutta* L.). *Institute of Freshwater Research, Drottningholm, Report No. 39*, 55–98.

Kazimirov NI and Morozova RN (1973). *Biological Cycling of Matter in Spruce Forests of Karelia*. Nauka Publishing House, Leningrad, Russia.

Keddy PA (1990). Why don't ecologists study positive interactions? *Bulletin of the Ecological Society of America* **71**, 101–102.

Keeling CD and Whorf TP (2004). Atmospheric CO_2 records from sites in the Scripps Institute of Oceanography air sampling network In *Trends: A Compendium of Data on Global Change*. Carbon Dioxide Information Analysis Center, U.S. Department of Energy, Oak Ridge, Tennessee.

Keeling CD, Whorf TP, Wahlen M and van der Plicht J (1995). Interannual extremes in the rate of rise of atmospheric carbon dioxide since 1980. *Nature* **375**, 666–670.

Keith H (1997). Nutrient cycling in eucalypt ecosystems. In *Eucalypt Ecology: Individuals to Ecosystems*. (Eds JE Williams and JCZ Woinarski) pp. 197–226. Cambridge University Press, Cambridge, UK.

Keller LF and Waller DM (2002). Inbreeding effects in wild populations. *Trends in Ecology and Evolution* **17**, 230–242.

Ketterson ED and Nolan V, Jr. (1982). The role of migration and winter mortality in the life history of a temperate-zone migrant, the dark-eyed junco, as determined from demographic analyses of winter populations. *Auk* **99**, 243–259.

Knowlton N (2001). The future of coral reefs. *Proceedings of the National Academy of Sciences, USA* **98**, 5419–5425.

Knox EA (1970). Antarctic marine ecosystems. In *Antarctic Ecology*. (Ed. MW Holdgate) pp. 69–96. Academic Press, London, UK.

Koop K, Booth D, Broadbent A, Brodie J, Bucher D, Capone D, *et al.* (2001). ENCORE: The Effect of Nutrient Enrichment on Coral Reefs. Synthesis of Results and Conclusions. *Marine Pollution Bulletin* **42**, 91–120.

Korzukhin MD, Porter SD, Thompson LC and Wiley S (2001). Modeling temperature-dependent range limits for the fire ant *Solenopsis invicta* (Hymenoptera: Formicidae) in the United States. *Environmental Entomology* **30**, 645–655.

Kotler BP, Brown JS and Hasson O (1991). Factors affecting gerbil foraging behaviour and rates of owl predation. *Ecology* **72**, 2249–2260.

Kramer LD and Bernard KA (2001). West Nile virus in the Western Hemisphere. *Current Opinion in Infectious Diseases* **14**, 519–525.

Krebs CJ and Boonstra R (2001). The Kluane Region. In *Ecosystem Dynamics of the Boreal Forest*. (Eds CJ Krebs, S Boutin and R Boonstra) pp. 9–24. Oxford University Press, New York.

Krebs CJ, Boutin S and Boonstra R (Eds) (2001). *Ecosystem Dynamics of the Boreal Forest: the Kluane Project*. Oxford University Press, New York.

Krebs JR and Davies NB (1993). *An Introduction to Behavioural Ecology*. Blackwell Scientific Publications, Oxford, UK.

Krebs JW, Mandel EJ, Swerdlow DL and Rupprecht CE (2005). Rabies surveillance in the United States during 2004. *Journal of the American Veterinary Medicine Association* **227**, 1912–1925.

Krupa SV (2003). Effects of atmospheric ammonia (NH_3) on terrestrial vegetation: a review. *Environmental Pollution* **124**, 179–221.

Kurki S, Nikula A, Helle P and Lindén H (2000). Landscape fragmentation and forest composition effects on grouse breeding success in boreal forests. *Ecology* **81**, 1985–1997.

Kuussaari M, Saccheri I, Camara M and Hanski I (1998). Allee effect and population dynamics in the Glanville fritillary butterfly. *Oikos* **82**, 384–392.

Lambert B and Peferoen M (1992). Insecticidal promise of *Bacillus thuringiensis*: Facts and mysteries about a successful biopesticide. *Bioscience* **42**, 112–121.

Landsberg J (1990). Dieback of rural eucalypts: response of foliar dietary quality and herbivory to defoliation. *Australian Journal of Ecology* **15**, 89–96.

Landsberg J and Wylie R (1991). A review of rural dieback in Australia. In *Growback '91*. pp. 3–11. Growback Publications, University of Melbourne, Victoria.

Landsberg JH (2002). The effects of harmful algal blooms on aquatic organisms. *Reviews in Fisheries Science* **10**, 113–390.

Landsberg JJ and Cork SJ (1997). Herbivory: interactions between eucalypts and the vertebrates and invertebrates that feed on them. In *Eucalypt Ecology: Individuals to Ecosystems*. (Eds JE Williams and JCZ Woinarski) pp. 342–372. Cambridge University Press, Cambridge, UK.

Larsen J (2004). The sixth great extinction: a status report. *Earth Policy Institute Eco-economy Update* **35**, March 2, 2004, Washington D.C.

Laurance WF, Lovejoy TE, Vasconcelos HL, Bruna EM, Didham RK, Stouffer PC, *et al.* (2002). Ecosystem decay of Amazonian forest fragments: a 22-year investigation. *Conservation Biology* **16**, 605–618.

Lawlor TE (1986). Comparative biogeography of mammals on islands. *Biological Journal of the Linnean Society* **28**, 99–125.

Leopold A (1933). *Game Management*. Charles Scribner's Sons, New York.

Levin SA (Ed.) (2001). *Encyclopedia of Biodiversity*. Academic Press, San Diego.

Lewis JR (1972). *The Ecology of Rocky Shores*. English Universities Press, London, UK.

Lewis VR, Merrill LD, Atkinson T, H. and Wasbauer JS (1992). Imported fire ants: Potential risk to California. *California Agriculture* **46**, 29–31.

Likens GE, Bormann FH, Johnson NM, Fisher DW and Pierce RS (1970). Effects of forest cutting and herbicide treatment on nutrient budgets in the Hubbard Brook watershed-ecosystem. *Ecological Monographs* **40**, 23–47.

Lima SL (1998). Nonlethal effects in the ecology of predator–prey interactions. *Bioscience* **48**, 25–34.

Link J (2002). Does food web theory work for marine ecosystems? *Marine Ecology Progress Series* **230**, 1–9.

Lizotte MP (2001). The contributions of sea ice algae to Antarctic marine primary production. *American Zoologist* **41**, 57–73.

Lorius C, Jouzel J, Ritz C, Merlivat L, Barkov NI, Korotkevich YS and Kotlyakov VM (1985). A 150,000-year climatic record from antarctic ice. *Nature* **316**, 591–596.

Lubchenco J (1978). Plant species diversity in a marine intertidal community: Importance of herbivore food preference and algal competitive abilities. *American Naturalist* **112**, 23–39.

Lubchenco J (1986). Relative importance of competition and predation: Early colonization by seaweeds in New England. In *Community Ecology*. (Eds J Diamond and TJ Case) pp. 537–555. Harper & Row, New York.

Ludwig D, Hilborn R and Walters C (1993). Uncertainty, resource exploitation, and conservation: lessons from history. *Science* **260**, 17, 36.

MacArthur R and Wilson EO (1967). *The Theory of Island Biogeography*. Princeton University Press, Princeton, New Jersey.

MacCracken MC (1985). Carbon dioxide and climate change: background and overview. In *Projecting the Climatic Effects of Increasing Carbon Dioxide*. (Eds MC MacCracken and FM Luther) pp. 1–23. U.S. Department of Energy, Washington, D.C.

Macdonald DW and Service K (Eds) (2007). *Key Topics in Conservation Biology*. Blackwell Publishing: Oxford, UK.

Macdonald DW and Voigt DR (1985). The biological basis of rabies models. In *Population Dynamics of Rabies in Wildlife*. (Ed. PJ Bacon) pp. 71–108. Academic Press, London, UK.

MacLean DA and Wein RW (1977). Nutrient accumulation for post fire jack pine and hardwood successional patterns in New Brunswick. *Canadian Journal of Forest Research* **7**, 562–578.

Macnab J (1991). Does game cropping serve conservation? A reexamination of the African data. *Canadian Journal of Zoology* **69**, 2283–2290.

Mangin S, Gauthier-Clerc M, Frenot Y, Gendner J-P and Maho YL (2003). Ticks *Ixodes uriae* and the breeding performance of a colonial seabird, king penguin *Aptenodytes patagonicus*. *Journal of Avian Biology* **34**, 30–34.

Mantua NJ, Hare SR, Zhang Y, Wallace JM and Francis RC (1997). A Pacific interdecadal climate oscillation with impacts on salmon production. *Bulletin of the American Meteorological Society* **78**, 1069–1079.

Margules CR and Pressey RL (2000). Systematic conservation planning. *Nature* **405**, 243–253.

Marks PL (1983). On the origin of the field plants of the northeastern United States. *American Naturalist* **122**, 210–228.

Marrs RH, Johnson SW and LeDuc MG (1998). Control of bracken and restoration of heathland. VIII. The regeneration of the heathland community after 18 years of continued bracken control or 6 years of control followed by recovery. *Journal of Applied Ecology* **35**, 857–870.

McCarthy MA, Lindenmayer DB and Dreschler M (1997). Extinction debts and risks faced by abundant species. *Conservation Biology* **11**, 221–226.

McCook LJ (1994). Understanding ecological community succession: causal models and theories, a review. *Vegetatio* **110**, 115–147.

McGarigal K (2002). Landscape pattern metrics. In *Encyclopedia of Environmentrics Volume 2*. (Eds AH El-Shaarawi and WW Piegorsch) pp. 1135–1142. John Wiley and Sons, Chichester, UK.

McGarigal K and Cushman SA (2002). Comparative evaluation of experimental approaches to the study of habitat fragmentation effects. *Ecological Applications* **12**, 335–345.

McKilligan NG (1987). Causes of nesting losses in the cattle egret *Ardeola ibis* in eastern Australia with special reference to the pathogenicity of the tick *Argas (Persicargas) robertsi* to nestlings. *Australian Journal of Ecology* **12**, 9–16.

McNeill WH (1976). *Plagues and Peoples*. Anchor Press, Garden City, New York.

McQueen DJ, Johannes MRS, Post JR, Stewart TJ and Lean DRS (1989). Bottom-up and top-down impacts on freshwater pelagic community structure. *Ecological Monographs* **59**, 289–309.

Menge BA and Sutherland JP (1987). Community regulation: variation disturbance, competition, and predation in relation to environmental stress and recruitment. *American Naturalist* **130**, 730–757.

Menges ES (1990). Population viability analysis for an endangered plant. *Conservation Biology* **4**, 52–62.

Menon S and Bawa KS (1998). Deforestation in the tropics: reconciling disparities in estimates for India. *Ambio* **27**, 576–577.

Messina FJ, Durham SL, Richards JH and McArthur ED (2002). Trade-off between plant growth and defense? A comparison of sagebrush populations. *Oecologia* **131**, 43–51.

Moritz MA (1997). Analysing extreme disturbance events: fire in Los Padres National Forest. *Ecological Applications* **7**, 1252–1262.

Morrow PA and Fox LR (1980). Effects of variation in *Eucalyptus* essential oil yield on insect growth and grazing damage. *Oecologia* **45**, 209–219.

Moss R, Oswald J and Baines D (2001). Climate change and breeding success: decline of the capercaille in Scotland. *Journal of Animal Ecology* **70**, 47–61.

Moss R and Watson A (2001). Population cycles in birds of the grouse family (Tetraonidae). *Advances in Ecological Research* **32**, 53–111.

Mougeot F, Redpath SM, Moss R, Matthiopoulos J and Hudson PJ (2003). Territorial behaviour and population dynamics in red grouse *Lagopus lagopus scoticus*. I. Population experiments. *Journal of Animal Ecology* **72**, 1073–1082.

Murdoch WW (1994). Population regulation in theory and practice. *Ecology* **75**, 271–287.

Myers JH and Bazely D (2003). *Ecology and Control of Introduced Plants*. Cambridge University Press, Cambridge, UK.

Myers JH and Risley C (2000). Why reduced seed production is not necessarily translated into successful biological control. In *Proceedings of the X International Symposium on Biological Control of Weeds*. (Ed. N Spencer) pp. 569–581. Montana State University, Bozeman, Montana.

Myers K, Marshall ID and Fenner F (1954). Studies in epidemiology of infectious myxomatosis of rabbits. III. Observations on two succeeding epizootics in Australian wild rabbits on the Riverine Plain of south-eastern Australia. *Journal of Hygiene* **52**, 337–360.

Myers N, Mittermeier RA, Mittermeier CG, da Fonseca GAB and Kent J (2000). Biodiversity hotspots for conservation priorities. *Nature* **403**, 853–858.

Myers RA, Hutchings JA and Barrowman NJ (1996). Hypotheses for the decline of cod in the North Atlantic. *Marine Ecology Progress Series* **138**, 293–308.

Nakai S, Halliday AN and Rea DK (1993). Provenance of dust in the Pacific Ocean. *Earth and Planetary Science Letters* **119**, 143–157.

Nathan R, Safriel UN, Noy-Meir I and Schiller G (2000). Spatiotemporal variation in seed dispersal and recruitment near and far from *Pinus halepensis* trees. *Ecology* **81**, 2156–2169.

Nemani RR, Keeling CD, Hashimoto H, Jolly WM, Piper SC, Tucker CJ, Myneni RB and Running SW (2003). Climate-driven increases in global terrestrial net primary production from 1982 to 1999. *Science* **300**, 1560–1563.

Newman EI (1997). Phosphorus balance of contrasting farming systems, past and present. Can food production be sustainable? *Journal of Applied Ecology* **34**, 1334–1347.

Newmark WD (1985). Legal and biotic boundaries of Western North American National Parks: A problem of congruence. *Biological Conservation* **33**, 197–208.

Newmark WD (1995). Extinction of mammal populations in Western North American national parks. *Conservation Biology* **9**, 512–526.

Newton I (1994). Experiments on the limitation of bird breeding densities: a review. *Ibis* **136**, 397–411.

Nichols JD, Conroy MJ, Anderson DR and Burnham KP (1984). Compensatory mortality in waterfowl populations: a review of the evidence and implications for research and management. *Transactions of the North American Wildlife and Natural Resources Conference* **49**, 535–554.

Nicklas KJ, Tiffney BH and Knoll AH (1980). Apparent changes in the diversity of fossil plants. *Evolutionary Biology* **12**, 1–89.

Noble IR (1981). Predicting successional change. In *Fire Regimes and Ecosystem Properties*. (Ed. HA Mooney) pp. 278–300. U.S. Department of Agriculture, Forest Service, General Technical Report.

Norby RJ, Wullschleger SD, Gunderson CA, Johnson DW and Ceulemans R (1999). Tree responses to rising CO_2 in field experiments: implications for the future forest. *Plant, Cell and Environment* **22**, 683–714.

Noss RF (1987). Corridors in real landscapes: a reply to Simberloff and Cox. *Conservation Biology* **1**, 159–164.

Oesterheld M, Sala OE and McNaughton SJ (1992). Effect of animal husbandry on herbivore-carrying capacity at a regional scale. *Nature* **356**, 234–236.

Olesen JM and Jordano P (2002). Geographic patterns in plant-pollinator mutualistic networks. *Ecology* **83**, 2416–2424.

Olson JS (1958). Rates of succession and soil changes on southern Lake Michigan sand dunes. *Botanical Gazette* **119**, 125–170.

Ostfeld RS (1997). The ecology of Lyme-disease risk. *American Scientist* **85**, 338–346.

Ostfeld RS, Jones CG and Wolff JO (1996). Of mice and mast—ecological connections in eastern deciduous forests. *BioScience* **46**, 323–330.

Ovington JD (1978). *Australia's Endangered Species: Mammals, Birds, and Reptiles*. Cassell Australia, Melbourne.

Pac HI and Frey K (1991). *Some Population Characteristics of the Northern Yellowstone Bison Herd During the Winter of 1988–89*. Montana Department of Fish, Wildlife and Parks, Bozeman, Montana.

Packer C, Herbst L, Pusey AE, Bygott JD, Hanby JP, Cairns SJ and Mulder MB (1988). Reproductive success in lions. In *Reproductive Success*. (Ed. TH Clutton-Brock) pp. 363–383. University of Chicago Press, Chicago.

Packer C and Pusey AE (1997). Divided we fall: cooperation among lions. *Scientific American* **276**, 52–59.

Pagel MD, May RM and Collie AR (1991). Ecological aspects of the geographical distribution and diversity of mammalian species. *American Naturalist* **137**, 791–815.

Paine RT (1974). Intertidal community structure: Experimental studies on the relationship between a dominant competitor and its principal predator. *Oecologia* **15**, 93–120.

Paine RT, Tegner MJ and Johnson EA (1998). Compounded perturbations yield ecological surprises. *Ecosystems* **1**, 535–545.

Parer I, Conolly D and Sobey WR (1985). Myxomatosis: the effects of annual introductions of an immunizing strain and a highly virulent strain of myxoma virus into rabbit populations at Urana, N.S.W. *Australian Wildlife Research* **12**, 407–423.

Parker MA and Salzman AG (1985). Herbivore exclosure and competitor removal: effects on juvenile survivorship and growth in the shrub *Gutierrezia microcephala*. *Journal of Ecology* **73**, 903–913.

Parmesan C, Ryrholm N, Stefanescu C, Hill JK, Thomas CD, Descimon H, *et al.* (1999). Poleward shifts in geographical ranges of butterfly species associated with regional warming. *Nature* **399**, 579–583.

Parmesan C and Yohe G (2003). A globally coherent fingerprint of climate change impacts across natural systems. *Nature* **421**, 37–42.

Pauly D and Christensen V (1995). Primary production required to sustain global fisheries. *Nature* **374**, 255–257.

Pauly D, Christensen V, Guenette S, Pitcher TJ, Sumaila UR, Walters CJ, Watson R and Zeller D (2002). Towards sustainability in world fisheries. *Nature* **418**, 689–695.

Payette S, Fortin M-J and Gamache I (2001). The subarctic forest-tundra: the structure of a biome in a changing climate. *BioScience* **51**, 709–718.

Pech R, Hood GM, McIlroy J and Saunders G (1997). Can foxes be controlled by reducing their fertility? *Reproduction, Fertility and Development* **9**, 41–50.

Penn D (1999). Explaining the human demographic transition. *Trends in Ecology and Evolution* **14**, 32–33.

Peñuelas J and Estiarte M (1998). Can elevated CO_2 affect secondary metabolism and ecosystem function? *Trends in Ecology and Evolution* **13**, 20–24.

Peters RH (1983). *The Ecological Implications of Body Size*. Cambridge University Press, New York.

Peterson CH (2001). The 'Exxon Valdez' oil spill in Alaska: acute, indirect and chronic effects on the ecosystem. *Advances in Marine Biology* **39**, 1–103.

Peterson RO, Thomas NJ, Thurber JM, Vucetich JA and Waite T (1998). Population limitation and the wolves of Isle Royale. *Journal of Mammology* **79**, 828–841.

Petit JR, Jouzel J, Raynaud D, Barkov NI, Barnola JM, Basile I, *et al.* (1999). Climate and atmospheric history of the past 420,000 years from the Vostok Ice Core, Antarctica. *Nature* **399**, 429–436.

Phillips OL, Malhi Y, Higuchi N, Laurance WF, Nunez PV, Vasquez RM, *et al.* (1998). Changes in the carbon balance of tropical forests: evidence from long-term plots. *Science* **282**, 439–441.

Pickett STA and White PS (Eds) (1985). *The Ecology of Natural Disturbance and Patch Dynamics*. Academic Press, Orlando, Florida.

Pielou EC (1991). *After the Ice Age: The Return of Life to Glaciated North America*. University of Chicago Press, Chicago.

Pierson EA and Turner RM (1998). An 85-year study of saguaro (*Carnegiea gigantea*) demography. *Ecology* **79**, 2676–2693.

Pimentel D (1997). Pest management in agriculture. In *Techniques for Reducing Pesticide Use: Economic and*

Environmental Benefits. (Ed. D Pimentel) pp. 1–11. John Wiley & Sons, Chichester, UK.

Pimentel D, Acquay H, Biltonen M, Rice P, Silva M, Nelson J, *et al.* (1992). Assessment of environmental and economic impacts of pesticide use. In *The Pesticide Question: Environment Economics and Ethics.* (Eds PD and H Lehman) pp. 47–83. Chapman and Hall, New York.

Platt T, Fuentes-Yaco C and Frank KT (2003). Spring algal bloom and larval fish survival. *Nature* **423**, 398–399.

Population Reference Bureau (2006). World Population Data Sheet. http://www.prb.org/.

Porter WF and Underwood HB (1999). Of elephants and blind men: deer management in the U.S. National Parks. *Ecological Applications* **9**, 3–9.

Possingham HP (2001). The business of biodiversity: applying decision theory principles to nature conservation. *Environment, Economy and Society* **9**, 1–37.

Postel SL, Daily GC and Ehrlich PR (1996). Human appropriation of renewable fresh water. *Science* **271**, 785–788.

Poulson TL and Platt WJ (1996). Replacement patterns of beech and sugar maple in Warren Woods, Michigan. *Ecology* **77**, 1234–1253.

Power ME, Tilman D, Estes JA, Menge BA, Bond WJ, Mills LS, *et al.* (1996). Challenges in the quest for keystones. *BioScience* **46**, 609–620.

Prasifka JR, Heinz KM and Minzenmayer RR (2005). Relationships of landscape, prey and agronomic variables to the abundance of generalist predators in cotton (*Gossypium hirsutum*) fields. *Landscape Ecology* **19**, 709–717.

Prins HHT and Weyerhaeuser FJ (1987). Epidemics in populations of wild ruminants: anthrax and impala, rinderpest and buffalo in Lake Manyara National Park, Tanzania. *Oikos* **49**, 28–38.

Proches S (2001). Back to the sea: secondary marine organisms from a biogeographical perspective. *Biological Journal of the Linnean Society* **74**, 197–203.

Qian H (2002). Floristic relationships between Eastern Asia and North America: test of Gray's Hypothesis. *American Naturalist* **160**, 317–332.

Rabalais NN, Turner RE and Scavia D (2002). Beyond science into policy: Gulf of Mexico hypoxia and the Mississippi River. *BioScience* **52**, 129–142.

Rapport DJ, Costanza R and McMichael AJ (1998). Assessing ecosystem health. *Trends in Ecology and Evolution* **13**, 397–402.

Rapport DJ and Whitford WG (1999). How ecosystems respond to stress. *BioScience* **49**, 193–203.

Rich SM, Caporale DA, Telford SR, III, Kocher TD, Hartl DL and Spielman A (1995). Distribution of the *Ixodes ricinus*-like ticks of eastern North America. *Proceedings of the National Academy of Science USA* **92**, 6284–6288.

Ricker WE (1982). Size and age of British Columbia sockeye salmon (*Oncorhynchus nerka*) in relation to environmental factors and the fishery. *Canadian Technical Reports in Fisheries and Aquatic Sciences* **1115**, 1–117.

Roberts L (1989). How fast can trees migrate? *Science* **243**, 735–737.

Robinson WD (1999). Long-term changes in the avifauna of Barro Colorado Island, Panama, a tropical forest isolate. *Conservation Biology* **13**, 85–97.

Rodda GH, Fritts TH and Chiszar D (1997). The disappearance of Guam's wildlife. *BioScience* **47**, 565–574.

Rohde K (1999). Latitudinal gradients in species diversity and Rapoport's rule revisited: a review of recent work and what can parasites teach us about the causes of the gradients? *Ecography* **22**, 593–613.

Roland J (1993). Large-scale forest fragmentation increases the duration of tent caterpillar outbreak. *Oecologia* **93**, 25–30.

Root T (1988). Energy constraints on avian distributions and abundances. *Ecology* **69**, 330–339.

Root TL, Price JT, Hall KR, Schneider SH, Rosenzweig C and Pounds JA (2003). Fingerprints of global warming on wild animals and plants. *Nature* **421**, 57–60.

Roubik DW (2002). Tropical agriculture: The value of bees to the coffee harvest. *Nature* **417**, 708.

Ruesink JL, Parker IM, Groom MJ and Kareiva PM (1995). Reducing the risks of nonindigenous species introductions. *BioScience* **45**, 465–477.

Running SW, Nemani RR, Heinsch FA, Zhao M, Reeves M and Hashimoto H (2004). A continuous satellite-derived measure of global terrestrial primary production. *BioScience* **54**, 547–560.

Russell RC (2002). Ross River virus: ecology and distribution. *Annual Review of Entomology* **47**, 1–31.

Rutberg AT, Naugle RE, Thiele LA and Liu IKM (2004). Effects of immunocontraception on a suburban population of white-tailed deer *Odocoileus virginianus*. *Biological Conservation* **116**, 243–250.

Ryel LA (1981). Population change in the Kirtland's Warbler. *Jack-Pine Warbler* **59**, 76–91.

Ryther JH (1969). Photosynthesis and fish production in the sea. *Science* **166**, 72–76.

Sæther B-E, Ringsby TH, Bakke O and Solberg EJ (1999). Spatial and temporal variation in demography of a house sparrow metapopulation. *Journal of Animal Ecology* **68**, 628–637.

Sagarin RD and Gaines SD (2002). The 'abundant centre' distribution: to what extent is it a biogeographical rule? *Ecology Letters* **5**, 137–147.

Saugier B, Roy J and Mooney HA (Eds) (2000). *Terrestrial Global Productivity*. Academic Press, New York.

Schadt S, Revilla E, Wiegand T, Knauer F, Kaczensky P, Breitenmoser U, *et al.* (2002). Assessing the suitability of central European landscapes for the reintroduction of Eurasian lynx. *Journal of Applied Ecology* **39**, 189–203.

Schall JJ (1983). Lizard malaria: cost to vertebrate host's reproductive success. *Parasitology* **87**, 1–6.

Schindler DW (1997). Liming to restore acidified lakes and streams: a typical approach to restoring damaged ecosystems? *Restoration Ecology* **5**, 1–6.

Schlesinger WH (1997). *Biogeochemistry: An Analysis of Global Change*. Academic Press, San Diego, California.

Schluter D (1982). Seed and patch selection by Galapagos ground finches: relation to foraging efficiency and food supply. *Ecology* **63**, 1106–1120.

Schneider SH (1989). The greenhouse effect: science and policy. *Science* **243**, 771–781.

Schoenly K and Cohen JE (1991). Temporal variation in food web structure: 16 empirical cases. *Ecological Monographs* **61**, 267–298.

Secord D (2003). Biological control of marine invasive species: cautionary tales and land-based lessons. *Biological Invasions* **5**, 117–131.

Shears NT and Babcock RC (2002). Marine reserves demonstrate top-down control of community structure on temperate reefs. *Oecologia* **132**, 131–142.

Sheppard CRC (2003). Predicted recurrences of mass coral mortality in the Indian Ocean. *Nature* **425**, 294–297.

Sherman PW (1977). Nepotism and the evolution of alarm calls. *Science* **197**, 1246–1253.

Silbernagel J (2003). Spatial theory in early conservation design: examples from Aldo Leopold's work. *Landscape Ecology* **18**, 635–646.

Silva M, Brown JH and Downing JA (1997). Differences in population density and energy use between birds and mammals: a macroecological perspective. *Journal of Animal Ecology* **66**, 327–340.

Silvertown JW and Charlesworth D (2001). *Introduction to Plant Population Biology*. Blackwell Science, Oxford, UK.

Simberloff D, Schmitz D and Brown T (1997). *Strangers in Paradise: Impact and Management of Nonindigenous Species in Florida*. Island Press, Washington, D.C.

Singer FJ, Harting A, Symonds KK and Coughenour MB (1997). Density dependence, compensation, and environmental effects on elk calf mortality in Yellowstone National Park. *Journal of Wildlife Management* **61**, 12–25.

Singer FJ, Swift DM, Coughenour MB and Varley JD (1998). Thunder on the Yellowstone revisited: an assessment of management of native ungulates by natural regulation, 1968–1993. *Wildlife Society Bulletin* **26**, 375–390.

Skellam JG (1951). Random dispersal in theoretical populations. *Biometrika* **38**, 196–218.

Smeding FW and de Snoo GR (2003). A concept of food-web structure in organic arable farming systems. *Landscape and Urban Planning* **65**, 219–236.

Smith TB, Freed LA, Lepson JK and Carothers JH (1995). Evolutionary consequences of extinctions in populations of a Hawaiian honeycreeper. *Conservation Biology* **9**, 107–113.

Somero GN (2002). Thermal physiology and vertical zonation of intertidal animals: optima, limits, and costs of living. *Integrative and Comparative Biology* **42**, 780–789.

Springer AM, Estes JA, van Vliet GB, Williams TM, Doak DF, Danner EM, Forney KA and Pfister B (2003). Sequential megafaunal collapse in the North Pacific Ocean: An ongoing legacy of industrial whaling? *Proceedings of the National Academy of Sciences of the USA* **100**, 12223–12228.

Stachowicz JJ (2001). Mutualism, facilitation, and the structure of ecological communities. *BioScience* **51**, 235–246.

Stamp N (2003). Out of the quagmire of plant defense hypotheses. *Quarterly Review of Biology* **78**, 23–55.

Stauffer B (1999). A cornucopia of ice core results. *Nature* **399**, 412–413.

Steinberg CEW and Wright RF (Eds) (1994). *Acidification of Aquatic Ecosystems: Implications for the Future.* John Wiley and Sons, Chichester, UK.

Stephens SE, Koons DN, Rotella JJ and Willey DW (2003). Effects of habitat fragmentation on avian nesting success: a review of the evidence at multiple spatial scales. *Biological Conservation* **115**, 101–110.

Stohlgren TJ, Schell LD and Vanden Heuvel B (1999). How grazing and soil quality affect native and exotic plant diversity in Rocky Mountain grasslands. *Ecological Applications* **9**, 45–64.

Storfer A (2003). Amphibian declines: future directions. *Diversity & Distributions* **9**, 151–163.

Stratford JA and Stouffer PC (1999). Local extinctions of terrestrial insectivorous birds in a fragmented landscape near Manaus, Brazil. *Conservation Biology* **13**, 1416–1423.

Strauss SY, Rudgers JA, Lau JA and Irwin RE (2002). Direct and ecological costs of resistance to herbivory. *Trends in Ecology and Evolution* **17**, 278–285.

Straw N and Sheppard A (1995). The role of plant dispersion pattern in the success and failure of biological control. In *Proceedings of the VIII International Symposium on Biological Control of Weeds.* (Eds E Delfosse and R Scott) pp. 161–168. CSIRO, Melbourne.

Sutherland WJ (Ed.) (2006). *Ecological Census Techniques: A Handbook.* Cambridge University Press: Cambridge, UK.

Sutherst RW, Floyd RB and Maywald GF (1995). The potential geographical distribution of the cane toad, *Bufo marinus* L. in Australia. *Conservation Biology* **9**, 294–299.

Sutherst RW and Maywald G (2005). A climate model of the red imported fire ant, *Solenopsis invicta* Buren (Hymenoptera: Formicidae): implications for inva-

sion of new regions, particularly Oceania *Environmental Entomology* **34**, 317–335.

Swinton J, Harwood J, Grenfell BT and Gilligan CA (1998). Persistence thresholds for phocine distemper virus infection in harbour seal *Phoco vitulina* metapopulations. *Journal of Animal Ecology* **67**, 54–68.

Symmons PM and Cressman K (2001). *Desert Locust Guidelines: 1. Biology and Behaviour.* Food and Agriculture Organization of the United Nations, Rome.

Szmant AM (2002). Nutrient enrichment on coral reefs: is it a major cause of coral reef decline? *Estuaries* **25**, 743–766.

Tanner JE, Hughes TP and Connell JH (1996). The role of history in community dynamics: a modelling approach. *Ecology* **77**, 108–117.

Tewksbury JJ, Hejl SJ and Martin TE (1998). Breeding productivity does not decline with increasing fragmentation in a western landscape. *Ecology* **79**, 2890–2903.

Tewksbury JJ, Levey DJ, Haddad NM, Sargent S, Orrock JL, Weldon A, *et al.* (2002). Corridors affect plants, animals, and their interactions in fragmented landscapes. *Proceedings of the National Academy of Science U.S.A.* **99**, 12923–12926.

Thies C and Tscharntke T (1999). Landscape structure and biological control in agroecosystems. *Science* **285**, 893–895.

Thomas CD, Franco AMA, and Hill JK (2006). Range retractions and extinction in the face of climate warming. *Trends in Ecology & Evolution* **21**, 415–416.

Tiessen H, Cuevas E and Chacon P (1994). The role of soil organic matter in sustaining soil fertility. *Nature* **371**, 783–785.

Tilghman NG (1989). Impacts of white-tailed deer on forest regeneration in northwestern Pennsylvania. *Journal of Wildlife Management* **53**, 524–532.

Tilman D (1986). Resources, competition and the dynamics of plant communities. In *Plant Ecology.* (Ed. MJ Crawley) pp. 51–75. Blackwell, Oxford, UK.

Tilman D (1996). Biodiversity: population versus ecosystem stability. *Ecology* **77**, 350–363.

Tilman D, May RM, Lehman CL and Nowak MA (1994). Habitat destruction and the extinction debt. *Nature* **371**, 65–66.

Tinker DB, Romme WH and Despain DG (2003). Historic range of variability in landscape structure in subalpine forests of the Greater Yellowstone Area, USA. *Landscape Ecology* **18**, 427–439.

Tomialojc L (1999). A long-term study of changing predation impact on breeding woodpigeons. In *Advances in Vertebrate Pest Management*. (Eds DP Cowan and CJ Feare) pp. 205–218. Filander Verlag, Furth, Germany.

Torti SD, Coley PD and Kursar TA (2001). Causes and consequences of monodominance in tropical lowland forests. *American Naturalist* **157**, 141–153.

Towns DR and Broome KG (2003). From small Maria to massive Campbell: forty years of rat eradications from New Zealand islands. *New Zealand Journal of Zoology* **30**, 377–398.

Tsurumi M (2003). Diversity at hydrothermal vents. *Global Ecology & Biogeography* **12**, 181–190.

Turchin P (1999). Population regulation: a synthetic view. *Oikos* **84**, 153–159.

Turesson G (1930). The selective effect of climate upon the plant species. *Hereditas* **14**, 99–152.

Twigg LE and Williams CK (1999). Fertility control of overabundant species: can it work for feral rabbits? *Ecology Letters* **2**, 281–285.

van der Heijden MGA and Sanders IR (Eds) (2002). *Mycorrhizal Ecology*. Springer-Verlag, Berlin, Germany.

Van Dover CL, German CR, Speer KG, Parson LM and Vrijenhoek RC (2002). Evolution and biogeography of deep-sea vent and seep invertebrates. *Science* **295**, 1253–1257.

Vitousek PM (1984). Litterfall, nutrient cycling, and nutrient limitation in tropical forests. *Ecology* **65**, 285–298.

Vitousek PM (1994). Beyond global warming: ecology and global change. *Ecology* **75**, 1861–1876.

Vitousek PM (2004). *Nutrient Cycling and Limitation: Hawai'i as a Model System*. Princeton University Press, Princeton, New Jersey.

Vitousek PM, D'Antonio CM, Loope LL and Westbrooks R (1996). Biological invasions as global environmental change. *American Scientist* **84**, 468–478.

Vitousek PM, Mooney HA, Lubchenco J and Melillo JM (1997). Human domination of earth's ecosystems. *Science* **277**, 494–499.

Wackernagel M, Onisto L, Bello P, Callejas Linares A, Lopez Falfan IS, Mendez Garcia J, *et al.* (1999). National natural capital accounting with the ecological footprint concept. *Ecological Economics* **29**, 375–390.

Wackernagel M and Rees WE (1996). *Our Ecological Footprint: Reducing Human Impact on the Earth*. New Society Publishers, Gabriola Island, B.C.

Walk JW and Warner RE (1999). Effects of habitat area on the occurrence of grassland birds in Illinois. *American Midland Naturalist* **141**, 339–344.

Walker BH, Steffen W, Canadell J and Ingram J (Eds) (1999). *The Terrestrial Biosphere and Global Change: Implications for Natural and Managed Ecosystems*. Cambridge University Press, Cambridge.

Walker LR and del Moral R (2003). *Primary Succession and Ecosystem Rehabilitation*. Cambridge University Press, Cambridge, UK.

Walkinshaw LH (1983). *Kirtland's Warbler: The Natural History of an Endangered Species*. Cranbrook Institute of Science, Bloomfield Hills, Michigan.

Wallace AR (1876). *The Geographical Distribution of Animals*. Macmillan, London, UK.

Walters CJ, Robinson CE and Northcote TG (1990). Comparative population dynamics of *Daphnia rosea* and *Holopedium gibberum* in four oligotrophic lakes. *Canadian Journal of Fisheries and Aquatic Sciences* **47**, 401–409.

Waring RH and Schlesinger WH (1985). *Forest Ecosystems: Concepts and Management*. Academic Press, Orlando, Florida.

Warming E (1896). *Lehrbuch der okologischen Pflanzengeographie*. Gebruder Borntraeger, Berlin.

Watson A, Hewson R, Jenkins D and Parr R (1973). Population densities of mountain hares compared with red grouse on Scottish moors. *Oikos* **24**, 225–230.

Watson R (1999). Common themes for ecologists in global issues. *Journal of Applied Ecology* **36**, 1–10.

Watt AS (1940). Contributions to the ecology of bracken (*Pteridium aquilinum*). I. The rhizome. *New Phytologist* **39**, 401–422.

Watt AS (1955). Bracken versus heather, a study in plant sociology. *Journal of Ecology* **43**, 490–506.

Way MJ and Heong KL (1994). The role of biodiversity in the dynamics and management of insect pests of tropical irrigated rice—a review. *Bulletin of Entomological Research* **84**, 567–587.

Wedin DA and Tilman D (1996). Influence of nitrogen loading and species composition of the carbon balance of grasslands. *Science* **274**, 1720–1723.

Weiss RA (2002). Virulence and pathogenesis. *Trends in Microbiology* **10**, 314–317.

West SA, Kiers ET, Simms EL and Denison RF (2002). Sanctions and mutualism stability: why do rhizobia fix nitrogen? *Proceedings of the Royal Society of London, Series B* **269**, 685–694.

Westemeier RL, Brawn JD, Simpson SA, Esker TL, Jansen RW, Walk JW, Kershner EL, *et al.* (1998). Tracking the long-term decline and recovery of an isolated population. *Science* **282**, 1695–1697.

Western D and Pearl MC (Eds) (1989). *Conservation for the Twenty-first Century*. Oxford University Press: New York.

White GG (1981). Current status of prickly pear control by *Cactoblastis cactorum* in Queensland. In *Proceedings of the 5th International Symposium on the Biological Control of Weeds*. (Ed. E Del Fosse) pp. 609-616. Department of Primary Industries, Brisbane.

White TCR (1974). A hypothesis to explain outbreaks of looper caterpillars, with special reference to populations of *Selidosema suavis* in a plantation of *Pinus radiata* in New Zealand. *Oecologia* **16**, 279–310.

Whittaker RH (1975). *Communities and Ecosystems*. Macmillan, New York.

Wiens JA (1984). On understanding a non-equilibrium world: myth and reality in community patterns and processes. In *Ecological Communities*. (Eds DR Strong Jr., D Simberloff, LG Abele and AB Thistle) pp. 439–457. Princeton University Press, Princeton, New Jersey.

Wiens JA (1999). The science and practice of landscape ecology. In *Landscape Ecological Analysis: Issues and Applications*. (Eds JM Klopatek and RH Gardner) pp. 371–383. Springer-Verlag, New York.

Wiens JA (2002). Riverine landscapes: taking landscape ecology into the water. *Freshwater Biology* **47**, 501–515.

Williams CB (1964). *Patterns in the Balance of Nature*. Academic Press, London, UK.

Williams TM, Estes JA, Doak DF and Springer AM (2004). Killer appetites: assessing the role of predators in ecological communities. *Ecology* **85**, 3373–3384.

Willig MR, Kaufman DM and Stevens RD (2003). Latitudinal gradients of biodiversity: pattern, process, scale, and synthesis. *Annual Review of Ecology, Evolution and Systematics* **34**, 273–309.

Wills C (1996). *Yellow Fever, Black Goddess: The Coevolution of People and Plagues*. Addison Wesley, Reading, Massachusetts.

Wilson EO (2001). *The Diversity of Life*. 2nd edn. Penguin, London, UK.

Wilson EO (2002). *The Future of Life*. Random House, New York.

Wilson MA and Carpenter SR (1999). Economic valuation of freshwater ecosystem services in the United States: 1971–1997. *Ecological Applications* **9**, 772–783.

Woinarski JCZ, Milne DJ and Wanganeen G (2001). Changes in mammal populations in relatively intact landscapes of Kakadu National Park, Northern Territory, Australia. *Austral Ecology* **26**, 360–370.

Wolff JO (1997). Population regulation in mammals: an evolutionary perspective. *Journal of Animal Ecology* **66**, 1–13.

Woodford J (2003). *The Dog Fence*. Text Publishing Company, Melbourne.

Woods KD and Davis MB (1989). Paleoecology of range limits: Beech in the upper peninsula of Michigan. *Ecology* **70**, 681–696.

Wootton JT, Parker MS and Power ME (1996). Effects of disturbance on river food webs. *Science* **273**, 1558–1561.

Worm B, Barbier EB, Beaumont N, Duffy JE, Folke C, Halpern BS, *et al.* (2006). Impacts of biodiversity loss on ocean ecosystem services. *Science* **314**, 787–790.

Worm B and Duffy JE (2003). Biodiversity, productivity and stability in real food webs. *Trends in Ecology and Evolution* **18**, 628–632.

Yeakley A and Weishampel JF (2000). Multiple source pools and dispersal barriers for Galapagos plant species distribution. *Ecology* **81**, 893–898.

Yoda K, Kira T, Ogawa H and Hozumi K (1963). Self-thinning in overcrowded pure stands under cultivated and natural conditions. *Journal of the Institute of Polytechnics, Osaka City University, Series D* **14**, 107–129.

Zanette L, Doyle P and Tremont SM (2000). Food shortage in small fragments: evidence from an area-sensitive passerine. *Ecology* 81, 1654–1666.

INDEX